AF566500

# FINITE MATHEMATICS with CALCULUS

## FOR THE MANAGEMENT, LIFE, AND SOCIAL SCIENCES

# FINITE MATHEMATICS with CALCULUS

## FOR THE MANAGEMENT, LIFE, AND SOCIAL SCIENCES

WILLIAM R. DERRICK
University of Montana

JUDITH L. DERRICK

ADDISON-WESLEY PUBLISHING COMPANY
Reading, Massachusetts | Menlo Park, California
London | Amsterdam | Don Mills, Ontario | Sydney

SPONSORING EDITOR: Stephen H. Quigley
PRODUCTION EDITOR: Martha K. Morong
DESIGNER: Catherine L. Dorin
ILLUSTRATOR: B. J. and F. W. Taylor
ANCO/Boston
COVER DESIGN: Marshall Henrichs

Photographs on pages 16 and 28 by John K. Long.
All other photographs by Marshall Henrichs.

**Library of Congress Cataloging in Publication Data**

Derrick, William R
Finite mathematics with calculus for the management, life, and social sciences.

Includes index.
1. Mathematics--1961- 2. Management --Mathematics. 3. Life sciences--Mathematics. 4. Social sciences--Mathematics. I. Derrick, Judith L., 1936- joint author. II. Title.
QA37.2.D467 510 78-18639
ISBN 0-201-01400-9



ISBN 0-201-01400-9
ABCDEFGHIJK-HA-79

# PREFACE

The topics presented in this book are an extension of those ideas of basic algebra and geometry that have application in the economic, social, and life sciences. The text is balanced to provide an equal treatment of the concepts needed by students of management, biology, and the behavioral sciences.

**APPROACH**

The main purpose of any first-year course in mathematics for nonmajors is to provide an introduction to quantitative ideas and methods, and to convince the student that these techniques will be useful in his or her discipline. In order to satisfy this last goal, this text contains a wide selection of examples from a diversity of fields. The finance and management student will find examples on marginal analysis, consumer's surplus, simple and compound interest, annuities, amortization, equity, trend analysis, taxes, wages, prices, and indifference curves; the life scientist will encounter examples on DNA, heredity, blood types, gestation, and other medical and biological topics; the behavioral scientist will meet examples in ballistics, learning curves, politics, and demography.

**MODULARIZED TEXT**

It is important that any first-year text be sufficiently flexible to meet the differing requirements at various colleges and universities. To provide this flexibility we have made each chapter as independent as possible from the others. The dependency chart at the end of this preface shows how the various chapters are related. Because of this flexibility, the book may be taught in three modules permitting a variety of course syllabi to be developed. The following core sections for each module are recommended.

*Module 1:* Elementary Functions and Linear Systems

Chapter 2 Sections 1–7
Chapter 3 Sections 1, 3
Chapter 4 Sections 1–4
Chapter 5 Sections 1–3

This module covers linear functions and systems, polynomials, exponential and logarithmic functions, matrices, and linear programming. Linear systems are solved both simultaneously and by the augmented matrix procedure, matrix inversion is accomplished by Gaussian elimination, and both the geometric method and simplex algorithm are used in solving linear programming problems.

*Module 2:* Probability, Statistics, and Decision Theory

Chapter 6 Sections 2–5, 7
Chapter 7 Sections 2–6

This core covers the essential concepts of set theory, probability, elementary statistics, Markov chains, and $2 \times 2$ games. Tree diagrams are emphasized in both chapters, and matrix methods are employed in the entirely self-contained sections on Markov chains and on $2 \times 2$ games. Either the section on Markov chains or that on $2 \times 2$ games may be omitted if time is pressing.

*Module 3:* Calculus—a short course

Chapter 8 Sections 1, 3–5
Chapter 9 Sections 1–6, 8
Chapter 10 Sections 1–5

These core sections present the basic concepts of limits, differentiation, and integration. The geometrical nature of these concepts is emphasized, while formal proofs are minimized. Limits are presented as a natural extension of geometric progressions, and derivatives as a consequence of marginal change and average growth rates. Integration is introduced as "reverse" differentiation; definite integrals are presented after the integration techniques have been mastered.

**ENRICHMENT SECTIONS** Ample supplementary material is provided in the remaining sections of this book and in Appendix 1 to enable the instructor to meet the requirements of his or her students. These include:

1. A self-contained chapter (Appendix 1) presenting the main topics of *mathematics of finance* including compound interest, annuities, and

amortization. Although these topics are also covered as examples in various sections throughout the book, some instructors may prefer to present this material as an organized whole.

2. A *review* of basic algebra, factoring, and binomials in Chapter 1. This material is not designed to provide a course in intermediate algebra, but simply to highlight the algebraic procedures that are assumed in this book. A one-year course in high school algebra is a prerequisite for this text.
3. A careful explanation of curve fitting by means of *least squares approximation* in Section 2.8. This method is very useful in management (trend analysis), as well as in the life and social sciences (linear regression). Exponential and power curve fitting are considered in Section 3.5.
4. A section (3.4) on *graphical methods* for plotting and determining exponential and power curve fits. These techniques are of importance to life science majors.
5. A section (9.9) on *partial derivatives* for functions involving two independent variables, including the usual second derivative test for maxima and minima and a brief development of Lagrange multipliers.
6. A summary (Appendix 2) of the main identities of trigonometry, intended for those instructors who wish to include trigonometric functions in their development of calculus. Sections on the differentiation and integration of trigonometric functions are also included (see 9.7 and 10.2); problems that involve the use of trigonometry are identified with the symbol ▶. A self-contained chapter on trigonometry, generously supplied with examples, applications, and exercises, is also available from Addison-Wesley.
7. A short introduction to *differential equations* (Sections 10.7 and 10.8), emphasizing their connection with integration and their applications in economics and the life sciences.
8. Several sections stressing applications, modeling, and detailed case studies of topics in the three main areas of the management, social, and life sciences. These sections generally contain one or more examples in each of the three main areas, allowing the instructor to select the examples that best fit the interests of the class.

**COURSE SYLLABI** While many adequate syllabi are possible, we recommend consideration of the following outlines.

*Finite Mathematics*
(If mathematics of finance is desired, include Appendix 1 in each course description.)

One-quarter course: Module 1 except Sections 2.5–2.7 and 3.3, plus Sections 6.2–6.7
One-semester course: Module 1 plus Sections 6.2–6.7 and 7.5 or 7.6
Two-quarter course: Modules 1 and 2

*Short Calculus*
One-quarter course: Module 3
One-semester course: Module 3 plus Sections 9.7 and 10.6–10.8

**EXAMPLES AND EXERCISES**

This text contains a large number of practical examples. Their locations by section number and topic are given below.

*Finance and Management Sciences*
Interest: Sections 3.1, 3.2, 8.1, 8.2, 9.6, A1.1
Annuities and amortization: Sections 8.1, 8.2, A1.2, A1.3
Margin: Sections 9.1, 9.5, 9.6
Consumer's surplus: Section 10.5
Trend analysis: Sections 2.8, 3.2, 3.4, 3.5, 7.5
Economics: Sections 2.7, 9.1, 9.5, 9.9, 10.5
Other topics: Sections 2.1, 2.2, 2.3, 2.4, 4.1, 4.5, 5.2, 5.3, 7.2, 7.5, 9.8, 10.7

*Behavioral Sciences*
Demography: Sections 2.1, 3.2, 8.2, 9.6, 10.7, 10.8
Psychology: Sections 3.2, 7.1, 7.6
Sociology: Sections 4.1, 4.5
Other topics: Sections 2.1, 4.5, 6.6, 6.7, 7.2

*Biological Sciences*
Ecology: Sections 4.5, 5.2, 5.3, 8.5, 10.5, 10.6, 10.8
Pollution and concentration problems: Sections 2.4, 8.2, 8.5, 10.5, 10.6, 10.8
Gestation: Sections 2.4, 2.8
DNA: Section 6.9
Heredity: Sections 6.8, 6.9, 8.1, 8.3
Blood types: Sections 6.2, 6.3
Medicine: Sections 6.1, 6.5, 6.8, 10.7, 10.8
Other topics: Sections 2.4, 3.4, 3.5, 9.5, 9.9

*Physical Sciences*
Free fall: Sections 9.1, 9.5, 10.5
Temperature: Sections 2.4, 7.4, 9.8
Meteorology: Sections 2.8, 7.1, 7.4, 10.8
Decay and other topics: Sections 3.2, 3.3, 9.1, 9.5, 10.5, 10.7

Great care has been taken in the construction of the exercises so that they will reflect the diversity of fields present in the examples. Exercise sets are generally arranged in matching pairs with answers *only for the odd-numbered exercises.* This allows the instructor to assign exercises in whichever way will best fit the needs of the class. Exercises, examples, and sections in which the use of a hand-calculator will greatly simplify the computation are identified by the symbol □. Difficult or challenging exercises are indicated by the notation ★.

**ACKNOWLEDGMENTS** We wish to express our appreciation to the many people who helped with the development of this book; to our students and colleagues for providing suggestions and criticisms while class testing this material; and to the following reviewers for their valuable comments, which significantly improved the organization and presentation of this material: Bill Bompart, Augusta College; Richard M. Crownover, University of Missouri at Columbia; Richard D. Derderian, Providence College; Carroll E. Flanagan, University of Wisconsin at Whitewater; Donald R. Horner, Eastern Washington State College; Frank Kocher, The Pennsylvania State University at University Park; Carl D. Meyer, Jr., North Carolina State University at Raleigh; Margaret Miller, Skyline College; Roger B. Nelsen, Lewis and Clark College; Donald F. Reynolds, West Virginia University; and Frederick R. Ward, Boise State University. We also thank especially the editorial and production staff at Addison-Wesley for their hard work and encouragement.

*Missoula, Montana* W. R. D.
*January 1979* J. L. D.

# CHAPTER DEPENDENCY CHART

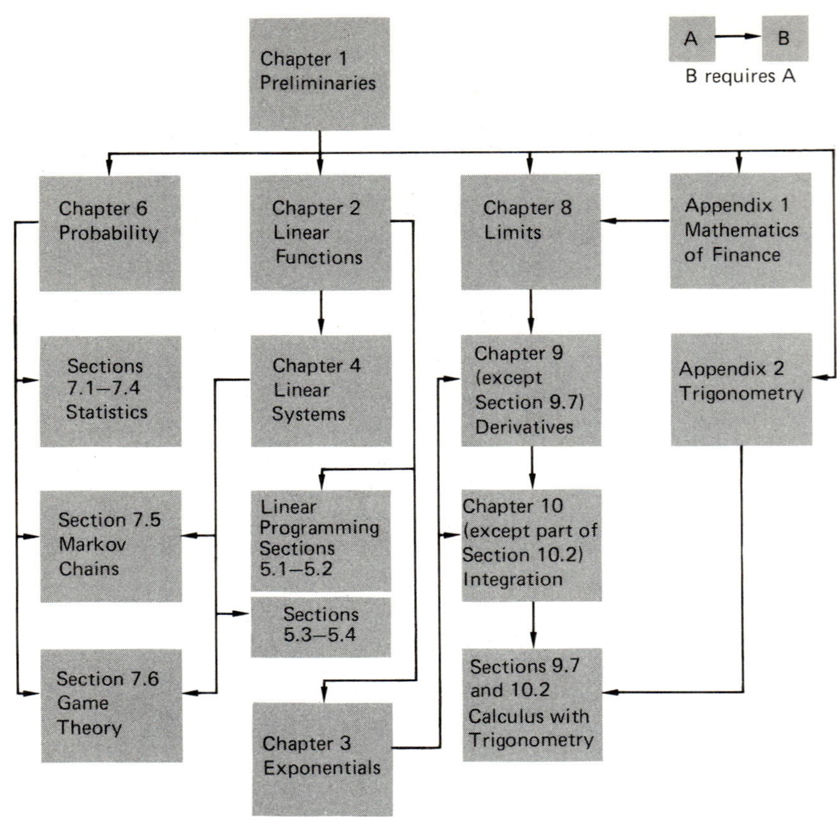


Location of Mathematics of Finance Topics

| | |
|---|---|
| Compound interest | Sections 3.1, 3.2, 8.1, 8.2, 9.6, A1.1 |
| Annuities | Sections 8.1, A1.2 |
| Amortization | Sections 8.2, A1.3 |
| Equity and other topics | Sections 8.2, A1.4 |

# CONTENTS

*□ indicates an optional section that requires the use of a hand-held calculator*

## 1 PRELIMINARIES

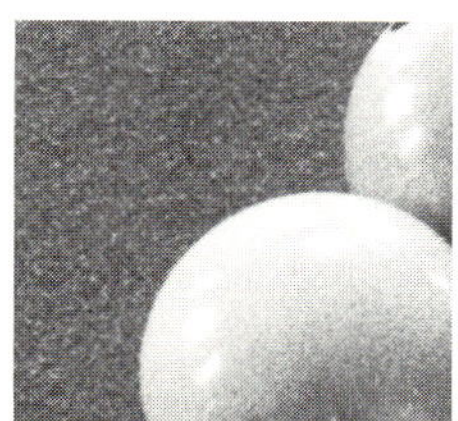


Algebraic Laws and Operations 2
Factoring 12
Pascal's Triangle 14


## PART I ELEMENTARY FUNCTIONS AND SYSTEMS

## 2 LINEAR AND POLYNOMIAL FUNCTIONS

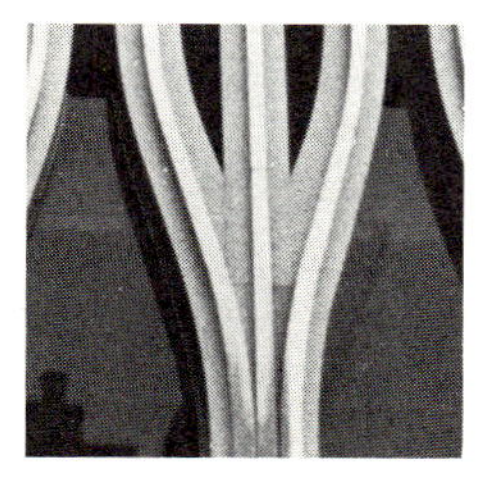


Functions and Graphs 22
Linear Equations 31
Simultaneous Linear Equations 38
Applications of Linear Equations 47
Parabolas 59
Circles 69
Polynomial and Rational Functions 72
□ Modeling by Least Squares Approximation 80

## 3 EXPONENTIAL AND LOGARITHMIC FUNCTIONS

Exponential Functions 91
Exponential Growth and Decay, Learning Models, and Logistic Curves 98
Logarithms and Their Properties 109
Graphical Methods: Double and Semi-logarithmic Plots (Optional) 119
□ Modeling by Exponential Curve Fitting 138

## 4 LINEAR SYSTEMS

Matrices and Matrix Multiplication 146
Writing a Linear System in Matrix Notation 158
The Augmented Matrix Procedure 163
Matrix Inversion by Gaussian Elimination 170
Some Applications of Matrices 177

## 5 LINEAR PROGRAMMING

Linear Inequalities 198
Linear Programming: Geometric Method 204
Linear Programming: The Simplex Algorithm 215
Nonstandard Linear Programming Problems 228

# PART II
# PROBABILITY AND STATISTICS

## 6 PROBABILITY

Introduction 240
Sets and Venn Diagrams 246
Events and Their Probabilities 256
Uniform Sample Spaces 266
Conditional Probability 269
Bayes's Formula 280

Counting Procedures 286
Examples and Applications 292
A Case Study: Counting and the Genetic Code 304

7 STATISTICS AND DECISION THEORY

Sampling a Population 312
Mean, Variance, and Expected Value 318
The Binomial Distribution 330
The Normal Distribution 339
Markov Chains 344
Game Theory 353

**PART III**
**CALCULUS**

8 SEQUENCES AND LIMITS

Arithmetic and Geometric Sequences 368
Discrete Models in Business and the Biological Sciences 377
Limit of a Sequence 388
Limits of a Function 395
Continuity of Functions 404

9 DIFFERENTIAL CALCULUS

Difference Quotients, Growth Rates, and Marginal Change 414
The Derivative 423
Six Rules of Differentiation 432
The Chain Rule 445
Rates of Change 450
Derivatives of Exponential and Logarithmic Functions 461
Derivatives of Trigonometric Functions (Optional) 468
Optimization: The Two Derivative Tests 475
Partial Derivatives 486

## 10 INTEGRAL CALCULUS

The Antiderivative 504
Substitution Techniques in Integration 509
Integration by Parts 518
The Fundamental Theorem of Calculus 525
Applications of Integral Calculus 535
□ Approximation: The Trapezoidal Rule 547
The Differential Equation $y' = ay + b$ 552
Some Ecological Models 563

## APPENDIXES

**1. A Summary of Mathematics of Finance Topics** 578
Compound Interest 578
Annuities 581
Amortization 583
Equity and Other Topics 585
**2. Trigonometry Formulas** 590
**3. Tables of Derivatives and Integrals** 593
**4. Common Logarithms** 597
**5. Natural Logarithms** 599

## ANSWERS TO ODD-NUMBERED PROBLEMS A-1

## INDEX I-1

# CHAPTER 1 PRELIMINARIES

In this chapter we will briefly review some of the basic properties of algebra. This is not intended as a complete course in algebra; instead we will simply outline some of the more important facts and give some examples of their use. You should refer to this chapter whenever you are having difficulty with any algebraic manipulations.

---

## 1.1 ALGEBRAIC LAWS AND OPERATIONS

Modern mathematics is characterized by the frequent use of different sets (or collections) of numbers. Four of these sets of numbers are so common that special names have been given to them. They are the integers, rational numbers, real numbers, and complex numbers.

The set of *integers* consists of all the counting numbers,

$$1, 2, 3, 4, 5, \ldots,$$

together with the number zero and the negatively signed counting numbers,

$$-1, -2, -3, -4, -5, \ldots,$$

Thus the set of integers contains all whole numbers:

$$\{\ldots, -3, -2, -1, 0, 1, 2, 3, \ldots\}. \tag{1.1}$$

We have learned that whenever we add, multiply, or subtract two integers we again obtain an integer. Division, however, causes problems, since the result of dividing two integers is often not an integer. Hence, in order to be able to divide (except by zero), we need a larger number system (or set of numbers).

The set of *rational numbers* consists of all quotients $a/b$, where $a$ and $b$ are integers and $b \neq 0$. We often call these numbers fractions and identify equivalent fractions such as

$$6/3 = 4/2 = 2/1$$

and

$$1/3 = 2/6 = 3/9 = 5/15.$$

Clearly, the rational number system includes all integers, since any integer can be written as a fraction:

$$17 = 17/1 \quad \text{and} \quad -12 = -12/1. \tag{1.2}$$

Thus, the integers are a subset of the rationals. We have learned that performing all four arithmetic operations on a pair of rationals leads to a rational number, *with the exclusion of division by zero*. However, the roots of rational numbers often are not rationals. For example, the square root of 2 is not a rational number, because if it were we could write it as a fraction

$$p/q = \sqrt{2}, \tag{1.3}$$

with $p$ and $q$ having only the integers $\pm 1$ as common factors. Squaring both sides of (1.3) yields

$$p^2/q^2 = 2,$$

and multiplying both sides by $q^2$, we obtain

$$p^2 = 2q^2. \tag{1.4}$$

Since the right side of Eq. (1.4) is even, the left side must also be even, and only even integers $p$ have even squares. Therefore, $p$ is even so that $p$ equals two times some integer $k$:

$$p = 2k \quad \text{and} \quad p^2 = (2k)^2 = 4k^2.$$

Substituting $4k^2$ for $p^2$ in Eq. (1.4) yields

$$4k^2 = 2q^2,$$

which may be divided by 2 on each side to obtain

$$2k^2 = q^2. \tag{1.5}$$

But the left side of Eq. (1.5) is even, so the right side must also be even, implying that $q$ is even. This is a contradiction, since the only integers that are common factors of $p$ and $q$ are $\pm 1$, but we have shown that $p$ and $q$ are even, hence both are divisible by 2. Thus, our assumption that $\sqrt{2}$ is rational must be false.

The existence of numbers that are not rationals leads to the construction of a larger number system in which root extraction of positive numbers is possible. The *real number system* consists of all numbers that may be geometrically represented by the points on an infinitely long straight line (see Fig. 1.1). The line is marked off in equally spaced intervals representing each of the integers, with the positive real numbers to the right of zero and the negative real numbers to the left. For emphasis, an arrow is often drawn at the right end of the line to indicate the positive direction. Every real number is represented by a

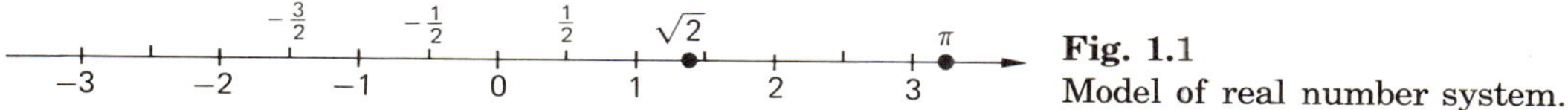


**Fig. 1.1**
Model of real number system.

single point on the line. All rational numbers $p/q$ can easily be located by subdividing the interval between 0 and 1 into $q$ equal subintervals and measuring $p$ such subintervals to the right or left of 0, depending on whether $p$ is positive or negative.

Although root extraction of positive numbers is now possible, the square roots of negative numbers are not real, since the product of any real number multiplied by itself is always nonnegative. To remedy this remaining defect, a larger set of numbers is required. This is the set of *complex numbers*, consisting of all numbers of the form

$$a + bi,$$

where $a$ and $b$ are real numbers and $i$ is a number that satisfies the property

$$i^2 = -1. \tag{1.6}$$

(Roughly, we can think of $i$ as the square root of $-1$.)

Since $i$ is clearly not a real number, it is called the *imaginary unit*, but this does not mean that it is any less useful than other numbers in the complex number system. Indeed, complex numbers play a fundamental role in science and engineering. The geometric model for the complex numbers is shown in Fig. 1.2. It consists of a plane of points spanned by two axes: a real axis from which the real numbers $a$ are selected and an imaginary axis from which the imaginary numbers $bi$ are chosen. The complex number $a + bi$ is located by drawing a vertical line through the number $a$ on the real axis and a horizontal line through the number $bi$ on the imaginary axis and finding where the two

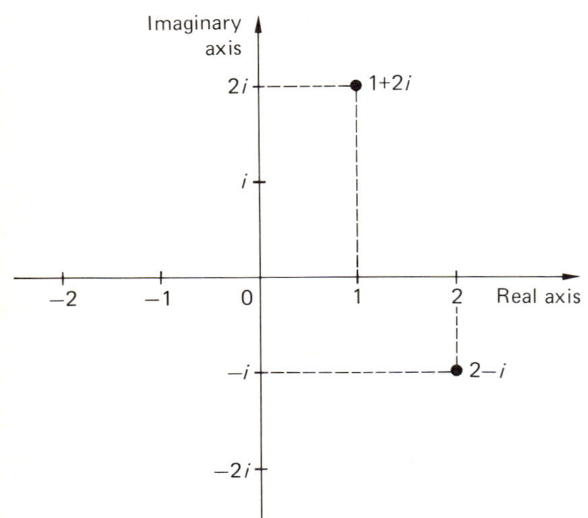


**Fig. 1.2**
Model of complex number system.

lines intersect. Clearly, the set of complex numbers contains the real numbers since every real number satisfies

$$a = a + 0i.$$

Figure 1.3 shows the relationships among these four number systems.

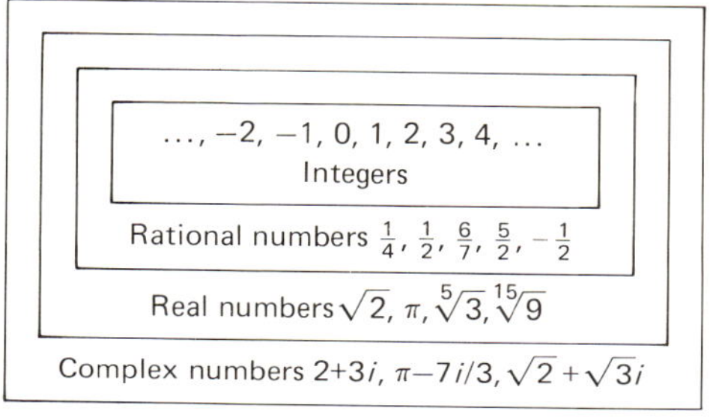


**Figure 1.3**

### Arithmetic Operations

We will briefly review some of the arithmetic operations as they apply to these four number systems. You should be familiar with all of these rules, since they form the basis of elementary algebra. The proof of any of these rules can be found in most high-school algebra texts.

The *properties of the number zero* are the ones that are most commonly misapplied.

> $a + 0 = a$; $a - 0 = a$; $a \cdot 0 = 0$; but division by zero is prohibited.

For example, $5 + 0 = 5$, $2 - 0 = 2$, $\pi \cdot 0 = 0$, but $7 \div 0$ is not allowed.

The *arithmetic of real signed numbers* is another frequent area of difficulty. We have the following rules:

> $a + (-b) = a - b$; $a - (-b) = a + b$;
>
> $a \cdot (-b) = (-a) \cdot b = -ab$; $(-a) \cdot (-b) = ab$;
>
> $a/(-b) = (-a)/b = -a/b$; $(-a)/(-b) = a/b$, provided $b \neq 0$.

For example,

$$7 + (-3) = 7 - 3 = 4; \qquad 7 - (-3) = 7 - 3 = 10;$$

$$7 \cdot (-3) = -21; \qquad (-7) \cdot (-3) = 21;$$

$$(-7)/3 = -7/3; \qquad (-7)/(-3) = 7/3.$$

The *arithmetic of fractions* is often a source of error. Recall that in adding or subtracting fractions a common denominator must first be found. Common denominators are not needed in multiplying or dividing fractions. Provided $b \neq 0$ and $d \neq 0$ we have the following:

$$\frac{a}{b} + \frac{c}{d} = \frac{ad + bc}{bd}; \qquad \frac{a}{b} - \frac{c}{d} = \frac{ad - bc}{bd};$$

$$\frac{a}{b} \cdot \frac{c}{d} = \frac{ac}{bd}; \qquad \frac{a}{b} \div \frac{c}{d} = \frac{a}{b} \cdot \frac{d}{c} = \frac{ad}{bc}, \quad (c \neq 0).$$

For example,

$$\frac{1}{2} + \frac{1}{3} = \frac{3}{6} + \frac{2}{6} = \frac{3+2}{6} = \frac{5}{6}; \qquad \frac{3}{4} - \frac{2}{3} = \frac{9}{12} - \frac{8}{12} = \frac{9-8}{12} = \frac{1}{12};$$

$$\frac{3}{5} \cdot \frac{4}{7} = \frac{12}{35}; \qquad \frac{3}{5} \div \frac{4}{7} = \frac{3}{5} \cdot \frac{7}{4} = \frac{21}{20}.$$

The *arithmetic of complex numbers* is different from that of real numbers, since the real and imaginary parts are handled separately. We have:

$$(a + bi) + (c + di) = (a + c) + (b + d)i,$$

$$(a + bi) - (c + di) = (a - c) + (b - d)i,$$

$$(a + bi) \cdot (c + di) = (ac - bd) + (bc + ad)i,$$

$$\frac{a + bi}{c + di} = \frac{(a + bi)(c - di)}{(c + di)(c - di)} = \frac{(ac + bd) + (bc - ad)i}{c^2 + d^2}.$$

The last rule can be verified by showing the multiplication:

$$\begin{array}{r} a + bi \\ \times\ c - di \\ \hline ac + bci \\ -\ adi - bdi^2 \\ \hline (ac + bd) + (bc - ad)i \end{array} \qquad \begin{array}{r} c + di \\ \times\ c - di \\ \hline c^2 + cdi \\ -\ cdi - d^2i^2 \\ \hline c^2 + d^2 \end{array}$$

since $i^2 = -1$. For example,

$$(2 + 3i) + (4 - 5i) = 6 - 2i,$$

$$(2 + 3i) - (4 - 5i) = -2 + 8i,$$

$$(2 + 3i) \cdot (4 - 5i) = (8 + 15) + (12 - 10)i = 23 + 2i,$$

and

$$\frac{2 + 3i}{4 - 5i} = \frac{(2 + 3i)(4 + 5i)}{(4 - 5i)(4 + 5i)} = \frac{(8 - 15) + (12 + 10)i}{16 + 25} = \frac{-7}{41} + \frac{22i}{41}.$$

The laws of exponents and radicals are commonly misapplied. Briefly, they are as follows.

$$a^m \cdot a^n = a^{m+n};$$

$$(a^m)^n = a^{mn};$$

$$a^m \div a^n = a^m/a^n = a^{m-n}, \quad a \neq 0;$$

$$a^0 = 1 \quad \text{and} \quad a^{-n} = 1/a^n, \quad a \neq 0;$$

$$(ab)^n = a^n b^n;$$

$$(a/b)^n = a^n/b^n, \quad b \neq 0.$$

For example,

1. $2^3 \cdot 2^4 = 2^7$, because $2^3 = 8$, $2^4 = 16$, and $8 \cdot 16 = 128 = 2^7$.
2. $(2^4)^2 = 2^8$, because $(16)^2 = 256 = 2^8$.
3. $2^5 \div 2^3 = 2^5/2^3 = 2^2$, since $2^5 = 32$ and $32/8 = 4 = 2^2$.
4. $(2 \cdot 3)^2 = 2^2 \cdot 3^2$, since $6^2 = 36 = 4 \cdot 9 = 2^2 \cdot 3^2$.
5. $\left(\frac{2}{3}\right)^3 = \frac{2^3}{3^3} = \frac{8}{27}$.

The laws of exponents are valid not only for integers but for any real numbers $m$ and $n$. When the exponent is a rational number $p/q$, the exponential

$$a^{p/q} = (a^{1/q})^p = (\sqrt[q]{a})^p,$$

that is, $a^{p/q}$ is the $p$th power of the $q$th root of $a$. Thus, square roots are indicated by an exponent of 1/2, cube roots by an exponent of 1/3, and so forth. For instance,

$$2^{1/2} = \sqrt{2},$$
$$3^{1/3} = \sqrt[3]{3},$$

and

$$\left(\frac{5}{4}\right)^{2/3} = \left(\sqrt[3]{\frac{5}{4}}\right)^2 = \left(\frac{\sqrt[3]{5}}{\sqrt[3]{4}}\right)^2.$$

The laws of exponents apply in any problem involving roots or radicals. Therefore,

$$\sqrt[q]{a} \cdot \sqrt[q]{b} = a^{1/q} \cdot b^{1/q} = (ab)^{1/q} = \sqrt[q]{ab};$$

$$\frac{\sqrt[q]{a}}{\sqrt[q]{b}} = \frac{a^{1/q}}{b^{1/q}} = \left(\frac{a}{b}\right)^{1/q} = \sqrt[q]{\frac{a}{b}}, \quad b \neq 0.$$

Thus,

$$\sqrt[3]{4} \cdot \sqrt[3]{2} = \sqrt[3]{8} = 2$$

and

$$\frac{\sqrt[3]{4}}{\sqrt[3]{2}} = \sqrt[3]{\frac{4}{2}} = \sqrt[3]{2}.$$

However, *no such rules apply to the addition and subtraction of radicals*:

$$\sqrt[q]{a+b} \neq \sqrt[q]{a} + \sqrt[q]{b};$$
$$\sqrt[q]{a-b} \neq \sqrt[q]{a} - \sqrt[q]{b}, \quad a, b \neq 0.$$

For example, $\sqrt{9+16} \neq \sqrt{9} + \sqrt{16}$, since $\sqrt{9+16} = \sqrt{25} = 5$, $\sqrt{9} = 3$, and $\sqrt{16} = 4$, but $5 \neq 3 + 4$. Similarly, $\sqrt{25-9} \neq \sqrt{25} - \sqrt{9}$.

The following observation may prevent many embarrassing algebraic errors.

> If you are uncertain about an algebraic rule, see whether it works using numbers to replace the variables. If it doesn't work with arbitrary numbers, it can't be valid.

**Inequalities and Absolute Values**

There is frequent need in mathematics to compare two quantities. If they are not equal, we often can show one quantity to be less than the other. In this case, we use *inequalities* to express the relationship. The following notation is used.

> $x < y$ means that $x$ is *less than* $y$;
>
> $x \leq y$ means that $x$ is *less than or equal to* $y$;
>
> $x > y$ means that $x$ is *greater than* $y$;
>
> $x \geq y$ means that $x$ is *greater than or equal to* $y$.

For example,

$$2 < 3, \quad \text{and} \quad 5 > 4;$$

and the set

$$\{x \mid x \leq -\sqrt{2}\}$$

consists of all real numbers equal to or to the left of the number $-\sqrt{2}$ on the number line (see the darker line in Fig. 1.4).

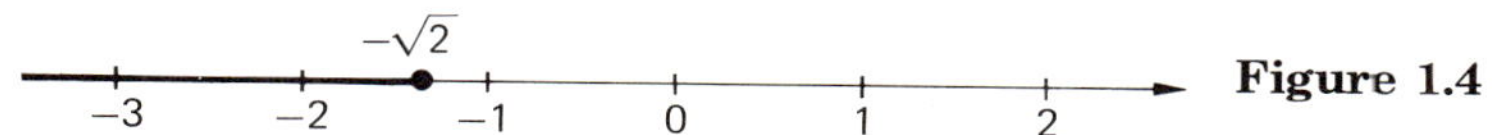


**Figure 1.4**

Two or more inequalities can be used in the same mathematical expression. For example,

$$2 < \sqrt{9} \leq 3;$$

and the set

$$\{x \mid 2 \leq x < \sqrt{10}\}$$

consists of all real numbers greater than or equal to 2, but less than the square root of 10 (see the darker line in Fig. 1.5).

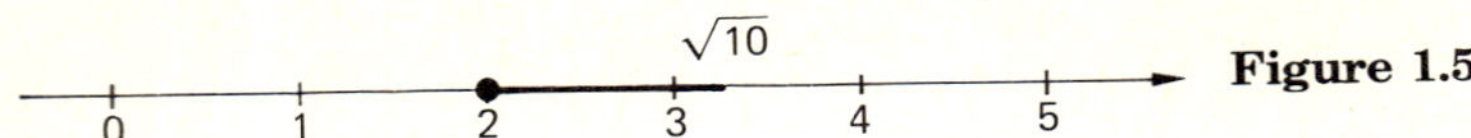

**Figure 1.5**

The *absolute value* of a number $a$ is the distance from $a$ to 0 on the number line, and is indicated by the symbol $|a|$. For example, the distances from 6 to 0 and from $-3$ to 0 are

$$|6| = 6 \quad \text{and} \quad |-3| = 3.$$

Observe that all the absolute value does is remove the sign from a signed number. We can also use absolute values in defining sets. For instance, the set

$$\{x \mid |x| \leq 2\}$$

consists of all numbers $x$ within at most 2 units of zero, and is shown as the darker line in Fig. 1.6.

**Figure 1.6**

We should note that the absolute value inequality $|x| \leq 2$ is equivalent to the two inequalities

$$-2 \leq x \leq 2.$$

Similarly, the set

$$\{x \mid |x + 3| < 5\} \tag{1.7}$$

consists of all points $x$ on the real line that, when added to the number 3, are less than 5 units from zero. To find what set of points this is we rewrite the absolute value inequality as the two inequalities

$$-5 < x + 3 < 5.$$

Now we subtract 3 from each of the three terms

$$-5 - 3 < x + 3 - 3 < 5 - 3,$$

obtaining

$$-8 < x < 2. \tag{1.8}$$

Thus, any number $x$ satisfying the inequality (1.8) lies in the set (1.7) (see the darker line in Fig. 1.7).

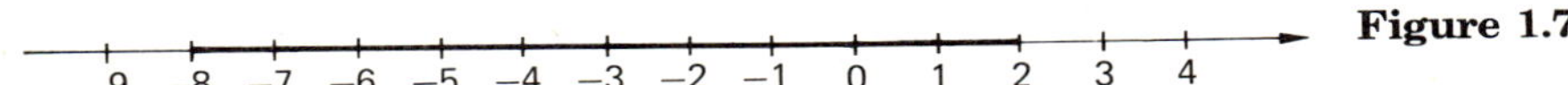

**Figure 1.7**

## EXERCISES 1.1

*Perform the indicated computation in Exercises 1–24.*

**1.** $2 + (-3)$

**2.** $3 - (-4)$

**3.** $(-2) + (-5)$

**4.** $(-2) - (-5)$

**5.** $(7)(-3)$

**6.** $(-3)(-7)$

**7.** $(-2)(5)$

**8.** $(-2) \div 5$

**9.** $(-8) \div (-2)$

**10.** $(8) \div (-2)$

**11.** $(3/4) \div (6/5)$

**12.** $(-3/2) \div (4/5)$

**13.** $(-3/2) \cdot (-4/5)$

**14.** $(5/6) \cdot (3/10)$

**15.** $(-2/7) \cdot (-14/15)$

**16.** $(-a/b) \cdot (-b/c)$

**17.** $(-a/b) \div [b/(-c)]$

**18.** $[x^2/(-a)] - (-x/a^2)$

**19.** $[x^2/(-a)] \cdot (-x/a^2)$

**20.** $(a^3a^4) \div a^5$

**21.** $(a^2 \cdot a^4) \div a^6$

**22.** $(a^4b) \div (a^2b^2)$

**23.** $(a^{1/2}a^{2/3}) \div a^{5/6}$

**24** $(a^3/b^2) \div (a^2/b^3)$

*Solve the following exercises involving complex numbers.*

**25.** $(2 + 3i) + (5 - i)$

**26.** $(3 - 4i) - (5 - 10i)$

**27.** $(3 + 4i)(2 + 3i)$

**28.** $(2 - i)(5 + 2i)$

**29.** $(-1 + i)(5 + 2i)$

**30.** $i(3i - 1)$

**31.** $(2 - i) \div i$

**32.** $(2 - i) \div 5i$

**33.** $(2 - i) \div (1 + i)$

**34.** $(2 - i) \div (1 - i)$

**35.** $(3 + 4i) \div (2 - i)$

**36.** $(5 + 2i) \div (3 + 4i)$

*Graph each of the following sets.*

**37.** $\{x \mid x < 3\}$

**38.** $\{x \mid x \geq -1\}$

**39.** $\{x \mid -1 \leq x < 4\}$

**40.** $\{x \mid |x| < 2\}$

**41.** $\{x \mid |x - 1| < 2\}$

**42.** $\{x \mid |x + 1| \leq 3\}$

**43.** Show that $\sqrt{3}$ is not a rational number. (*Hint*: Replace the word "even" by "divisible by 3" in the proof that $\sqrt{2}$ is not rational.)

**44.** Show that $\sqrt{5}$ is not a rational number.

**45.** Show that there is a rational number between any two different rational numbers.

**46.** Show that the number

$$x = 0.1\,2\,3\,2\,3\,2\,3\,2\,3\,2\,3\ldots,$$

where the group of digits 2 3 keeps repeating indefinitely, is a rational number. (*Hint*: Multiply $x$ by 100 and subtract.)

**47.** Show that any number whose decimal representation contains a repeated group of digits must be a rational number. (*Hint*: If $n$ is the number of digits in the repeating group, multiply the number by $10^n$ and subtract.)

**48.** Show the *triangle inequality*

$$|a + b| \leq |a| + |b|$$

by considering four possible cases:

a) $a < 0, b < 0$;
b) $a < 0, b \geq 0$;
c) $a \geq 0, b < 0$;
d) $a \geq 0, b \geq 0$.

---

## 1.2 FACTORING

This section is intended merely as a *review* of factoring. If you have never factored an algebraic expression before, you should consult any textbook on intermediate algebra.

The process of rewriting an algebraic expression as a product of two or more quantities is called *factoring*. Factoring is a very useful tool in simplifying algebraic operations, since common factors of the numerator and denominator of a fraction may be canceled. In other words, *cancellation is permissible only when the terms to be canceled are factors of the entire numerator and the entire denominator*. For instance, the $x$ terms may be canceled in the expression

$$\frac{x(y + 2)}{x(y - 2)}$$

but the $y$ terms and the twos may *not* be canceled.

Some common facts about factoring are listed below.

$$a^2 - b^2 = (a - b)(a + b) \qquad \textbf{(1.9)}$$

$$a^2 + 2ab + b^2 = (a + b)(a + b) = (a + b)^2 \qquad \textbf{(1.10)}$$

$$a^2 - 2ab + b^2 = (a - b)(a - b) = (a - b)^2 \qquad \textbf{(1.11)}$$

$$a^3 - b^3 = (a - b)(a^2 + ab + b^2) \qquad \textbf{(1.12)}$$

$$a^3 + b^3 = (a + b)(a^2 - ab + b^2) \qquad \textbf{(1.13)}$$

Here $a$ and $b$ may each stand for a more complicated expression. For example, if $a = 4x$ and $b = 3y$, Eq. (1.9) becomes

$$16x^2 - 9y^2 = (4x)^2 - (3y)^2 = (4x - 3y)(4x + 3y),$$

and Eq. (1.11) becomes

$$16x^2 - 24xy + 9y^2 = (4x)^2 - 2(4x)(3y) + (3y)^2$$
$$= (4x - 3y)^2.$$

An expression of the form $ax^2 + bx + c$ can always be factored by using the *quadratic formula*

$$x = \frac{-b \pm \sqrt{b^2 - 4ac}}{2a} \qquad \textbf{(1.14)}$$

to find the two roots

$$r_1 = \frac{-b + \sqrt{b^2 - 4ac}}{2a} \quad \text{and} \quad r_2 = \frac{-b - \sqrt{b^2 - 4ac}}{2a}$$

of the equation $ax^2 + bx + c = 0$. (A proof of this result can be found in Section 2.5.) Once we have computed $r_1$ and $r_2$, we may write

$$ax^2 + bx + c = a(x - r_1)(x - r_2). \qquad \textbf{(1.15)}$$

***Example 1*** Factor the expression $6x^2 + 13x + 6$.

We set $a = 6$, $b = 13$, and $c = 6$ and use the quadratic formula (1.14) to find the roots $r_1$ and $r_2$:

$$x = \frac{-13 \pm \sqrt{(13)^2 - 4(6)6}}{2(6)} = \frac{-13 \pm \sqrt{169 - 144}}{12}$$

$$= \frac{-13 \pm \sqrt{25}}{12} = \frac{-13 \pm 5}{12} = \begin{cases} -8/12 = -2/3, \\ -18/12 = -3/2. \end{cases}$$

Then

$$6x^2 + 13x + 6 = 6(x - (-2/3))(x - (-3/2))$$
$$= 3 \cdot 2(x + 2/3)(x + 3/2)$$
$$= (3x + 2)(2x + 3). \qquad \textbf{(1.16)}$$

Alternatively, we note that $6 = 3 \cdot 2$ and $2 \cdot 2 + 3 \cdot 3 = 13$, hence,

$$(3x + 2)(2x + 3) = (3x)(2x) + (3x)3 + 2(2x) + 2 \cdot 3$$
$$= 6x^2 + 13x + 6.$$

## EXERCISES 1.2

*Factor the following algebraic expressions.*

1. $9x^2 - 4$
2. $27y^3 + 8x^3$
3. $16x^2 - 49y^2$
4. $x^2 - x - 2$
5. $16x^2 + 56xy + 49y^2$
6. $x^2 - 2x - 3$
7. $9x^2 - 24xy + 16y^2$
8. $2x^2 - 5x + 3$
9. $8x^3 - y^3$
10. $2x^2 - 7x + 3$
11. $27x^3 - 8y^3$
12. $2x^2 - 5x - 3$

*Use the quadratic formula to find the factors of the following equations.*

13. $-x^2 - 3x + 2$
14. $x^2 - 2x - 1$
15. $x^2 - 3x + 1$
16. $x^2 + x + 1$
17. $x^2 - 3x - 5$
18. $x^2 + x + 2$
19. $4x^2 - 5x + 9$
20. $7x^2 - 2x - 4$
21. $3x^2 + 7x + 2$
22. $9x^2 + 4x - 7$

---

## 1.3 PASCAL'S TRIANGLE

We will encounter several important concepts in this book that will require us to consider positive integral powers of a binomial expression, that is, an algebraic expression of the form $(a + b)^n$. While it is not difficult to calculate the first few such powers by repeated multiplication, it is clear that we would not be willing to perform such a procedure for much larger values of $n$. For example, for $n = 1, 2, 3$, and 4 we have

$$(a + b)^1 = a + b; \tag{1.17}$$

$$(a + b)^2 = a^2 + 2ab + b^2; \tag{1.18}$$

$$\begin{aligned}(a + b)^3 &= (a + b)(a^2 + 2ab + b^2)\\ &= a^3 + 2a^2b + ab^2 + a^2b + 2ab^2 + b^3\\ &= a^3 + 3a^2b + 3ab^2 + b^3;\end{aligned} \tag{1.19}$$

and

$$\begin{aligned}(a + b)^4 &= (a + b)(a^3 + 3a^2b + 3ab^2 + b^3)\\ &= a^4 + 3a^3b + 3a^2b^2 + ab^3 + a^3b + 3a^2b^2 + 3ab^3 + b^4\\ &= a^4 + 4a^3b + 6a^2b^2 + 4ab^3 + b^4.\end{aligned} \tag{1.20}$$

There is, however, a much simpler way to obtain these expansions by a procedure known as *Pascal's triangle.* To form Pascal's triangle we construct a triangular array topped by the single entry 1, with ones down the two sides which meet at the top (Fig. 1.8). To fill in the interior numbers we simply add the two numbers in the array immediately above and to the left and right of the number we wish to find. There is nothing to find in the zeroth and first rows, but we see that the two in the middle of the second row is the sum of the two ones above it. Similarly, the threes in the third row are the sums of the ones and twos immediately above them.

If we compare the coefficients in the rows of Pascal's triangle with the coefficients of the expressions $(a + b)^n$ that we computed in Eqs. (1.17)–(1.20), we see that

$$(a + b)^0 = 1;$$

$$(a + b)^1 = 1 \cdot a + 1 \cdot b;$$

$$(a + b)^2 = 1 \cdot a^2 + 2 \cdot ab + 1 \cdot b^2,$$

$$(a + b)^3 = 1 \cdot a^3 + 3 \cdot a^2b + 3 \cdot ab^2 + 1 \cdot b^3;$$

and

$$(a + b)^4 = 1 \cdot a^4 + 4 \cdot a^3b + 6 \cdot a^2b^2 + 4 \cdot ab^3 + 1 \cdot b^4.$$

Thus, the coefficients in Pascal's triangle are exactly the same numbers as those we found in the expansion of $(a + b)^n$. (This fact is not surprising if we carefully examine what happens when we multiply the

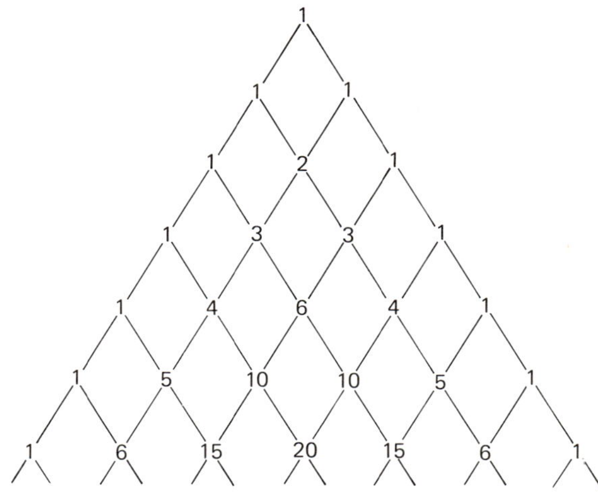


**Fig. 1.8**
Pascal's triangle.

expansion of any binomial by $a + b$.) Therefore, we may use Pascal's triangle to quickly write the result of raising a binomial to an integral power.

***Example 1*** Compute $(a + b)^7$.

In this expression the sum of the powers of the $a$'s and $b$'s must equal 7. Starting with $a^7b^0$ we decrease the exponent of $a$ by one while increasing the exponent of $b$ by one, and multiply each consecutive term by the next entry in the 7th row of Pascal's triangle. Finally, we add all the terms together. That is, we compute

$$\begin{aligned}(a + b)^7 &= 1 \cdot a^7b^0 + 7 \cdot a^6b^1 + 21 \cdot a^5b^2 + 35 \cdot a^4b^3 + 35 \cdot a^3b^4 \\ &\quad + 21 \cdot a^2b^5 + 7a^1b^6 + 1 \cdot a^0b^7 \\ &= a^7 + 7a^6b + 21a^5b^2 + 35a^4b^3 + 35a^3b^4 + 21a^2b^5 \\ &\quad + 7ab^6 + b^7.\end{aligned}$$

***Example 2*** Calculate $(2x + 3y)^5$ using Pascal's triangle.

Here we let $a = 2x$ and $b = 3y$ and compute $(a + b)^5$:

$$(a + b)^5 = 1 \cdot a^5b^0 + 5 \cdot a^4b^1 + 10 \cdot a^3b^2 + 10 \cdot a^2b^3 + 5 \cdot a^1b^4 + 1 \cdot a^0b^5$$

$$= a^5 + 5a^4b + 10a^3b^2 + 10a^2b^3 + 5ab^4 + b^5. \tag{1.21}$$

Replacing each $a$ in Eq. (1.21) with $2x$ and each $b$ with $3y$, we have

$$(2x + 3y)^5 = (2x)^5 + 5(2x)^4(3y) + 10(2x)^3(3y)^2 + 10(2x)^2(3y)^3 + 5(2x)(3y)^4 + (3y)^5$$

$$= 32x^5 + 240x^4y + 720x^3y^2 + 1080x^2y^3 + 810xy^4 + 243y^5.$$

***Example 3*** Calculate $(7x - 2y^2)^4$.

Here we set $a = 7x$ and $b = -2y^2$, and calculate:

$$(a + b)^4 = a^4 + 4a^3b + 6a^2b^2 + 4ab^3 + b^4,$$

so that

$$(7x - 2y^2)^4 = (7x)^4 + 4(7x)^3(-2y^2) + 6(7x)^2(-2y^2)^2 + 4(7x)(-2y^2)^3 + (-2y^2)^4$$

$$= 2401x^4 - 2744x^3y^2 + 1176x^2y^4 - 224xy^6 + 16y^8.$$

---

The entries in Pascal's triangle are called *binomial coefficients*, since they arise in the expansion of binomials. The $(k + 1)$st entry in the $n$th row of Pascal's triangle is often denoted by $C_{n,k}$.* Using this symbol we can rewrite the expression $(a + b)^n$ in the form

$$(a + b)^n = C_{n,0}a^n + C_{n,1}a^{n-1}b + C_{n,2}a^{n-2}b^2 + \cdots + C_{n,n-1}ab^{n-1} + C_{n,n}b^n.$$

It is easy to remember the symbol that is associated with each term since *the second number $k$ equals the exponent of the $b$ term.*

---

* Symbols other than $C_{n,k}$ that are sometimes used are

$${}_nC_k, \quad {}^nC_k, \quad C(n, k), \quad \binom{n}{k}.$$

## EXERCISES 1.3

*Expand the following binomials.*

| | | |
|---|---|---|
| **1.** $(a - b)^3$ | **2.** $(a - b)^5$ | **3.** $(2a + b)^4$ |
| **4.** $(3a - 2b)^4$ | **5.** $(3a + 2b)^6$ | **6.** $(2a - 4b)^6$ |
| **7.** $(2x - 4y)^5$ | **8.** $(\frac{1}{2}x - 3y)^4$ | **9.** $(\frac{1}{2}x + \frac{1}{3}y)^5$ |
| **10.** $(x + 1/x)^4$ | **11.** $(x - 1/x)^5$ | **12.** $(x - 2/x)^4$ |
| **13.** $[(x + 1) - y]^5$ | **14.** $(x - 4)^7$ | **15.** $(2x - 3)^5$ |
| **16.** $(9x - 2)^4$ | **17.** $(9 + 0.7)^3$ | **18.** $(8 - 0.3)^3$ |
| **19.** $(1 - 0.2)^5$ | **20.** $(100 - 1)^5$ | **21.** $(1 + 0.02)^4$ |

---

## CHAPTER VOCABULARY

**absolute value** $|a|$
**binomial** $(a + b)^n$
**binomial coefficient** $C_{n,k}$
**complex arithmetic**
**complex numbers** $a + ib$
**exponent** $a^n$
**factoring**
**imaginary unit** $i$
**inequalities** $<, \leq, >, \geq$
**integers** (= whole numbers)
**Pascal's triangle**
**quadratic formula** $x = \dfrac{-b \pm \sqrt{b^2 - 4ac}}{2a}$
**radicals** $\sqrt[n]{a}$
**rational numbers** (= fractions) $p/q$
**real numbers**
**repeating decimals**
**roots of an equation**
**roots of a number** $\sqrt[n]{a}$
**set** (= collection)
**signed numbers** $\pm a$
**triangle inequality** $|a + b| \leq |a| + |b|$

# PART I ELEMENTARY FUNCTIONS and SYSTEMS

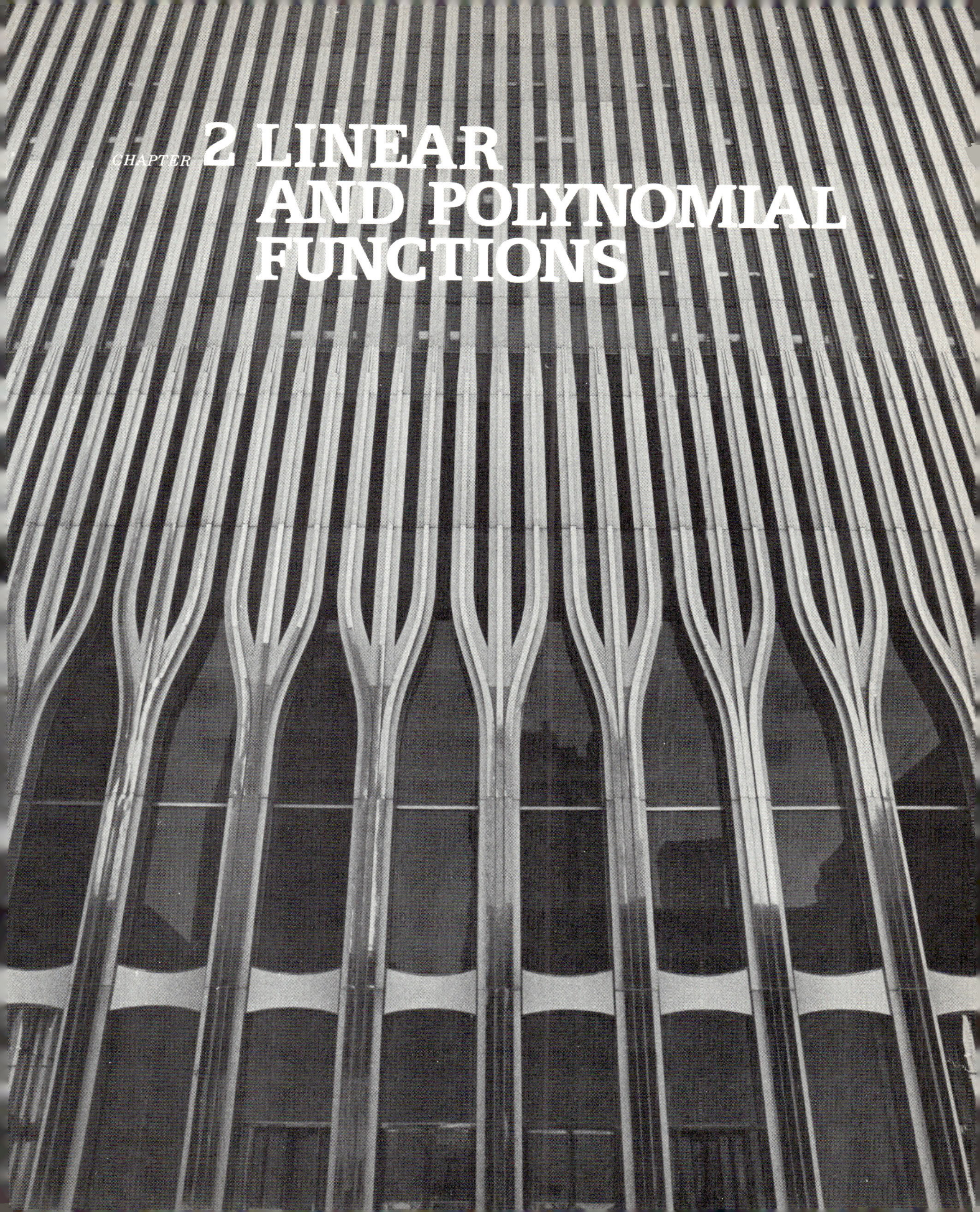

CHAPTER 2 LINEAR AND POLYNOMIAL FUNCTIONS

## 2.1 FUNCTIONS AND GRAPHS

In our modern world we are constantly assaulted by large quantities of numerical information. Football scores and rankings, stock market reports, and new and used car prices are a few of the more common lists or *tables* found in every daily newspaper. Often this data has been arranged as a visual display, such as the daily close of the Dow-Jones averages over the period of a week (see Fig. 2.1). Displays of this kind are *graphs*, and are useful to the investor in determining the trend of the market. Every student is also familiar with mathematical formulas, called *functions* and *relations*, which convey numerical information in a compact manner. For example, the formula $A = \pi r^2$ determines the area of any circle of radius $r$. It might seem that no connection exists among tables, graphs, and formulas, but actually all three concepts are simply different ways of describing the correspondence between two (or more) variables. We will devote several sections in this and the next chapter to discussing the methods used to transform any one of these concepts into the other two. First, however, we must explain some basic terminology.

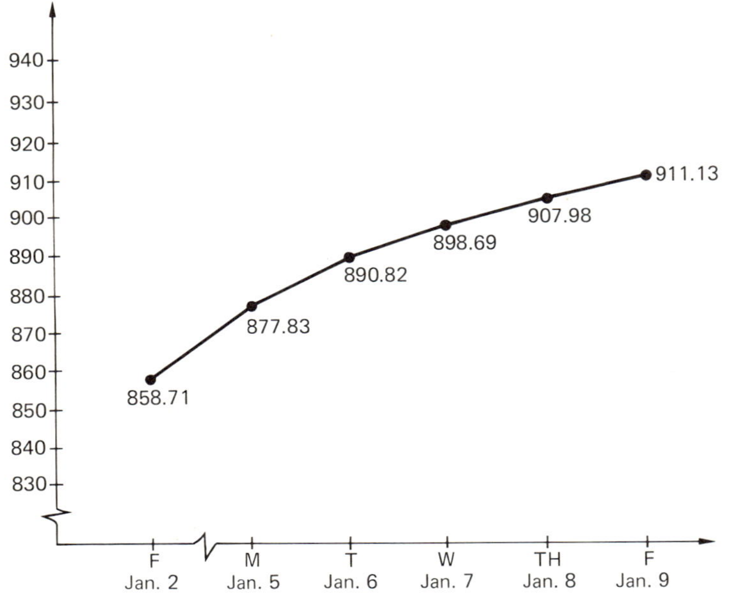


**Fig. 2.1** Dow-Jones average of 30 industrials, January 2–9, 1976.

### Graphs and Coordinate Systems

We will begin by discussing the simplest model of the plane, which is called the *Cartesian coordinate system*.* It consists of the set of all points on and surrounding a pair of perpendicular lines called the

* Named after the French mathematician and philosopher René Descartes (1596–1650).

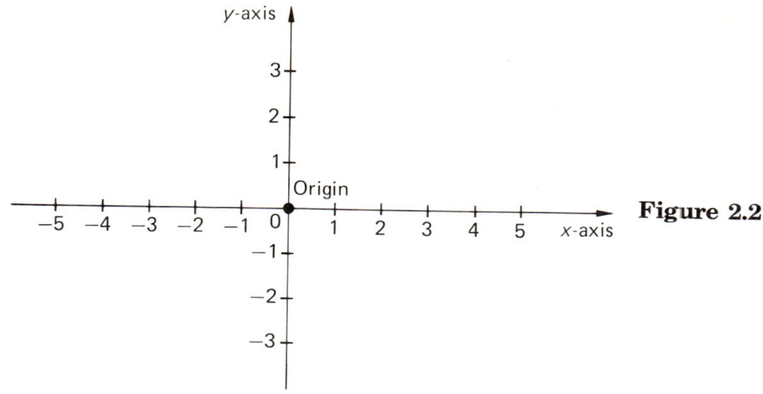


**Figure 2.2**

coordinate axes: The horizontal axis is called the *abscissa* or *x-axis*, while the vertical axis is the *ordinate* or *y-axis*. The point at which the two axes intersect is called the *origin* of the coordinate system. Starting with zero at the origin, each axis is marked off in equally spaced units in increasing order as we move right for the $x$-axis and up for the $y$-axis (see Fig. 2.2). Each point in the plane is represented by a pair of real numbers $(x, y)$ called the *coordinates* of the point. By convention, *the x-coordinate is always given first.* We can locate, or *plot*, the point $(x_0, y_0)$ by erecting perpendiculars to the axes at the points $x_0$ on the $x$-axis and $y_0$ on the $y$-axis. The point at which the two perpendiculars intersect is the point $(x_0, y_0)$ (see Fig. 2.3). Actually, it is simpler to plot $(x_0, y_0)$ by moving first to the value $x_0$ on the $x$-axis, and then proceeding vertically a distance of $y_0$.

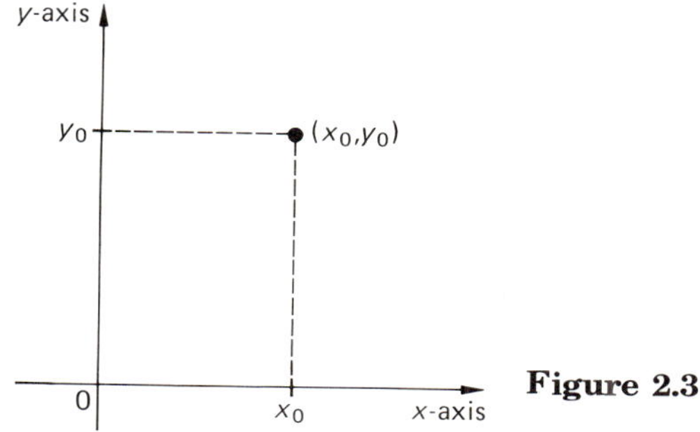


**Figure 2.3**

***Example 1*** Plot the points $(2, 0)$, $(3, 2)$, $(-2, 4)$, $(3, -2)$, and $(-3, -1)$. These points appear in Fig. 2.4. Observe that if the first coordinate is negative we move left of the origin, while if the second coordinate is negative we move down from the origin. Clearly, since the origin is the point $(0, 0)$,

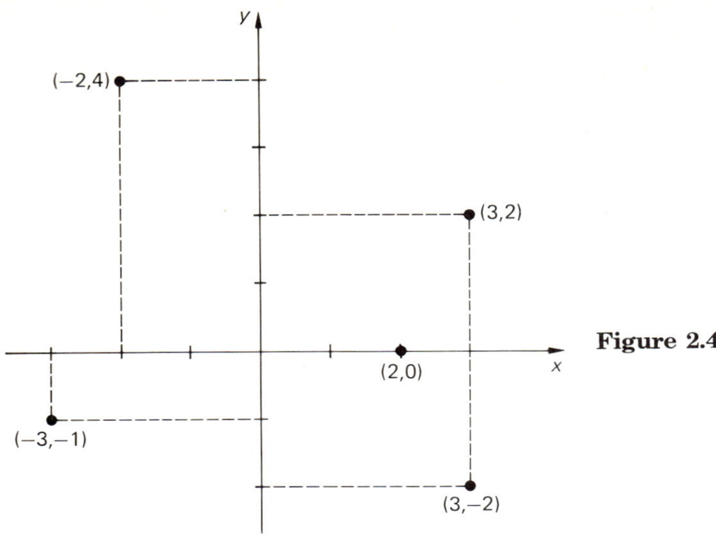


Figure 2.4

points on the $x$-axis have a $y$-coordinate equal to zero, and points on the $y$-axis have an $x$-coordinate equal to zero.

Actually the concept of plotting is a familiar one. Frequently in asking directions we are given locations such as 21st South and 28th East. This address indicates that we must go 21 blocks south of the main East–West avenue and 28 blocks east of the main North–South avenue. The two main avenues are the coordinate axes of the system, and their point of intersection (the origin) is usually the center of the town.

After we have plotted the data on the coordinate system, we call the resulting set of points the *graph* of the data. For example, Fig. 2.4 is the graph of the set of points $(2, 0)$, $(3, 2)$, $(-2, 4)$, $(3, -2)$, and $(-3, -1)$. Of course, we are not limited to graphs of finite sets of points. For example, the diagonal line shown in Fig. 2.5(a) is the graph of the set of all points whose $x$- and $y$-coordinates are equal, while the shaded region in Fig. 2.5(b) is the graph of all points whose $x$- and $y$-coordinates are both positive.

***Example 2*** As we indicated at the beginning of this section, graphs are a useful way of describing the relationship between two variables. Figure 2.6 shows the percentage of the United States population living in urban areas from 1800 to 1960. The definite trend toward urban living is easily recognizable, even though we have only a rough idea as to what the exact percentage was in any given year. In general, graphical methods pro-

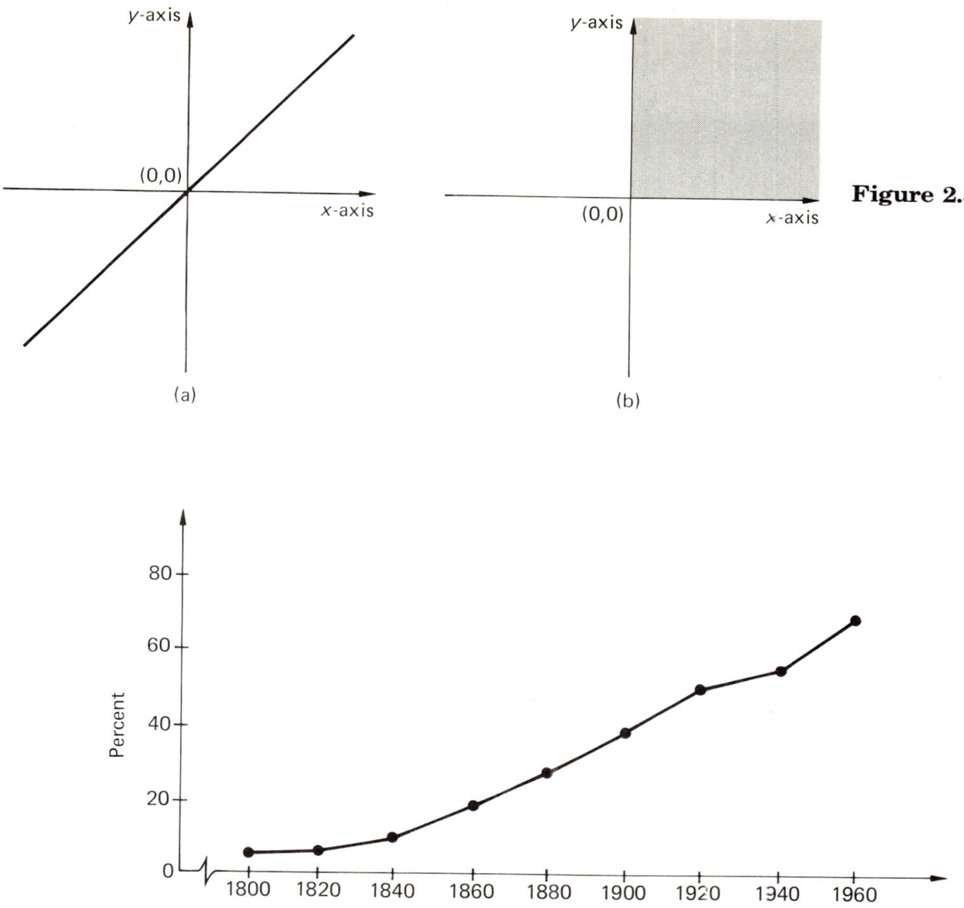


**Figure 2.5**

**Fig. 2.6**
Percent of U.S. population living in urban areas (data obtained from *Information Please Almanac, 1969*).

vide the quickest way of understanding the relationship between two variables, but they often do so at the expense of accuracy.

---

It is not necessary that either axis in a coordinate system be the set of real numbers. In Fig. 2.1, the units on the $x$-axis consist of the days of the week. The following example shows that both axes can be nonnumerical.

***Example 3*** Suppose we wish to make a graph of the final letter grades that are obtained in your class. Each point can then be described by the pair (name of the student, grade in the class). To construct the graph, we space the names of the students equally on the $x$-axis, and mark the letters A, B, C, D, F at equally spaced intervals on the $y$-axis. (If your school uses a different grading scale, you will have to mark the $y$-axis accordingly.) Now we merely have to plot the points (name, grade) to obtain the desired graph.

---

**Tables** Tables are a convenient method of collecting information when greater accuracy is required than can be provided by graphs. Essentially, a *table* is a list that associates a number $y$ with each given number $x$. Thus, a table really consists of a list of points $(x, y)$. Should we wish to construct a graph from a table, we need only plot the given points. On the other hand, we can easily construct a table from a graph by selecting a set of $x$-values and measuring their corresponding $y$-values from the graph. For example, Table 2.1 lists a few of the points on the graph of Fig. 2.6.

**Table 2.1**
**Percentage of population living in urban areas. Data obtained from *Information Please Almanac, 1969.*)**

| *Year* ($x$) | *Percent* ($y$) |
|---|---|
| 1800 | 6.1 |
| 1820 | 7.2 |
| 1840 | 10.8 |
| 1860 | 19.8 |
| 1880 | 28.2 |
| 1900 | 39.7 |
| 1920 | 51.2 |
| 1940 | 56.5 |
| 1960 | 69.9 |

In later sections of this book we will encounter a number of different tables: common logarithms, tables of integrals, etc. At that time we will explain carefully how each is used.

**Functions and Relations** The most accurate way of describing the correspondence between two variables—say, $x$ and $y$—is through a mathematical formula, called a *relation*. Unfortunately, it is often very difficult to find such relations; their discovery constitutes the main goal of most scientific investigations.

To be precise, a *relation is simply any subset of the set of all pairs* $(x, y)$.

***Example 4*** The formula

$$xy = 0 \tag{2.1}$$

determines a relation, since its graph consists of the set of all points $(x, y)$ in the Cartesian coordinate system lying on either of the coordinate axes. It is obvious that every point on one of the coordinate axes satisfies Eq. (2.1), since one of its coordinates must equal zero. That these are the only points that satisfy Eq. (2.1) is also clear, since neither coordinate of any point not on the coordinate axes is zero; thus their product is not zero.

---

As we can see by this example, graphs play an important role in helping us understand the properties of a given relation. It is usually easy to construct a graph from a relation. First we consider the set of all $x$-values (first components) of the pairs $(x, y)$ defining the relation. This set of $x$-values is called the *domain* of the relation. Then we use the relation to compute the $y$-values that correspond to each $x$ in the domain. The set of $y$-values (second components) defining a relation is usually called the *range* of the relation. Once "enough" of these points have been calculated, we may plot them to obtain the graph of the relation.

One type of relation is so important that it is called by a special name. A *function is a relation having the property that exactly one $y$-element corresponds to each $x$-element in its domain.* Note that this means that each vertical line in the $xy$-plane meets the set of points of the function either not at all or at only a single point $(x, y)$. It is easiest to remember the concept of a function as a *rule or formula that assigns exactly one value $y$ to each value $x$ in the domain.*

We have already seen an example of a relation that is a function and an example of one that is not a function. The diagonal line in Fig. 2.5(a) of all points $(x, y)$ whose $x$- and $y$-coordinates are identical is a function, since only one possible $y$-coordinate will equal a given $x$-coordinate. A simple formula describes this function completely:

$$y = x.$$

This expression indicates that whatever value is selected for $x$ is also assigned to $y$. On the other hand, the relation $xy = 0$, in Example 4, is definitely not a function, since the entire $y$-axis corresponds to the value $x = 0$.

In these two examples the domain has consisted of all real numbers $x$. The expression

$$y = \sqrt{1-x^2}$$

provides a single $y$-value for each value of $x$ between $-1$ and 1. When $x$ lies outside of this range no real value of $y$ can be found. Thus the domain of this function is the set $-1 \leq x \leq 1$ and the range is $0 \leq y \leq 1$. The graph of this function is the semicircle shown in Fig. 2.7.

Since a function is a rule assigning exactly one $y$-value to each $x$-value in the domain, it is frequently possible to obtain a formula of the form

$$y = f(x),$$

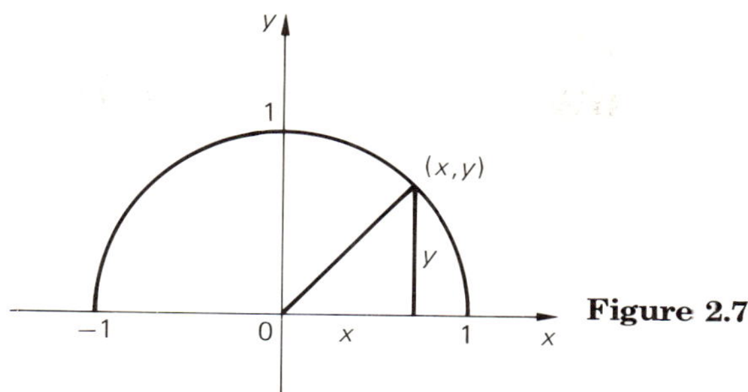


Figure 2.7

which precisely describes the correspondence. This equation is read "y is a function of x" and provides a convenient mathematical shorthand called *functional notation.* To understand how it works, consider the following example.

***Example 5*** Suppose that

$$f(x) = x^2 + 2.$$

If we write $f(0)$ we mean that each $x$ in the right side of the equation is to be replaced by 0. Hence,

$$f(0) = 0^2 + 2 = 2.$$

Similarly,

$$f(2) = 2^2 + 2 = 6,$$

$$f(-3) = (-3)^2 + 2 = 11.$$

and

$$f(12) = (12)^2 + 2 = 146.$$

It is not necessary that the values of $x$ be numbers. For example,

$$f(a) = a^2 + 2,$$

$$f(a + b) = (a + b)^2 + 2,$$

and

$$f(x + h) = (x + h)^2 + 2.$$

To reiterate, the notation $f(z)$ means that every $x$ in the definition of $f(x)$ is to be replaced by $z$, regardless of what $z$ might be.

In the definition of a function, it is important to observe that the $x$- and $y$-values play quite different roles. No harm is done if the same $y$-value is assigned to two different values of $x$. For instance, consider the semicircle $y = \sqrt{1 - x^2}$. The value $y = 0$ is assigned to both $x = -1$ and $x = 1$. Thus, the variables are generally not interchangeable in the construction of a function. We indicate this by calling $x$ the *independent* variable and $y$ the *dependent* variable. The following example illustrates the care that must be exercised in defining a function.

***Example 6*** The science of ballistics is based on the fact that each bullet fired from a given revolver acquires a distinctive set of markings induced by the rifling of the barrel. Suppose we have five bullets $b_1$, $b_2$, $b_3$, $b_4$, and $b_5$, each of which has been fired from one of a set of three revolvers $r_1$, $r_2$, and $r_3$. Which of the variables (the $b$'s or the $r$'s) should be the independent variable if we wish to construct a function showing which bullet came from which revolver? Clearly, more than one bullet must have been fired from at least one of the revolvers. Therefore, if we consider the revolvers the independent variable, we will not obtain a function. (Remember that each $x$ can have at most one $y$). Choosing the bullets as the independent variable clearly yields a function, since each bullet could have been fired from only one revolver.

---

**EXERCISES 2.1**

**1.** Plot the points $(2, 5)$, $(-3, 4)$, $(4, 1)$, $(0, 5)$, and $(-1, 0)$.

**2.** Plot the points $(2, 3)$, $(4, 1)$, $(-1, -3)$, and $(-3, 0)$.

**3.** Draw the straight-line graph that passes through the points $(2, 3)$ and $(4, 1)$. What $y$-value corresponds to the value $x = 3$? to the value $x = 7$?

**4.** Answer Exercise 3 for the points $(-1, -3)$ and $(5, 9)$.

*In Exercises 5–12, draw the graph of the given relation and decide whether the relation is a function*

**5.** $x + y = 0$

**6.** $y = 0$

**7.** $y/x = 0$

**8.** $x^2 + y^2 = 0$

**9.** $x^2 = y^2$

**10.** $y = \sqrt{x}$

**11.** $x/y = 0$

**12.** $x^2 + y^2 = 16$

*In Exercises 13–20, find the values $f(0)$, $f(2)$, $f(5)$, $f(-1)$, $f(-3)$, $f(a)$, and $f(a + b)$ of the given function.*

**13.** $f(x) = x + 2$

**14.** $f(x) = 3x - 1$

**15.** $f(x) = x^2$

**16.** $f(x) = 2x^2 - 3$

**17.** $f(x) = x^2 - x$

**18.** $f(x) = x^3$

**19.** $f(x) = \sqrt{100 - x^2}$

**20.** $f(x) = \frac{x}{1+x}$

**21.** Can the relation between the human population and the set of all their fingerprints be chosen in such a way that it is a function?

*In Exercises 22–25, draw a graph of the given relation*

**22.** $y \geq x$

**23.** $y < 0$

**24.** $x^2 + y^2 \leq 1$

**25.** $xy < 0$

**26.** The ABO blood groups are theoretically explained by the presence of a combination of two alleles or genes at the same gene locus on a person's chromosomes. The three types of alleles are represented by the letters a, b, and o. The alleles a and b are known to be dominant over o. Each individual obtains one allele from each parent and the gene combination received determines the individual's blood type. Individuals carrying oo have neither antigen A nor B in their blood, so they are said to have type O blood. Either of the combinations aa or ao produces an individual with type A blood, while bb and bo lead to type B blood and ab to type AB blood. If we call the allele carried by the sperm cell the $x$-variable and that carried by the egg cell the $y$-variable, is the relation of all individuals with type B blood a function? Which blood types do lead to functions? Which do not?

**27.** Plot the following data concerning the number of mass transit riders between 1940 and 1953. Construct a curve that passes through all the data points by drawing a straight line joining each point to the data point with the next largest date. Call this curve the graph of the data. Which variable should be chosen as the independent variable to guarantee that the graph is a function?

| Year | 1941 | 1943 | 1944 | 1946 | 1948 | 1950 | 1953 |
|---|---|---|---|---|---|---|---|
| Passengers (billions) | 14 | 22 | 22.5 | 24 | 21.5 | 17.5 | 14 |

*Source*: American Public Transit Association.

## 2.2 LINEAR EQUATIONS

Jack works as a chemical technician in a dye factory. He earns \$6 per hour plus double pay for any overtime work beyond the normal 40 hours per week. Is there an equation that describes his pay scale?

The base pay for Jack's normal 40-hour week is \$240. In addition, he earns \$12 per hour for each hour of overtime he works. If we let $x$

represent the number of hours of overtime work, then the pay he should receive is

$$y = 12x + 240,$$

or \$240 plus \$12 times the number of hours of overtime. Thus, a 50-hour week would result in earnings of

$$y = 12(10) + 240 = \$360$$

(less taxes) since only 10 of these hours are overtime.

We might prefer to draw a graph of the equation

$$y = 12x + 240.$$

To do so we construct a table by substituting several different values for $x$ into this equation and calculating the resulting values of $y$. Some typical values are shown in Fig. 2.8(a). Plotting these values in Fig. 2.8(b) we see that the points that we have plotted all lie on a straight line. If we were to continue to plot points for all possible values of $x$, including fractions of an hour, we would discover that all these points would also lie on the straight line shown in Fig. 2.8(b). Thus, the equation

$$y = 12x + 240$$

has a straight line for its graph.

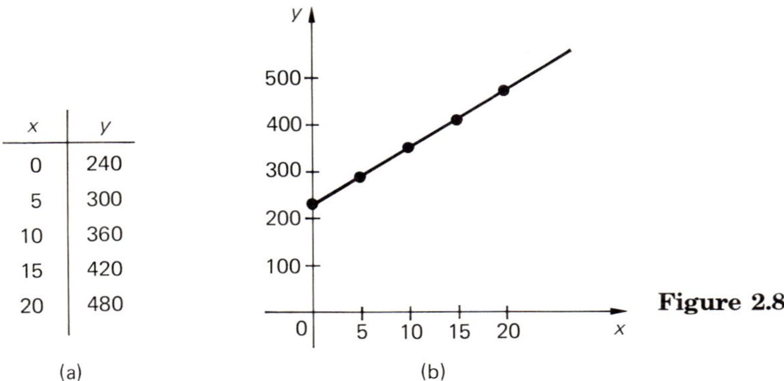

| x | y |
|---|---|
| 0 | 240 |
| 5 | 300 |
| 10 | 360 |
| 15 | 420 |
| 20 | 480 |

(a)



**Figure 2.8**

If we wish to generalize the work we have done above, we study the properties of the equation

$$y = mx + b, \tag{2.2}$$

where $m$ and $b$ are constant (real) numbers. To see what this function looks like, we again graph a number of points that satisfy Eq. (2.2). We can begin by setting $x = 0$; then $y = m \cdot 0 + b = b$, so the point $(0, b)$ lies on the graph. Setting $x = 1$, we obtain $y = m \cdot 1 + b = m + b$, so that $(1, m + b)$ lies on the graph. Continuing in this fashion, we would again discover that all solutions to Eq. (2.2) lie on a straight line that crosses the $y$-axis at $y = b$ and rises by $m$ units each time we increase the value of $x$ by 1 (see Fig. 2.9). We call the value $b$, where the line crosses the $y$-axis, the *y-intercept* of the line. The number $m$, which equals the number of units that the line rises as it moves one unit to the right, is called the *slope* of the line, so that

$$\text{slope} = m = \frac{\text{rise}}{\text{run}}. \tag{2.3}$$

For these reasons, Eq. (2.2) is called the *slope-intercept* formula of the line.

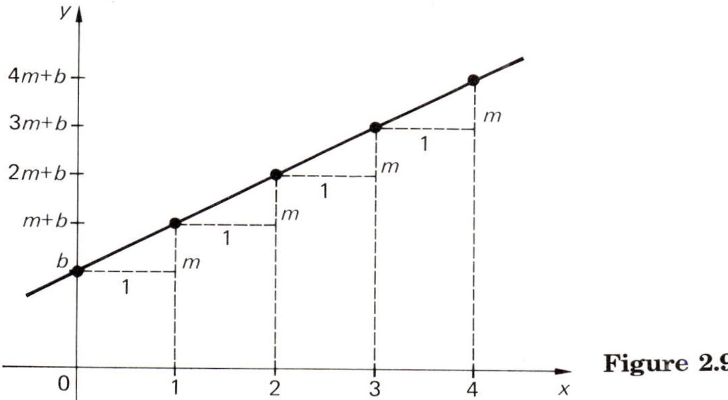


**Figure 2.9**

Since a line is determined by any two of its points, we can easily graph any equation of the form (2.2) by drawing the line that passes through the $y$-intercept $b$ and any other point of the line—say, $(1, m + b)$.

***Example 1*** Graph the line

$$y = -3x + 4. \tag{2.4}$$

Since $-3(1) + 4 = -3 + 4 = 1$, we draw the straight line passing through the $y$-intercept 4 and the point (1, 1) obtained by substituting the value $x = 1$ in Eq. (2.4) (see Fig. 2.10a).

---

Suppose, instead, that we wish to obtain the equation associated with the graph in Fig. 2.10(b). Here we know that the $y$-intercept is $b = 2$ and that the line passes through the point $(4, 0)$. To write Eq. (2.2) we need only calculate the number $m$ (since $b = 2$ is already known). Recalling that the slope equals the rise over the run (see Eq. 2.3) we note that the line *falls* two units over a run of four units. Thus, the rise is $-2$ and the run is 4, and we have

$$m = \frac{\text{rise}}{\text{run}} = \frac{-2}{4} = \frac{-1}{2}.$$

Therefore, by the slope-intercept formula, the equation of this line is

$$y = -\tfrac{1}{2}x + 2.$$

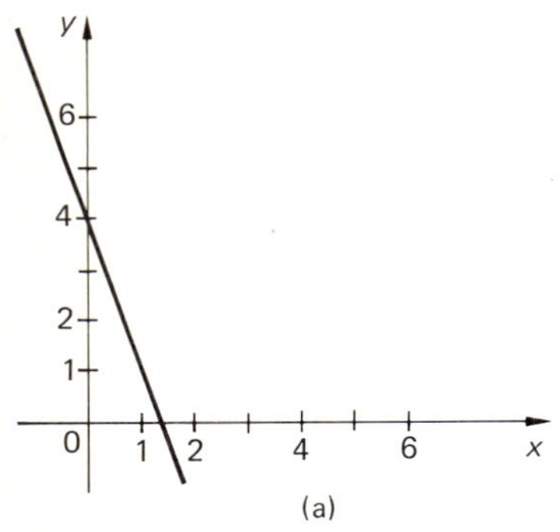


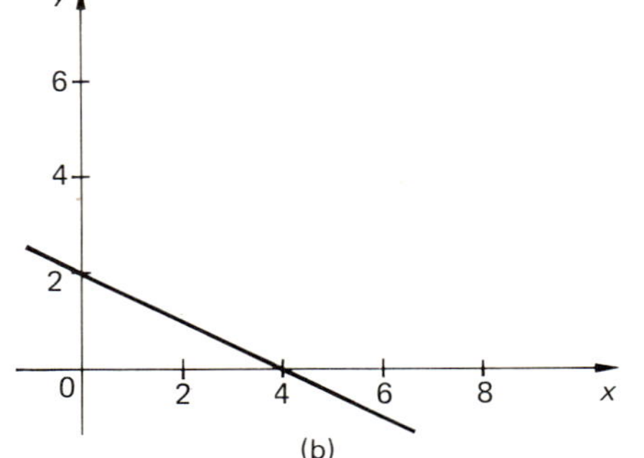


Figure 2.10

**Point-slope formula** Suppose we know that a line has slope $m$ and passes through the point $(x_0, y_0)$. Although in this case the $y$-intercept is unknown, we can easily compute $b$ by substituting the numbers $m$, $x_0$, and $y_0$ in Eq. (2.2):

$$y_0 = mx_0 + b, \qquad \textbf{(2.5)}$$

and solving Eq. (2.5) for $b$. We illustrate this process with an example.

***Example 2*** Find the equation of the line with slope $m = 3$ passing through the point $(4, 7)$. Here $x_0 = 4$ and $y_0 = 7$ and Eq. (2.2) yields

$$7 = 3(4) + b$$

or, subtracting 12 from each side of the equation,

$$b = 7 - 12 = -5.$$

Thus, $y = 3x - 5$ is the desired equation.

Equation (2.5) indicates that $b = y_0 - mx_0$. Substituting this value into Eq. (2.2) we have

$$y = mx + (y_0 - mx_0).$$

Collecting the terms having a factor of $m$ yields the *point-slope formula*

$$y = m(x - x_0) + y_0$$

or, equivalently,

$$\frac{y - y_0}{x - x_0} = m. \tag{2.6}$$

However, we suggest that you learn the simple procedure shown in Example 2 rather than memorize the formula in Eq. (2.6).

**Two-point formula** This time we wish to find the equation of the line that passes through two given points $(x_0, y_0)$ and $(x_1, y_1)$, $x_0 \neq x_1$. Since the slope is not given, we must find it by using its definition as rise over run (see Fig. 2.11). From Fig. 2.11 we see that the rise equals $y_1 - y_0$, while the run equals $x_1 - x_0$, hence,

$$m = \frac{y_1 - y_0}{x_1 - x_0}. \tag{2.7}$$

We can now apply the point-slope formula (2.6), using the slope we have just calculated and the point $(x_0, y_0)$, to obtain the *two-point formula*

$$\frac{y - y_0}{x - x_0} = \frac{y_1 - y_0}{x_1 - x_0} \tag{2.8}$$

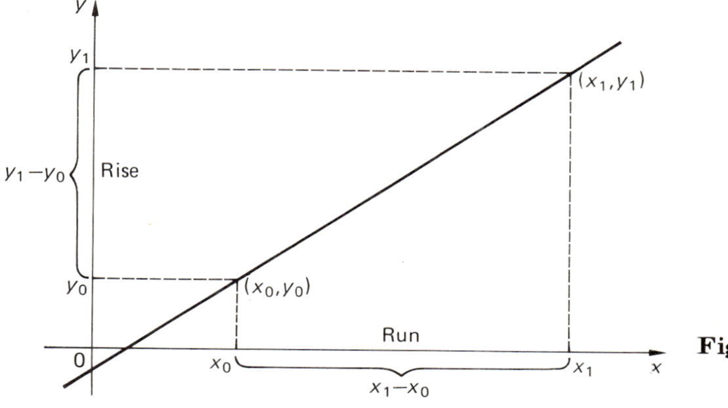


**Figure 2.11**

or, equivalently,

$$y = \frac{y_1 - y_0}{x_1 - x_0}(x - x_0) + y_0.$$

Again, we recommend learning the method rather than memorizing the formula in Eq. (2.8).

***Example 3*** Find the equation of the straight line that passes through the points (2, 3) and (4, 1).

Here we let $(x_0, y_0) = (2, 3)$ and $(x_1, y_1) = (4, 1)$. Then since the slope is the rise over the run, we have

$$m = \frac{y_1 - y_0}{x_1 - x_0} = \frac{1 - 3}{4 - 2} = \frac{-2}{2} = -1.$$

Substituting this value and the point (2, 3) in the slope-intercept formula (2.2), we obtain

$$3 = y_0 = mx_0 + b = (-1)(2) + b,$$

so $b = 5$ and the equation of the line is

$$y = -x + 5.$$

---

We now consider the equation

$$Ax + By = C, \tag{2.9}$$

where $A$, $B$, and $C$ are real numbers, and at least one of the numbers $A$ and $B$ is not zero. Subtracting $Ax$ from both sides, we have

$$By = -Ax + C. \tag{2.10}$$

If $B \neq 0$ we can divide both sides of Eq. (2.10) by $B$, obtaining

$$y = \frac{-A}{B}x + \frac{C}{B}.$$

Setting $m = -A/B$ and $b = C/B$, we see that Eq. (2.9) has been transformed into the slope-intercept formula of the line. If $B = 0$, the left side of Eq. (2.9) reduces to $Ax = C$ or, since $A \neq 0$,

$$x = C/A. \tag{2.11}$$

Equation (2.11) yields the set of all points whose $x$-coordinate equals $C/A$ (see Fig. 2.12); their graph is the vertical line passing through the

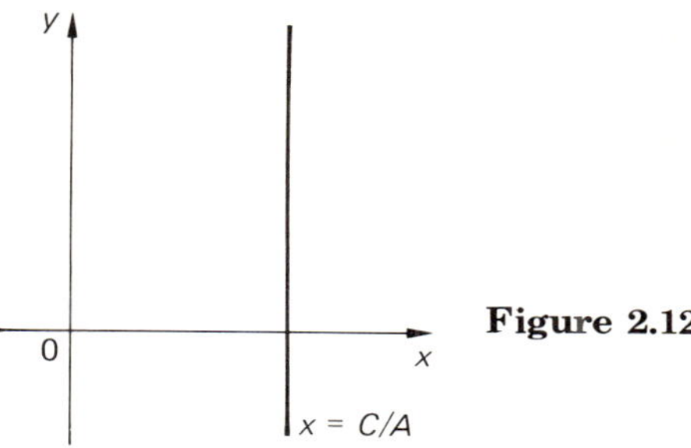


**Figure 2.12**

point $C/A$ on the $x$-axis. Thus, whatever the value of $B$, Eq. (2.9) yields a straight line. For this reason, Eq. (2.9) is called a *linear equation.*

This definition has been extended to higher dimensions. If $x_1, x_2, \ldots, x_n$ are variables and $a_1, a_2, \ldots, a_n$ and $b$ are constants, we also call

$$a_1x_1 + a_2x_2 + \cdots + a_nx_n = b \tag{2.12}$$

a *linear equation.*

We summarize the above facts in Table 2.2.

**Table 2.2**
**Lines**

| Name | Formula | Remarks |
|---|---|---|
| Slope-intercept | $y = mx + b$ | $m$ is the slope, $b$ is the $y$-intercept |
| Point-slope | $y = m(x - x_0) + y_0$ | $m$ is the slope, $(x_0, y_0)$ is the point |
| Two-point | $y = \dfrac{y_1 - y_0}{x_1 - x_0}(x - x_0) + y_0$ | Points are $(x_0, y_0)$, $(x_1, y_1)$ |
| Linear equation | $Ax + By = C$ | At least one of the numbers $A$ and $B$ is not zero |

## EXERCISES 2.2

*In Exercises 1–6, draw the graph of each given line.*

**1.** $y = 2x + 3$

**2.** $y = 5x - 2$

**3.** $y = x + 4$

**4.** $y = -3x - 2$

**5.** $y = 6$

**6.** $y = -4$

**7.** Graph the lines $y = 2x + 4$ and $y = 2x - 1$. What effect did the change in $y$-intercept $b$ have on the lines?

8. Show that any two lines with the same slope but different $y$-intercepts are parallel. (*Hint:* Use the fact that the sides of a parallelogram are parallel.)

*In Exercises 9–12, find the equation and draw the graph of the line having the given slope and y-intercept.*

9. $m = 2,\ b = 3$
10. $m = -3,\ b = 1$
11. $m = 4,\ b = -2$
12. $m = -5,\ b = -4$

*In Exercises 13–16, find the equation of the line having the given slope and passing through the indicated point.*

13. $m = 1,\ (2, 3)$
14. $m = -2,\ (-1, 4)$
15. $m = 3,\ (4, -2)$
16. $m = -4,\ (-1, -3)$

*In Exercises 17–20, find the equation of the line passing through the two given points.*

17. $(1, 5),\ (3, 11)$
18. $(0, -3),\ (2, 5)$
19. $(1, 3),\ (-2, 6)$
20. $(-1, -3),\ (-5, 5)$
21. Sue works as a seamstress in a nonunion sweatshop. She is expected to work 12-hour days, at $1.60 per hour, six days a week. If she works on Sundays, she receives time-and-a-half for each hour worked. Find a linear equation that determines Sue's weekly wage before taxes.
22. A TV technician charges a flat fee of $20 per service call plus $10 per hour. Find a linear equation that determines the total charges, not including the cost of parts.
23. A taxicab company charges a flat rate of one dollar plus 15¢ for each mile traveled. Find a linear equation that determines the total fare. Assume that the meter changes continually.
24. The downward velocity of an object falling freely near the earth's surface increases by approximately 32.2 ft per second. If the initial downward velocity of the object is 50 ft per second, find a linear equation that determines the object's velocity at any time $x$.

---

## 2.3 SIMULTANEOUS LINEAR EQUATIONS

Larry is offered two jobs. One pays $3.20 per hour plus double pay for overtime, the other $3.60 per hour plus time-and-a-half for overtime. Both jobs require a normal 40-hour week. How many hours must Larry work to make the same pay regardless of which job he chooses?

Let $x$ be the number of overtime hours that Larry works. In the first job, Larry receives $6.40 for each hour of overtime so that his weekly pay becomes

$$y = 6.40x + 3.20(40). \tag{2.13}$$

Similarly, the weekly pay equation for the second job is

$$y = 5.40x + 3.60(40), \tag{2.14}$$

since $3.60 + (3.60)/2 = 5.40$. Thus, we have the two linear equations

$$\begin{aligned} y &= 6.40x + 128 \\ y &= 5.40x + 144, \end{aligned} \tag{2.15}$$

and we wish to find the number $x$ that yields the same value $y$ for both linear equations *simultaneously*. Graphically, we wish to find the point at which the two lines (2.13) and (2.14) intersect (see Fig. 2.13). Whatever the number $x$ is, it yields the same value for $y$ in each equation, so that both sides of both equations in (2.15) must be equal. Thus,

$$6.40x + 128 = 5.40x + 144,$$

and we may subtract $5.40x + 128$ from each side to obtain

$$6.40x + 128 - (5.40x + 128) = 5.40x + 144 - (5.40x + 128)$$

or

$$x = 16.$$

Hence, Larry must work 16 hours of overtime in the first job before he breaks even with the second job. If he anticipates working only a

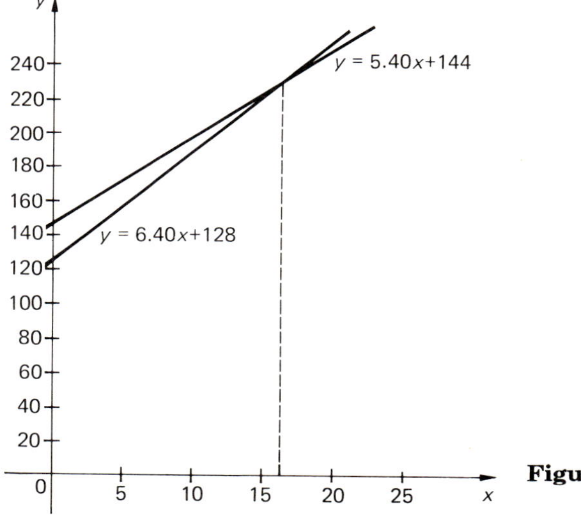


**Figure 2.13**

50-hour week, he should choose the second job, all other factors being equal.

Generally, suppose we are given two linear equations

$$\begin{aligned} A_1x + B_1y &= C_1 \\ A_2x + B_2y &= C_2, \end{aligned} \tag{2.16}$$

and are asked to find all points lying simultaneously on both lines. Such points satisfy *both* of the equations in (2.16). The situation is shown graphically in Fig. 2.14. In Fig. 2.14(a) we have two lines that are not parallel. These two lines have a unique point of intersection $(x_0, y_0)$, called the *solution of the system of equations*. If the two given lines are parallel, there are two possibilities: Either they are distinct parallel lines (Fig. 2.14b) or the *same* line (Fig. 2.14c). In Fig. 2.14(b) there can be no solution of the system of equations, since parallel lines never meet. However, in Fig. 2.14(c) *every* point on the line is a solution, since the system of two equations reduces to only one line.

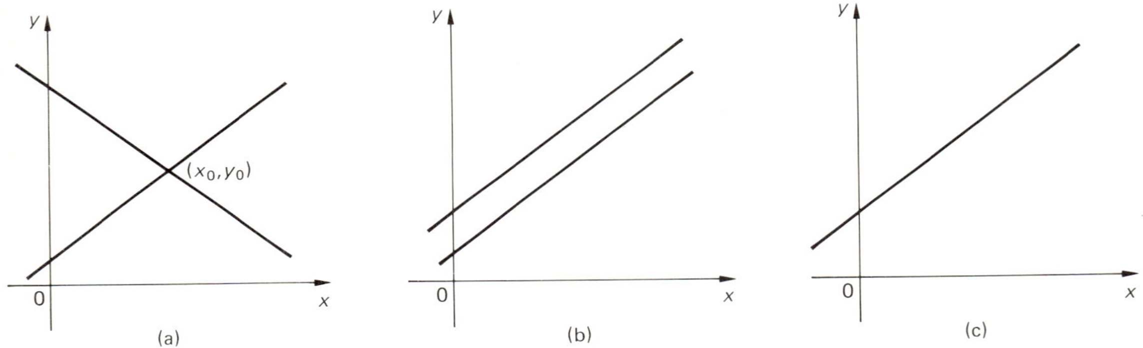


**Fig. 2.14**
(a) Intersecting lines: one solution; (b) parallel lines: no solution; (c) identical lines: infinitely many solutions.

We will see very shortly that it is quite easy to distinguish among the three cases. First, however, we must discuss the effect of multiplying every term in a linear equation

$$Ax + By = C \tag{2.17}$$

by a fixed *nonzero* constant $k$:

$$kAx + kBy = kC. \tag{2.18}$$

Suppose the point $(x_1, y_1)$ satisfies Eq. (2.17). We now show that it also satisfies Eq. (2.18). Substituting $(x_1, y_1)$ into Eq. (2.18) we obtain

$$kAx_1 + kBy_1 = kC,$$

or, factoring $k$ on the left side of this equation,

$$k(Ax_1 + By_1) = kC. \tag{2.19}$$

But the term in parentheses in Eq. (2.19) equals $C$ because the point $(x_1, y_1)$ satisfies Eq. (2.17). Thus, Eq. (2.19) holds since $k(C) = kC$. The converse also follows, since Eq. (2.17) can be obtained by multiplying each term of Eq. (2.18) by $1/k$. Thus, *the graphs of the two straight lines are identical*, so we may use this property of lines to our advantage.

We return to our system of linear equations:

$$A_1x + B_1y = C_1$$

$$A_2x + B_2y = C_2.$$

Multiplying each term in the first equation by $B_2$ and in the second by $B_1$, we obtain

$$\begin{aligned} A_1B_2x + B_1B_2y &= C_1B_2 \\ A_2B_1x + B_1B_2y &= C_2B_1. \end{aligned} \tag{2.20}$$

Subtracting $A_1B_2x$ from each side of the first equation and $A_2B_1x$ from each side of the second yields the equations

$$\begin{aligned} B_1B_2y &= C_1B_2 - A_1B_2x \\ B_1B_2y &= C_2B_1 - A_2B_1x. \end{aligned} \tag{2.21}$$

Since we are searching for points $(x, y)$ that lie on both lines and since the left sides of both equations in (2.21) are identical, we may equate the right sides of the two equations, obtaining

$$C_1B_2 - A_1B_2x = C_2B_1 - A_2B_1x. \tag{2.22}$$

Adding $A_2B_1x$ and subtracting $C_1B_2$ from both sides of Eq. (2.22) yields

$$A_2B_1x - A_1B_2x = C_2B_1 - C_1B_2. \tag{2.23}$$

*Note that* (2.23) *could have been obtained directly by subtracting the first equation of* (2.20) *from the second.* Factoring the value $x$ from the left side of (2.23) we have

$$x(A_2B_1 - A_1B_2) = C_2B_1 - C_1B_2. \tag{2.24}$$

If $A_2B_1 - A_1B_2 \neq 0$, we may divide both sides of (2.24) by $A_2B_1 - A_1B_2$ obtaining a unique value $x$. Substituting this value of $x$ into the first equation of (2.21) yields the corresponding unique value for $y$. Since only one solution is obtained, the two lines are not parallel.

***Example 1*** Solve the simultaneous system of equations

$$\begin{aligned} 3x + 2y &= 6 \\ 4x - y &= 2. \end{aligned} \tag{2.25}$$

Multiplying every term in the second equation by 2 yields the system

$$\begin{aligned} 3x + 2y &= 6 \\ 8x - 2y &= 4. \end{aligned}$$

Adding the two equations together, we obtain

$$11x = 10 \quad \text{or} \quad x = 10/11.$$

Now, subtracting $3x$ from both sides of the first equation, we have

$$2y = 6 - 3x,$$

or, dividing by 2,

$$y = 3 - \frac{3}{2}x. \tag{2.26}$$

Substituting the value $x = 10/11$ in Eq. (2.26) yields

$$y = 3 - \frac{3}{2}\left(\frac{10}{11}\right) = \frac{66 - 30}{22} = \frac{36}{22} = \frac{18}{11}.$$

Thus, the unique solution to the system is the point $\left(\frac{10}{11}, \frac{18}{11}\right)$.

---

If $A_2B_1 - A_1B_2 = 0$, then the left side of Eq. (2.24) vanishes and we have the equation

$$0 = C_2B_1 - C_1B_2.$$

Then if the number $C_2B_1 - C_1B_2$ is zero it follows that the two lines are identical, and if $C_2B_1 - C_1B_2$ is *not* zero the two lines are parallel. The proofs of these two assertions are left as exercises (see Exercises 16 and 17 on p. 47).

***Example 2*** Consider the system of equations

$$\begin{aligned} 3x - 9y &= 6 \\ -x + 3y &= -2. \end{aligned} \tag{2.27}$$

Multiplying each term in the second equation by 3, we obtain the system

$$\begin{aligned} 3x - 9y &= 6 \\ -3x + 9y &= -6. \end{aligned}$$

Adding the two equations, we have

$$0 = 0,$$

which we have asserted implies that the two lines are identical. Indeed, we see that if we multiply each term in the second equation of (2.27) by −3 we will obtain the first equation. Thus every point on this line is a solution (see Fig. 2.15).

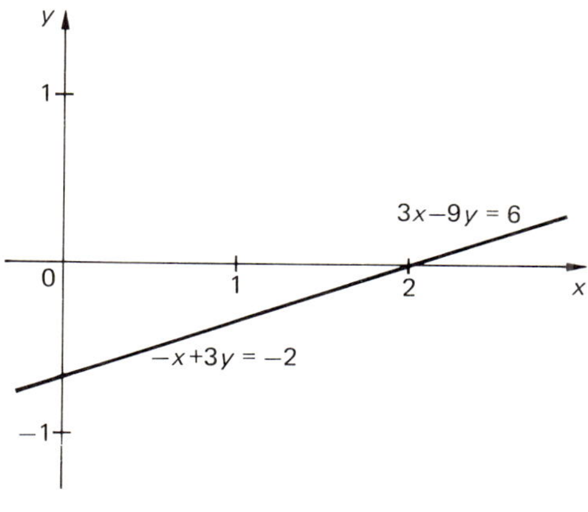


**Figure 2.15**

***Example 3*** We are given the system

$$\begin{aligned} 4x - 2y &= 6 \\ -6x + 3y &= 8. \end{aligned} \tag{2.28}$$

Multiplying each term in the first equation by 3 and each term in the second equation by 2, we have

$$\begin{aligned} 12x - 6y &= 18 \\ -12x + 6y &= 16. \end{aligned}$$

Adding these two equations yields

$$0 = 34,$$

so we know that the two lines are parallel and there is no solution.

---

This method may be extended to solving $n$ linear equations in $n$ unknowns simultaneously. The basic idea is to eliminate one of the unknowns from $(n - 1)$ of the equations by adding appropriate multiples of one of the equations to the rest. The process is then repeated, obtaining a system of $(n - 2)$ equations in $(n - 2)$ unknowns and so forth, until one finally arrives at one equation with one unknown. The same difficulties that arose in the case of two equations in two unknowns also occur here: If we obtain a false equation (for example, $0 = 2$) there is no solution, but if the system reduces to an identity (that is, $2 = 2$, etc.) then there are infinitely many solutions. We illustrate the procedure using a system of three linear equations in three unknowns.

***Example 4*** We solve the system

$$\begin{aligned} 4x - 2y + z &= 9 \\ 3x + y - 2z &= 9 \\ x - 4y + 5z &= 0. \end{aligned} \tag{2.29}$$

We will eliminate the unknown $z$ first. Add 2 times the first equation to the second equation and $-5$ times the first equation to the third equation, obtaining the system of two equations in $x$ and $y$:

$$\begin{aligned} 11x - 3y &= 27 \\ -19x + 6y &= -45. \end{aligned} \tag{2.30}$$

We now eliminate $y$ by adding 2 times the first equation of (2.30) to the second equation, obtaining

$$3x = 9.$$

Thus, $x = 3$ and substituting this value in the first equation of (2.30), we get

$$33 - 3y = 27,$$

or

$$3y = 33 - 27 = 6.$$

Therefore, $y = 2$. Substituting $x = 3$ and $y = 2$ in the first equation of (2.29) yields

$$12 - 4 + z = 9,$$

or $z = 1$. Thus the point (3, 2, 1) is the unique solution of system (2.29).

***Example 5*** Consider the system

$$\begin{aligned} 4x - 2y + z &= 9 \\ 3x + y - 2z &= 9 \\ x - 3y + 3z &= 1. \end{aligned} \tag{2.31}$$

Eliminating the $z$-variable by adding 2 times the first equation to the second and $(-3)$ times the first equation to the third, we obtain the two-equation system

$$\begin{aligned} 11x - 3y &= 27 \\ -11x + 3y &= -26. \end{aligned} \tag{2.32}$$

Adding both of these equations yields $0 = 1$, which is false. Thus, there is *no* solution in this case.

***Example 6*** Consider the system

$$\begin{aligned} 4x - 2y + z &= 9 \\ 3x + y - 2z &= 9 \\ x - 3y + 3z &= 0. \end{aligned} \tag{2.33}$$

Eliminating $z$ by adding 2 times the first equation to the second and $-3$ times the first equation to the third, we have

$$\begin{aligned} 11x - 3y &= \phantom{-}27 \\ -11x + 3y &= -27. \end{aligned} \tag{2.34}$$

Clearly, the second equation in (2.34) is $(-1)$ times the first so the two lines are identical. If we added both equations we would get the identity $0 = 0$. Thus every point on the line $11x - 3y = 27$ is a solution of system (2.34). Solving this equation for $y$ in terms of $x$ we have

$$3y = 11x - 27,$$

or

$$y = \frac{11}{3}x - 9. \tag{2.35}$$

Substituting (2.35) in the first equation of (2.33) and solving for $z$ in terms of $x$, we obtain

$$\begin{aligned} z &= 9 - 4x + 2y \\ &= 9 - 4x + 2\left(\frac{11}{3}x - 9\right), \end{aligned}$$

or

$$z = \frac{10}{3}x - 9. \tag{2.36}$$

Therefore, the solution set consists of all points

$$\left(x, \frac{11}{3}x - 9, \frac{10}{3}x - 9\right)$$

for all real numbers $x$, so we have infinitely many solutions.

***Example 7*** If we are given the system

$$\begin{aligned} 4x - 2y + \phantom{2}z &= 9 \\ 8x - 4y + 2z &= 18 \\ -4x + 2y - \phantom{2}z &= -9, \end{aligned}$$

we might observe that multiplying each term of the first equation by two will yield the second equation, while the third equation is simply the negative of the first equation. Subtracting twice the first equation from the second and adding the first equation to the third will reduce system (2.37) to a single equation:

$$4x - 2y + z = 9.$$

Then any point that satisfies this equation is a solution of system (2.37). For example, the points $(1, -2, 1)$, $(2, 0, 1)$, and $(0, -4, 1)$ all satisfy this equation. In fact, there are infinitely many solutions since

$$(0, t - 4, 2t + 1)$$

satisfies the first equation of (2.37) for all values of $t$:

$$4(0) - 2(t - 4) + (2t + 1) = -2t + 8 + 2t + 1 = 9.$$

Actually, in this case, there is an entire *plane* of solutions to system (2.37) consisting of all points of the form

$$(s, t, 9 - 4s + 2t),$$

where $s$ and $t$ are any real numbers.

---

## EXERCISES 2.3

*In Exercises 1–15, find all simultaneous solutions to the given system of equations.*

**1.** $3x - y = 1$
$2x + 3y = 8$

**2.** $4x + y = -5$
$-x + 3y = 11$

**3.** $3x - 5y = 12$
$x + 4y = 4$

**4.** $3x - y = 2$
$2x + 4y = -8$

**5.** $x + y = 2$
$x - y = 1$

**6.** $2x - 3y = 1$
$6x + 6y = -7$

**7.** $4x - 2y = 8$
$-2x + y = 4$

**8.** $-3x + 2y = 1$
$6x - 4y = -2$

**9.** $6x - 15y = 1$
$-2x + 5y = 2$

**10.** $x + y + z = 6$
$2x + y - z = 3$
$3x - y + 2z = 4$

**11.** $x + y + z = 6$
$2x + y - z = 3$
$4x + 3y + z = 15$

**12.** $x + y + z = 6$
$2x + y - z = 3$
$-4x - 2y + 2z = 4$

**13.** $2x + y + 3z = 8$
$4x - 2y + 3z = 3$
$-6x + 7y - 6z = 1$

**14.** $2x + y + 3z = 8$
$4x - 8y + 3z = -9$
$-6x + 7y - 6z = 1$

**15.** $2x + y + 3z = 8$
$4x - 8y + 3z = 9$
$-6x + 7y - 6z = 1$

**16.** Prove that if both

$$A_2B_1 - A_1B_2 = 0 \quad \text{and} \quad C_2B_1 - C_1B_2 = 0,$$

then the graphs of the two equations in (2.16) are identical. (*Hint*: Multiply all terms of the first equation of (2.16) by $B_2/B_1$ and use the two given identities.)

**17.** Prove that if

$$A_2B_1 - A_1B_2 = 0 \quad \text{and} \quad C_2B_1 - C_1B_2 \neq 0,$$

then the graphs of the two equations in (2.16) are parallel lines. (See the hint in Exercise 16.)

---

## 2.4 APPLICATIONS OF LINEAR EQUATIONS

In this section we will consider a number of applications in which linear equations and systems of linear equations may be used.

***Example 1*** A simple application of linearity is represented by the mercury thermometer, invented by G. D. Fahrenheit in 1717. The device consists of a sealed spherical bulb with a long tubular extension on the top, the lower

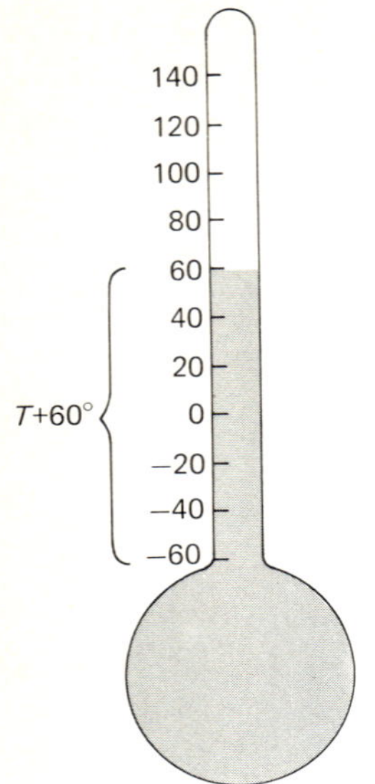

Figure 2.16

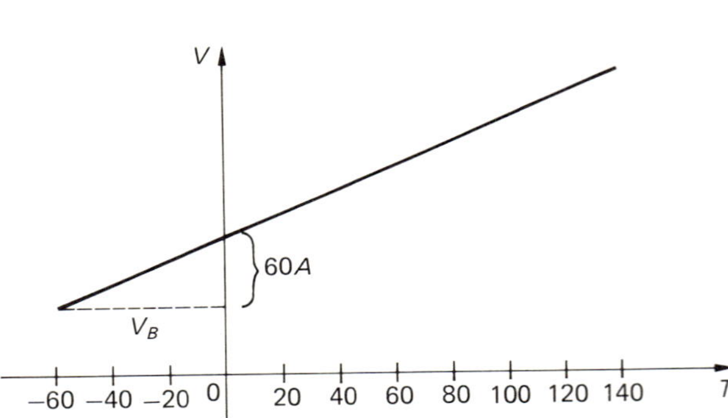


**Fig. 2.17**
Temperature versus volume.

part of which is filled with mercury (see Fig. 2.16). As the temperature increases, the mercury expands and occupies a greater volume, causing the level of the mercury in the tube to rise. Let $T$ denote the temperature, $V_B$ the volume of mercury in the bulb, and $A$ the cross-sectional area of the tube. Then the volume that the mercury occupies at any temperature $T$ is given by the formula

$$V(T) = A(T + 60) + V_B, \tag{2.38}$$

since the volume of the cylinder is obtained by multiplying the height $(T + 60)$ by the cross-sectional area $A$. Letting $T$ be the independent variable and $V$ the dependent variable, we see that Eq. (2.38) has the straight-line graph shown in Fig. 2.17. Since Eq. (2.38) may be rewritten as

$$V = AT + (60A + V_B), \tag{2.39}$$

$A$ is the slope and $(60A + V_B)$ is the $y$-intercept. Although the device is simple, the principle is ingenious, because it limits the volumetric expansion to a single direction.

***Example 2*** A portion of the 1970 income tax rate schedule for married taxpayers filing joint returns is given in Table 2.3. The tax for each income bracket can easily be written as a linear equation. Suppose that the taxable income $\$x$ lies in the \$4000–\$8000 bracket. Then the tax $\$y$ is

**Table 2.3**

| If taxable income is | the tax is |
|---|---|
| 0–\$1000 | 14% of taxable income |
| \$1000–\$2000 | \$140 + 15% of excess over \$1000 |
| \$2000–\$3000 | \$290 + 16% of excess over \$2000 |
| \$3000–\$4000 | \$450 + 17% of excess over \$3000 |
| \$4000–\$8000 | \$620 + 19% of excess over \$4000 |
| \$8000–\$12000 | \$1380 + 22% of excess over \$8000 |

obtained by adding 19% of $\$(x - 4000)$ to \$620, that is,

$$y = (0.19)(x - 4000) + 620. \tag{2.40}$$

Recalling the point-slope formula (Eq. 2.6) discussed in Section 2.2, we observe that Eq. (2.40) is the equation of a line of slope 0.19 (= 19%) passing through the point (4000, 620). Note that when $x = \$8000$ we obtain $y = 0.19(4000) + 620 = \$1380$, the minimum amount of tax in the next bracket. We may convert Table 2.3 into a graph by drawing straight line segments joining each of the points (0, 0), (1000, 140), (2000, 290), (3000, 450), (4000, 620), and (8000, 1380) to the next (see Fig. 2.18). To graph the \$8000–\$12,000 bracket, we calculate the tax when $x = 12{,}000$ by using the equation

$$y = (0.22)(x - 8000) + 1380,$$

so that $y = 0.22(4000) + 1380 = \$2260$. Thus this graph consists of a sequence of abutting straight line segments.

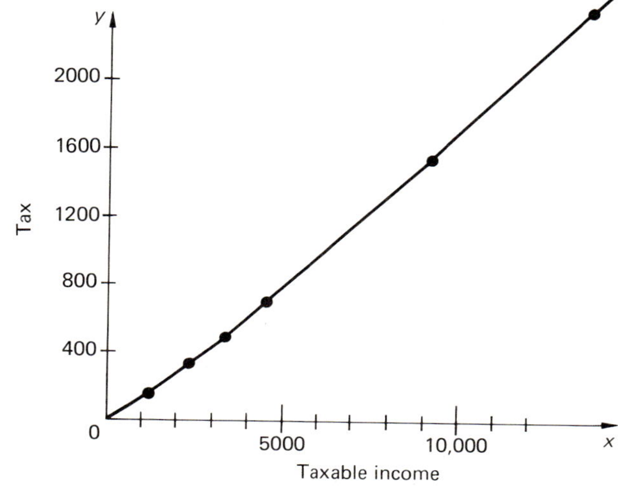


**Fig. 2.18**
1970 income tax rate schedule for married taxpayers filing joint returns.

Curves composed of straight line segments are frequently encountered in practice. This method of "fitting" a curve to a given set of data points is useful when very little data are available and each given value $x$ has only one corresponding value of $y$.

The graph in Fig. 2.19 shows the gill-to-bone ratios of zinc observed in the autopsies of bullheads exposed for 96 hours to lethal concentrations of zinc in their water. Experiments of this nature are performed to determine acceptable levels of water pollution and to prevent catastrophic fish kills.

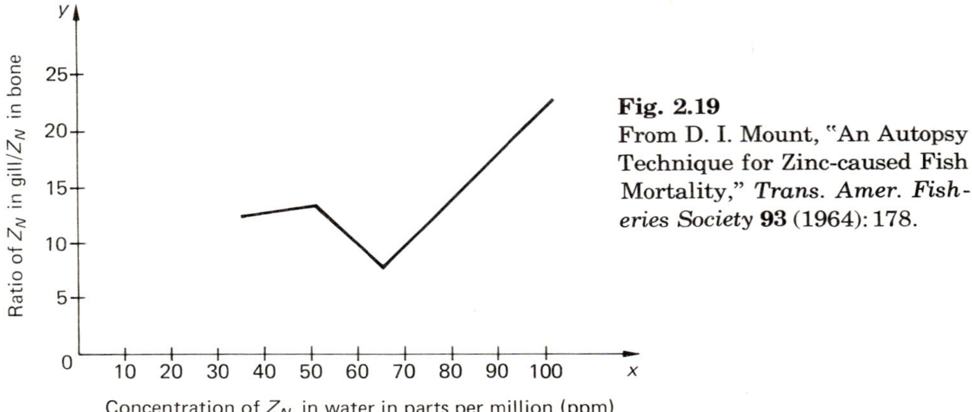


**Fig. 2.19**
From D. I. Mount, "An Autopsy Technique for Zinc-caused Fish Mortality," *Trans. Amer. Fisheries Society* **93** (1964): 178.

A different problem occurs when larger quantities of data are collected. It is then very likely that the data points will be quite scattered. When this occurs we select a single curve (very often a straight line) that passes "as close as possible" to the data points. The most common technique of fitting a straight line to data, called the *method of least squares*, will be discussed in Section 2.9. Until then we shall be content simply to draw the straight line that looks best to us, as illustrated below.

***Example 3*** In comparing gestation periods (shorter than 120 days) to longevity for animals, the data given in Table 2.4 are available. The information for kangaroos and opossums is omitted since the marsupial reproductive cycle includes a lengthy stay in the mother's pouch. Plotting gestation versus longevity in Fig. 2.20, we find that the line

$$L = \frac{3}{20}G + 1 \tag{2.41}$$

provides a good fit of the data. We can use Eq. (2.41) to predict the longevity of a lion (gestation period, 100 days) and a baboon (gestation

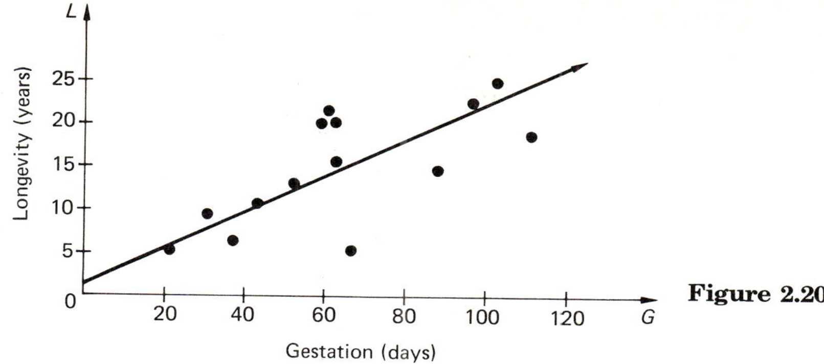


Figure 2.20

period, 187 days). For the lion we have

$$L(100) = \frac{3}{20}(100) + 1 = 16,$$

and for the baboon

$$L(187) = \frac{3}{20}(187) + 1 = 28.05.$$

These two values coincide very well with the values of 15–19 years and 27 years given in the same table of the *World Almanac, 1971.*

---

In this example we used Eq. (2.41) to make two kinds of predictions. The predicted value for the longevity of the lion is called a (*linear*) *interpolation* since the gestation period of 100 days is within the 120-day range of the data given in Table 2.4. The predicted longevity of the baboon is an example of (*linear*) *extrapolation* since the gestation period of 187 days is outside this range. Generally, more faith is placed

**Table 2.4**

| Animal | Gestation (days) | Longevity (years) | Animal | Gestation (days) | Longevity (years) |
|---|---|---|---|---|---|
| Badger | 60 | 15 | Mouse | 21 | 4 |
| Cat | 63 | 15 | Pig | 112 | 14 |
| Chipmunk | 31 | 7 | Puma | 90 | 11 |
| Dog | 61 | 16 | Rabbit | 37 | 5 |
| Fox | 52 | 8 | Squirrel | 44 | 8 |
| Guinea pig | 68 | 4 | Tiger | 105 | 19 |
| Leopard | 98 | 17 | Wolf | 63 | 12 |

*Source: World Almanac, 1971,* p. 142.

in values obtained by interpolation than in those obtained by extrapolation. The lack of reliance on extrapolation is a consequence of the fact that if the data are *not* linear, our linear approximation is good only within a short interval $I$, and may become progressively worse as we move further from $I$ (see Fig. 2.21).

It is not necessary to obtain the equation of the line in order to interpolate linearly between two given data points. We illustrate the procedure in the following example.

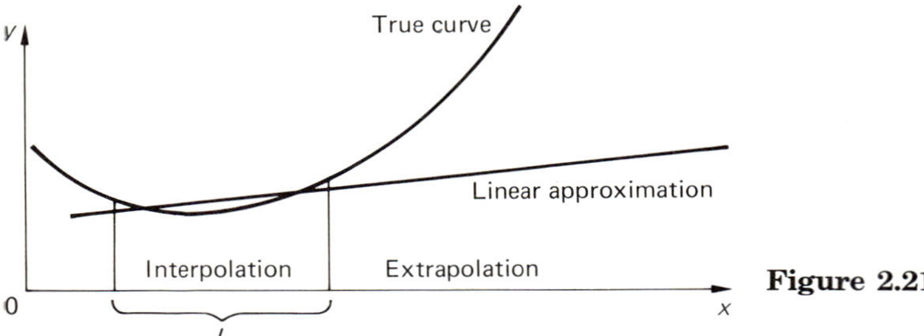


**Figure 2.21**

***Example 4*** Consider the following table of two values obtained from Fig. 2.19:

| $x$ | $y$ |
|---|---|
| 65 | 7 |
| 100 | 22 |

Suppose we wish to find the $y$-value corresponding to $x = 85$. We enter the values $x = 85$ and $y$ to form the following:

| | | $x$ | $y$ | | |
|---|---|---|---|---|---|
| Difference = 35 | Difference = 20 | 65 | 7 | Difference = $y - 7$ | Difference = 15 |
| | | 85 | $y$ | | |
| | | 100 | 22 | | |

After finding the differences indicated above, we *set the ratios of like differences equal* and solve for $y$:

$$\frac{y - 7}{20} = \frac{15}{35}. \tag{2.42}$$

Note that we placed the $y$-differences in the numerator and the $x$-differences in the denominator. *Do not mix positions.* Multiplying both sides of Eq. (2.42) by 20, we have

$$y - 7 = \frac{300}{35} \quad \text{or} \quad y = \frac{545}{35} = 15.57.$$

That is, we expect a concentration of 85 ppm to lead to a gill-to-bone ratio of 15.57.

---

Generalizing, suppose we have the table

| | | x | y | | |
|---|---|---|---|---|---|
| Difference $= x_1 - x_0$ | Difference $= x - x_0$ | $x_0$ | $y_0$ | Difference $= y - y_0$ | Difference $= y_1 - y_0$ |
| | | $x$ | $y$ | | |
| | | $x_1$ | $y_1$ | | |

Setting ratios of like differences equal yields the equation

$$\frac{y - y_0}{x - x_0} = \frac{y_1 - y_0}{x_1 - x_0}. \tag{2.43}$$

Observe that if we multiply both sides by $x - x_0$ we have

$$y - y_0 = \frac{y_1 - y_0}{x_1 - x_0}(x - x_0).$$

Adding $y_0$ to both sides yields the two-point formula of a line (Eq. 2.8):

$$y = \frac{y_1 - y_0}{x_1 - x_0}(x - x_0) + y_0.$$

Thus, linear interpolation essentially consists of finding the straight line through the two given points $(x_0, y_0)$ and $(x_1, y_1)$ and evaluating $y$ for a given value $x$.

The same procedure will also work for extrapolation.

***Example 5*** Given the data points (4, 7) and (5.12), extrapolate the $y$-value corresponding to $x = 8$. Here we have the table

| | | x | y | | |
|---|---|---|---|---|---|
| Difference = 4 | Difference = 1 | 4 | 7 | Difference = 5 | Difference $= y - 7$ |
| | | 5 | 12 | | |
| | | 8 | $y$ | | |

so that

$$\frac{y - 7}{4} = \frac{5}{1},$$

and

$$y - 7 = 20 \quad \text{or} \quad y = 27.$$

---

We now turn to problems involving systems of simultaneous equations.

***Example 6*** An electric utility proposes to change its rate structure for residential users from:

| **Present Rate Structure** |
|---|
| First 50 kWh or less per month for \$5.00 |
| All additional kWh per month @ 5.5¢ per kWh |

to a new rate structure:

| **Proposed Rate Structure** |
|---|
| First 40 kWh or less per month for \$3.00 |
| All additional kWh per month @ 8.5¢ per kWh |

Although the proposed rate is favorable to low users of electric power, where is the break-even point? We can write a linear equation for each rate structure. The present rate structure begins at the point (50, 5.00), where the $x$ entry is the number of kWh and the $y$ entry is the rate in dollars. The slope is 0.055 dollars, so by the point-slope formula, the rate structure is given by

$$y = 0.055(x - 50) + 5.00. \qquad \textbf{(2.44)}$$

Similarly, the proposed rate structure begins at (40, 3.00) with slope = 0.085, or

$$y = 0.085(x - 40) + 3.00. \qquad \textbf{(2.45)}$$

Subtracting $0.055x$ from both sides of Eq. (2.44) and $0.085x$ from both sides of Eq. (2.45), we obtain the system

$$\begin{aligned} -0.055x + y &= 2.25 \\ -0.085x + y &= -0.40. \end{aligned} \tag{2.46}$$

Subtracting the second equation from the first yields

$$0.03x = 2.65,$$

or,

$$3x = 265.$$

Thus $x = 88\frac{1}{3}$ kWh is the break-even point. All those who use more than $88\frac{1}{3}$ kWh will pay more under the proposed rate structure. Using 90 kWh per month costs \$7.20 under the old structure and \$7.25 under the new structure.

***Example 7*** A wholesaler wishes to package 10-lb bags of kidney beans having a price of \$1.35 per bag. The top grade of kidney beans costs 18 cents a pound and the next grade of kidney beans costs 12 cents a pound. How many pounds of each grade should be used in making up the packages?

Let $x$ represent the number of pounds of the top grade of kidney beans and $y$ the number of pounds of the next grade of kidney beans that must be used in making the mixture. Then

$$x + y = 10,$$

since we wish to obtain 10 lb of beans. We now balance the costs of the beans to find the desired cost for the 10-lb bag by multiplying each pound of each type of bean by its cost, obtaining the equation

$$0.18x + 0.12y = 1.35. \tag{2.47}$$

We obtain the system

$$\begin{aligned} x + y &= 10 \\ 18x + 12y &= 135 \end{aligned} \tag{2.48}$$

by multiplying each term of Eq. (2.47) by 100. Subtracting 12 times the first equation from the second equation in (2.48) yields $6x = 15$ or $x = 2.5$. Thus $y = 7.5$, implying that we need to use $2\frac{1}{2}$ lbs of the best beans and $7\frac{1}{2}$ lb of the next grade of beans in making up the mixture.

## EXERCISES 2.4

1. A surveyor's chain is a graduated coiled strip of metal, often approximately 100 ft in length. Chains are accurate when the temperature is 68°F, but they expand or contract uniformly if the temperature is higher or lower than 68°F, respectively. Suppose the chain is 101 ft long when the temperature is 80°F. Find an equation for the length of the chain at any temperature $T$ (°F).

2. Water freezes when the temperature is 0°C (= 32°F) and boils when the temperature is 100°C (= 212°F). Find an equation that converts °F into °C. What temperature in °C corresponds to 68°F?

3. Find an equation that converts °C into °F. What temperature in °F corresponds to 15°C?

4. If 87-octane gasoline costs 56.9¢ per gallon and 93-octane gasoline costs 65.9¢ per gallon, determine a formula for the price of gasoline in terms of its octane rating. How much should a gallon of 97-octane gasoline cost?

5. A rental agency advertises a rental rate for compact cars of $13.95 per day plus 11¢ per mile. Find a formula for determining the cost $C$ of the rental in terms of the number of days $x$ and the miles traveled $y$. What does this formula become if the car is rented for only one day? What is the charge for renting a car for one day and driving 320 miles?

6. The 1971 first-class postage rate for letters was 6¢ for each ounce or fraction of an ounce. Graph this rate schedule up to and including 6 oz. Do the straight lines in this graph abut?

7. The 1971 third-class postage rate for single mailings was 6¢ for the first two ounces and 2¢ for each additional ounce or fraction of an ounce. Graph this rate schedule up to a weight of 8 oz. Do the straight lines in the graph abut?

8. A portion of the 1970 income tax rate schedule for single taxpayers is given in Table 2.5. Graph the data in this table and find a linear equation for each straight line segment in the graph. Do the lines abut?

**Table 2.5**

| If taxable income is | the tax is |
|---|---|
| $14,000–$16,000 | $3550 plus 39% of excess over $14,000 |
| $16,000–$18,000 | $4330 plus 42% of excess over $16,000 |
| $18,000–$20,000 | $5170 plus 45% of excess over $18,000 |
| $20,000–$22,000 | $6070 plus 48% of excess over $20,000 |
| $22,000–$26,000 | $7030 plus 50% of excess over $22,000 |

9. Find a linear equation for each straight line segment in Table 2.3.

10. The 1971 residential electric service rate for the Montana Power Company is given in Table 2.6. Draw a graph of the data in this table, and find a

linear equation for each straight line segment in the graph. How much would a user of 180 kWh have to pay? Of 360 kWh?

**Table 2.6**

| | |
|---|---|
| First 20 kWh or less per month | $1.50 |
| Next 80 kWh per month | 4.65¢ per kWh |
| Next 100 kWh per month | 3.30¢ per kWh |
| Next 200 kWh per month | 1.75¢ per kWh |
| All additional kWh per month | 1.55¢ per kWh |

**11.** On the basis of a four-year study of 5,000,000 persons, the Society of Actuaries reports that the average weight of a 20–24-year-old man according to his height is as given below:

| Height | 5′ | 5′2″ | 5′4″ | 5′6″ | 5′8″ | 5′10″ | 6′ | 6′2″ | 6′4″ |
|---|---|---|---|---|---|---|---|---|---|
| Weight (lb) | 122 | 128 | 136 | 142 | 149 | 157 | 166 | 174 | 181 |

Draw a straight line that you feel best fits this data, and calculate an equation for this line. Use your equation to predict the weight of a man who is 5′11″ tall. (According to the Society of Actuaries, the average weight is 161 lb.) How much should a basketball player who is 7′ tall weigh?

**12.** The following data were gathered by the Society of Actuaries for women between 20 and 24 years of age:

| Height | 4′10″ | 5′ | 5′2″ | 5′4″ | 5′6″ | 5′8″ | 5′10″ | 6′ |
|---|---|---|---|---|---|---|---|---|
| Weight (lb) | 102 | 108 | 115 | 121 | 129 | 136 | 144 | 154 |

Draw a straight line that best fits this data, find its equation, and use it to predict the average weight of a woman who is 5′3″ tall (118 lb). How much should a woman who is 6′4″ weigh?

**13.** *Mysis relicta* are freshwater crustaceans commonly called the opossum shrimp. They usually inhabit deep cold-water lakes in the northern United States and Canada, and are a valuable food source for lake trout. They are very intolerant to exposure to light. A study of their tolerance to light was conducted by W. E. Smith [*Transactions of the American Fisheries Society* **99** (1970): 418–422]; approximate values of the data are given in the following table:

| Exposure to light (weeks) | 1 | 2 | 3 | 4 | 5 | 6 | 7 | 8 | 9 | 10 | 11 | 12 |
|---|---|---|---|---|---|---|---|---|---|---|---|---|
| Percent survival | 99 | 96 | 89 | 82 | 78 | 65 | 57 | 48 | 40 | 33 | 25 | 19 |

Draw a straight line that best fits this data and find its equation.

**14.** In comparing body length to the heart weight of 723 male children, the following table of averages was obtained by D. M. Schultz and D. A. Giordano [*Archives of Pathology* **74** (1962): 464]:

| Length (cm) | 51.4 | 54.0 | 57.7 | 64.2 | 69.4 | 73.8 | 84 | 90 | 101 | 109 | 114 | 121 | 127 |
|---|---|---|---|---|---|---|---|---|---|---|---|---|---|
| Weight (g) | 23 | 27 | 30 | 40 | 45 | 50 | 60 | 70 | 88 | 94 | 105 | 110 | 119 |

Draw a straight line that best fits this data and find its equation. How much would you expect the hearts of two children who are 62 cm and 138 cm long to weigh? (Actual average weights are 35 g and 150 g.) Was interpolation more effective than extrapolation?

**15.** Use interpolation in Table 2.3 to find the taxable income of a married taxpayer filing a joint return whose tax is $500.

**16.** Do Exercise 15 using a tax of $800.

**17.** Using Eq. (2.41) estimate the longevity of a grizzly bear (gestation period 225 days).

*In Exercises* 18–20, *find the missing value in each table.*

**18.**

| $x$ | $y$ |
|---|---|
| 2 | 9 |
| 3 | |
| 5 | 17 |

**19.**

| $x$ | $y$ |
|---|---|
| 4 | 12 |
| 5 | |
| 7 | −3 |

**20.**

| $x$ | $y$ |
|---|---|
| 3 | 11 |
| | 6 |
| 8 | 2 |

**21.** If 87-octane gasoline costs 56.9¢ per gallon and 93-octane gasoline costs 65.9¢ per gallon, how much should 91-octane gasoline cost?

**22.** Find the break-even point for the service rate in Table 2.6 and for the service rate in the table below:

| | |
|---|---|
| First 50 kWh or less per month | $5.00 |
| Next 150 kWh per month | 2.8¢ per kWh |
| All additional kWh per month | 1.4¢ per kWh |

**23.** Salted peanuts cost $1.20 per pound, while cashews cost $2.80 per pound. How many ounces of peanuts and cashews are required per pound for a mixture that sells at $2.30 per pound?

**24.** How many gallons of 87-octane gasoline and 93-octane gasoline should be mixed to obtain 100 gallons of 91-octane gasoline?

**25.** Suppose that a gorilla and a puma are born on the same day and weigh 5 lb and 1 lb, respectively. A year later they are reweighed and now weigh 20 lb and 60 lb, respectively. Assuming their growth rate was linear with time, at what age did they weigh the same?

## 2.5 PARABOLAS

Another function that often arises in applications is the *quadratic function*

$$y = ax^2 + bx + c. \tag{2.49}$$

The main difference between this equation and the linear equation (2.2) is the $ax^2$ term, which indicates that the dependent variable is related to the square of the independent variable. Of course, if $a = 0$, Eq. (2.49) reduces to a linear equation, so we shall assume throughout this section that $a$ is not zero.

To obtain an idea of the properties of quadratic functions, we begin by considering the simplest possible case

$$y = x^2, \tag{2.50}$$

in which $a = 1$ and $b = c = 0$. From Eq. (2.50) we easily derive the following table:

| $x$ | −3 | −2 | −1 | 0 | 1 | 2 | 3 |
|---|---|---|---|---|---|---|---|
| $y$ | 9 | 4 | 1 | 0 | 1 | 4 | 9 |

If we plot these and other points that satisfy Eq. (2.50), we will arrive at the graph shown in Fig. 2.22. Such a graph is called a *parabola*.

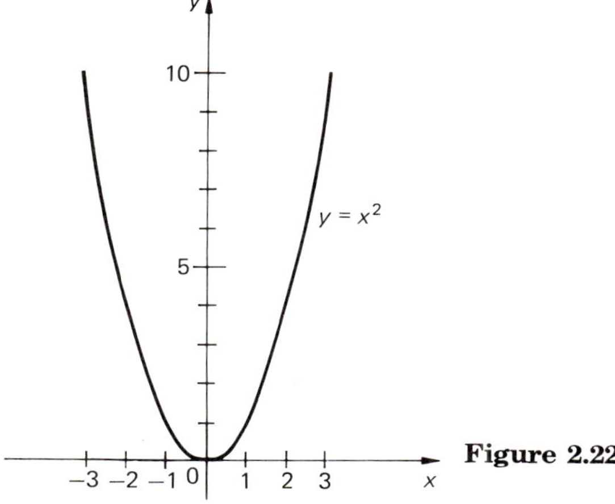


**Figure 2.22**

One property of this parabola is readily apparent: It is *symmetric about the y-axis*. By this we mean that if we fold the coordinate system along the $y$-axis, the part of the graph in the right half-plane will coincide exactly with the part of the graph in the left half-plane. Furthermore the minimum $y$-value lies on the $y$-axis, that is, on the axis of symmetry of the parabola. The point at which the parabola crosses its axis of symmetry is called the *vertex* of the parabola.

Suppose we change the value of $a$ and consider the parabola

$$y = 2x^2. \tag{2.51}$$

This equation yields the table

| $x$ | $-3$ | $-2$ | $-1$ | $0$ | $1$ | $2$ | $3$ |
|---|---|---|---|---|---|---|---|
| $y$ | $18$ | $8$ | $2$ | $0$ | $2$ | $8$ | $18$ |

and the graphs of Eqs. (2.50) and (2.51) are shown in Fig. 2.23. We see that increasing $a$ brings the parabola in closer to the $y$-axis. On the other hand, changing the sign of $a$ flips the parabola about the $x$-axis: The equation

$$y = -x^2 \tag{2.52}$$

yields the table

| $x$ | $-3$ | $-2$ | $-1$ | $0$ | $1$ | $2$ | $3$ |
|---|---|---|---|---|---|---|---|
| $y$ | $-9$ | $-4$ | $-1$ | $0$ | $-1$ | $-4$ | $-9$ |

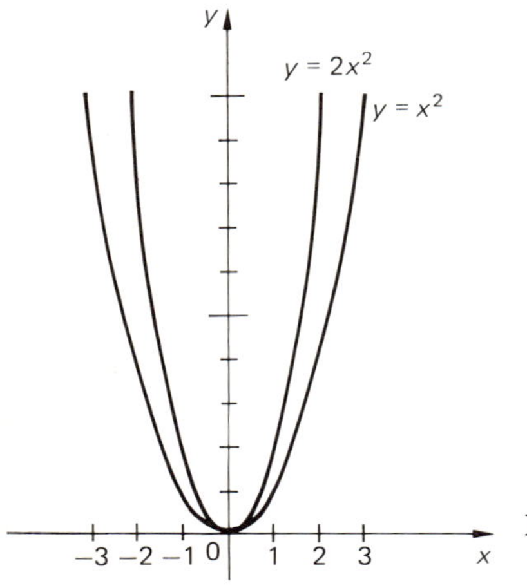


Figure 2.23

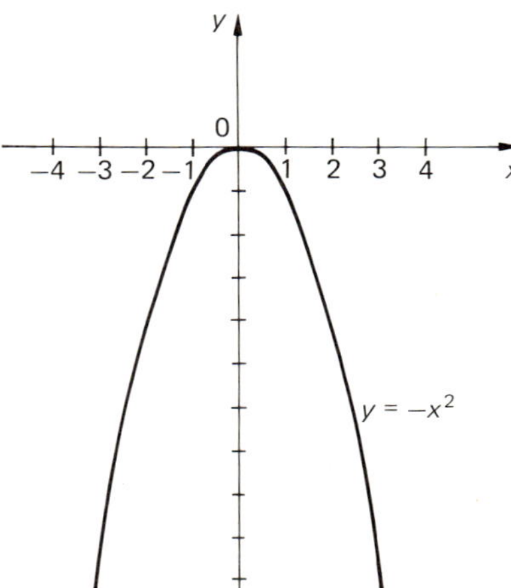


Figure 2.24

and the resulting graph shown in Fig. 2.24. The behavior of the parabolas

$$y = ax^2 \tag{2.53}$$

is shown in Fig. 2.25 for several values of $a$. Clearly, each of these parabolas has the $y$-axis as its axis of symmetry. Furthermore, *the vertex determines either the minimum or the maximum y-value of all the points of each parabola. The y-value is a minimum if $a > 0$ and a maximum if $a < 0$.*

If we add a number $c$ to the right-hand side of Eq. (2.53), obtaining the equation

$$y = ax^2 + c, \tag{2.54}$$

we simply shift the parabola vertically by the quantity $c$. For example,

$$y = x^2 + 1, \tag{2.55}$$

simply raises the graph in Fig. 2.22 by one unit, while

$$y = x^2 - 2 \tag{2.56}$$

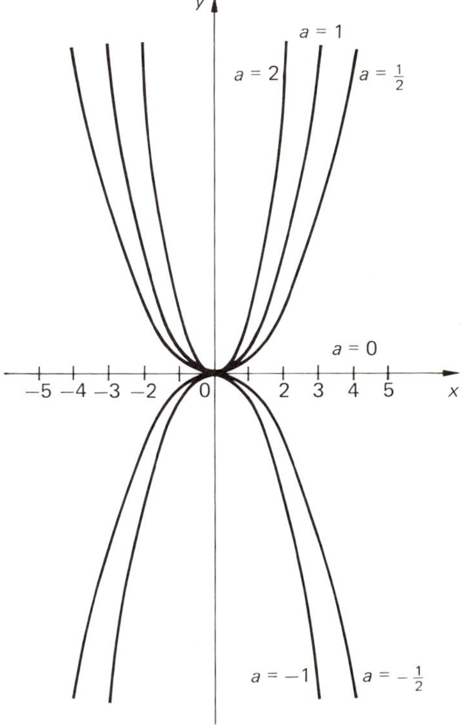


**Fig. 2.25**
Graph of $y = ax^2$ for different values of $a$.

lowers it by two units (see Fig. 2.26). Again, each of the parabolas given by Eq. (2.54) has the $y$-axis as its axis of symmetry and its maximum or minimum $y$-value at the vertex.

We now turn to the general quadratic function

$$y = ax^2 + bx + c, \tag{2.57}$$

and assume that neither $a$ nor $b$ is zero. To motivate our discussion, we consider the equation

$$y = x^2 + 2x + 1. \tag{2.58}$$

In this instance, the right side is a perfect square, so we may rewrite Eq. (2.58) in the form

$$y = (x + 1)^2. \tag{2.59}$$

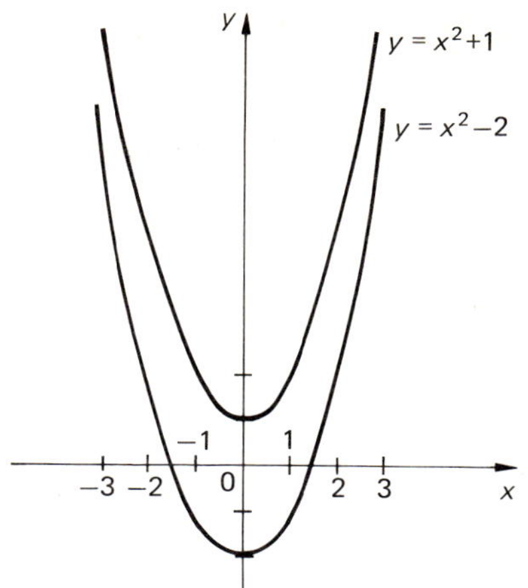


Figure 2.26

Using Eq. (2.59) we derive the table

| $x$ | −4 | −3 | −2 | −1 | 0 | 1 | 2 |
|---|---|---|---|---|---|---|---|
| $y$ | 9 | 4 | 1 | 0 | 1 | 4 | 9 |

from which we obtain the graph shown in Fig. 2.27. If we compare Fig. 2.27 with Fig. 2.22 and their respective tables of values, we see that the parabola has been shifted one unit to the left. The axis of symmetry is now the line $x = -1$. We now show that when $b$ is positive, the parabola is shifted to the left and when $b$ is negative it is shifted to the right.

If we write Eq. (2.57) in the form

$$y = a\left(x^2 + \frac{b}{a}x \qquad \right) + c, \tag{2.60}$$

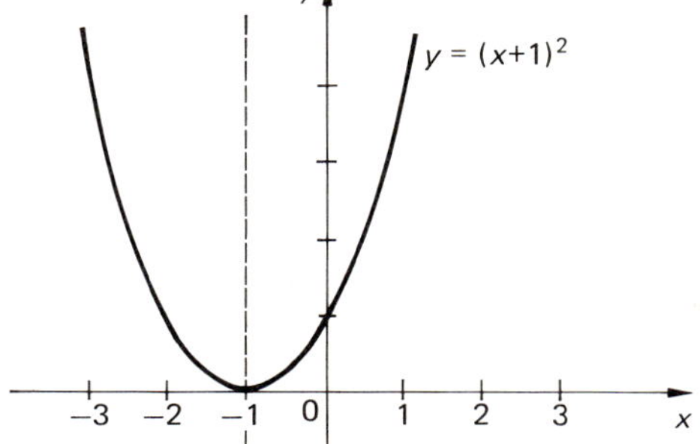


Figure 2.27

we can insert a constant inside the parentheses in order to make the terms within a perfect square. This process is called *completing the square.* Since

$$\left(x + \frac{b}{2a}\right)^2 = x^2 + \frac{b}{a}x + \frac{b^2}{4a^2}, \tag{2.61}$$

we add the term $b^2/4a^2$ inside the parentheses. To preserve the equality we must subtract $b^2/4a$ from $c$, since the term added within the parentheses is multiplied by $a$. Thus we obtain the equation

$$y = a\left(x^2 + \frac{b}{a}x + \frac{b^2}{4a^2}\right) + c - \frac{b^2}{4a}, \tag{2.62}$$

or

$$y = a\left(x + \frac{b}{2a}\right)^2 + \left(c - \frac{b^2}{4a}\right). \tag{2.63}$$

If we substitute

$$X = x + \frac{b}{2a}, \qquad C = c - \frac{b^2}{4a},$$

we can write Eq. (2.63) as

$$y = aX^2 + C.$$

This is a parabola in the $Xy$-plane whose axis of symmetry is the line $X = 0$ and whose graph has been shifted vertically by $C$ units. But the line $X = 0$ in the $Xy$-plane corresponds to the line

$$x + \frac{b}{2a} = 0$$

in the $xy$-plane, so the axis of symmetry in the $xy$-plane is the vertical line $x = -b/2a$.

We have proved the following fact.

> The quadratic function
>
> $$y = ax^2 + bx + c, \qquad a \neq 0,$$
>
> is a parabola whose axis of symmetry is the vertical line
>
> $$x = -b/2a.$$

***Example 1*** The axis of symmetry of the parabola

$$y = 3x^2 - 7x + 9$$

is the vertical line

$$x = -\frac{(-7)}{2(3)} = \frac{7}{6}.$$

If we wish to find the vertex of this parabola, we need only substitute $x = 7/6$ in the equation of the parabola, obtaining

$$y = 3\left(\frac{7}{6}\right)^2 - 7\left(\frac{7}{6}\right) + 9 = \frac{59}{12}$$

as the $y$-coordinate of the vertex. Since the coefficient of the $x^2$-term is positive, every $y$-value on the parabola is greater than or equal to 59/12.

---

Equation (2.63) can also be used to obtain the *quadratic formula*:

$$x = \frac{-b \pm \sqrt{b^2 - 4ac}}{2a}. \tag{2.64}$$

This formula, which was mentioned briefly in Chapter 1, is a convenient method of finding the solution or *roots* of the quadratic equation

$$ax^2 + bx + c = 0, \qquad a \neq 0. \tag{2.65}$$

Of course, it is often possible to find the roots of Eq. (2.65) by factoring, but when the roots are irrational or complex it is usually necessary to use the quadratic formula (2.64).

Since the right side of Eq. (2.63) equals $ax^2 + bx + c$, we can convert Eq. (2.63) into the quadratic equation (2.65) by setting $y = 0$. Subtracting $c - (b^2/4a)$ from both sides yields

$$a\left(x + \frac{b}{2a}\right)^2 = \frac{b^2}{4a} - c = \frac{b^2 - 4ac}{4a}$$

and dividing both sides by $a$, we obtain

$$\left(x + \frac{b}{2a}\right)^2 = \frac{b^2 - 4ac}{4a^2}.$$

Taking the square root of both sides yields

$$x + \frac{b}{2a} = \pm\sqrt{\frac{b^2 - 4ac}{4a^2}} = \pm\frac{\sqrt{b^2 - 4ac}}{2a}$$

and subtracting $b/2a$ from both sides, we have

$$x = \frac{-b}{2a} \pm \frac{\sqrt{b^2 - 4ac}}{2a},$$

or

$$x = \frac{-b \pm \sqrt{b^2 - 4ac}}{2a}.$$

***Example 2*** Suppose we wish to find the roots of the equation

$$x^2 - 7x + 5 = 0.$$

It is unlikely that we would be able to guess the correct factors for this equation since they are irrational. When we see that the most obvious methods of factoring fail, we immediately turn to the quadratic formula. In this case the coefficients are $a = 1$, $b = -7$, and $c = 5$; the quadratic formula (2.64) yields

$$x = \frac{-(-7) \pm \sqrt{(-7)^2 - 4(1)5}}{2(1)} = \frac{7 \pm \sqrt{29}}{2}.$$

Thus the equation $x^2 - 7x + 5$ has the roots

$$x = \frac{7 + \sqrt{29}}{2} \quad \text{and} \quad x = \frac{7 - \sqrt{29}}{2}.$$

---

Although every quadratic equation has two roots, in some instances these roots are equal and we say that we have a *double root.* This happens when the $(b^2 - 4ac)$-term in Eq. (2.64), called the *discriminant*, equals zero. Since adding and subtracting zero has no effect, the two roots then equal $-b/2a$.

***Example 3*** Find the roots of the quadratic equation

$$4x^2 + 4x + 1 = 0.$$

Since $a = 4$, $b = 4$, and $c = 1$, the quadratic formula (2.64) yields

$$x = \frac{-4 \pm \sqrt{(4)^2 - 4(4)1}}{2(4)} = \frac{-4 \pm \sqrt{16 - 16}}{8}$$

so that $x = -4/8 = -1/2$ is a double root.

---

Complex numbers are often the roots of a quadratic equation. This happens whenever the discriminant $b^2 - 4ac$ is negative. Each complex root is then obtained by changing the sign of the discriminant and multiplying the resulting positive square root by the imaginary unit $i$.

***Example 4*** Consider the equation

$$x^2 + x + 1 = 0. \tag{2.66}$$

Since $a = b = c = 1$, if we use the quadratic formula (2.64) we have

$$x = \frac{-1 \pm \sqrt{1 - 4(1)1}}{2(1)} = \frac{-1 \pm \sqrt{-3}}{2}.$$

Thus, the two complex numbers

$$x = \frac{-1 + i\sqrt{3}}{2} \quad \text{and} \quad x = \frac{-1 - i\sqrt{3}}{2}$$

are the roots of Eq. (2.66).

---

The fact that the vertex determines the minimum $y$-value when $a > 0$ and the maximum $y$-value when $a < 0$ may be used to optimize a quadratic function. The following example illustrates this procedure.

***Example 5*** A homeowner wishes to fence a rectangular area adjacent to her house. One side of the rectangle will lie along the side of the house. What is the largest area she can fence with 100 ft of fencing?

This situation is illustrated in Fig. 2.28. If we let $x$ represent the length of each vertical side, then the length of the horizontal side is $100 - 2x$. Hence, the area $y$ is given by the product

$$y = x(100 - 2x)$$

or

$$y = -2x^2 + 100x. \tag{2.67}$$

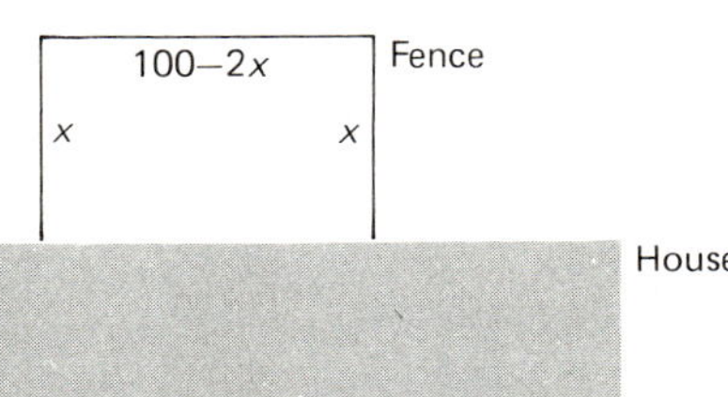


**Figure 2.28**

The axis of symmetry of this parabola is the vertical line

$$x = \frac{-100}{2(-2)} = 25,$$

and, since the coefficient of the $x^2$-term in Eq. (2.67) is negative, the maximum area is obtained by letting $x = 25$ ft. To find the area we set $x = 25$ and solve Eq. (2.67) for $y$:

$$y = -2(25)^2 + 100(25) = 1250 \text{ ft}^2.$$

---

## EXERCISES 2.5

*In Exercises 1–8, find the axis of symmetry and the vertex of the given quadratic function, and use this information to graph each function.*

**1.** $y = -2x^2$

**2.** $y = 3x^2 - 5$

**3.** $y = -2x^2 + 7$

**4.** $y = x^2 + 2x + 2$

**5.** $y = 2x^2 - 4x - 5$

**6.** $y = 3x^2 + 5x - 8$

**7.** $y = 7x^2 - x + 4$

**8.** $y = 2x^2 + 2x - 1$

*In Exercises 9–20, find the roots of the given quadratic equation by using the quadratic formula.*

**9.** $x^2 - x - 2 = 0$

**10.** $x^2 - 6x + 9 = 0$

**11.** $2x^2 + 3x - 2 = 0$

**12.** $21x^2 - 4x - 1 = 0$

**13.** $4x^2 - 4x + 1 = 0$

**14.** $x^2 - 5x + 3 = 0$

**15.** $2x^2 + 3x - 4 = 0$

**16.** $3x^2 - 5x - 1 = 0$

**17.** $3x^2 - x + 4 = 0$

**18.** $2x^2 + 8x + 9 = 0$

**19.** $\sqrt{8}x^2 + x + \sqrt{2} = 0$

**20.** $\sqrt{3}x^2 - \sqrt{5}x - \sqrt{12} = 0$

**21.** Show that the sum of the two roots of the quadratic equation

$$ax^2 + bx + c = 0$$

equals $-b/a$, and that the product of the two roots equals $c/a$.

**22.** A barber gives 120 haircuts a week if he charges \$2.50 per haircut. For each increase of 50¢ in price he loses 20 customers per week. How much should he charge to maximize his revenue?

**23.** Jo's Taco Shop can sell 300 tacos a day at a price of 50¢. Each increase of 10¢ in the price cuts the sales by 50.
a) How much should be charged to maximize the revenue?
b) If the cost of making each taco is 20¢, how much should be charged to maximize profits?

**24.** A butcher can sell 200 lb of sirloin every day if he charges \$1.89 per lb. Each increase of 10¢ in price decreases sales by 25 lb. His costs are \$1.50 per lb. What should he charge to maximize profits?

---

## 2.6 CIRCLES

Since a *circle* is the set of all points whose distance from a given point (called the *center*) is a fixed positive number, we need to determine a method for finding the distance between two points $(x_0, y_0)$ and $(x_1, y_1)$. We plot these two points and consider the shaded right triangle shown in Fig. 2.29. The Pythagorean theorem of geometry states that *the square of the length of the hypotenuse equals the sum of the squares of the lengths of the other two sides*:

$$d^2 = (x_1 - x_0)^2 + (y_1 - y_0)^2. \tag{2.68}$$

Taking the square root of both sides of the equation, we arrive at the *distance formula*:

$$d = \sqrt{(x_1 - x_0)^2 + (y_1 - y_0)^2}\,. \tag{2.69}$$

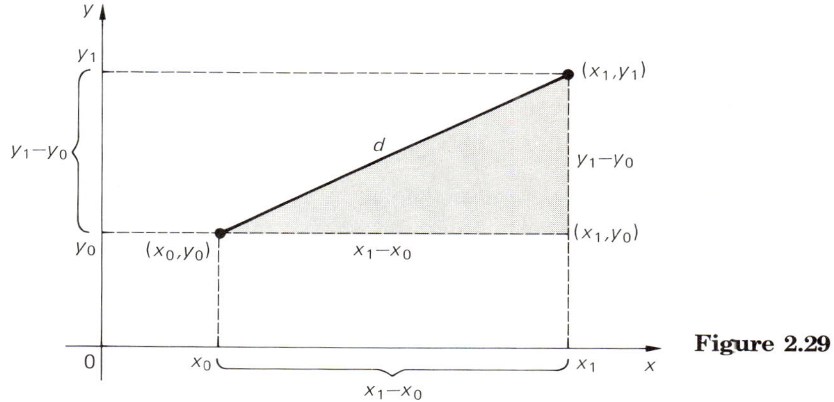


**Figure 2.29**

***Example 1*** Find the distance between the points (2, 3) and (4, 1). Letting $(x_0, y_0) = (2, 3)$ and $(x_1, y_1) = (4, 1)$, and applying the distance formula (2.69), we have

$$d = \sqrt{(4-2)^2 + (1-3)^2} = \sqrt{2^2 + (-2)^2} = \sqrt{8}.$$

---

The *equation of a circle of radius r centered at the point* $(x_0, y_0)$ (see Fig. 2.30) can be obtained as an immediate consequence of the distance formula. Clearly, every point $(x, y)$ on this circle has a distance

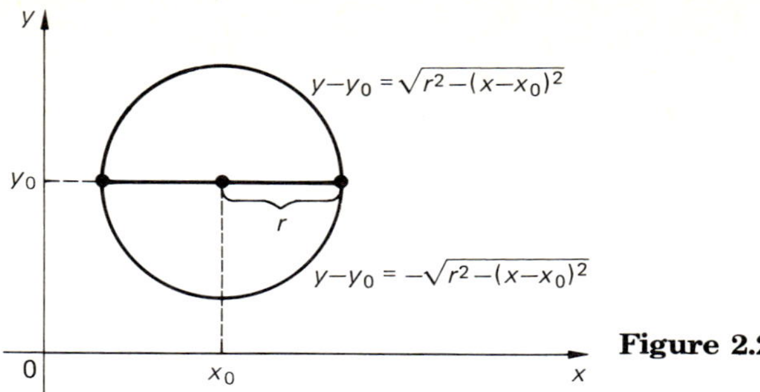


**Figure 2.28**

$r$ from the center $(x_0, y_0)$, so squaring the distance formula yields

$$(x - x_0)^2 + (y - y_0)^2 = r^2. \tag{2.70}$$

Equation (2.70) is easily applied, as is seen in the following example.

***Example 2*** What is the equation of a circle of radius 3 centered on the point $(-1, 2)$?

Using $r = 3$ and $(x_0, y_0) = (-1, 2)$, we can determine the equation of the circle (2.70):

$$[x - (-1)]^2 + (y - 2)^2 = 3^2$$

or

$$(x + 1)^2 + (y - 2)^2 = 9.$$

---

As can be seen in Fig. 2.29, circles are not functions, since two $y$-values will correspond to all $x$-values satisfying $-r < x - x_0 < r$. Nevertheless, the same example indicated that we can "break" this relation into two functions: the upper and lower semicircles. To obtain the equation of these functions, we subtract $(x - x_0)^2$ from both sides of Eq. (2.70). Then, taking the square roots of both sides, we find that

$$y - y_0 = \sqrt{r^2 - (x - x_0)^2}$$

is the equation of the upper semicircle, while

$$y - y_0 = -\sqrt{r^2 - (x - x_0)^2}$$

is the equation of the lower semicircle (see Fig. 2.30). In terms of Example 2, the upper semicircle is the function

$$y - 2 = \sqrt{9 - (x + 1)^2}$$

or

$$y = 2 + \sqrt{8 - 2x - x^2},$$

while the lower semicircle is the function

$$y - 2 = -\sqrt{9 - (x + 1)^2}$$

or

$$y = 2 - \sqrt{8 - 2x - x^2}.$$

***Example 3*** Describe the graph of the equation

$$x^2 - 4x + y^2 + 6y = 7.$$

Whenever we have an equation with quadratic terms in $x$ and $y$, we should check to see whether the coefficients of the $x^2$- and $y^2$-terms are identical. If so, we can convert the equation into the equation of a circle by completing the squares. We may write the equation above in the form

$$(x^2 - 4x \qquad) + (y^2 + 6y \qquad) = 7.$$

If we add 4 to the first group of parentheses and 9 to the second, each term in parentheses will become a perfect square. Thus, adding 4 + 9 to both sides yields

$$(x^2 - 4x + 4) + (y^2 + 6x + 9) = 7 + 4 + 9,$$

or

$$(x - 2)^2 + (y + 3)^2 = 20.$$

This is the equation of a circle of radius $\sqrt{20}$ centered at the point $(2, -3)$.

---

## EXERCISES 2.6

*In Exercises 1–16, find the distance between the given pair of points using the distance formula.*

**1.** $(1, 2), (3, 4)$

**2.** $(1, 2), (4, 3)$

**3.** $(-1, 2), (3, 4)$

**4.** $(1, -2), (4, 3)$

**5.** $(-1, 2), (-3, 4)$

**6.** $(-1, 2), (3, -4)$

**7.** $(2, 7), (-1, -3)$

**8.** $(-2, 7), (-1, -3)$

**9.** $(12, -2), (-5, 3)$

**10.** $(1, 5), (4, 1)$

**11.** $(4, 2), (6, 2)$

**12.** $(7, 3), (7, 18)$

**13.** $(2, 0), (0, 5)$

**14.** $(3, 1), (-3, -1)$

**15.** $(2, 4), (2, -6)$

**16.** $(9, -8), (-7, 6)$

*In Exercises 17–30, find the equation of the circle with the given center and radius.*

**17.** $(0, 0), r = 2$

**18.** $(1, 0), r = 3$

**19.** $(2, -1), r = 2$

**20.** $(4, 2), r = 5$

**21.** $(-1, 0), r = 4$

**22.** $(1, 2), r = 3$

**23.** $(9, 5), r = 6$

**24.** $(4, 2), r = 2$

**25.** $(0, 0), r = \sqrt{2}$

**26.** $(2, -1), r = \sqrt{2}$

**27.** $(1, 0), r = \sqrt{3}$

**28.** $(4, 2), r = \sqrt{5}$

**29.** $(0, -1), r = 1 + \sqrt{2}$

**30.** $(-1, -3), r = \sqrt{7}$

*In Exercises 31–36, graph the following circles.*

**31.** $(x - 1)^2 + (y + 1)^2 = 4$

**32.** $(x + 3)^2 + (y - 5)^2 = 16$

**33.** $(x - 3)^2 + (y - 4)^2 = 1$

**34.** $x^2 + (y - 1)^2 = 1$

**35.** $(x + 1)^2 + y^2 = 1$

**36.** $x^2 + y^2 = 1$

---

## 2.7 POLYNOMIAL AND RATIONAL FUNCTIONS

Let $a_0, a_1, a_2, \ldots, a_{n-1}, a_n$ be $(n + 1)$ constants. The expression

$$a_0 + a_1x + a_2x^2 + \cdots + a_{n-1}x^{n-1} + a_nx^n \tag{2.71}$$

is called a *polynomial of nth degree in x* if $a_n \neq 0$ and all the exponents of $x$ are positive integers. In other words, the degree of a polynomial is determined by the highest (integer) power of $x$ present in the equation. In this section we will consider equations in which the dependent and independent variables are connected by means of a polynomial or a quotient of polynomials, that is, equations of the form

$$y = a_nx^n + a_{n-1}x^{n-1} + \cdots + a_2x^2 + a_1x + a_0, \tag{2.72}$$

or the form

$$y = \frac{a_nx^n + a_{n-1}x^{n-1} + \cdots + a_2x^2 + a_1x + a_0}{b_mx^m + b_{m-1}x^{m-1} + \cdots + b_2x^2 + b_1x + b_0}. \tag{2.73}$$

Actually, we have already studied two such cases: the linear equation

$$y = a_1x + a_0,$$

which involves a polynomial of first degree, and the quadratic equation

$$y = a_2x^2 + a_1x + a_0,$$

involving a polynomial of second degree.

Since each value of $x$ produces only one value of $y$ when substituted into Eq. (2.72), it follows that the equation defines a function. The situation is slightly more complicated in Eq. (2.73): All zeros in the denominator must be avoided, since division by zero is not allowed. Thus, Eq. (2.73) also defines a function but only for those real numbers that are not zeros of the denominator. The following four examples illustrate these properties.

***Example 1*** Graph the cubic equation

$$y = x^3 + 1. \tag{2.74}$$

As we saw in Section 2.1, we may graph Eq. (2.74) by constructing a sufficiently complete table of points satisfying that equation. Note that Eq. (2.74) involves a polynomial of third degree

$$x^3 + 1 = 1 \cdot x^3 + 0 \cdot x^2 + 0 \cdot x + 1.$$

Substituting the values of $x$ given in the first column of Fig. 2.31(a) into Eq. (2.74), we obtain the $y$-values given in the second column. If we graph these and other points similarly generated, we arrive at the graph shown in Fig. 2.31(b).

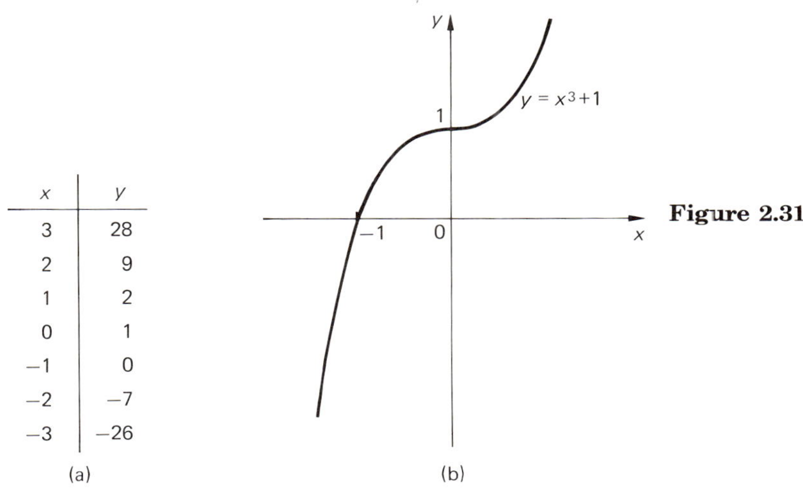

| $x$ | $y$ |
|---|---|
| 3 | 28 |
| 2 | 9 |
| 1 | 2 |
| 0 | 1 |
| −1 | 0 |
| −2 | −7 |
| −3 | −26 |



**Figure 2.31**

---

Sometimes a polynomial function is given as a product of polynomials.

***Example 2*** Graph the third-degree polynomial

$$y = (x + 1)(x + 2)(x + 3). \tag{2.75}$$

We can verify that this is a third-degree polynomial by multiplying the terms on the right side of the equation:

$$(x + 1)(x + 2)(x + 3) = (x^2 + 3x + 2)(x + 3)$$
$$= x^3 + 6x^2 + 11x + 6.$$

Thus, Eq. (2.75) could also have been written in the form

$$y = x^3 + 6x^2 + 11x + 6.$$

Actually, it is more convenient to know the individual factors of the cubic equation, as given in Eq. (2.75), because then we know the zeros of the right side of the equation: $x = -1, -2, -3$. For these values the right side becomes zero, since one of the factors equals zero, so the graph must meet the $x$-axis at these points. These and other points on this graph are given in Fig. 2.32(a) and the graph of Eq. (2.75) is shown in Fig. 2.32(b).

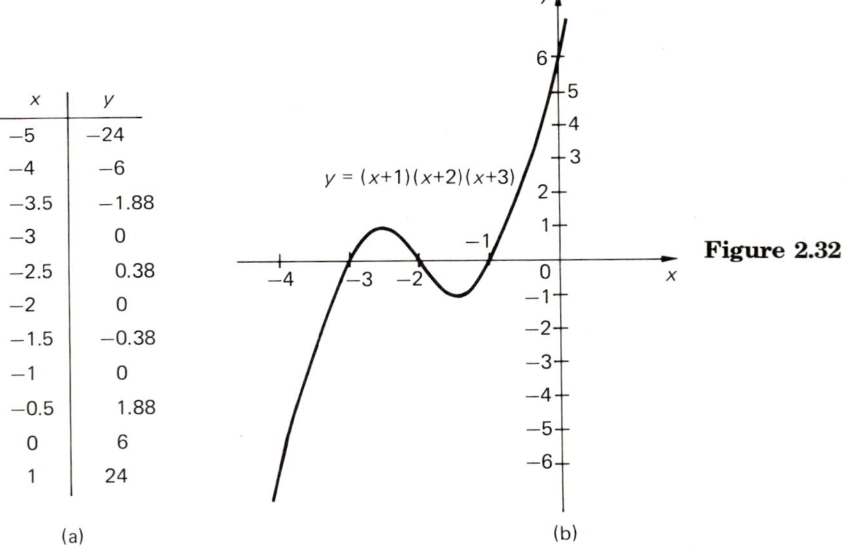

| $x$ | $y$ |
|---|---|
| −5 | −24 |
| −4 | −6 |
| −3.5 | −1.88 |
| −3 | 0 |
| −2.5 | 0.38 |
| −2 | 0 |
| −1.5 | −0.38 |
| −1 | 0 |
| −0.5 | 1.88 |
| 0 | 6 |
| 1 | 24 |



(a) (b)

**Figure 2.32**

Polynomial functions are easy to work with because the $y$-values are simply obtained by means of multiplications and additions. In particular, we will see in Chapters 9 and 10 that the operations of differential and integral calculus are especially simple for polynomials.

We now turn to functions that are defined as the quotient of two polynomials (see Eq. 2.73). Any such function is called a *rational function* and is defined for all real values of $x$ that are not roots of the denominator.

***Example 3*** Graph the rational function

$$y = \frac{1}{x}. \tag{2.76}$$

(Note that the numerator of this function is a polynomial of zeroth degree and the denominator is a polynomial of first degree.)

Since $x = 0$ is a root of the denominator of the right side of Eq. (2.76), the function will not be defined at this value of $x$. However, for all other values of $x$ we can obtain a corresponding value for $y$. Figure 2.33(a) is a short list of points on the graph of this function, which is shown in Fig. 2.33(b). Note that as $x$ gets larger in either direction, the values of $y$ tend to zero, while the reverse is true as $x$ approaches zero. Thus, the graph of the function gets arbitrarily close to the lines $x = 0$ and $y = 0$ (the coordinate axes) without ever meeting them. We say that the lines $x = 0$ and $y = 0$ are *asymptotes* of the function $y = 1/x$.

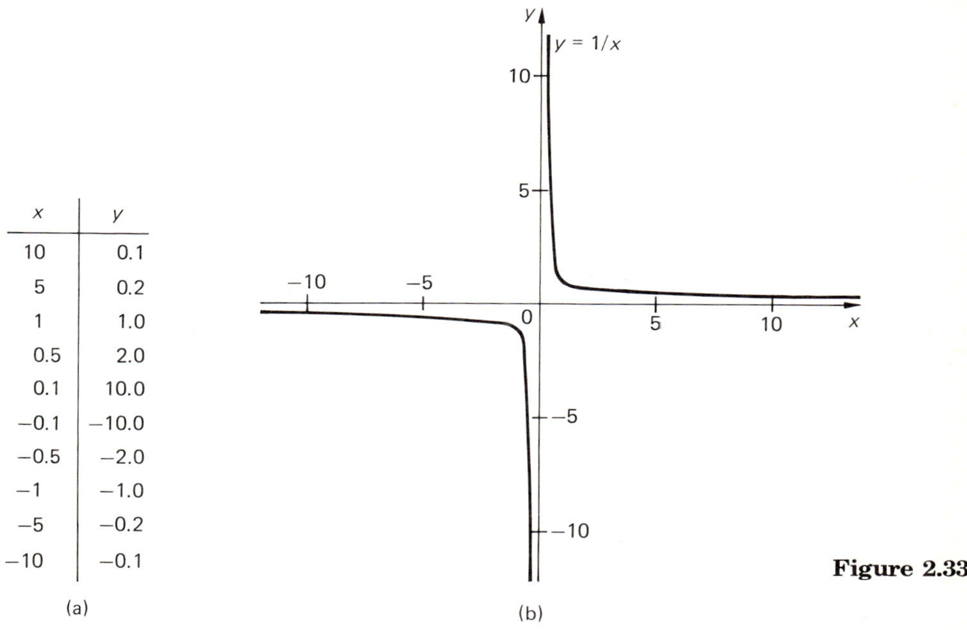

| x | y |
|---|---|
| 10 | 0.1 |
| 5 | 0.2 |
| 1 | 1.0 |
| 0.5 | 2.0 |
| 0.1 | 10.0 |
| −0.1 | −10.0 |
| −0.5 | −2.0 |
| −1 | −1.0 |
| −5 | −0.2 |
| −10 | −0.1 |



**Figure 2.33**

***Example 4*** Graph the rational function

$$y = \frac{x}{1 + x^2}. \tag{2.77}$$

Since the denominator never vanishes, because $x^2 \geq 0$, this rational function is defined for all values of $x$. Substituting several different values of $x$ into Eq. (2.77) we obtain the results shown in Fig. 2.34(a). Note that as we select larger and larger values of $x$, the corresponding $y$-values diminish, since there is an $x^2$-term in the denominator that is much larger than the value of $x$. Thus, the graph will approach the $x$-axis ever closer as $x$ increases without bound. Therefore, the $x$-axis is an asymptote of Eq. (2.77) (see Fig. 2.34b).

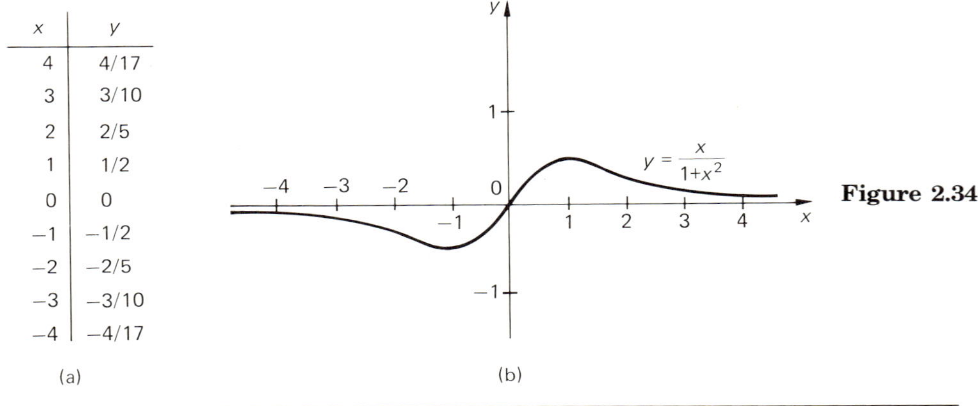

| $x$ | $y$ |
|---|---|
| 4 | 4/17 |
| 3 | 3/10 |
| 2 | 2/5 |
| 1 | 1/2 |
| 0 | 0 |
| −1 | −1/2 |
| −2 | −2/5 |
| −3 | −3/10 |
| −4 | −4/17 |

(a)



(b)

**Figure 2.34**

Rational functions have important applications in the fields of economics and sociology, as the following example illustrates.

***Example 5*** Ground beef and potatoes are two well-liked menu items in Mary's home. Consequently, she always stocks up on each item during her weekly visit to the supermarket. The amount that she buys each time varies according to each item's price and Mary's relative preferences.

Suppose that on one particular shopping trip Mary buys 15 lb of potatoes and 4 lb of ground beef. Obviously, if the store manager offered to let her have 20 lb of potatoes and 5 lb of ground beef for the same price, Mary would certainly accept (although she might be somewhat suspicious). Hence, Mary would definitely prefer more of everything. If the manager had offered her 25 lb of potatoes and 3 lb of ground beef, rather than her 15 lb of potatoes and 4 lb of ground beef, she might have been willing to exchange 1 lb of ground beef for 10 lb of potatoes. However, if the offer had been 20 lb of potatoes and 3 lb of ground beef (an exchange of 5 lb of potatoes for 1 lb of ground beef), Mary might have thought this bundle to be no better than the one she already had. Thus, it is possible to have two different bundles for which Mary has no preference *if cost is not a factor.*

Consider this situation more systematically. We let $x$ represent the number of lb of potatoes and $y$ the number of lb of ground beef in a given bundle. If two different combinations of the items are perceived as being equivalent, it is apparent that each must contain more of one item than the other. Thus, each increase in $x$ is offset by a decrease in $y$. The graph of all combinations that Mary judges to be equivalent to (15, 4) might resemble the curve shown in Fig. 2.35. Such a curve is called an *indifference curve* because Mary finds each combination equally appealing and is therefore indifferent to any particular selection.

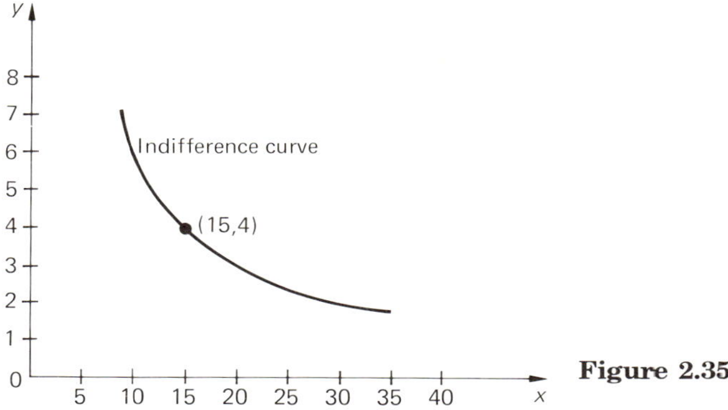


**Figure 2.35**

We see that the given indifference curve closely resembles the graph of the function in Example 3. Indeed, the rational function

$$y = \frac{60}{x} \tag{2.78}$$

yields the graph shown in Fig. 2.35. Writing Eq. (2.78) as $xy = 60$, we see that it has the form

$$xy = u, \tag{2.79}$$

where $u$ is some real number. If we increase $u$, we obtain a curve that is further away from the origin; conversely, decreasing $u$ yields a curve that is closer to the origin (see Fig. 2.36). Suppose that each of these curves is an indifference curve, because Mary finds each combination on each curve to be equally appealing. Mary will prefer those bundles lying on the curve $xy = 100$ to those on the curve $xy = 60$. To see this, select any bundle on the indifference curve $xy = 60$—say, (15, 4). If we continue along the horizontal line $y = 4$, we find that it meets the curve $xy = 100$ at the point (25, 4). Each bundle has 4 lb of ground beef, but the second has 10 more lb of potatoes. The same process will work for any

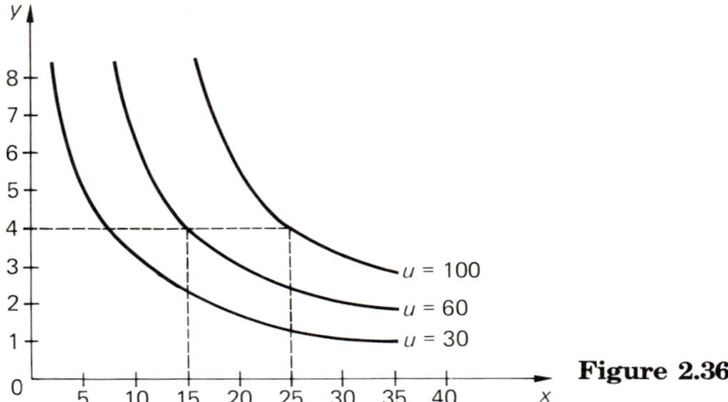


Figure 2.36

other bundle on the indifference curve $xy = 60$. Therefore, increasing the value of $u$ in Eq. (2.79) will increase the value or usefulness of the bundles. Economists and sociologists describe this by saying that the *utility* increases as one moves to an indifference curve that is further away from the origin. The set of all indifference curves (which may be infinite) is called an *indifference map* and represents Mary's particular preferences.

Mary would certainly like to be on the indifference curve with the highest utility that she can reach. She is limited only by what she can afford to spend for the commodities. Thus, her budget determines the utility that she can afford. Suppose that she can spend only \$10 and that potatoes cost 25¢ per lb, while ground beef costs \$1 per lb. The straight line in Fig. 2.37 is her *budget line* and shows all possible combinations that she can afford. For example, she can buy 40 lb of potatoes or 10 lb of ground beef, etc. Observe that the indifference curve $xy = 100$ just

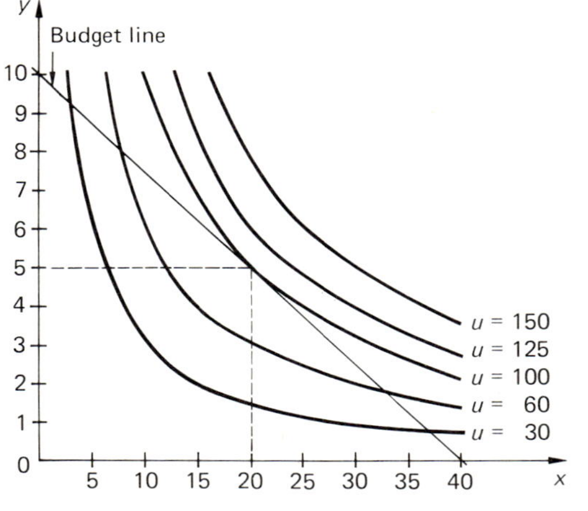


Figure 2.37

touches the budget line at (20, 5). This is the highest utility that she can reach and is called the *equilibrium point*. Thus, to maximize utility on a $10 budget, she should buy 20 lb of potatoes and 5 lb of ground beef.

Two factors influence the budget line: Mary may decide to spend more on these items, or the prices of the items may change. Either situation may yield a different equilibrium point and maximum utility (see Exercises 27–30 on p. 80).

---

Although this example seems quite artificial, the principles that it illustrates are quite common to many important situations. Most important decisions involve trade-offs of this type. For example, reducing the pollution emitted by a coal-burning power-generating plant raises the cost of the power generated; increasing the cost of an athletic event can decrease the attendance. Such situations often have indifference curves that can be expressed as rational functions. The analysis of the problem may be technically more difficult, but the principles are the same.

## EXERCISES 2.7

*In Exercises 1–24, graph the given function (by constructing a sufficiently long table of values) and find any asymptotes that it may have.*

**1.** $y = x^3$

**2.** $y = 2x^4$

**3.** $y = x^4 - 1$

**4.** $y = 3x^4 + 2x + 1$

**5.** $y = x^4 - 2x + 1$

**6.** $y = 4x^4 - 4x^2 + 1$

**7.** $y = (x - 1)(x - 2)(x - 3)$

**8.** $y = (x - 1)(x + 1)(x - 3)$

**9.** $y = (x^2 - 1)(x^2 + 1)$

**10.** $y = (x^2 - 4)(x^2 - 9)$

**11.** $y = \dfrac{1}{x + 1}$

**12.** $y = \dfrac{2}{x - 1}$

**13.** $y = \dfrac{3}{2 - x}$

**14.** $y = \dfrac{1}{x^2 + 1}$

**15.** $y = \dfrac{1}{x^2 - 1}$

**16.** $y = \dfrac{x}{x^2 - 1}$

**17.** $y = \dfrac{x + 1}{x - 1}$

**18.** $y = \dfrac{x - 1}{x + 1}$

**19.** $y = \dfrac{2x - 1}{x - 3}$

**20.** $y = \dfrac{1 + x^2}{1 - x^2}$

**21.** $y = \dfrac{x}{x^3 - 1}$

**22.** $y = \dfrac{x}{x^3 + 1}$

**23.** $y = (x - 1)^2(x - 2)^3$

**24.** $y = (x - 1)^2(x - 2)^3(x - 3)^4$

**25.** Construct a nonnegative polynomial function that touches the $x$-axis at $x = 1$, 2, and 3.

**26.** Construct a polynomial function that crosses the $x$-axis at $x = 1$, 2, and 3, but merely touches it at $x = 4$.

**27.** Suppose that Mary decides to spend $12 in purchasing potatoes and ground beef. Using the indifference curves given in Example 5 in Section 2.7, find the equilibrium point.

**28.** Suppose that the price of potatoes in Example 5 in Section 2.7 drops to 20¢ per lb, all other facts remaining the same. What is Mary's equilibrium point?

**29.** An unannounced special lowers the price of ground beef in Example 5 in Section 2.7 to 80¢ per lb. What is the equilibrium point?

**30.** Find the equilibrium point if the price of ground beef in Example 5 in Section 2.7 is raised to $1.25 per lb.

---

## ☐ 2.8 MODELING BY LEAST SQUARES APPROXIMATION

In Example 3 in Section 2.4 we compared the gestation periods and the longevities of a number of mammals. The data was plotted in Fig. 2.20 and showed a typical situation affecting large quantities of experimental data: The data points are quite scattered. In that example we used the graph of the data points to sketch in a straight line that we hoped would "best" fit the data. The linear model we obtained,

$$L = \frac{3}{20}G + 1,$$

was given in Eq. (2.41). We shall now see if we made a good choice.

The method of *least squares approximation* is a commonly used technique for fitting a straight line through a collection of data points. The vertical distance between the data point and the point on the line *having the same x-coordinate* is called the *error* of the approximation for that data point (see Fig. 2.38). The principal idea behind the method is to find the straight line that *minimizes the square of these errors.* Since the squares of the errors are least for this line, it is natural to call it the *least squares line* and to name the method after this fact.

What is the reason behind "squaring" the errors rather than merely adding them? One justification is that if we square a small error $e < 1$ the square $e^2 < e$ is often much smaller than the error $e$, so that the sum of all such errors is less likely to outweigh a large error. After all, the least squares line is already doing a good job of approximating those

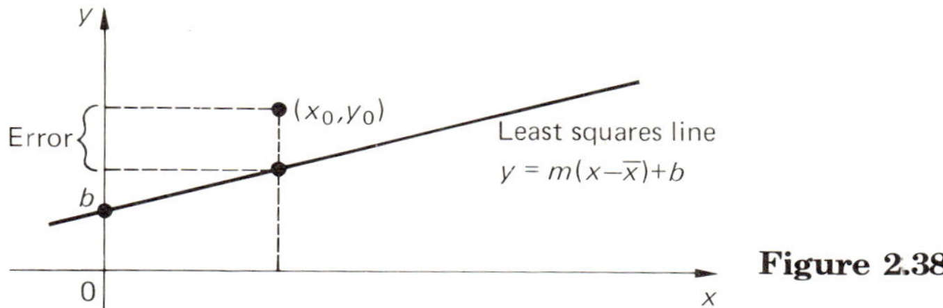


**Figure 2.38**

data points having a small error. A second reason is that the formula for the least squares line is easy to obtain.

Suppose we have the set of $n$ data points

$$(x_1, y_1), (x_2, y_2), (x_3, y_3), \ldots, (x_{n-1}, y_{n-1}), (x_n, y_n) \tag{2.80}$$

Denote the average $x$-value by $\bar{x}$ and the average $y$-value by $\bar{y}$; that is,

$$\bar{x} = \frac{1}{n}(x_1 + x_2 + \cdots + x_n),$$

$$\bar{y} = \frac{1}{n}(y_1 + y_2 + \cdots + y_n).$$

Then we have the following theorem.

**LEAST SQUARES THEOREM 2.1** **The least squares line for the data points (2.80) is given by the equation**

$$y = m(x - \bar{x}) + b, \tag{2.81}$$

**where the coefficients $m$ and $b$ are given by $b = \bar{y}$ and**

$$m = \frac{(x_1y_1 + x_2y_2 + \cdots + x_ny_n) - n\bar{x}\bar{y}}{(x_1^2 + x_2^2 + \cdots + x_n^2) - n(\bar{x})^2}. \tag{2.82}$$

Rather than prove this theorem, we will illustrate how it works by using the data given in Example 3 in Section 2.4. (Since we will be doing a number of additions and multiplications, a hand-held calculator will save us a tremendous amount of time. Although some of the more sophisticated calculators do the computations automatically, even the least expensive calculator can quickly give the answer.) First we relabel each gestation period by $x_i$ and the corresponding longevity by $y_i$ and gather the data in the first three columns of Table 2.7. The first column numbers the data points; in this case, $n = 14$. The second and third columns contain the gestation periods and longevities of the 14 species of animals. Column four contains the product of the gestation and the longevity, while column five contains the squares of the gestation

**Table 2.7**

| $i$ | $x_i$ | $y_i$ | $x_i y_i$ | $x_i^2$ |
|---|---|---|---|---|
| 1 | 60 | 15 | 900 | 3600 |
| 2 | 63 | 15 | 945 | 3969 |
| 3 | 31 | 7 | 217 | 961 |
| 4 | 61 | 16 | 976 | 3721 |
| 5 | 52 | 8 | 416 | 2704 |
| 6 | 68 | 4 | 272 | 4624 |
| 7 | 98 | 17 | 1666 | 9604 |
| 8 | 21 | 4 | 84 | 441 |
| 9 | 112 | 14 | 1568 | 12,544 |
| 10 | 90 | 11 | 990 | 8100 |
| 11 | 37 | 5 | 185 | 1396 |
| 12 | 44 | 8 | 352 | 1936 |
| 13 | 105 | 19 | 1995 | 11,025 |
| 14 | 63 | 12 | 756 | 3969 |
| Sum | 905 | 155 | 11,322 | 68,567 |
| Average | $\bar{x} = 64.64$ | $\bar{y} = 11.07$ | | |

periods for each species. We sum the quantities in columns two through five and divide columns two and three by $n = 14$ to get their averages. The resulting numbers can now be substituted in Eq. (2.82):

$$m = \frac{(x_1y_1 + \cdots + x_{14}y_{14}) - 14\bar{x}\bar{y}}{(x_1^2 + \cdots + x_{14}^2) - 14\bar{x}^2}$$

$$= \frac{11322 - 14(64.64)(11.07)}{68567 - 14(64.64)^2}$$

$$= \frac{11322 - 10017.91}{68567 - 58496.61} = \frac{1304.19}{10070.39} = 0.1295.$$

Equation (2.81) becomes

$$y = 0.1295\,(x - 64.64) + 11.07$$

or

$$y = 0.1295x + 2.7,$$

which is fairly close to the formula derived in Example 3, in Section 2.4,

$$y = 0.15x + 3,$$

using our current notation. If we use Eq. (2.83) to predict the longevity of a lion and a baboon (gestation periods $x$ equaling 100 and 187 days, respectively), we obtain for the lion

$$y = 0.1295(100) + 2.7 = 15.65$$

and for the baboon

$$y = 0.1295(187) + 2.7 = 26.92.$$

These values are slightly better than those obtained in Example 3 in Section 2.4.

***Example 1*** While studying the wind field for a convective storm, R. A. Kropfli and L. J. Miller [*Journal of the Atmospheric Sciences* **33** (1976): 528] discovered a significant connection between the vertical velocity of the updrafts and downdrafts and the cyclonic and anticyclonic vorticity, or whirling motion, of the wind. The values they obtained are approximately as follows:

| Vertical velocity (m/sec) | −6 | −2 | 2 | 7 | 12 |
|---|---|---|---|---|---|
| Vorticity ($\text{sec}^{-1} \times 10^{-3}$) | −0.28 | −0.33 | 0 | 0.8 | 1.1 |

(A negative wind velocity is a downdraft and a negative vorticity is anticyclonic.)

Again arranging our work in a table (see Table 2.8), we let $x_i$ represent the values of the vertical wind velocity and $y_i$ the values of the vorticity. In columns four and five we calculate the products $x_i y_i$ and $x_i^2$, and we sum the columns at the bottom.

**Table 2.8**

| $i$ | $x_i$ | $y_i$ | $x_i y_i$ | $x_i^2$ |
|---|---|---|---|---|
| 1 | −6 | −0.28 | −1.68 | 36 |
| 2 | −2 | −0.33 | −0.66 | 4 |
| 3 | 2 | 0.00 | 0.00 | 4 |
| 4 | 7 | 0.80 | 5.60 | 49 |
| 5 | 12 | 1.10 | 13.20 | 144 |
| Sum | 13 | 1.29 | 16.46 | 237 |
| Average | $\bar{x} = 2.6$ | $\bar{y} = 0.26$ | | |

Using the least squares theorem with $n = 5$, we obtain $b = \bar{y} = 0.26$ and

$$m = \frac{(x_1y_1 + \cdots + x_5y_5) - 5\bar{x}\bar{y}}{(x_1^2 + \cdots + x_5^2) - 5\bar{x}^2}$$
$$= \frac{16.46 - 5(2.6)(0.26)}{237 - 5(2.6)^2}$$
$$= \frac{13.08}{203.2} = 0.0644.$$

Thus, the least squares line is given by

$$y = 0.0644(x - 2.6) + 0.26.$$

We can now use this equation to predict vorticities at other vertical wind velocities. For example, a vertical updraft of 5m/sec should produce a cyclonic vorticity of

$$y = 0.0644(5 - 2.6) + 0.26 = 0.4146\ \text{sec}^{-1} \times 10^{-3}.$$

---

As we have seen, least squares approximation is an easy method to apply in constructing a linear model that fits a collection of data points. It requires only systematic routine calculations, which is the reason for its popularity. We will encounter it again in Section 3.5 and Chapter 9. Least squares approximation is often called *linear regression*, especially in statistics. A proof of the least squares theorem is given in Example 7 in Section 9.9. That proof requires a knowledge of calculus. A different proof, based on optimizing quadratic functions, is outlined in Exercise 19 on p. 86. This is a very challenging exercise.

## EXERCISES 2.8

*In Exercises 1–12, find the least squares line for the given data points. Round m and b to three decimal places.*

**1.**

| x | y |
|---|---|
| −4 | 2 |
| 1 | 5 |

**2.**

| x | y |
|---|---|
| 2 | 4 |
| 3 | 1 |

**3.**

| x | y |
|---|---|
| 2 | 7 |
| 5 | 10 |

**4.**

| x | y |
|---|---|
| −3 | 6 |
| −1 | 2 |

**5.**

| x | y |
|---|---|
| −1 | 2 |
| 1 | 5 |
| 3 | 4 |

**6.**

| x | y |
|---|---|
| −2 | 3 |
| 0 | 7 |
| 2 | 5 |

**7.**

| x | y |
|---|---|
| −3 | 6 |
| 1 | 2 |
| 5 | 4 |

**8.**

| x | y |
|---|---|
| 1 | 7 |
| 3 | 2 |
| 7 | 5 |

**9.**

| x | y |
|---|---|
| −4 | 8 |
| −2 | 5 |
| 0 | 1 |
| 2 | 3 |

**10.**

| x | y |
|---|---|
| −3 | 2 |
| 1 | 4 |
| 5 | 6 |
| 8 | 8 |

**11.**

| x | y |
|---|---|
| −5 | −2 |
| −2 | 0 |
| 3 | 2 |
| 4 | 4 |
| 5 | 6 |
| 9 | 8 |

**12.**

| x | y |
|---|---|
| −10 | −30 |
| −5 | −10 |
| 0 | 5 |
| 5 | 20 |
| 10 | 50 |
| 20 | 80 |

**13.** Find the best least squares approximation to the data in Exercise 11 in Section 2.4. Use the equation so obtained to answer the questions posed in that exercise.

**14.** Find the best least squares approximation to the data in Exercise 12 in Section 2.4, and answer the questions posed in that exercise.

**15.** Find the best least squares approximation to the data in Exercise 13 in Section 2.4. How many weeks of exposure to light are required to ensure that only 50% of the *Mysis relicta* will survive?

**16.** Find the best least squares approximation to the data in Exercise 14 in Section 2.4, and answer the questions posed in that exercise.

**17.** In an experiment to determine the energy expended by an adult male while walking at various velocities, E. Atzler and R. Herbst [*Pfluegers Archiv.* **215** (1972): 391] obtained the following data:

| | | | | | | | | | | | |
|---|---|---|---|---|---|---|---|---|---|---|---|
| $x$ (km/hr) Walking speed | 1.25 | 1.75 | 2.00 | 2.75 | 3.4 | 4.5 | 4.7 | 5.5 | 5.75 | 6.75 | 8.9 |
| $y$ (kcal/min) Energy expenditure | 2.1 | 2.75 | 2.4 | 2.75 | 3.3 | 3.9 | 4.1 | 4.9 | 5.2 | 6.6 | 8.75 |

Find the least squares approximation to these data and estimate the energy expenditure at a velocity of 6 km/hr.

**★18.** Prove that the least squares line for two data points is the straight line passing through these two points.

★**19.** A proof of the least squares theorem may be based on the technique of optimizing quadratic functions, which we discussed in Section 2.5. The proof is divided into the following steps.

a) Let $y = m(x - \bar{x}) + b$ be the least squares line and $(x_i, y_i)$ be any data point. Explain (see Fig. 2.38) why the line causes an error of

$$y_1 - [m(x_i - \bar{x}) + b].$$

b) Squaring the errors at each data point $(x_1, y_1), (x_2, y_2), \cdots, (x_n, y_n)$, show that the sum of these squares equals

$$\begin{aligned}&[y_1^2 + y_2^2 + \cdots + y_n^2]\\ &\quad + m^2[(x_1 - \bar{x})^2 + (x_2 - \bar{x})^2 + \cdots + (x_n - \bar{x})^2]\\ &\quad - 2m[y_1(x_1 - \bar{x}) + y_2(x_2 - \bar{x}) + \cdots + y_n(x_n - \bar{x})]\\ &\quad + 2mb[(x_1 - \bar{x}) + (x_2 - \bar{x}) + \cdots + (x_n - \bar{x})] - 2nb\bar{y} + nb^2.\end{aligned}$$

c) Show that

$$[(x_1 - \bar{x}) + (x_2 - \bar{x}) + \cdots + (x_n - \bar{x})] = 0.$$

d) Show that

$$\begin{aligned}&y_1(x_1 - \bar{x}) + y_2(x_2 - \bar{x}) + \cdots + y_n(x_n - \bar{x})\\ &\qquad = (x_1y_1 + x_2y_2 + \cdots + x_ny_n) - n\bar{x}\bar{y}.\end{aligned}$$

e) Show that

$$(x_1 - \bar{x})^2 + (x_2 - \bar{x})^2 + \cdots + (x_n - \bar{x})^2 = (x_1^2 + x_2^2 + \cdots + x_n^2) - n\bar{x}^2.$$

f) Using the results in parts (c), (d), and (e) and completing the square, show that the expression in (b) can be written in the form

$$\begin{aligned}&[(x_1^2 + x_2^2 + \cdots + x_n^2) - n\bar{x}^2]\left[m - \frac{(x_1y_1 + x_2y_2 + \cdots + x_ny_n) - n\bar{x}\bar{y}}{(x_1^2 + x_2^2 + \cdots + x_n^2) - n\bar{x}^2}\right]^2\\ &\qquad + n(b - \bar{y})^2\\ &\quad + \left[(y_1^2 + y_2^2 + \cdots + y_n^2) - n\bar{y}^2 - \frac{((x_1y_1 + x_2y_2 + \cdots + x_ny_n) - n\bar{x}\bar{y})^2}{(x_1^2 + x_2^2 + \cdots + x_n^2) - n\bar{x}^2}\right].\end{aligned}$$

g) Minimize the expression in part (f) by selecting $m$ and $b$ so that all terms but the last one (in brackets) will vanish. This proves the least squares theorem, since it shows that the sum of the squares of the error terms is least for these values of $m$ and $b$.

## CHAPTER VOCABULARY

**abscissa** (= $x$-axis)

**asymptote**

**axis**

**Cartesian coordinates**

**circle** $(x - x_0)^2 + (y - y_0)^2 = r^2$

**completing the square**

**dependent variable**

**discriminant** $b^2 - 4ac$

**distance formula** $d = \sqrt{(x_0 - x_1)^2 + (y_0 - y_1)^2}$

**domain (of a function)**

**extrapolation**

**function**

**functional notation** $f(x)$

**graph**

**hypotenuse**

**independent variable**

**indifference curve**

**indifference map**

**interpolation**

**least squares line**

**least squares theorem**

**linear equation** $Ax + By = C$

**linear regression**

**ordinate** (= $y$-axis)

**parabola** $y = ax^2 + bx + c$

**plotting**

**point-slope formula** $y - y_0 = m(x - x_0)$

**polynomial** $y = a_n x^n + \cdots + a_1 x + a_0$

**Pythagorean theorem** $a^2 + b^2 = c^2$

**quadratic formula** $x = \dfrac{-b \pm \sqrt{b^2 - 4ac}}{2a}$

**quadratic function** $ax^2 + bx + c$

**radius**

**range (of a function)**

**rational function** $p(x)/q(x)$

**relation**

**root (of an equation)**

**simultaneous linear equations**

**slope** $m$

**slope-intercept formula** $y = mx + b$

**symmetry (about an axis)**

**system of equations**

**table**

**two-point formula** $\dfrac{y - y_0}{x - x_0} = \dfrac{y_1 - y_0}{x_1 - x_0}$

**utility (economic)**

**vertex (of a parabola)**

**$y$-intercept** $b$

# CHAPTER 3 EXPONENTIAL AND LOGARITHMIC FUNCTIONS

## 3.1 EXPONENTIAL FUNCTIONS

We saw in Chapter 2 that linear and polynomial functions can be quite useful in describing and analyzing many practical situations. In this chapter we will study another class of functions, the exponential functions and the closely related logarithmic functions, which are equally important in practical applications. Exponential functions can be used, for example, to describe the growth of populations or money with time or the decay of radioactivity in a given sample. Before considering a number of these applications, we shall digress by telling two stories that illustrate the behavior of exponential functions.

In a kingdom, long ago, there lived a noble knight who spent his time doing all he could to help the peasants. The peasants lived in near starvation due to the greed of their king. It came to pass that the kingdom was overrun by 60 dragons that destroyed most of the crops. The king was very upset because it was the law of the kingdom that half of every crop belonged to him. He summoned the knight and begged him to slay the dragons. The knight agreed, saying that since this would help the peasants he would only charge the king one grain of wheat for the first dragon, two for the second, four for the third, and so on, doubling for each additional dragon he slew. The greedy king was only too happy to accept these terms. The knight slew the dragons and the king, for all his wealth, was unable to pay the agreed amount. In a rage he threw the knight in the dungeon, but the peasants rebelled, assassinated the king, and installed the knight as their new king. Why was the king *unable* to pay the agreed amount?

To discover the answer, we need to determine how many grains of wheat were required. We arrange this information in Table 3.1. How much is 2 multiplied by itself 59 times? If you have some time you will discover that it is the enormous number 576,460,752,303,423,488, or

**Table 3.1**

| Dragon slain | Grains of wheat to be paid for that dragon |
|---|---|
| 1 | $1$ |
| 2 | $2 = 1 \cdot 2$ |
| 3 | $4 = 1 \cdot 2 \cdot 2$ |
| 4 | $8 = 1 \cdot 2 \cdot 2 \cdot 2$ |
| 5 | $16 = 1 \cdot \underbrace{2 \cdot 2 \cdot 2 \cdot 2}_{\text{4 times}}$ |
| . | . |
| . | . |
| . | . |
| 60 | $1 \cdot \underbrace{2 \cdot 2 \cdot \cdots \cdot 2}_{\text{59 times}}$ |

576 quadrillion grains of wheat for the sixtieth dragon. Indeed, the total amount for all the dragons is one grain less than double this amount (see Exercise 31 on p. 98). Since 7000 grains of wheat weigh one pound and 2000 lb equal a ton, 14 million grains of wheat weigh a ton; thus, 80 billion tons of wheat were required as payment for slaying the 60 dragons. This is far more than the world's total production of wheat.

Now let's move on to our second story. One thousand years ago a Swiss clockmaker deposited one gram of gold (now worth about four dollars) with a Swiss banker. Each year the bank increased the amount of gold in the account by five percent. The account was to be equally divided among all the clockmaker's heirs after 1000 years. The bank has located 1200 heirs. Will each heir receive enough gold to buy a new Cadillac?

To find the answer to this question we need to determine the effect of adding 5% to a given amount $x$:

$$x + 5\%x = x + 0.05x = 1.05x.$$

Thus, we multiply each year's amount by 1.05 to get the next year's amount. Beginning with 1 gram of gold, we multiply this by 1.05 one thousand times to obtain the number of grams of gold in the account after 1000 years. Any calculator with a $y^x$ key will quickly supply the answer of

$$1.546319 \times 10^{21}$$

grams of gold. Each heir will receive

$$1.288599 \times 10^{18}$$

grams of gold worth

$$\$5{,}154{,}396{,}600{,}000{,}000{,}000,$$

enough to buy General Motors lock, stock, and barrel.

These two stories illustrate the fact that if a number greater than one is multiplied by itself a large number of times, the result is an astronomically large number. Indeed, the number need not be much greater than one, as was shown in the second story. This process, which we shall now study systematically, is called *exponential growth.*

A review of the *laws of exponents* (see Section 1.1) will be helpful. Recall that the expression $a^n$, where $a$ is a real number and $n$ is an integer, is used to denote the multiplication of $a$ by itself $n$ times:

$$a^n = \underbrace{a \cdot a \cdot \cdots \cdot a}_{n \text{ times}} \tag{3.1}$$

We call $a^n$ an *exponential* with *base a* and *exponent* (or *power*) $n$. Then, it follows that

$$a^m a^n = a^{m+n} \tag{3.2}$$

since each side consists of $m + n$ multiples of $a$. Similarly, if $m > n$,

$$\frac{a^m}{a^n} = a^{m-n}, \qquad a \neq 0, \tag{3.3}$$

since we can cancel $n$ multiples of $a$ from the numerator of the left side. Of course, $a$ cannot equal zero since division by zero is not allowed. Note further that if $m = n$ in Eq. (3.3), there are just as many factors of $a$ in the numerator as there are in the denominator. If we cancel all these factors, the left side of Eq. (3.3) equals 1, so that

$$a^0 = 1, \qquad a \neq 0. \tag{3.4}$$

If we extend this definition to the case in which $m < n$, we obtain an exponential with a *negative* power. This simply means that there are more factors of $a$ in the denominator:

$$a^{-n} = \frac{1}{a^n}, \qquad a \neq 0. \tag{3.5}$$

Thus, *we may transfer an exponential from the denominator to the numerator (or vice versa) by simply changing the sign of the exponent.*

Returning to the basic definition, we observe that

$$(a^m)^n = a^{mn}, \tag{3.6}$$

since the left side consists of $n$ factors of $a^m$, and each of these consists of $a$ multiplied by itself $m$ times. Thus, $a$ multiplies itself $mn$ times. Using Eq. (3.6), we can assign a meaning to *rational* (or fractional) powers. Since

$$(a^{1/n})^n = a^{n/n} = a^1 = a, \tag{3.7}$$

it follows that $a^{1/n}$ must be the $n$th *root of a*. Thus, $a^{1/2}$ is the *square root* of $a$, $a^{1/3}$ is the *cube root* of $a$, and so on, or

$$a^{1/2} = \sqrt{a}, \qquad a^{1/3} = \sqrt[3]{a}, \cdots.$$

Continuing this process, we define

$$a^{m/n} = (a^{1/n})^m = (\sqrt[n]{a})^m; \tag{3.8}$$

that is, $a^{m/n}$ is the $m$th power of the $n$th root of $a$. At this point some care is necessary for three reasons:

**1.** If $a$ is positive and $n$ is even, there are *two* real numbers that are both $n$th roots of $a$. For example, 2 and $-2$ are both square roots of 4, and 3 and $-3$ are both fourth roots of 81. Fortunately, they differ only in sign, so we define $a^{1/n} = \sqrt[n]{a}$ to be the positive $n$th root of $a$ and denote the other $n$th root by $-\sqrt[n]{a} = -(a^{1/n})$.

**2.** If $a$ is negative and $n$ is even, there is *no* real number that is the $n$th root of $a$. This is a consequence of Eq. (3.7), since any real number multiplied by itself an even number of times is nonnegative.

**3.** Although the laws of exponents allow us to rewrite the middle term in Eq. (3.8) as

$$a^{m/n} = (a^{1/n})^m = (a^m)^{1/n}, \tag{3.9}$$

this last term may lead to inconsistencies when $a$ is negative. For example, we know that $(-1)^{1/2}$ is not a real number, but if we apply the last term of Eq. (3.9) to an equivalent expression we incorrectly obtain

$$(-1)^{1/2} = (-1)^{2/4} = [(-1)^2]^{1/4} = [1]^{1/4} = 1.$$

In evaluating $a^{m/n}$ for negative values of $a$, we should first reduce $m/n$ to its lowest form—say, $p/q$—and then use only Eq. (3.8):

$$a^{p/q} = (a^{1/q})^p.$$

If $a < 0$ and $q$ is even, then $a^{p/q}$ is not a real number.

We illustrate these laws in the following example.

***Example 1*** a) $2^3 2^2 = 2^5 = 32$

b) $\dfrac{3^4}{3^7} = 3^{4-7} = 3^{-3} = \dfrac{1}{3^3} = \dfrac{1}{27}$

c) $\dfrac{4^2}{2^4} = \dfrac{(2^2)^2}{2^4} = \dfrac{2^{2\cdot 2}}{2^4} = \dfrac{2^4}{2^4} = 2^{4-4} = 2^0 = 1$

d) $5^{-1} + 2^{-2} = \dfrac{1}{5} + \dfrac{1}{2^2} = \dfrac{1}{5} + \dfrac{1}{4} = \dfrac{9}{20}$

e) $(64)^{1/2} = 8$, but $-(64)^{1/2} = -8$

f) $(27)^{1/3} = 3$

g) $(8)^{2/3} = (8^{1/3})^2 = (2)^2 = 4,$ or

$$= (8^2)^{1/3} = (64)^{1/3} = 4$$

h) $\dfrac{3^{5/3}}{9^{1/3}} = \dfrac{3^{5/3}}{(3^2)^{1/3}} = \dfrac{3^{5/3}}{3^{2/3}} = 3^{(5/3)-(2/3)} = 3$

i) $(-32)^{2/5} = [(-32)^{1/5}]^2 = (-2)^2 = 4$

j) $(-8)^{2/6} = (-8)^{1/3} = -2$

k) $(-4)^{2/4} = (-4)^{1/2}$, which is not a real number

---

By using the laws of exponents, we have successfully defined expressions having a rational exponent, at least for positive values of $a$. This fundamental fact allows us to define the exponential functions. We will illustrate the process with a particular example.

***Example 2*** We wish to draw the graph of the function

$$y = 2^x$$

in the $xy$-plane. Obviously, we can calculate $2^x$ whenever $x$ is a rational number. If we were to stop here, our graph would consist of a series of points corresponding only to the rational values of $x$. We need to extend our definition to real values of $x$, hopefully in such a way that a smooth curve will result from the extension. To do this we define $2^x$ as the real number that satisfies the inequalities

$$2^r < 2^x < 2^s$$

for all rational numbers $r$ and $s$ satisfying

$$r < x < s.$$

Then $2^x$ will be larger than $2^r$ for all rational numbers $r$ smaller than $x$, and $2^x$ will be less than $2^s$ for all rational numbers $s$ larger than $x$. Since $2^r$ and $2^s$ approach each other as $r$ and $s$ become closer, this guarantees that the graph of $y = 2^x$ will be a smooth curve having no jumps or gaps.

A similar process can be used to define the *exponential function*

$$y = a^x$$

for any *positive* real number $a$. It should also be apparent why the process will fail for negative values of $a$, since gaps will occur whenever the rational number $x$ (in lowest form) has an even denominator.

Furthermore, the numbers $a^r$ and $a^s$ need not be close, even when $r$ and $s$ are close together:

$$(-1)^{1/11} = -1 \quad \text{but} \quad (-1)^{2/21} = [(-1)^{1/21}]^2 = (-1)^2 = 1.$$

Now that we know that the graph of an exponential function

$$y = a^x, \qquad a > 0 \tag{3.10}$$

is smooth, it is an easy task to draw the graph. For example, to graph $y = 2^x$, we make a table of $x$- and $y$-values until we have enough points to determine the shape of the curve. We then draw a smooth curve through them and obtain the graph shown in Fig. 3.1.

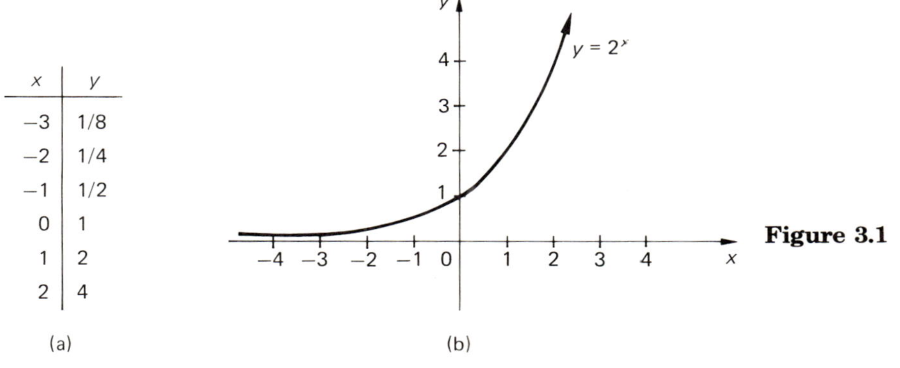

| x | y |
|---|---|
| −3 | 1/8 |
| −2 | 1/4 |
| −1 | 1/2 |
| 0 | 1 |
| 1 | 2 |
| 2 | 4 |



**Figure 3.1**

Similarly, the graphs of $y = 3^x$ and $y = (1/2)^x$ are both shown in Fig. 3.2. Observe that the function $y = a^x$ increases (as $x$ increases) whenever $a > 1$, and decreases if $0 < a < 1$. If $a = 1$, the curve becomes $y = 1$.

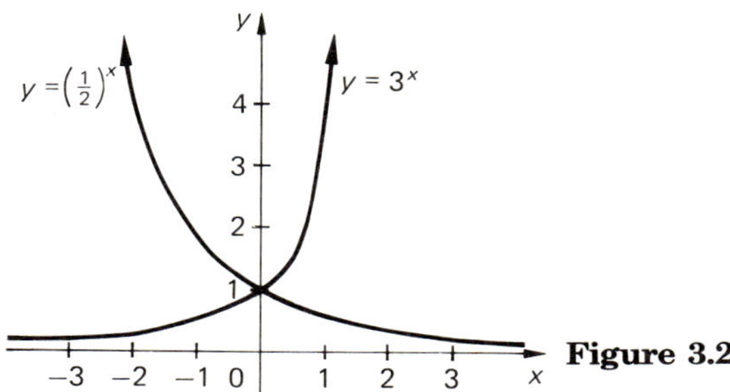


**Figure 3.2**

***Example 3*** In practice, one of the most frequently encountered exponential functions is the function

$$y = e^x. \tag{3.11}$$

The letter $e$ is used to denote a certain irrational number that arises naturally in calculus. Although the number $e$ can be obtained to whatever accuracy is desired, we rarely require more than ten decimal places:

$$e = 2.7182818285 \cdots .$$

Figure 3.3(a) lists some values of this function and the graph of Eq. (3.11) is shown in Fig. 3.3(b). If $f(x)$ is any function in $x$, we also call the function

$$y = a^{f(x)}, \qquad a > 0 \tag{3.12}$$

an exponential function. Functions of this type are plotted in the same fashion: We construct a table, graph the points in the table, and draw a smooth curve through these points. The following examples illustrate this procedure.

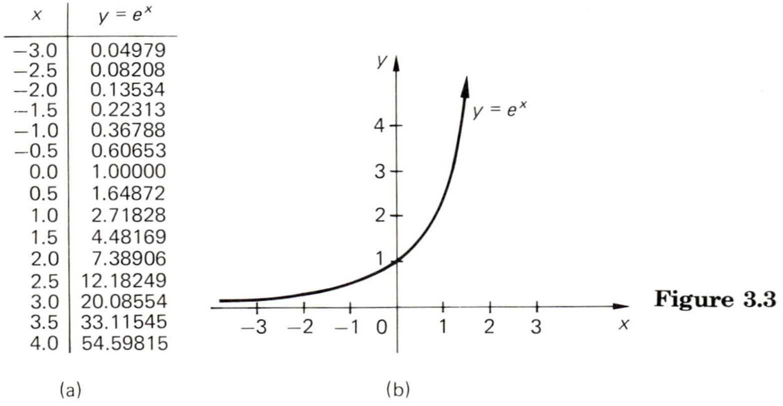

| $x$ | $y = e^x$ |
|---|---|
| −3.0 | 0.04979 |
| −2.5 | 0.08208 |
| −2.0 | 0.13534 |
| −1.5 | 0.22313 |
| −1.0 | 0.36788 |
| −0.5 | 0.60653 |
| 0.0 | 1.00000 |
| 0.5 | 1.64872 |
| 1.0 | 2.71828 |
| 1.5 | 4.48169 |
| 2.0 | 7.38906 |
| 2.5 | 12.18249 |
| 3.0 | 20.08554 |
| 3.5 | 33.11545 |
| 4.0 | 54.59815 |



(a) (b)

**Figure 3.3**

***Example 4*** Graph the function

$$y = e^{-x^2}. \tag{3.13}$$

We may rewrite Eq. (3.13) in the form

$$y = \frac{1}{e^{x^2}},$$

and use the table in Fig. 3.3(a) to obtain the values in Fig. 3.4(a), which yield the "bell-shaped" curve in Fig. 3.4(b). This function is very important in statistics as we shall see in Section 7.4, where it is associated with the Normal distribution.

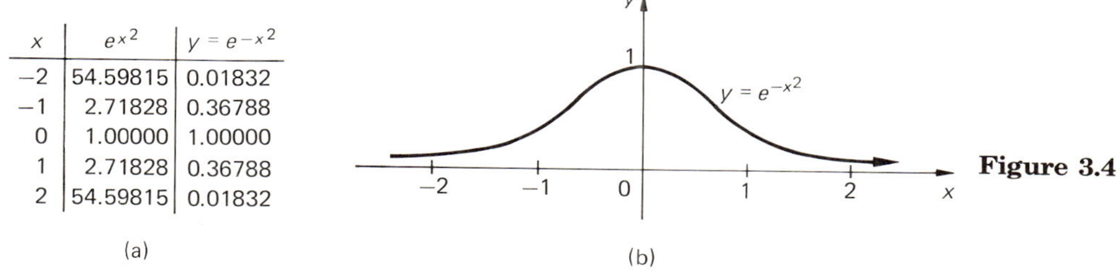

| $x$ | $e^{x^2}$ | $y = e^{-x^2}$ |
|---|---|---|
| −2 | 54.59815 | 0.01832 |
| −1 | 2.71828 | 0.36788 |
| 0 | 1.00000 | 1.00000 |
| 1 | 2.71828 | 0.36788 |
| 2 | 54.59815 | 0.01832 |



**Figure 3.4**

***Example 5*** Graph the function

$$y = 2^{-1/(1+|x|)}$$

We construct a table of values (Fig. 3.5a) from which the graph of Fig. 3.5(b) is obtained.

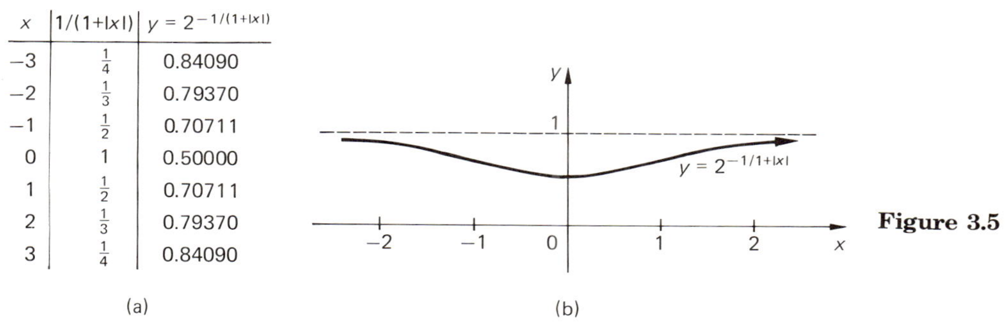

| $x$ | $1/(1+\|x\|)$ | $y = 2^{-1/(1+\|x\|)}$ |
|---|---|---|
| −3 | $\frac{1}{4}$ | 0.84090 |
| −2 | $\frac{1}{3}$ | 0.79370 |
| −1 | $\frac{1}{2}$ | 0.70711 |
| 0 | 1 | 0.50000 |
| 1 | $\frac{1}{2}$ | 0.70711 |
| 2 | $\frac{1}{3}$ | 0.79370 |
| 3 | $\frac{1}{4}$ | 0.84090 |



**Figure 3.5**

---

In the following section on exponential functions, we will see several examples that arise quite naturally in everyday situations. Indeed, as we shall see, exponential functions have far-reaching consequences in business, politics, and science.

## EXERCISES 3.1

*Evaluate each of the following expressions.*

**1.** $3^2$

**2.** $2^3$

**3.** $(-2)^0$

**4.** $5^0$

**5.** $5^{-1}$

**6.** $6^{-2}$

**7.** $1^{-3}$

**8.** $2^3 \cdot 2^4$

**9.** $3^2 \cdot 3^{-4}$

**10.** $(-1)^2 \cdot (-1)^3$

**11.** $5^3/5^2$

**12.** $5^5/5^3$

**13.** $3^2/3^5$

**14.** $(4^2)^3$

**15.** $[(-1)^3]^2$

**16.** $8^{1/3} \cdot 8^{4/3}$

**17.** $9^{3/2} \cdot 9^{-1/2}$

**18.** $(8^2 \cdot 8^{-3})/(8^3 \cdot 8^{1/3})$

**19.** $(9^{2/3} \cdot 9^{5/3})/9^{1/3}$

**20.** $(4^{3/2} \cdot 4^{-5/2})/(4^{5/2} \cdot 4^{-7/2})$

*In Exercises* 21–30, *find the values of* $y$ *for* $x = -3, -2, -1, 0, 1, 2,$ *and* 3 *and graph the given function.*

**21.** $y = 4^x$

**22.** $y = (3/2)^x$

**23.** $y = 3^{-x}$

**24.** $y = (1.1)^x$

**25.** $y = (1.25)^x$

**26.** $y = 2^{x^2}$

**27.** $y = 3^{1-x}$

**28.** $y = 4^{1+x^2}$

**29.** $y = 2^{1+x}$

**30.** $y = 2^{2^x}$

**31.** Show that the total number of grains of wheat earned by the knight for slaying all 60 dragons is $2^{60} - 1$. [*Hint:* $1 + z + z^2 + z^3 + \cdots + z^n = (z^{n+1} - 1)/(z - 1)$.]

**32.** How much grain would the knight have earned if the remaining dragons fled after he slew the 40th?

□ **33.** How much gold was in the clockmaker's account after 200 years? After 500 years?

□ **34.** Suppose the Swiss banker increased the amount of gold in the account by only 1% each year. Would each of the 1200 heirs be a millionaire?

□ **35.** How much money would you have after 30 years if you initially deposited \$1000 at 6% annual interest and made no withdrawals from the account?

---

## □ 3.2 EXPONENTIAL GROWTH AND DECAY, LEARNING MODELS, AND LOGISTIC CURVES

The notion of exponential growth was first popularized by the English social philosopher Thomas Robert Malthus (1766–1834) in a pamphlet, "Essay on the Principles of Population," in 1798. In that pamphlet he noted that unrestrained population growth was multiplicative in nature, that is, that the population would double every so many years. Malthus believed that the growth rate of the population would be much more rapid than the increase in food production. Eventually, he felt that widespread famines would become commonplace and that our destiny is to live perpetually at a starvation level. Whether the recent famines in Africa's Sahel and Bangladesh are indications of the validity of Malthus's prediction only time will tell.

That populations exhibit such growth is undeniable. Figure 3.6 gives estimates of the world's population from 1650 to 1960. The exponential nature of this graph is unmistakable. In this section we will present six models in which exponential functions play a significant role.

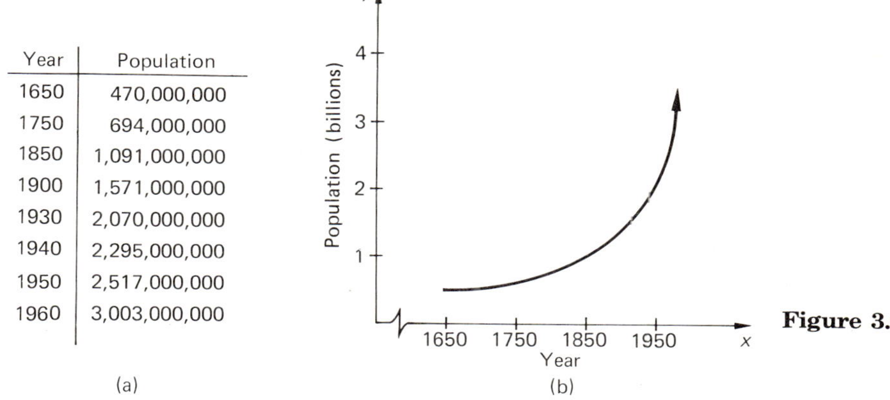

| Year | Population |
|---|---|
| 1650 | 470,000,000 |
| 1750 | 694,000,000 |
| 1850 | 1,091,000,000 |
| 1900 | 1,571,000,000 |
| 1930 | 2,070,000,000 |
| 1940 | 2,295,000,000 |
| 1950 | 2,517,000,000 |
| 1960 | 3,003,000,000 |



**Figure 3.6**

***Example 1*** Compound interest is one of the simplest examples of exponential growth. Consider a specific situation. Suppose we deposit \$100 in a savings account that pays 6% interest compounded annually. At the end of the first year we have our initial deposit plus the interest on that deposit, or

$$100 + 0.06(100).$$

Factoring the 100 from both terms, we see that

$$100 + 0.06(100) = 100(1 + 0.06) = 100(1.06),$$

so that the amount in the account at the end of the first year is obtained by multiplying the initial deposit by 1.06. Thus, at the end of the first year we have \$106.00 in the account.

To find the amount at the end of the second year, we add the interest, 6% of \$106, to the 106 dollars in the account, obtaining

$$106 + 0.06(106).$$

Again we may factor the common term 106, obtaining

$$106 + 0.06(106) = 106(1 + 0.06) = 106(1.06).$$

But $106 = 100(1.06)$ so the right-hand term becomes

$$106(1.06) = 100(1.06)(1.06) = 100(1.06)^2. \tag{3.14}$$

Thus, the value of the account at the end of the second year is simply $(1.06)^2$ times the original amount. Continuing in this fashion we discover that the value of the account at the end of $x$ years is simply the initial deposit of \$100 multiplied by (1.06) $x$ times, or

$$y = 100(1.06)^x. \tag{3.15}$$

As an example, we may use Eq. (3.15) to find the value of the account after 20 years. Setting $x = 20$, we use a calculator to get

$$y = 100(1.06)^{20} = \$320.71.$$

Equation (3.15) is an example of a discrete exponential function since it holds only for whole values of $x$. This is due to the fact that the interest is paid only at the end of each year. The graph of Eq. (3.15) is the sequence of points shown in Fig. 3.7.

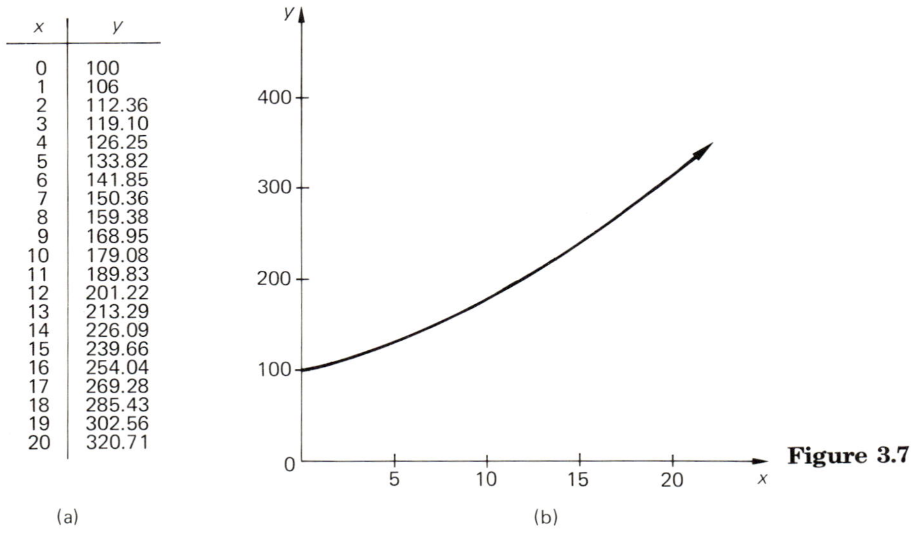

| x | y |
|---|---|
| 0 | 100 |
| 1 | 106 |
| 2 | 112.36 |
| 3 | 119.10 |
| 4 | 126.25 |
| 5 | 133.82 |
| 6 | 141.85 |
| 7 | 150.36 |
| 8 | 159.38 |
| 9 | 168.95 |
| 10 | 179.08 |
| 11 | 189.83 |
| 12 | 201.22 |
| 13 | 213.29 |
| 14 | 226.09 |
| 15 | 239.66 |
| 16 | 254.04 |
| 17 | 269.28 |
| 18 | 285.43 |
| 19 | 302.56 |
| 20 | 320.71 |



**Figure 3.7**

***Example 2*** Many chemical isotopes emit radiant energy in the form of particles or rays. We say that these isotopes are *radioactive*. As the energy is emitted, the atomic nucleus disintegrates and other isotopes are formed. When this happens, we say that the original isotope has decayed and we call the process *radioactive decay*. A common method of measuring radioactive decay is to determine the period of time required for half of a given quantity of the isotope to decay. This is called the *half-life* of the isotope.

Archaeologists and geologists have found that the process of radioactive decay provides them with a useful method of determining the age of artifacts and rocks. We will focus our attention on the use of carbon-14, $^{14}C$, in dating organic samples.

Carbon dating rests on the assumption that the level of the radioactive isotope $^{14}C$ in the atmosphere (primarily in the form of carbon dioxide, carbon monoxide, and methane) has remained approximately constant for many thousands of years. New $^{14}C$ is constantly produced by the cosmic bombardment of nitrogen and balances the amount of $^{14}C$ that disintegrates into the stable form $^{12}C$. All living objects absorb a fixed percentage of $^{14}C$ in their tissues since they are unable to distinguish between $^{12}C$ and $^{14}C$. Thus, it is reasonable to assume that a cubic centimeter of old wood (for example) contained the same amount of $^{14}C$ when it was living as a comparable piece would now. However, when the wood was used to make an artifact it died and was unable to absorb any more $^{14}C$. Over the years, the $^{14}C$ contained in the artifact has gradually disintegrated, so that the amount of $^{14}C$ in the sample of the artifact should be substantially different from that in an identical sample of newly cut wood. The ratio of $^{14}C$ in the old wood to $^{14}C$ in the new wood can be obtained by comparing their levels of radioactivity with a Geiger counter.

The best available estimate of the half-life of $^{14}C$ is 5730 years. Suppose a sample of freshly cut pine exhibits a radioactivity measurement of 15.3 dpm/gc (disintegrations per minute per gram of carbon). After 5730 years, it will read $\frac{1}{2}(15.3) = 7.56$ dpm/gc; after $2(5730) = 11{,}460$ years, it will read only $(\frac{1}{2})^2(15.3) = 3.825$ dpm/gc. Thus, the radioactivity present in the sample at any time $x$ is given by

$$y = 15.3(\tfrac{1}{2})^{x/5730} \text{ dpm/gc}, \tag{3.16}$$

since the exponential term equals 1 when $x = 0$, $\frac{1}{2}$ when $x = 5730$, $(\frac{1}{2})^2$ when $x = 11{,}460$, and so forth. Equation (3.16) is valid for all values of $x$, since radioactive decay occurs continually. In particular, a sample of pine 1000 years old has a radioactivity of

$$y = 15.3(\tfrac{1}{2})^{1000/5730} = 15.3(0.886) = 13.56 \text{ dpm/gc}.$$

Figure 3.8 is the graph of Eq. (3.16). We can use Fig. 3.8 to find the radioactivity of a sample of pine 2000 years old or the age of a sample of pine with a radioactivity of 2.5 dpm/gc. The 2000-year-old sample would have a reading of approximately 12 dpm/gc, while the sample with a reading of 2.5 dpm/gc, would be approximately 15,000 years old.

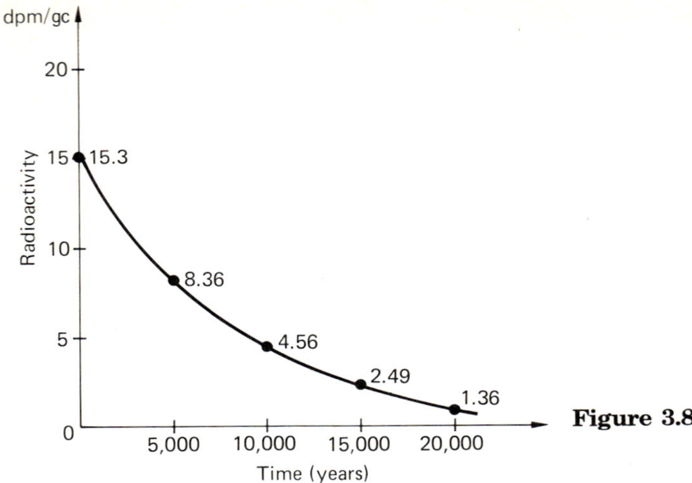


**Figure 3.8**

It is possible to use Eq. (3.16) to directly determine the age of a sample of wood with known radioactivity. However, the calculation involves logarithms, so further discussion is delayed until Section 3.3.

The reader may have noticed that Eqs. (3.15) and (3.16) share certain characteristics. Both equations have the form

$$y = a(b)^{kx}, \qquad b > 0 \tag{3.17}$$

where $a$, $b$, and $k$ are constants. In Eq. (3.15) $a = 100$, $b = 1.06$, and $k = 1$, while in Eq. (3.16) $a = 15.3$, $b = 1/2$, and $k = 1/5730$. Both equations are certainly exponential functions (although the domain of Eq. 3.15 is restricted to whole values of $x$). Note also that in both cases $a$ equals the initial amount of money or radioactivity. This is not surprising since Eq. (3.17) yields the solution $y = a$ when the time $x = 0$. Thus $a$ is always the initial amount.

We may apply this observation in the following way. The potassium isotope $^{42}K$, having a half-life of 12.5 hr, is often used to trace the movement of drugs in the human body. Suppose that a researcher receives a one-gram sample of potassium containing 10% $^{42}K$. Four hours are required to prepare for the experiment. How much of the sample is still $^{42}K$ at the beginning of the experiment?

The initial amount of $^{42}K$ is 10% of 1 g, or $a = 0.1$ g. Since the half-life is 12.5 hr, $b = 1/2$ and $k = 1/12.5$; thus Eq. (3.17) becomes

$$y = 0.1(\tfrac{1}{2})^{x/12.5},$$

where $x$ is in hours and $y$ measures the number of grams of $^{42}$K. Setting $x = 4$ yields

$$y = 0.1(\tfrac{1}{2})^{4/12.5} = 0.1(0.8) = 0.08 \text{ g of } {}^{42}\text{K}.$$

If the constant $b$ in Eq. (3.17) equals the irrational number $e = 2.7182\ldots$ (see Example 3 in Section 3.1), the constant $k$ represents the *instantaneous growth* (or *decay*) *rate per unit* of $y$. This fact will not be proved now, since it requires the use of calculus. It will be demonstrated in Section 9.6. Nevertheless, we can use this result to our advantage in the following example.

***Example 3*** The population of the United States, according to the census of 1 April 1960, was 179,323,175. The (annual) birth and death rates per thousand estimated on 1 July 1960 were 23.7 and 9.5, respectively. Assuming that the birth and death rates remain constant, predict the United States population in the years 1970, 1980, 1990, and 2000, ignoring all such factors as immigration and emigration.

To find the birth and death rates per individual we simply divide 23.7 and 9.5 by 1000, since these facts were given per 1000 individuals. Thus, the birth and death rates per individual become 0.0237 and 0.0095, respectively. The difference $(0.0237 - 0.0095 = 0.0142)$ is the growth rate per individual for the United States population. We can now use Eq. (3.17) with $b = e$, $k = 0.0142$, and $a = 179{,}323{,}175$ to obtain a formula for the United States population

$$y = 179{,}323{,}175e^{0.0142x}, \tag{3.18}$$

where $x$ is the number of years after 1960, since the initial population $a = 179{,}323{,}175$ refers to the year 1960. Therefore, to find the population in the year 1970, we substitute $x = 1970 - 1960 = 10$ into Eq. (3.18), obtaining

$$y = 179{,}323{,}175e^{0.142} = 206{,}683{,}704.$$

This compares favorably with the 1970 census figure of 200,251,326 (*World Almanac, 1971,* p. 66,409), the difference being due to the drop-off in birth rates during the late 1960s. Using Eq. (3.18) we can predict a population of

$$y = 179{,}323{,}175e^{0.0142(20)} = 238{,}218{,}811$$

in 1980,

$$y = 179{,}323{,}175e^{0.0142(30)} = 274{,}565{,}439$$

in 1990, and

$$y = 179{,}323{,}175e^{0.0142(40)} = 316{,}457{,}713$$

in 2000. Of course, the fact that the birth rates did not remain constant will have a significant effect on these predictions.

Our next example shows a situation in which an exponential function is superimposed on a constant function. In particular, we will consider a function of the form

$$y = A + a(b)^{kx}. \tag{3.19}$$

***Example 4*** Commercial products often enjoy large initial sales, followed by a gradual tapering off with time. Some recent examples include ten-speed bicycles, frisbees, and CB radios. Using Eq. (3.19) one can often obtain an approximation to the *sales curve* of such a product. Figure 3.9 describes the sales of Pennsylvania anthracite from 1945 to 1965. We are pretty safe in assuming that sales of anthracite will not totally vanish, so we might assume that sales will never fall below ten million tons. We might decide that the formula

$$y = 10{,}000 + 54{,}000(0.8865)^x \quad \text{(thousands of tons)}, \tag{3.20}$$

where $x$ is the number of years after 1945, is satisfactory for our purposes, since it fits the most recent production figures quite well (see Fig. 3.9b). Using Eq. (3.20) we might predict the production figures for the year 1970 by setting $x = 25$. We obtain

$$y = 10{,}000 + 54{,}000(0.8865)^{25} = 12{,}657.$$

***Example 5*** *Learning curves* can also be described by an equation of the form (3.19). Such curves are used by psychologists in explaining how rats (or humans) learn to make a particular response.

Consider a T-maze (shown in Fig. 3.10). A rat is placed at the base of the T. It is always rewarded if it turns left and never rewarded if it turns right. Initially there is a 50-percent chance that the rat will turn left. As the number of trials increases and the rat learns that turning left is rewarded and turning right is not, the chances that it will turn left increase, until the rat turns left consistently. Thus, the graph of the rat's chances of turning left should closely approximate that shown in Fig. 3.11. Hence, we can model this situation by an equation of the form

$$y = 100 - 50(b)^x,$$

where $0 < b < 1$ is some suitable constant reflecting the speed with which the rat learns by trial and error. Thus, the constant $b$ is some

*Pennsylvania anthracite*

| Year | Thousands of tons |
|---|---|
| 1945 | 54,934 |
| 1950 | 44,077 |
| 1955 | 26,205 |
| 1960 | 18,817 |
| 1965 | 14,866 |

(a)

*Thousands of tons of Pennsylvania anthracite*

| Year | $x$ | $y = 10{,}000 + 54{,}000\,(0.8865)^x$ |
|---|---|---|
| 1945 | 0 | 64,000 |
| 1950 | 5 | 39,566 |
| 1955 | 10 | 26,188 |
| 1960 | 15 | 18,863 |
| 1965 | 20 | 14,853 |

(b)

**Figure 3.9**

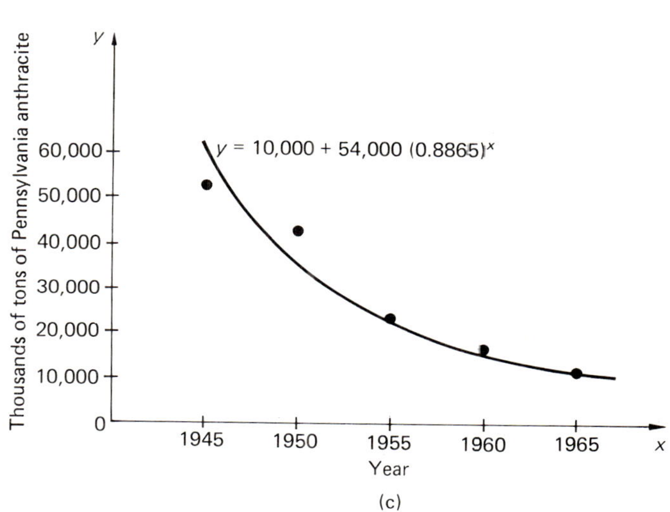


(c)

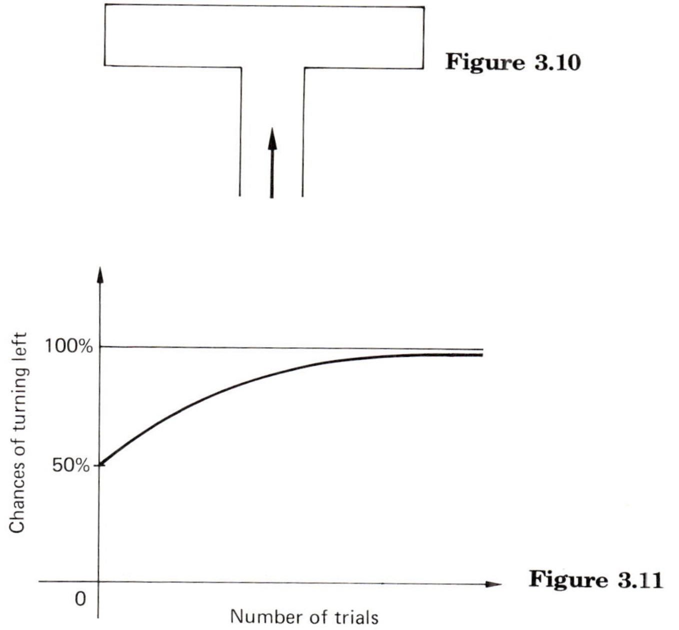


**Figure 3.10**

**Figure 3.11**

measure of the intelligence of the animal: If $b$ is close to zero the animal is very intelligent, since $y$ approaches 100 percent very quickly; if $b$ is close to one the animal is quite stupid and learns very slowly from its mistakes.

Our final example considers a situation in which the exponential function occurs in the denominator.

***Example 6*** In Example 3 we studied the effects of unrestricted population growth. This model is not valid when a lack of food or resources plays a part in controlling the population. If a population is restricted from growing beyond a certain limit by starvation or some other condition, it exhibits a graph similar to that shown in Fig. 3.12. We see there that the curve shows exponential growth when the population is small, but tapers off as we approach the *carrying capacity*, that is, the maximum population that the resources will allow. The S-shaped curve in Fig. 3.12 is called a *logistic* (or *sigmoidal*) curve. Such curves occur quite frequently, as in the case of a timber plantation. At first there is very little timber in a recently reseeded tract. As the timber matures, the trees put on a growth spurt. Eventually the tract is covered with mature trees that grow very slowly from then on.

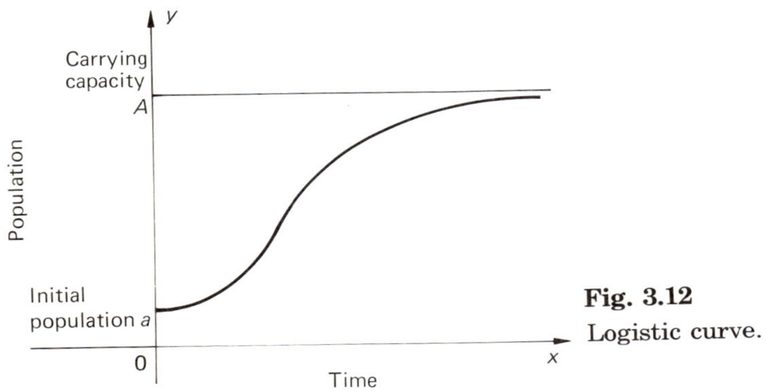


**Fig. 3.12**
Logistic curve.

The formula that is used for logistic growth is (see Section 10.8)

$$y = \frac{Aa}{a + (A - a)e^{-kx}}, \tag{3.21}$$

where $k$ is the growth rate per individual, $a$ is the initial population, and $A$ is the carrying capacity. Observe that if we set $x = 0$ in Eq. (3.21) we have

$$y = \frac{Aa}{a + (A - a)e^{0}} = \frac{Aa}{a + (A - a)} = \frac{Aa}{A} = a.$$

Similarly, if we allow $x$ to increase, the term $e^{-kx}$ will get closer and closer to zero. For very large values of $x$ we can ignore the term $(A - a)e^{-kx}$ since it is very close to zero. Thus Eq. (3.21) becomes

$$y = \frac{Aa}{a + (A - a)e^{-kx}} \approx \frac{Aa}{a + 0} = \frac{Aa}{a} = A.$$

That is, as $x$ increases, $y$ approaches the carrying capacity.

We illustrate these facts with the following situation. The population of India in 1967 was estimated to be 511,115,000. The birth and death rates per thousand were 25.8 and 10.3, respectively. Suppose the carrying capacity (the maximum possible population) of India is 750,000,000. Predict India's population in the year 1985.

In this case, $a = 511{,}115{,}000$, $A = 750{,}000{,}000$, and the birth and death rates per individual are 0.0258 and 0.0103, respectively. Assuming that these rates remain constant (which need not be the case) the growth rate per individual is

$$k = 0.0258 - 0.0103 = 0.0155.$$

Using Eq. (3.12) we have

$$y = \frac{(750{,}000{,}000)(511{,}115{,}000)}{511{,}115{,}000 + (750{,}000{,}000 - 511{,}115{,}000)e^{-0.0155x}}; \tag{3.22}$$

dividing numerator and denominator by 511,155,000 reduces Eq. (3.22) to

$$y = \frac{750{,}000{,}000}{1 + \left[\dfrac{750{,}000{,}000}{511{,}115{,}000} - 1\right]e^{-0.0155x}}$$

$$= \frac{750{,}000{,}000}{1 + 0.4674e^{-0.0155x}}. \tag{3.23}$$

Since $x = 0$ in 1967, we must set $x = 18$ for the year 1985. Replacing $x$ by 18 in Eq. (3.23) we have

$$y = \frac{750{,}000{,}000}{1 + 0.4674e^{-0.279}}$$

$$= \frac{750{,}000{,}000}{1 + 0.4674(0.7565)}$$

$$= \frac{750{,}000{,}000}{1.3536} = 554{,}075{,}251.$$

## EXERCISES 3.2

☐ *In Exercises 1–15, find the value of an account in which P dollars have been deposited at i% interest compounded annually for n years.*

**1.** $P = 100, i = 5\%, n = 3$

**2.** $P = 100, i = 7\%, n = 4$

**3.** $P = 1000, i = 10\%, n = 3$

**4.** $P = 1000, i = 12\%, n = 2$

**5.** $P = 1000, i = 8\%, n = 4$

**6.** $P = 500, i = 9\%, n = 10$

**7.** $P = 500, i = 12\%, n = 10$

**8.** $P = 500, i = 6\%, n = 20$

**9.** $P = 5000, i = 7\%, n = 8$

**10.** $P = 5000, i = 5\%, n = 15$

**11.** $P = 3800, i = 7\%, n = 20$

**12.** $P = 3800, i = 6\%, n = 25$

**13.** $P = 2400, i = 5\frac{1}{2}\%, n = 10$

**14.** $P = 2400, i = 5\frac{1}{4}\%, n = 8$

**15.** $P = 3750, i = 6\frac{1}{2}\%, n = 15$

☐ *In Exercises 16–21, use the data given in Example 2 in Section 3.2 to find the radioactivity of a sample of pine that is*

**16.** 2000 years old

**17.** 2800 years old

**18.** 1550 years old

**19.** 1800 years old

**20.** 6000 years old

**21.** 16,000 years old

*Use the graph in Fig. 3.8 to determine the age (within 5%) of the samples of pine having the radioactivities given in Exercises 22–27.*

**22.** 10 dpm/gc

**23.** 8 dpm/gc

**24.** 5 dpm/gc

**25.** 2 dpm/gc

**26.** 1.5 dpm/gc

**27.** 14 dpm/gc

☐ **28.** Radium-228 has a half-life of 6.7 years. Determine how great a reduction of the original amount occurs in two years.

☐ **29.** The half-life of phosphorus-32, $^{32}P$, is 14.2 days. A sample of phosphorus containing 20 percent $^{32}P$ is received and used two days later. What is the percentage of $^{32}P$ in the sample when it is used?

☐ **30.** Uranium-238 has a half-life of four billion years. How great a reduction of the original amount occurs in one million years?

☐ *In Exercises 31–38, predict the population of the given country in 1985, given its 1967 population a, birth rate per thousand B, and death rate per thousand D. Assume that the birth and death rates remain constant and that no other conditions influence the population.*

**31.** Canada: $a = 20{,}441{,}000$, $B = 18.0$, $D = 7.3$

**32.** Australia: $a = 11{,}751{,}000$, $B = 19.5$, $D = 8.8$

**33.** United Kingdom: $a = 54{,}909{,}000$, $B = 17.4$, $D = 11.2$

**34.** Japan: $a = 99{,}920{,}000$, $B = 19.3$, $D = 6.7$

35. France: $a = 49{,}890{,}000$, $B = 16.8$, $D = 10.8$

36. Sweden: $a = 7{,}869{,}000$, $B = 15.5$, $D = 10.1$

37. Mexico: $a = 45{,}671{,}000$, $B = 42.7$, $D = 8.9$

38. Venezuela: $a = 9{,}352{,}000$, $B = 44.4$, $D = 7.0$

☐ **39.** Using Eq. (3.21) find the chances that a rat will turn left in the maze on the tenth trial $x$ given that the rat's intelligence factor $b$ is 0.4. Repeat the exercise for $b = 0.8$ and $b = 0.9$.

☐ *In Exercises 40–45, use the facts given in Exercises 33–38 and the carrying capacities A listed below to predict the population in 1985 for each of the given countries.*

40. United Kingdom: $A = 100{,}000{,}000$

41. Japan: $A = 120{,}000{,}000$

42. France: $A = 80{,}000{,}000$

43. Sweden: $A = 25{,}000{,}000$

44. Mexico: $A = 100{,}000{,}000$

45. Venezuela: $A = 40{,}000{,}000$

---

## 3.3 LOGARITHMS AND THEIR PROPERTIES

In Sections 3.1 and 3.2 we discussed the behavior of exponential functions. In this section we will study the *logarithmic functions*, which are intimately related to the exponential functions.

We saw in Section 3.1 that the graph of the exponential function

$$y = a^x, \qquad a > 0 \tag{3.24}$$

is smooth for all real values of $x$. If $a = 1$, the expression $a^x$ takes on all positive values as $x$ ranges over the real numbers (see Fig. 3.2). Thus, *for each fixed positive number $y$ there is a unique number $x$ such that $y = a^x$*. That number $x$ is called the *logarithm of $y$ to the base $a$* and is written as

$$x = \log_a y. \tag{3.25}$$

For example, since

$$100 = 10^2$$

if we let $y = 100$ and $a = 10$, the number $x = 2$ is the logarithm of 100 to the base 10 or

$$2 = \log_{10} 100.$$

Similarly,

$$3 = \log_2 8$$

since $8 = 2^3$, and

$$\tfrac{1}{2} = \log_4 2$$

since $2 = 4^{1/2}$. Another way of expressing the relationship is to say that *the logarithm of* $y$ *to the base* $a$ *is that power to which we must raise* $a$ *in order to get the number* $y$ or, symbolically,

$$y = a^{\log_a y}. \tag{3.26}$$

Note that Eq. (3.26) implies that $\log_a a = 1$ and $\log_a 1 = 0$, since $a = a^1$ and $1 = a^0$.

If we take another look at Eqs. (3.24) and (3.25) we may notice that each equation reverses the action of the other. That is, suppose we start with some fixed number $x$ and substitute it into Eq. (3.24), obtaining the number $y$. If we substitute $y$ into Eq. (3.25), we will obtain the value $x$ with which we originally started. The same behavior occurs if we begin with some fixed number $y$ and substitute it first into Eq. (3.25). Using the resulting $x$ in Eq. (3.24) will reverse the process and yield the original value $y$. Functions that behave in this fashion are said to be *inverses* of each other. We can express these steps symbolically by writing the identities

$$x = \log_a (a^x) \tag{3.27}$$

and

$$y = a^{\log_a y}. \tag{3.28}$$

Figure 3.13 shows the graphs of the two functions $y = a^x$ and $y = \log_a x$. Note that if we fold the $xy$-plane along the line $y = x$, the two curves coincide. This geometrical property is a direct consequence of the functions being inverses of each other.

Although it is possible to use any positive number $a \neq 1$ as a base, logarithms to the base 10 are the most convenient for numerical calculations. This is due to the fact that our number system is based on the number 10 and we are already accustomed to thinking in terms of

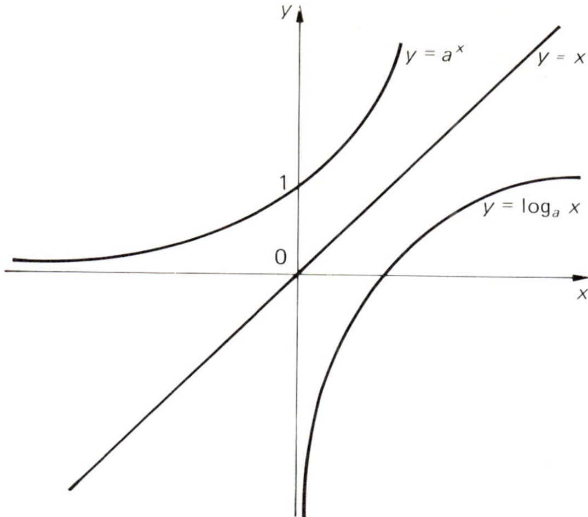


**Fig. 3.13**
Graphs of $y = a^x$ and $y = \log_a x$, $a > 1$.

powers of 10. Logarithms to the base 10 are called *common logarithms* and are often represented symbolically by "log $x$" with no base specified. Thus,

$$\log x \quad \text{means} \quad \log_{10} x$$

for any positive number $x$.

Almost every other practical application of logarithms uses the number $e = 2.71828\ldots$ as its base. We will see in Chapters 9 and 10 that logarithms to the base $e$ occur naturally in calculus. For this reason such logarithms are called *natural logarithms* and are usually written "ln $x$." Thus,

$$\ln x \quad \text{means} \quad \log_e x$$

for every positive number $x$.

Until the recent advent of calculators, it was usually necessary to refer to a logarithm table in determining logarithms. Tables such as those in Appendixes 4 and 5 were used in computation. We will explain briefly how these two tables can be used to find common and natural logarithms. First, however, we will show how these numbers can be obtained on a calculator.

If your calculator has a [LOG] key, simply key in the number $x$ and then push the [LOG] key. The calculator will display the common logarithm of $x$; that is, it gives $\log x$. For example, if $x = 54.6$, then the sequence of keys is

| Press | Display |
|---|---|
| 54.6 | 54.6 |
| [LOG] | 1.73719264 (to eight decimal places). |

Hence,

$$\log 54.6 = 1.73719264. \tag{3.29}$$

Similarly, if your calculator has an [LN] key, keying in the number $x$ and then pushing the [LN] key will give the natural logarithm of $x$, $\ln x$. Thus, the sequence of keys is

| Press | Display | |
|---|---|---|
| 54.6 | 54.6 | |
| [LN] | 4.00003388 | (to eight places), |

indicating that

$$\ln 54.6 = 4.00003388.$$

***Example 1*** The following logarithms (to four decimal places) can be found by keying in the number $x$ and then pushing either the [LOG] key or the [LN] key:

| For the logarithm | key | obtaining |
|---|---|---|
| ln 1.7 | 1.7 [LN] | 0.5306 |
| ln 25 | 25 [LN] | 3.2189 |
| ln 345.67 | 345.67 [LN] | 5.8455 |
| log 345.67 | 345.67 [LOG] | 2.5387 |
| log 0.0078 | 0.0078 [LOG] | −2.1079 |
| log 140 | 140 [LOG] | 2.1461 |

To see how the tables are used turn first to Appendix 5. We can easily see that the calculator obtained the correct answers to the first two natural logarithms. However, ln 345.67 is not in this table, so we must use a different table to determine this logarithm. Convenience is an important advantage of calculators. A second advantage arises from the fact that ln 1.78 can easily be determined on a calculator:

$$\ln 1.78 = 0.5766 \quad \text{(to four places)},$$

but linear interpolation is required when using the table. To find ln 1.78 using Appendix 5, we construct the table

| $x$ | $y = \ln x$ |
|---|---|
| 1.7 | 0.5306 |
| 1.78 | $y$ |
| 1.8 | 0.5878 |

obtaining the ratio of differences

$$\frac{y - 0.5306}{0.5878 - 0.5306} = \frac{1.78 - 1.7}{1.8 - 1.7} = \frac{0.08}{0.10} = 0.8.$$

Hence

$$y - 0.5306 = 0.8(0.5878 - 0.5306) = 0.04576$$

or

$$y = 0.57636.$$

Note that a great deal of work is required and that the final answer, when rounded to four decimal places, is inaccurate. The error of 0.0002 is caused by trying to fit a straight line to a curve $y = \ln x$ similar to that shown in Fig. 3.13. Clearly, the line passing through the two points $(1.7, \ln 1.7)$ and $(1.8, \ln 1.8)$ will always lie *below* the curve $y = \ln x$, so linear interpolation will always underestimate the correct value.

---

Suppose we wish to find the common logarithm of the number 54.6. Our first step is to rewrite 54.6 in *scientific notation*, that is, as *a real number $x$ greater than or equal to* 1 *but less than* 10 *and multiplied by an integral power of* 10.

In our example, we write

$$54.6 = 5.46 \times 10^1.$$

Similarly, we would have

$$4{,}520 = 4.52 \times 10^3,$$

$$32{,}500{,}000 = 3.25 \times 10^7,$$

$$3.472 = 3.472 \times 10^0,$$

$$0.003986 = 3.986 \times 10^{-3}.$$

An easy way to find the exponent of 10 is to count the number of places that you must move the decimal point, and use a negative sign if the original number is less than 1.

Once the number has been written in scientific notation, the common logarithm is obtained in two steps:

**1.** The integer part of the logarithm, called the *characteristic*, is the exponent of 10 in the number's scientific notation. Thus, in the number $54.6 = 5.46 \times 10^1$, the characteristic is 1.

**2.** The decimal part of the logarithm, called the *mantissa*, is the entry in the common logarithm table (see Appendix 4) corresponding to the number $x$. First we find 5.4 in the column headed by the $x$, and 6 in the top row. The place at which the row labeled 5.4 intersects the column headed by 6 has an entry of 0.7372, which is the mantissa of our original number.

The common logarithm of 54.6 is now obtained by adding the characteristic and mantissa together. Hence,

$$\log 54.6 = 1 + 0.7372 = 1.7372.$$

Note that this number agrees (to four decimal places) with the value we obtained with the calculator in Eq. (3.29) by keying in 54.6 and pushing the LOG key.

Until the recent advent of calculators, logarithms were a common necessity for three reasons: they made it possible to

1. Multiply numbers by adding their logarithms.
2. Divide numbers by subtracting their logarithms, and
3. Find the exponential of a number by multiplying its logarithm by the exponent.

Although the first two properties were not truly necessary, they saved a great deal of time in lengthy calculations. The third property was essential, since the only convenient method of calculating exponentials was to use logarithms or a slide rule (which is really a "logarithm table on a stick"). These properties are proved in the following theorem.

**THEOREM 3.1** **Let $x$ and $y$ represent any positive numbers, $a$ a positive number such that $a \neq 1$, and $k$ an arbitrary constant. Then the following are true.**

1. $\log_a (xy) = \log_a x + \log_a y$ **(multiply by adding)**
2. $\log_a (x/y) = \log_a x - \log_a y$ **(divide by subtracting)**
3. $\log_a (x^k) = k \log_a x$ **(exponentiate by multiplying)**

***Proof*** The verification of Properties 1–3 depends on Eq. (3.26) and the laws of exponents. We have

**1.** $a^{\log_a(xy)} = xy = a^{\log_a x} a^{\log_a y} = a^{\log_a x + \log_a y}$,

where the first and second equalities follow from Eq. (3.26) and the last equality holds by the laws of exponents (see Eq. 3.2). Since the values on both ends of the string of equalities agree, their exponents must also agree, because each value of the exponential function corresponds to a unique value of the exponent. Thus, Property 1 is proved.

**2.** $a^{\log_a(x/y)} = \dfrac{x}{y} = \dfrac{a^{\log_a x}}{a^{\log_a y}} = a^{\log_a x - \log_a y}$

Here Eq. (3.26) gives the first two equalities and Eq. (3.3) the last. Thus, Property 2 holds.

**3.** $a^{\log_a(x^k)} = x^k = (a^{\log_a x})^k = a^{k \log_a x}$,

where again the first two equalities follow from Eq. (3.26) and the last from Eq. (3.6). Hence, Property 3 is proved. ■

***Example 2*** Calculate, using common logarithms and Appendix 4:

$$\frac{5370 \times 0.0512}{(54.1)^{1.1}}.$$

Using the common logarithm table in Appendix 4 we calculate the logarithms of the three numbers involved in the problem

$$\log 5370 = 3.7300,$$

$$\log 0.0512 = 8.7093 - 10 = -1.2907,$$

and

$$\log 54.1 = 1.7332.$$

Since we wish to multiply the numbers 5370 and 0.0512, we add their logarithms (see Property 1). Since the second logarithm is negative, this amounts to subtracting these two numbers:

$$\begin{array}{rcr} \log 5370 & = & 3.7300 \\ +\log 0.0512 & = & -1.2907 \\ \hline & = & 2.4393 \end{array}$$

The number 2.4393 is the logarithm of the product 5370 × 0.0512. We wish to divide this product by $(54.1)^{1.1}$, which by Property 3 is

$$1.1 \log 54.1 = 1.1(1.7332) = 1.9065 \quad \text{(to four places)}.$$

By Property 2, we subtract the logarithm of the denominator from that of the numerator. Thus the logarithm of the answer is

$$2.4394 - 1.9065 = 0.5328.$$

Glancing at the table in Appendix 4, we see that

$$\log 3.41 = 0.5328.$$

Therefore,

$$\frac{5370 \times 0.0512}{(54.1)^{1.1}} = 3.41.$$

Actually, if the problem is solved on a calculator we obtain 3.40978703 to eight decimal places. Thus, the common logarithms were effective in obtaining the desired result.

---

It is easy to convert logarithms from one base into another by using the following result.

**THEOREM 3.2** $$\log_a x = (\log_a b)(\log_b x). \tag{3.30}$$

***Proof*** We again use the identity (3.26) and the laws of exponents:

$$a^{\log_a x} = x = b^{\log_b x} = (a^{\log_a b})^{\log_b x} = a^{(\log_a b)(\log_b x)}.$$

As in Theorem 3.1, this proves the identity. ■

We now illustrate how Theorem 3.2 can be used to convert common logarithms into natural logarithms and vice versa. If $a = e$ and $b = 10$, Eq. (3.30) becomes

$$\ln x = (\ln 10)(\log x).$$

Using a calculator we find that $\ln 10 = 2.302585$ (to six decimal places), yielding the identity

$$\ln x = 2.302585 \log x. \tag{3.31}$$

***Example 3*** We can use this identity to find ln 345.67, which is not in the table in Appendix 5. The common logarithm of 345.67 is (see Example 1)

$$\log 345.67 = 2.5387.$$

Thus, multiplying the common logarithm by 2.302585 gives the natural logarithm

$$\ln 345.67 = (2.302585)(2.5387) = 5.8456.$$

The difference between this answer and that given in Example 1 is due entirely to round-off error.

***Example 4*** Use Eq. (3.16) to determine the age of a sample of pine whose radioactivity measures 10 dpm/gc.

In this problem we know that $y = 10$ dpm/gc and we need to find the number of years $x$ required for the decay of $^{14}C$ to reach this level (see Example 2 in Section 3.2). Using Eq. (3.16) we have

$$y = 15.3(\tfrac{1}{2})^{x/5730} = 15.3(2)^{-x/5730},$$

since we may transfer the exponential $2^{x/5730}$ from the denominator to the numerator by changing the sign (see Eq. 3.5). Dividing both sides by 15.3 and taking common logarithms of each side, we obtain

$$\log(y/15.3) = \log(2)^{-x/5730}.$$

Applying Properties 2 and 3 of Theorem 3.1 to each side, respectively,

yields

$$\log y - \log 15.3 = \frac{-x}{5730} \log 2.$$

Multiplying both sides by $-5730/\log 2$, we have

$$x = \frac{5730}{\log 2}(\log 15.3 - \log y). \tag{3.32}$$

Setting $y = 10$ in Eq. (3.32) and using Property 2 of Theorem 3.1, we obtain

$$x = \frac{5730}{\log 2} \log\left(\frac{15.3}{10}\right) = \frac{5730 \log 1.53}{\log 2}.$$

We may now use a calculator or Appendix 4 to obtain the values of the logarithms:

$$\log 1.53 = 0.1847, \qquad \log 2 = 0.3010.$$

Hence,

$$x = 5730 \frac{(0.1847)}{(0.3010)} = 3515.5 \text{ years.}$$

---

## EXERCISES 3.3

*Find each of the following common logarithms.*

**1.** $\log 3.91$ **2.** $\log 4.24$ **3.** $\log 97.2$

**4.** $\log 87.7$ **5.** $\log 654$ **6.** $\log 5800$

**7.** $\log 90{,}000$ **8.** $\log 0.003$ **9.** $\log 0.871$

**10.** $\log 30{,}300$ **11.** $\log 0.00104$ **12.** $\log 0.00005$

**13.** $\log 0.0702$ **14.** $\log 1000$ **15.** $\log 999$

*Find the numbers that have the following common logarithms.*

**16.** 0.6937 **17.** 0.4669 **18.** 2.8876

**19.** 4.5276 **20.** 6.8021 **21.** 1.8976

**22.** 2.9243 **23.** 3.6021 **24.** $6.8451 - 10$

**25.** 8.8069 − 10 **26.** 9.2625 − 10 **27.** 2.6770

**28.** 3.9200 **29.** 8.6460 − 10 **30.** 7.4660 − 10

*Use logarithms to evaluate the following expressions.*

**31.** (28.2)(9.43)

**32.** $\dfrac{(37.6)(0.0248)}{58{,}200}$

**33.** $\dfrac{(463)(0.00887)}{46.2}$

**34.** $\dfrac{(23.4)(1830)}{(0.0215)(54.2)}$

**35.** $\dfrac{(1200)(0.0051)}{(21.05)(4755)}$

**36.** $\dfrac{(237.5)(0.001054)}{12.34}$

**37.** $\sqrt{47.3}(18.2)^{1.3}$

**38.** $\dfrac{(15.46)^{5/4}}{\sqrt{88.32}(754.8)}$

*Use Eq. (3.32) to determine the age x of a sample of pine whose radioactivity is*

**39.** $y = 7.15$ dpm/gc **40.** $y = 4.55$ dpm/gc **41.** $y = 3.12$ dpm/gc

**42.** $y = 0.4$ dpm/gc **43.** $y = 0.058$ dpm/gc **44.** $y = 10.5$ dpm/gc

## 3.4 GRAPHICAL METHODS: DOUBLE- AND SEMI-LOGARITHMIC PLOTS (OPTIONAL)

In Section 2.4 we discussed the methods that can be used to graph experimental data on ordinary graph paper. In particular, we saw that a mathematical equation can easily be derived from a straight-line graph by using the point-slope formula for a line. But what can a scientist do if the experimental data are clearly nonlinear? How can formulas be obtained for nonlinear graphs?

In this section we will show two extremely useful methods for determining whether a table of experimental data satisfies either an exponential function or a power function. Furthermore, we will show how to obtain this exponential function or a power function directly from the graph.

In order that you may fully profit from this section we recommend that you purchase at least six sheets of *semilogarithmic* paper and six sheets of *full-* or *double-logarithmic* paper, which should be available in any college bookstore.

### Semilogarithmic Plots

Graphing paper is called semilogarithmic if one of the scales (edges) of the sheet is equally spaced, while the other scale is logarithmic. For an example of a logarithmic scale, see Fig. 3.14. Note that the logarithmic scale starts at 1 rather than 0, and that the spacing is uneven. The reason it is called logarithmic becomes clear when we study the common

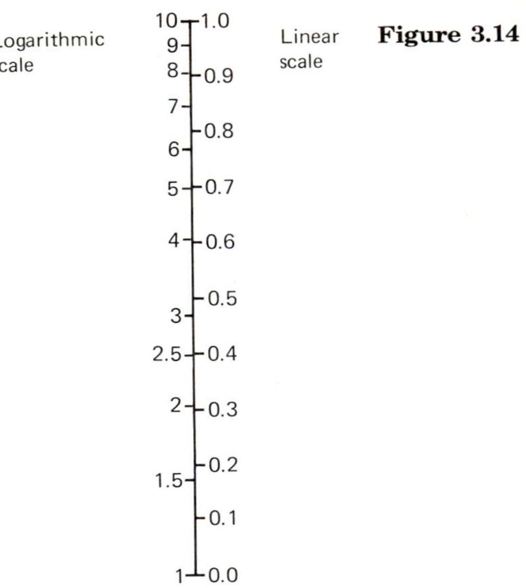


**Figure 3.14**

logarithm table (see Appendix 4). A brief table of common logarithms is given below.

| $x$ | 1 | 2 | 3 | 4 | 5 | 6 | 7 | 8 | 9 |
|---|---|---|---|---|---|---|---|---|---|
| $\log x$ | 0.0000 | 0.3010 | 0.4771 | 0.6020 | 0.6990 | 0.7782 | 0.8451 | 0.9031 | 0.9542 |

Now consider Fig. 3.14 very carefully. Note that the value 2 on the logarithmic scale is very close to the value 0.3 on the linear scale. Similarly, the value 5 on the logarithmic scale is very close to 0.7 on the linear scale. Hence, it is clear that the common logarithm table has been used to space the entries on the logarithmic scale. Note, further, that the entry 50 on the logarithmic scale is close to 1.7 on the linear scale, which is what we would expect since

$$\log 50 = \log (5.0 \times 10^1) = 1.6990;$$

the characteristic is 1 and the mantissa is $\log 5.0 = 0.6990$. A sample of semilogarithmic paper is shown in Fig. 3.15(b). Observe that the vertical scale is logarithmic, while the horizontal scale is linear.

We shall now see what happens if we plot some values of the exponential equation

$$y = 3(2)^x. \tag{3.33}$$

| $x$ | $y$ |
|---|---|
| −1 | 1.5 |
| 0 | 3.0 |
| 1 | 6.0 |
| 2 | 12.0 |
| 3 | 24.0 |
| 4 | 48.0 |

(a)

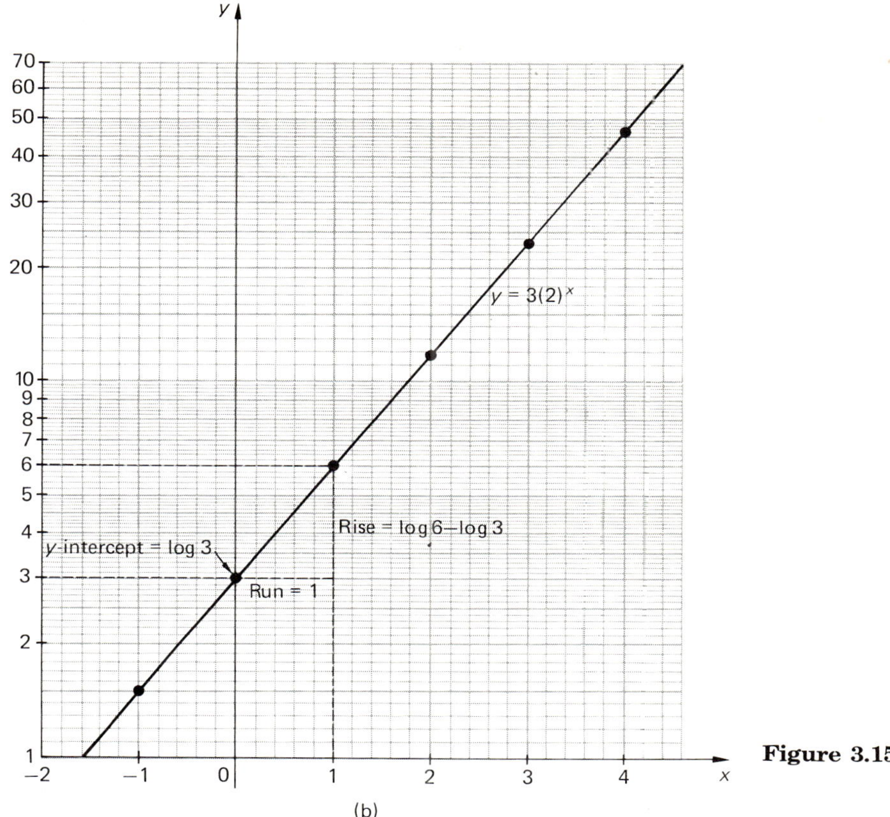


**Figure 3.15**

If we give $x$ the values in the left-hand column of Fig. 3.15(a), we get the $y$-values given in the right-hand column of that table. We plot these values in Fig. 3.15(b). Note that the points all lie on a *straight line* on semilogarithmic paper. This illustrates the most important property of semilogarithmic paper:

> Exponential functions have straight-line graphs on semilogarithmic paper.

The reason for this is clear if we take the common logarithm of both sides of Eq. (3.33). Then, since the logarithm of a product is the sum of the logarithms, we have

$$\log y = \log 3 + x \log 2,$$

or

$$\log y = (\log 2)x + \log 3. \tag{3.34}$$

If we set $Y = \log y$, and remember that $\log 2 = 0.3010$ and $\log 3 = 0.4771$, we see that Eq. (3.34) becomes

$$Y = 0.3010x + 0.4771, \tag{3.35}$$

which is the equation of a straight line in the $xY$-plane. Thus, to get a straight line we need a piece of graph paper whose $x$-scale is linear and whose $Y$ (or $\log y$)-scale is logarithmic in $y$.

In general, however, we must reverse this process: We have plotted some experimental data on a piece of semilogarithmic paper and we discover, to our delight, that the data lie on a straight line. How can we find the exponential equation that corresponds to this graph? To see how the equation may be read from the graph we again consider Fig. 3.15. Note that the graph intersects the $y$-axis at the value $y = 3$ *on the logarithmic scale.* Thus, in the slope-intercept formula (Eq. 2.2) for lines

$$Y = mx + b, \tag{3.36}$$

the value $b = \log 3$ while $Y = \log y$, so that Eq. (3.36) becomes

$$\log y = mx + \log 3. \tag{3.37}$$

Now, how do we find the slope? Recall that

$$m = \frac{\text{rise}}{\text{run}}.$$

Here, for a run of $x = 1$ we have a rise of $(\log 6 - \log 3)$ (see Fig. 3.15). Thus,

$$m = \frac{\log 6 - \log 3}{1} = \frac{\log (6/3)}{1} = \log 2.$$

Substituting this value of $m$ into Eq. (3.37), we get

$$\log y = (\log 2)x + \log 3,$$

which is exactly the same as Eq. (3.34). Now reversing the steps we used in getting to Eq. (3.34), we have

$$\log y = \log 2^x + \log 3 = \log 3(2)^x,$$

or

$$y = 3(2)^x.$$

We do not actually need to do all these calculations to obtain the exponential equation if we just note what happens to the slope $m = \log 2$ and the $y$-intercept $b = \log 3$. We simply "removed the logarithms" and inserted the remaining numbers ($m = 2$ and $b = 3$) into the exponential equation

$$y = b(m)^x. \tag{3.38}$$

To verify this procedure, observe that

$$\log y = \log b(m)^x = \log (m)^x + \log b,$$

or

$$\log y = (\log m)x + \log b. \tag{3.39}$$

Thus, the number $b$ is the *value* of the $y$-intercept, while $m$ is found by *dividing the $y$-value at $x = 1$ by the $y$-value at $x = 0$* (see Fig. 3.16); that

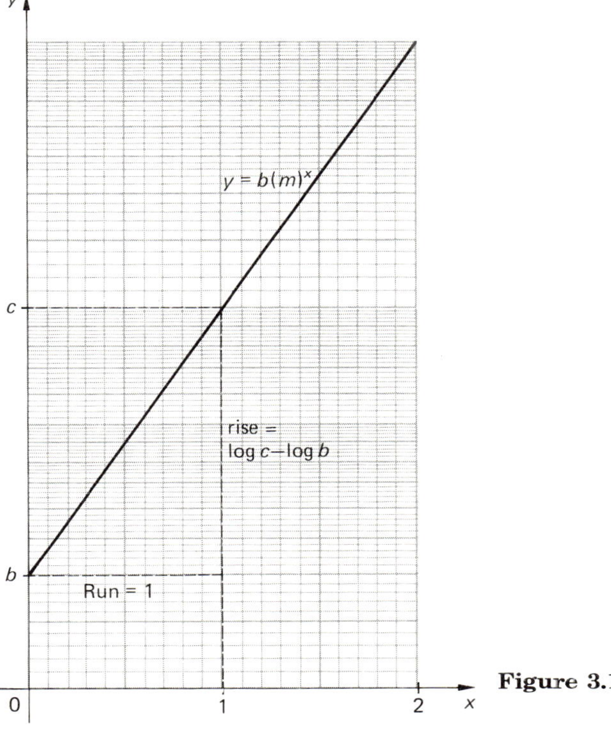


**Figure 3.16**

is,

$$m = \frac{c}{b}. \tag{3.40}$$

This relationship can be determined by recalling that the slope of the line (3.39) equals

$$\log m = \frac{\text{rise}}{\text{run}} = \frac{\text{rise}}{1} = \log c - \log b = \log (c/b),$$

from which Eq. (3.40) follows.

In the next two examples we will find the exponential formula that corresponds to the given data. You will be asked to repeat this process in the exercises.

***Example 1*** Find the exponential function that yields the following data:

| $x$ | $-3/2$ | $-1$ | $-1/2$ | $0$ | $1/2$ | $3/4$ |
|---|---|---|---|---|---|---|
| $y$ | 833.41 | 315 | 119.06 | 45 | 17.01 | 10.46 |

Plotting these points on the semilogarithmic plot of Fig. 3.17, we see that they lie on a straight line. Before we find the corresponding exponential equation, two observations are in order.

1. Compare the scaling on the logarithmic axis in Fig. 3.15 with that in Fig. 3.17. The $y$-values in Fig. 3.15(a) range from 1.5 to 48 so we have scaled the $y$-values in Fig. 3.15(b) from $10^0 = 1$ to $10^2 = 100$. For this problem, the $y$-values lie between 10 and 1000, so we scale the $y$-values from $10^1 = 10$ to $10^3 = 1000$. Thus, we may select whatever range is most suitable for graphing the data points providing we always begin each scale with consecutive powers of 10.
2. As in Fig. 3.15(b), the $x$-axis is not shown since the $y$-values in all semilogarithmic plots are always positive. This, however, causes no difficulties.

Now, in this example, the $y$-intercept $b = 45$. However in this case, rather than use the values of $y$ at $x = 0$ and $x = 1$ to find $m$, we use the values for the run between $x = -1$ and $x = 0$. Hence,

$$\log m = \frac{\text{rise}}{\text{run}} = \frac{\log 45 - \log 315}{0 - (-1)},$$

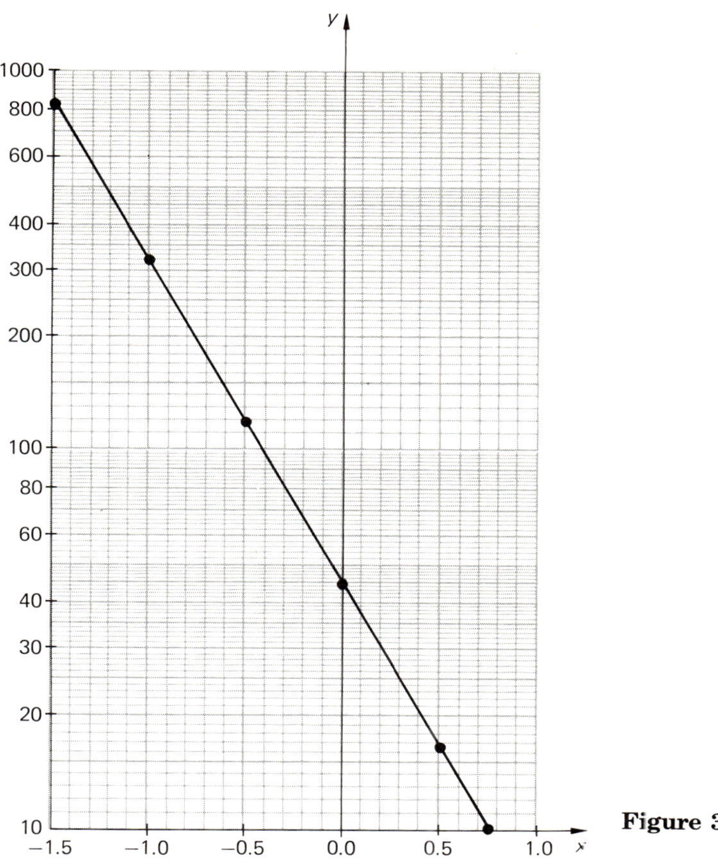


**Figure 3.17**

or

$$\log m = \log(45/315) = \log(1/7).$$

Thus $m = 1/7$, and the exponential equation for this problem is

$$y = 45\,(1/7)^x.$$

***Example 2*** Consider the table shown in Fig. 3.18(a). These points are plotted in Fig. 3.18(b). Note in this case that it would be hard to obtain a run equal to 1. However, if we extend the line we have

$$\log m = \frac{\text{rise}}{\text{run}} = \frac{\log 0.050 - \log 0.135}{0.1},$$

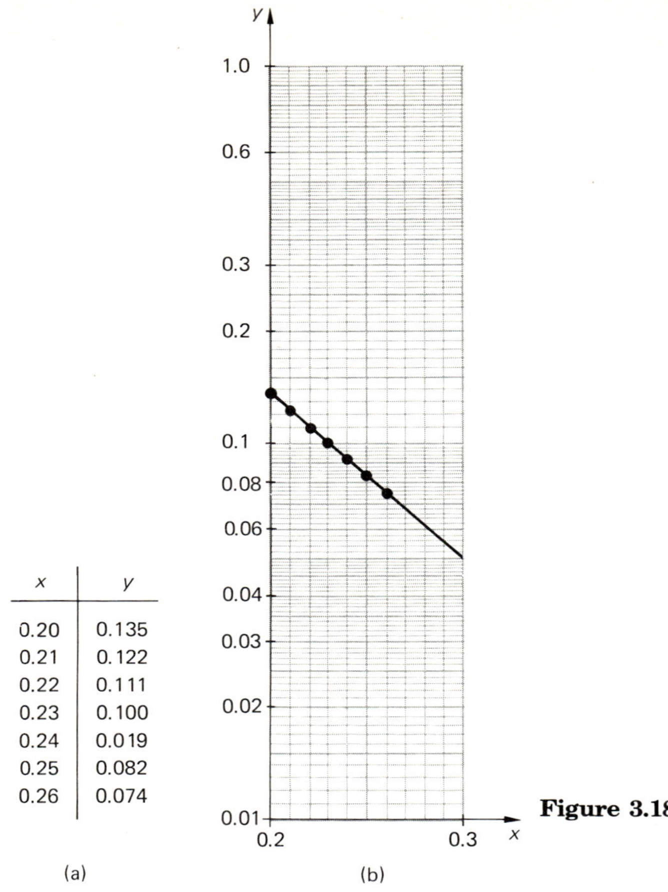

| $x$ | $y$ |
|---|---|
| 0.20 | 0.135 |
| 0.21 | 0.122 |
| 0.22 | 0.111 |
| 0.23 | 0.100 |
| 0.24 | 0.019 |
| 0.25 | 0.082 |
| 0.26 | 0.074 |



**Figure 3.18**

so that

$$\log m = \frac{\log (0.05/0.135)}{(1/10)} = 10 \log 0.37$$

or

$$\log m = \log (0.37)^{10}.$$

Hence we obtain $m = (0.37)^{10}$. Now we also face a problem in determining the $y$-intercept since the $y$-axis is not present in Fig. 3.18(b). To find $b$, we substitute the value of $m$ that we have obtained into Eq. (3.38):

$$y = b[(0.37)^{10}]^x. \tag{3.41}$$

Then we substitute the coordinates of *any* point in Fig. 3.18(a) into Eq. (3.14) to obtain the value of $b$. For example, using the point (0.20, 0.135), we obtain

$$0.135 = b[(0.37)^{10}]^{0.20}$$

or, dividing both sides by $[(0.37)^{10}]^{0.2} = (0.37)^2$, we have

$$b = \frac{0.135}{(0.37)^2} = 0.986.$$

Therefore, we end up with the exponential equation

$$y = 0.986(0.37)^{10x},$$

which can easily be checked with a calculator to give the values shown in Fig. 3.18(a).

***Example 3*** In Example 4 in Section 3.2 we obtained a formula (Eq. 3.20) for the production of Pennsylvania anthracite (see Fig. 3.9a). We shall now determine whether we can use the methods above to obtain Eq. (3.20). In Example 4 in Section 3.2, we assumed that production would never fall below 10 million tons, so we seek an equation of the form

$$y - 10{,}000 = b(m)^x, \tag{3.42}$$

where $x$ is the number of years after 1945 and $y$ is in thousands of tons of anthracite. Subtracting 10,000 thousand tons from each entry in Fig. 3.9(a) we obtain the data shown in Fig. 3.19(a). These entries are plotted in Fig. 3.19(b). As we can see, the points do not lie on a straight line, but are reasonably close to one. This procedure of sketching in a line that "fits" the data well is the same technique that we used in Example 3 in Section 2.4 for ordinary graphs. Indeed, if you studied the discussion in Section 2.8 on least squares approximations, you will note that a similar procedure is being used here. We will consider the numerical procedure for determining the least squares "line" in Section 3.5.

Returning to our present problem, we note that the $y$-intercept of the line that we have sketched in is $b = 54{,}000$, while the slope satisfies

$$\log m = \frac{\log 16{,}200 - \log 54{,}000}{10} = \frac{1}{10}\log\frac{16{,}200}{54{,}000}.$$

Thus, $m = (16{,}200/54{,}000)^{0.1} = (0.3)^{0.1} = 0.8865$, so that we obtain

$$y = 10{,}000 + 54{,}000(0.8865)^x,$$

which yields Eq. (3.20).

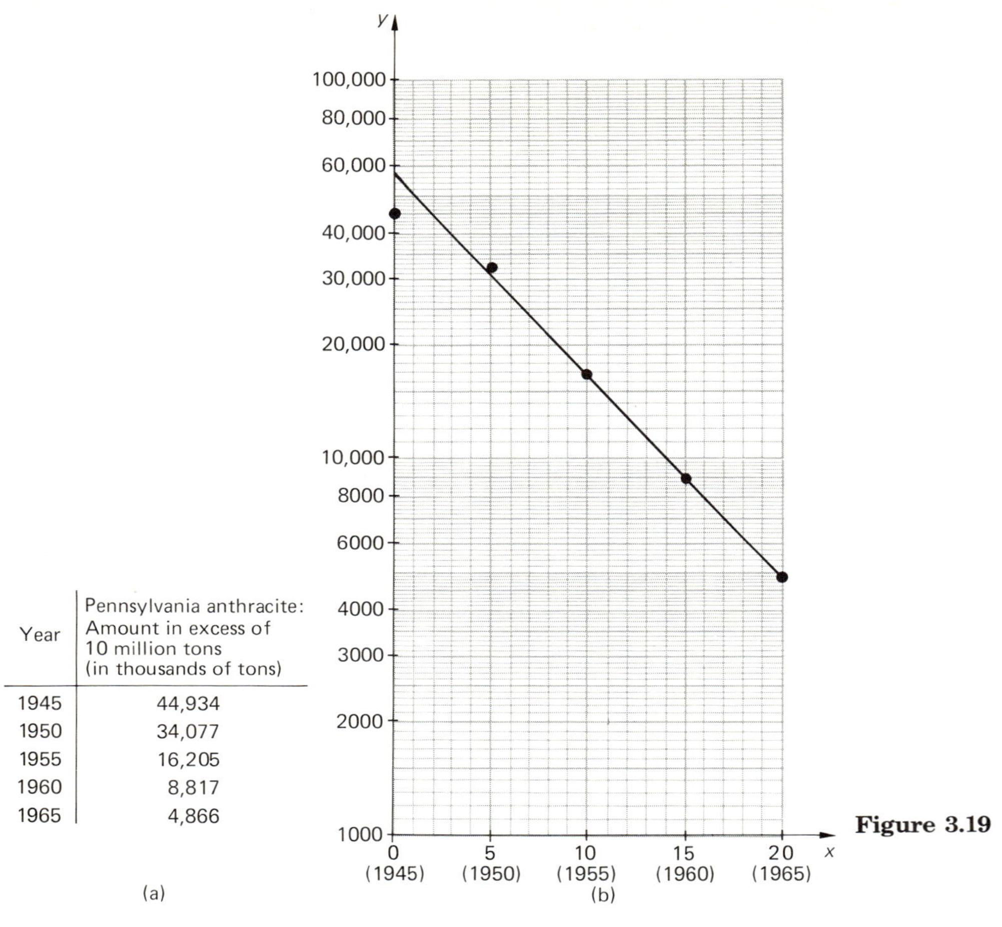

| Year | Pennsylvania anthracite: Amount in excess of 10 million tons (in thousands of tons) |
|---|---|
| 1945 | 44,934 |
| 1950 | 34,077 |
| 1955 | 16,205 |
| 1960 | 8,817 |
| 1965 | 4,866 |



**Figure 3.19**

---

**Double-Logarithmic Plots** Graph paper is called full or *double logarithmic* if both of its scales are logarithmic. As we shall see immediately, this type of graphing paper is very useful when the function involved is a *power function* of the form

$$y = b(x)^m. \tag{3.43}$$

Note the difference between this equation and the exponential equation (3.38). Here the constant $m$ is the *power* of $x$, while in Eq. (3.38) $m$ is the *base* of the exponent.

***Example 4*** Figure 3.20(a) gives a few points that satisfy the power function

$$y = 2x^3. \tag{3.44}$$

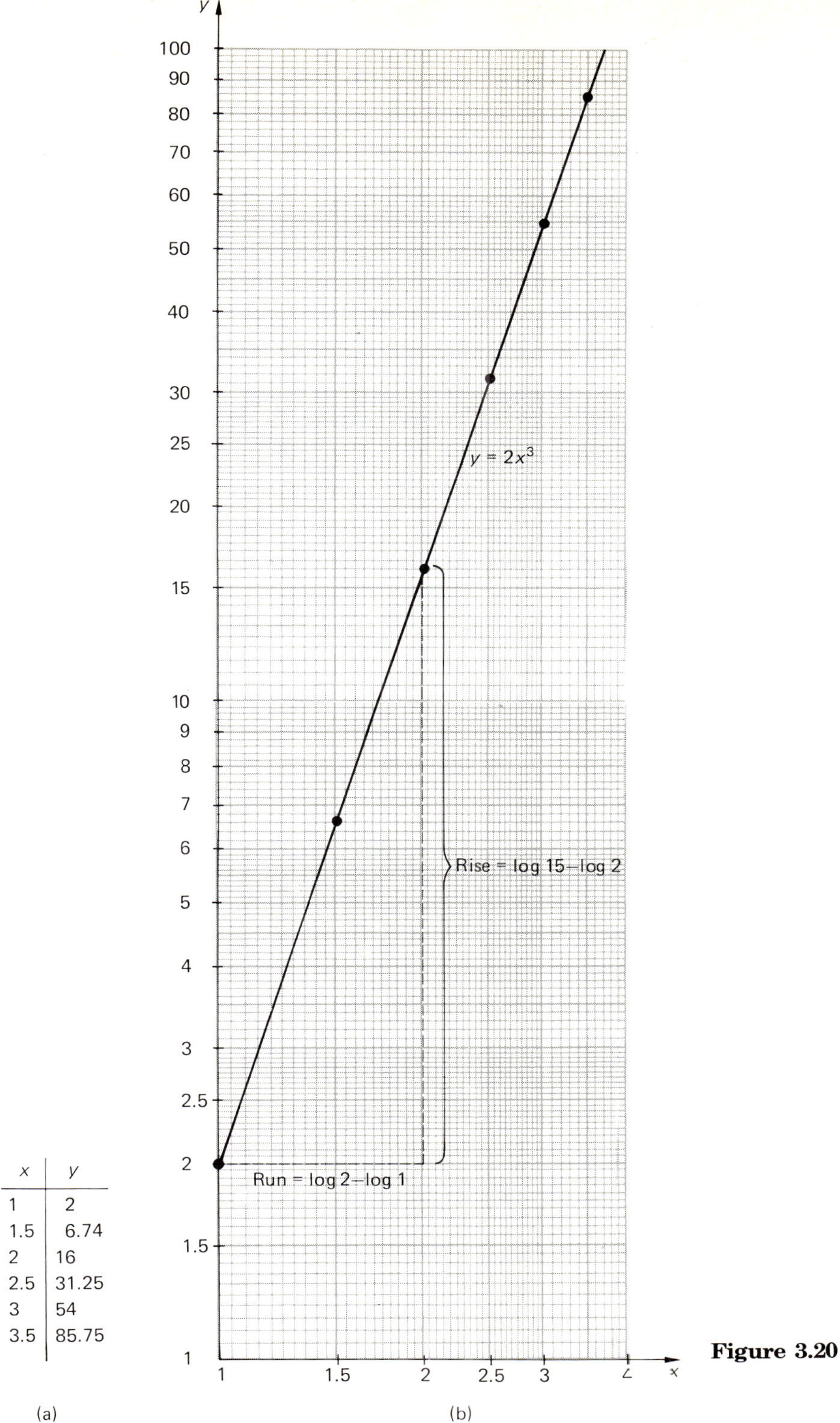


| $x$ | $y$ |
|---|---|
| 1 | 2 |
| 1.5 | 6.74 |
| 2 | 16 |
| 2.5 | 31.25 |
| 3 | 54 |
| 3.5 | 85.75 |

**Figure 3.20**

These points are plotted on the double-logarithmic graph in Fig. 3.20(b), and all lie on a straight line.

| Power functions have straight line graphs on double-logarithmic paper. |
|---|

We can see why double-logarithmic paper works this way by taking the common logarithm of both sides of Eq. (3.44):

$$\begin{aligned} \log y &= \log(2x^3) \\ &= \log 2 + \log x^3 \\ &= 3 \log x + \log 2. \end{aligned}$$

Now, if we set $Y = \log y$ and $X = \log x$, and remember that $\log 2 = 0.3010$, we have the equation of the straight line

$$Y = 3X + 0.3010$$

in the $XY$-plane. Thus, Eq. (3.44) yields a straight line graph on any piece of paper whose $X\,(=\log x)$ scale is logarithmic in $x$ and whose $Y\,(=\log y)$ scale is logarithmic in $y$.

---

A general formula for recovering the coefficients $b$ and $m$ from a graph on double-logarithmic paper can quickly be obtained by taking the common logarithm of both sides of Eq. (3.43). We have

$$\begin{aligned} \log y &= \log(bx^m) \\ &= \log b + \log(x^m), \end{aligned}$$

or

$$\log y = m \cdot \log x + \log b. \tag{3.45}$$

Now the $(\log x)$-term will vanish when $x = 1$, so the value $b$ can be determined immediately by noting at what point the straight line intersects the line $x = 1$. In Fig. 3.20(b) we see that this occurs for $b = 2$. To find $m$, which in this case is the slope of the straight line, we pick any two points on the line. Then, for example,

$$m = \frac{\text{rise}}{\text{run}} = \frac{\log 16 - \log 2}{\log 2 - \log 1},$$

so that

$$m = \frac{\log (16/2)}{\log (2/1)} = \frac{\log 8}{\log 2} = \frac{0.9030}{0.3010} = 3.$$

We will illustrate this procedure again in the following example.

***Example 5*** Most birds have access to fresh water, but those living in deserts must metabolize much of the water they need from the seeds they consume. The metabolic rate of birds is very high and the loss of water through respiration is relatively large. The percentage of body weight lost as water each day was studied by G. A. Bartholomew and T. C. Cade [*Auk* **80** (1963): 504–539]. The results of their study are illustrated in Fig. 3.21, and show that the percentage loss of body weight in small birds is much larger than in large ones.

As we can see, the straight line in Fig. 3.21 provides a reasonable fit to the data. Therefore, we can hope to obtain a formula relating the mean body weight $x$ to the percentage of body weight lost as water each day $y$.

One immediate problem that we must face is that Fig. 3.21 does not include the line $x = 1$, so we can not find $b$ directly. Instead, we will first find $m$ and then use the same technique that we used in Example 2 to obtain $b$. Since $m$ is the slope of the straight line, we have

$$m = \frac{\text{rise}}{\text{run}} = \frac{\log 3.3 - \log 32}{\log 100 - \log 10},$$

so that

$$m = \frac{\log (3.3/32)}{\log (100/10)} = \frac{\log 0.103}{\log 10} = -0.987.$$

Substituting this value into Eq. (3.43), we get

$$y = bx^{-0.987}. \tag{3.46}$$

Now we may use any point on the straight line in Fig. 3.21 to determine $b$. For example, substituting the point (100, 3.3) into Eq. (3.46) yields

$$3.3 = b(100)^{-0.987},$$

or

$$b = \frac{3.3}{(100)^{-0.987}} = 3.3(100)^{0.987} = 310.82.$$

Hence, we obtain the formula

$$y = 310.82x^{-0.987} \tag{3.47}$$

as the relationship of body weight to percentage of body weight lost as water each day.

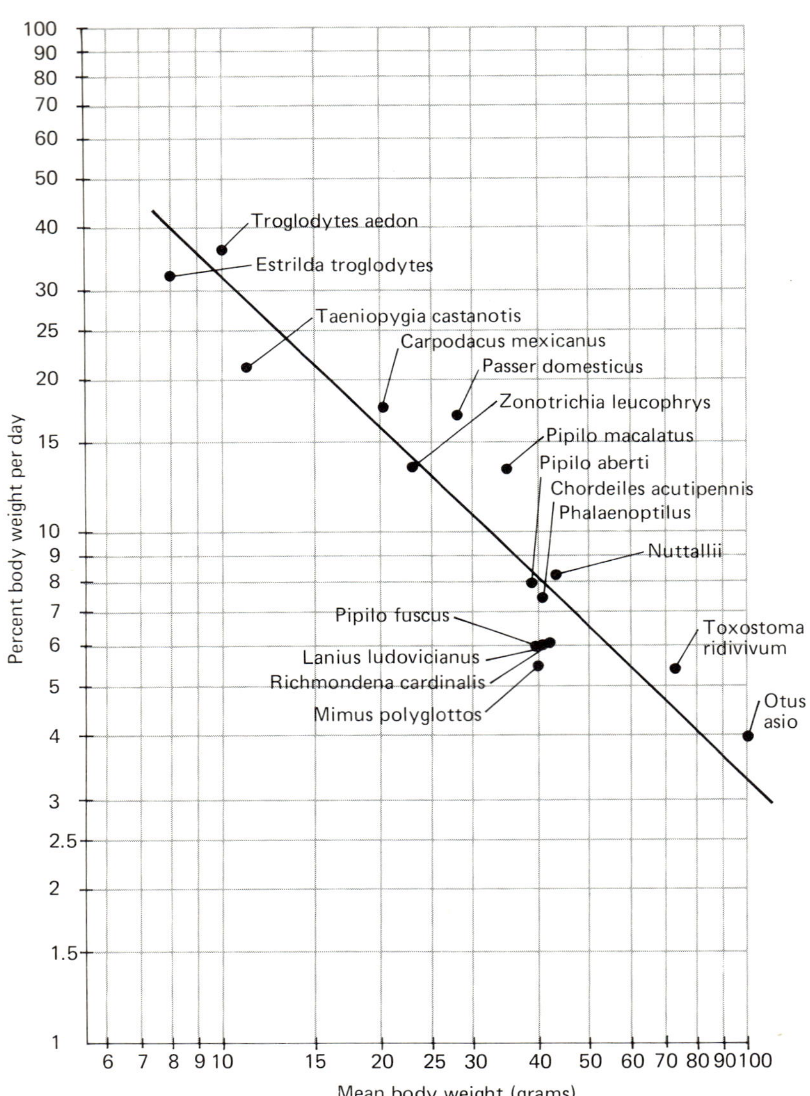


Figure 3.21

## EXERCISES 3.4

A small supply of semilogarithmic and double-logarithmic paper will be required to do the exercises in this set.

*In Exercises 1–16, use semi- and double-logarithmic paper to determine the equation that yields the given table of values.*

**1.**

| x | y |
|---|---|
| −2 | 1.75 |
| −1 | 3.5 |
| 0 | 7 |
| 1 | 14 |
| 2 | 28 |

**2.**

| x | y |
|---|---|
| 0.25 | 0.13 |
| 0.50 | 1.00 |
| 0.75 | 3.38 |
| 1.00 | 8.00 |
| 1.25 | 15.63 |

**3.**

| x | y |
|---|---|
| 2.00 | 28.00 |
| 2.25 | 35.44 |
| 2.50 | 43.75 |
| 2.75 | 52.94 |
| 3.00 | 63.00 |

**4.**

| x | y |
|---|---|
| 0.25 | 10.53 |
| 0.50 | 13.86 |
| 0.75 | 18.24 |
| 1.00 | 24.00 |
| 1.25 | 31.59 |

**5.**

| x | y |
|---|---|
| −1.0 | 0.07 |
| −0.5 | 0.12 |
| 0.0 | 0.20 |
| 0.5 | 0.35 |
| 1.0 | 0.60 |
| 1.5 | 1.04 |

**6.**

| x | y |
|---|---|
| 2 | 0.1600 |
| 3 | 0.0320 |
| 4 | 0.0064 |
| 5 | 0.0013 |
| 6 | 0.0003 |
| 7 | 0.0001 |

**7.**

| x | y |
|---|---|
| 2.0 | 3.20 |
| 2.5 | 7.81 |
| 3.0 | 16.20 |
| 3.5 | 30.01 |
| 4.0 | 51.20 |
| 4.5 | 82.01 |

**8.**

| x | y |
|---|---|
| 2 | 3.45 |
| 3 | 3.74 |
| 4 | 3.96 |
| 5 | 4.14 |
| 6 | 4.29 |
| 7 | 4.43 |

**9.**

| x | y |
|---|---|
| 1.0 | 3.14 |
| 1.2 | 5.43 |
| 1.4 | 8.62 |
| 1.6 | 12.87 |
| 1.8 | 18.32 |
| 2.0 | 25.13 |
| 2.2 | 33.45 |

**10.**

| x | y |
|---|---|
| 1.0 | 3.00 |
| 1.2 | 5.32 |
| 1.4 | 8.63 |
| 1.6 | 13.13 |
| 1.8 | 19.01 |
| 2.0 | 26.47 |
| 2.2 | 35.72 |

**11.**

| x | y |
|---|---|
| 0.1 | 3.36 |
| 0.2 | 3.77 |
| 0.3 | 4.23 |
| 0.4 | 4.74 |
| 0.5 | 5.32 |
| 0.6 | 5.96 |
| 0.7 | 6.69 |

**12.**

| x | y |
|---|---|
| 0.1 | 3.51 |
| 0.2 | 3.91 |
| 0.3 | 4.37 |
| 0.4 | 4.88 |
| 0.5 | 5.44 |
| 0.6 | 6.07 |
| 0.7 | 6.78 |

**13.**

| $x$ | $y$ |
|---|---|
| 2.00 | 7.39 |
| 2.25 | 9.49 |
| 2.50 | 12.18 |
| 2.75 | 15.64 |
| 3.00 | 20.09 |
| 3.25 | 25.79 |
| 3.50 | 33.12 |
| 3.75 | 42.52 |

**14.**

| $x$ | $y$ |
|---|---|
| 2.00 | 6.58 |
| 2.25 | 9.06 |
| 2.50 | 12.07 |
| 2.75 | 15.64 |
| 3.00 | 19.81 |
| 3.25 | 24.63 |
| 3.50 | 30.13 |
| 3.75 | 36.34 |

**15.**

| $x$ | $y$ |
|---|---|
| 2.0 | 26.92 |
| 2.5 | 45.99 |
| 3.0 | 71.23 |
| 3.5 | 103.12 |
| 4.0 | 142.07 |
| 4.5 | 188.49 |
| 5.0 | 242.72 |
| 5.5 | 305.10 |

**16.**

| $x$ | $y$ |
|---|---|
| 2.0 | 29.38 |
| 2.5 | 45.51 |
| 3.0 | 70.50 |
| 3.5 | 109.22 |
| 4.0 | 169.21 |
| 4.5 | 262.13 |
| 5.0 | 406.09 |
| 5.5 | 629.12 |

*The data in Exercises* 17–20 *consist of the Olympic Games records for the given events. Plot the points on semilogarithmic paper, find a straight line that fits the data fairly well, and determine an equation for the event in terms of the number of years after* 1900 A.D. *Then use this formula to predict what the record will be in the year* 2000.

**17.** Men's 400-meter run

| Year | 1896 | 1900 | 1904 | 1906 | 1908 | 1912 | 1920 | 1924 | 1928 |
|---|---|---|---|---|---|---|---|---|---|
| Time (sec) | 54.2 | 49.4 | 49.2 | 53.2 | 50.0 | 48.2 | 49.6 | 47.6 | 47.8 |

| Year | 1932 | 1936 | 1948 | 1952 | 1956 | 1960 | 1964 | 1968 |
|---|---|---|---|---|---|---|---|---|
| Time (sec) | 46.2 | 46.5 | 46.2 | 45.9 | 46.7 | 44.9 | 45.1 | 43.8 |

**18.** Men's metric mile (1500-meter run)

| Year | 1896 | 1900 | 1904 | 1906 | 1908 | 1912 | 1920 | 1924 | 1928 |
|---|---|---|---|---|---|---|---|---|---|
| Time (min) | 4.55 | 4.10 | 4.09 | 4.20 | 4.06 | 3.95 | 4.03 | 3.89 | 3.89 |

| Year | 1932 | 1936 | 1948 | 1952 | 1956 | 1960 | 1964 | 1968 |
|---|---|---|---|---|---|---|---|---|
| Time (min) | 3.85 | 3.80 | 3.83 | 3.75 | 3.69 | 3.59 | 3.64 | 3.58 |

**19.** Women's 100-meter freestyle

| Year | 1912 | 1920 | 1924 | 1928 | 1932 | 1936 | 1948 | 1952 | 1956 |
|---|---|---|---|---|---|---|---|---|---|
| Time (min) | 1.37 | 1.23 | 1.21 | 1.18 | 1.11 | 1.10 | 1.11 | 1.11 | 1.03 |

| Year | 1960 | 1964 | 1968 |
|---|---|---|---|
| Time (min) | 1.02 | 0.99 | 1.00 |

**20.** Women's 400-meter relay

| Year | 1928 | 1932 | 1936 | 1948 | 1952 | 1956 | 1960 | 1964 | 1968 |
|---|---|---|---|---|---|---|---|---|---|
| Time (sec) | 48.4 | 47.0 | 46.9 | 47.5 | 45.9 | 44.5 | 44.5 | 43.6 | 42.8 |

**21.** The following table lists the official census of the United States for the years 1900 to 1970. Plot these values on semilogarithmic paper, draw a straight line that fits the data fairly well, and devise a formula for the United States population in terms of the number of years after 1900. Use this formula to predict the population in the years 1980, 1990, and 2000.

| Year | 1900 | 1910 | 1920 | 1930 | 1940 |
|---|---|---|---|---|---|
| Pop. | 75,994,575 | 91,972,266 | 105,710,620 | 122,775,046 | 131,669,275 |

| Year | 1950 | 1960 | 1970 |
|---|---|---|---|
| Pop. | 150,697,361 | 179,323,175 | 200,251,326 |

**22.** The total electric energy output in the United States in millions of kilowatt hours is listed in the table below. Use the methods discussed in Section 3.4

to estimate the output in 1985. Let $x$ represent the number of years after 1940.

| Year | 1939 | 1943 | 1951 | 1956 | 1960 | 1965 | 1966 | 1967 |
|---|---|---|---|---|---|---|---|---|
| Output | 127,642 | 217,759 | 370,673 | 600,668 | 753,350 | 1,055,252 | 1,144,350 | 1,211,749 |

**23.** The United States motor fuel demand in thousands of 42-gallon barrels is given in the table below. Estimate the demand in the year 1980. (Let $x$ represent the number of years after 1950.)

| Year | 1950 | 1955 | 1960 | 1965 | 1969 |
|---|---|---|---|---|---|
| Demand | 994,290 | 1,329,788 | 1,511,670 | 1,750,028 | 2,072,068 |

**24.** The total amount of natural gas (in millions of cubic feet) marketed in the United States is given in the table below. Estimate the demand in the year 1980. (Let $x$ represent the number of years after 1940.)

| Year | 1940 | 1945 | 1950 | 1955 | 1960 | 1965 |
|---|---|---|---|---|---|---|
| Amount marketed | 2,660,222 | 3,918,686 | 6,282,060 | 9,405,351 | 12,771,038 | 16,039,753 |

**25.** Figure 3.22 shows a double-logarithmic plot of the average total heat production and average body weight of birds and mammals, as determined by experiments of F. G. Benedict [reported in the Carnegie Institute Washington Publication 503(1938): 1–215]. Using the methods discussed in Section 3.4, obtain a power function relating body weight to heat production.

**26.** The rate of oxygen consumption while running is higher for small animals than for large ones, according to the information gathered by C. R. Taylor, K. Schmidt-Nielsen, and J. L. Raab [*American Journal of Physiology* **219** (1970): 1104–1107], and illustrated in Fig. 3.23. Using the methods discussed in Section 3.4, determine a power function relating body weight to oxygen consumption per gram-kilometer. Use your formula and Fig. 3.23 to estimate the oxygen consumption per gram-kilometer for an elephant.

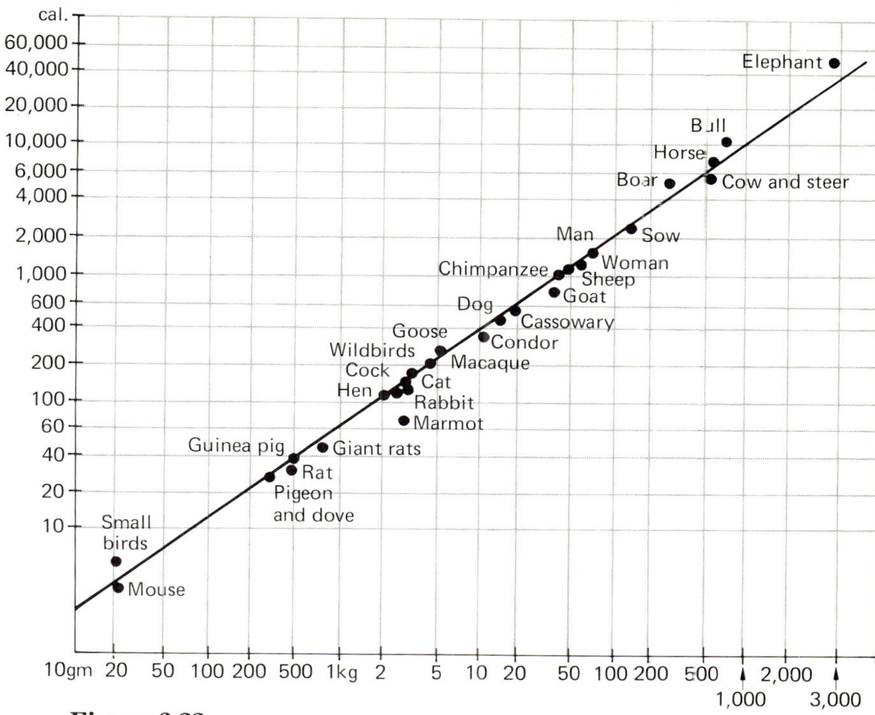


**Figure 3.22**
Double-logarithmic plot of average total heat production and average body weight of birds and mammals. (From F. G. Benedict, "Vital Energetics," *Carnegie Inst. Washington Publ.* **503** (1938): 171.)

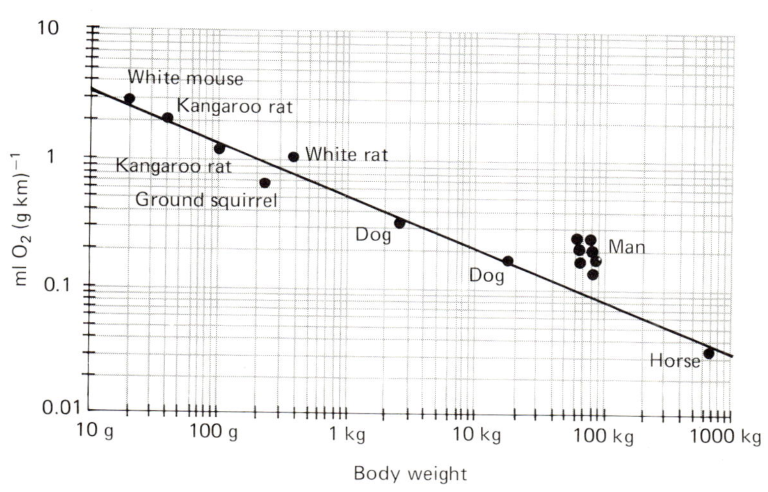


**Figure 3.23**

From C. R. Taylor, K. Schmidt-Nielsen, and J. L. Rabb, "Scaling of Energetic Cost of Running to Body Size in Mammals, *Am. J. Physiol.* **219** (1970): 1104–1107.

**27.** The Society of Actuaries obtained the following table of average heights and weights based on a four-year study of five million Americans.

Men, 50–59 years of age

| Height (in.) | 60 | 62 | 64 | 66 | 68 | 70 | 72 | 74 | 76 |
|---|---|---|---|---|---|---|---|---|---|
| Weight (lb) | 136 | 142 | 149 | 157 | 166 | 175 | 185 | 194 | 205 |

Use the methods discussed in Section 3.4 to find a formula relating height and weight. How much should a male who is 50–59 years of age and seven feet tall weigh?

## ☐ 3.5 MODELING BY EXPONENTIAL CURVE FITTING

In Section 3.4 we observed that exponential functions have straight line graphs on semilogarithmic paper. Similarly, power functions have straight line graphs on double-logarithmic paper. We also saw that by fitting a straight line to experimental data, we can obtain exponential and power functions describing real situations. In this section we will adopt the techniques shown in Section 2.8 to obtain *least squares approximations* for exponential and power functions. Actually, the method we shall apply is almost identical to that used in Section 2.8. The only difference is that we will use the *logarithm* of one (or both) of the variables, which effectively converts our problem to a semi- or double-logarithmic situation, and will find the *least squares line* of the resulting data. We illustrate the method with two examples.

***Example 1*** In Exercise 21 on p. 135 graphical techniques are used to obtain an exponential function describing the population of the United States in terms of the number of years after 1900. In this example we will use the least squares theorem (see Theorem 2.1) to derive numerically the exponential function

$$y = b(m)^x \tag{3.48}$$

that "best" fits the data in Exercise 21 on p. 135. Taking the common logarithm of both sides of Eq. (3.48) yields

$$\log y = \log b + \log (m^x),$$

or

$$\log y = (\log m)x + \log b. \tag{3.49}$$

Substituting $Y = \log y$, $M = \log m$, and $B = \log b$ into Eq. (3.49), we obtain the straight line

$$Y = Mx + B \tag{3.50}$$

in the $xY$-plane. We will make use of these substitutions in our calculations below.

**Table 3.2**

| Year | Population | $i$ | $x_i$ | $Y_i$ | $x_i Y_i$ | $x_i^2$ |
|---|---|---|---|---|---|---|
| 1900 | 75,994,575 | 1 | 0 | 7.8808 | 0 | 0 |
| 1910 | 91,972,266 | 2 | 10 | 7.9637 | 79.637 | 100 |
| 1920 | 105,710,620 | 3 | 20 | 8.0241 | 160.482 | 400 |
| 1930 | 122,775,046 | 4 | 30 | 8.0891 | 242.673 | 900 |
| 1940 | 131,669,275 | 5 | 40 | 8.1195 | 324.780 | 1600 |
| 1950 | 150,697,361 | 6 | 50 | 8.1781 | 408.905 | 2500 |
| 1960 | 179,323,175 | 7 | 60 | 8.2536 | 495.216 | 3600 |
| 1970 | 200,251,326 | 8 | 70 | 8.3016 | 581.112 | 4900 |
| | | Sum | 280 | 64.3105 | 2,292.805 | 14,000 |
| | | Average | | $\bar{x} = 35$, | $\bar{Y} = 8.1013$ | |

Consider Table 3.2. The first two columns contain the census data given in Exercise 21 on p. 135. In column 3, headed by the symbol $i$, we number the data points. Column 4, headed $x_i$, gives the number of years after 1900, since our final formula will be in terms of years after 1900. Column 5, headed $Y_i$, gives the *common logarithms* of the population figures in column 2. We use the logarithms because the least squares line (Eq. 3.50) involves the dependent variable $Y (= \log y)$, the logarithm of the populations. The entries in this column can easily be checked if your calculator has a [LOG] key. Columns 6 and 7 contain the results of multiplying the entries in columns 4 and 5 and squaring the entry in column 4, respectively. We will make use of these two columns in our calculations.

We now add the numbers in columns 4, 5, 6, and 7 divide the sum of the first two columns by 8 (the number of entries) to obtain the averages. Compare what we have done in Table 3.2 with the calculation given in Table 2.7 on p. 82.

After we have completed the entries in Table 3.2, we use the least squares theorem to obtain the coefficients $M$ and $B$ of the least squares line

$$Y = M(x - \bar{x}) + B. \tag{3.51}$$

Hence, we have $B = \bar{Y}$ and

$$M = \frac{(x_1 Y_1 + x_2 Y_2 + \cdots + x_n Y_n) - n\bar{x}\bar{Y}}{(x_1^2 + x_2^2 + \cdots + x_n^2) - n(\bar{x})^2}, \tag{3.52}$$

so $B = 8.1013$ and

$$M = \frac{2292.805 - 8(35)(8.1013)}{14{,}000 - 8(35)^2}$$

$$= \frac{24.44}{4200} = 0.00582.$$

Thus, substituting these values into Eq. (3.51) gives us

$$Y = 0.00582(x - 35) + 8.1013$$

$$= 0.00582x + [8.1013 - 35(0.00582)],$$

or

$$Y = 0.00582x + 7.89760. \tag{3.53}$$

Equation (3.53) is the least squares line in the $xY$-plane. We must now reverse the steps we took to obtain Eq. (3.50) to recover the exponential function (3.48). Hence, we let $\log y = Y$, $\log m = M = 0.00582$, and $\log b = B = 7.89760$, getting $m = 1.0135$ and $b = 78{,}995{,}072$. Substituting these values into Eq. (3.48) gives us the final formula

$$y = 78{,}995{,}072(1.0135)^x. \tag{3.54}$$

This equation is very close to our graphical solution:

$$y = 77{,}000{,}000\left(\frac{200}{77}\right)^{x/70}$$

$$= 77{,}000{,}000\left[\left(\frac{200}{77}\right)^{1/70}\right]^x$$

$$= 77{,}000{,}000(1.0137)^x.$$

We can use Eq. (3.54) to predict the population of the United States in 1980, 1990, and 2000:

$$y_{1980} = 78{,}995{,}072(1.0135)^{80} = 230{,}940{,}574;$$

$$y_{1990} = 78{,}995{,}072(1.0135)^{90} = 264{,}081{,}374;$$

$$y_{2000} = 78{,}995{,}072(1.0135)^{100} = 301{,}977{,}998.$$

***Example 2*** We will use the data given in Exercise 27 on p. 138 to derive a least squares approximation relating height to weight for men from 50 to 59 years of age. In this problem we seek an equation of the form

$$y = bx^m. \tag{3.55}$$

Taking common logarithms of both sides, we obtain

$$\log y = \log b + m \log x. \tag{3.56}$$

Substituting $Y = \log y$, $X = \log x$, and $B = \log b$, Eq. (3.56) yields the straight line

$$Y = mX + B \tag{3.57}$$

in the $XY$-plane. This is the line we wish to find by least squares methods.

The first two columns in Table 3.3 contain the data from Exercise 27 on p. 138. Column 3 counts the number of entries, while columns 4 and 5 give the common logarithms of the entries in the first two columns, respectively. This is done because the least squares line is in terms of $X$, the logarithm of the height, and $Y$, the logarithm of the weight. The entries in columns 6 and 7 are self-explanatory.

**Table 3.3**

| Height (in.) | Weight (lb) | $i$ | $X_i$ | $Y_i$ | $X_iY_i$ | $X_i^2$ |
|---|---|---|---|---|---|---|
| 60 | 136 | 1 | 1.7782 | 2.1335 | 3.7938 | 3.1620 |
| 62 | 142 | 2 | 1.7924 | 2.1523 | 3.8578 | 3.2127 |
| 64 | 149 | 3 | 1.8062 | 2.1732 | 3.9252 | 3.2624 |
| 66 | 157 | 4 | 1.8195 | 2.1959 | 3.9954 | 3.3106 |
| 68 | 166 | 5 | 1.8325 | 2.2201 | 4.0683 | 3.3581 |
| 70 | 175 | 6 | 1.8451 | 2.2430 | 4.1386 | 3.4044 |
| 72 | 185 | 7 | 1.8573 | 2.2672 | 4.2109 | 3.4496 |
| 74 | 194 | 8 | 1.8692 | 2.2878 | 4.2764 | 3.4939 |
| 76 | 205 | 9 | 1.8808 | 2.3118 | 4.3480 | 3.5374 |
| | | Sum | 16.4812 | 19.9848 | 36.6144 | 30.1911 |
| | | Average | $\bar{X} = 1.8312$, | $\bar{Y} = 2.2205$ | | |

By the least squares theorem, the least squares line

$$Y = m(X - \bar{X}) + B \tag{3.58}$$

satisfies $B = \bar{Y} = 2.2205$ and

$$m = \frac{(X_1Y_1 + X_2Y_2 + \cdots + X_nY_n) - n\bar{X}\bar{Y}}{(X_1^2 + X_2^2 + \cdots + X_n^2) - n(\bar{X})^2}$$

$$= \frac{36.6144 - 9(1.8312)(2.2205)}{30.1911 - 9(1.8312)^2} = \frac{0.0179}{0.0107} = 1.6729.$$

Thus, $Y = 1.6729(X - 1.8312) + 2.2205$, so we obtain

$$Y = 1.6729X - 0.8429; \tag{3.59}$$

substituting back in terms of $x$ and $y$ we see that $m = 1.6729$ and $\log b = B = -0.8429$ or $b = 10^{-0.8429} = 0.1436$. Therefore, the final equation is

$$y = 0.1436x^{1.6729}, \tag{3.60}$$

which agrees very closely with the answer to Exercise 27.

---

### EXERCISES 3.5

*In Exercises 1–8, find the least squares exponential function that best fits the data in the given table.*

**1.**

| x | y |
|---|---|
| 1.3 | 7.4 |
| 1.6 | 9.2 |
| 1.9 | 11.2 |
| 2.2 | 13.8 |

**2.**

| x | y |
|---|---|
| 0.9 | 17.0 |
| 1.3 | 32.5 |
| 1.7 | 61.0 |
| 2.1 | 117.0 |

**3.**

| x | y |
|---|---|
| 0.5 | 3.5 |
| 0.8 | 4.9 |
| 1.1 | 7.0 |
| 1.4 | 9.9 |

**4.**

| x | y |
|---|---|
| 1 | 9.4 |
| 2 | 29.7 |
| 3 | 93.1 |
| 4 | 292.0 |

**5.**

| x | y |
|---|---|
| 0.3 | 1.36 |
| 0.7 | 2.03 |
| 1.1 | 3.00 |
| 1.5 | 4.50 |
| 1.9 | 6.70 |

**6.**

| x | y |
|---|---|
| 1.0 | 0.38 |
| 1.2 | 0.30 |
| 1.4 | 0.25 |
| 1.6 | 0.20 |
| 1.8 | 0.18 |

**7.**

| $x$ | $y$ |
|---|---|
| 0.8 | 7.0 |
| 1.2 | 10.4 |
| 1.6 | 15.6 |
| 2.0 | 23.2 |
| 2.4 | 34.6 |

**8.**

| $x$ | $y$ |
|---|---|
| 1.00 | 8.55 |
| 1.25 | 11.34 |
| 1.50 | 15.15 |
| 1.75 | 20.12 |
| 2.00 | 26.84 |

*In Exercises 9–12, find the least squares power function that best fits the data in the given table.*

**9.**

| $x$ | $y$ |
|---|---|
| 2 | 12.2 |
| 3 | 26.9 |
| 4 | 48.1 |
| 5 | 74.7 |

**10.**

| $x$ | $y$ |
|---|---|
| 0.5 | 0.3 |
| 1.0 | 5.1 |
| 1.5 | 25.3 |
| 2.0 | 79.9 |

**11.**

| $x$ | $y$ |
|---|---|
| 1 | 3.2 |
| 2 | 12.6 |
| 3 | 28.2 |
| 4 | 50.2 |

**12.**

| $x$ | $y$ |
|---|---|
| 1.2 | 1.6 |
| 2.5 | 12.1 |
| 3.7 | 35.0 |
| 6.1 | 136.1 |

*Solve the following exercises (from pp. 134–136) using the methods discussed in Section 3.5.*

**13.** Exercise 17

**14.** Exercise 18

**15.** Exercise 19

**16.** Exercise 20

**17.** Exercise 22

**18.** Exercise 23

**19.** The average weight for women from 50 to 59 years of age, as determined by the Society of Actuaries, is given in the table below. Use the methods discussed in this section to determine a formula that relates weight to height.

| Height (in.) | 58 | 60 | 62 | 64 | 66 | 68 | 70 | 72 |
|---|---|---|---|---|---|---|---|---|
| Weight (lb) | 125 | 130 | 136 | 141 | 152 | 160 | 169 | 180 |

**20.** Rod and reel fishing for marlin is one of the most challenging sports. The world records (as of September 1970) were as follows.

| Species | Weight (lb) | Length | Girth |
|---|---|---|---|
| Black marlin | 1560 | 14′6″ | 81″ |
| Blue marlin | 845 | 13′1″ | 71″ |
| Pacific blue marlin | 1153 | 14′8″ | 73″ |
| Striped marlin | 430 | 10′8.5″ | 54.5″ |
| White marlin | 159.5 | 9′ | 36″ |

Determine the relationships between each pair of the three measurements: weight vs. length, weight vs. girth, and length vs. girth.

## CHAPTER VOCABULARY

**base (of an exponent)**
**base (of a logarithm)**
**bell-shaped curve**
**carrying capacity**
**characteristic (of a common logarithm)**
**compound interest** $A = P(1 + i)^n$
**common logarithm** $\log x$
**double-logarithmic plot**
**$e$ (= 2.71828 · · ·)**
**exponent**
**exponential decay**
**exponential function**
**exponential growth**
**half-life**
**instantaneous decay**
**instantaneous growth**
**learning curve**
**least squares approximation**
**least squares line**
**least squares theorem**
**logarithm** $\log_a x$
**logarithmic scale**
**logistic curve**
**mantissa (of a common logarithm)**
**natural logarithm** $\ln x$
**power function** $ax^b$
**radioactive decay**
**rational powers** $x^{p/q}$
**sales curve**
**scientific notation**
**semilogarithmic plot**
**sigmoidal curve**

# CHAPTER 4 LINEAR SYSTEMS

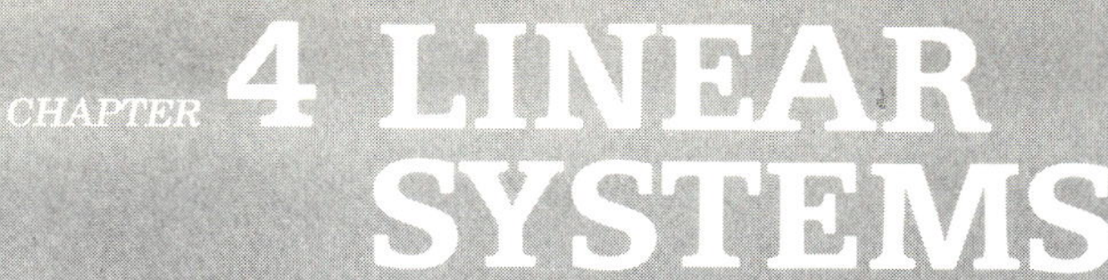

## 4.1 MATRICES AND MATRIX MULTIPLICATION

In this chapter we shall consider one of the most important techniques used in working with systems of linear equations: the use of *matrices*. We present, in this section, the basic definition of a *matrix* and discuss how to add and multiply matrices. In later sections we will see that matrix notation allows us to handle systems of linear equations almost as simply as single linear equations. In particular, we will note that matrix methods simplify the theoretical concepts and provide systematic procedures for obtaining the solution of a system of simultaneous linear equations or of a linear programming problem.

Although the term *matrix* may be new to you, you have already used some simple matrices in this book. For example, in Chapter 2 we discussed the rectangular coordinates of points in the Euclidean plane. We saw that the symbol $(2, 3)$ represents the point in the plane two units to the right of and three units up from the origin. This pair of numbers is a matrix consisting of one *row* and two *columns* of real numbers. Similarly, the triple of numbers $(1, 2, 3)$ representing a point in three-dimensional Euclidean space (see Fig. 4.1) is a matrix consisting of one row and three columns of numbers.

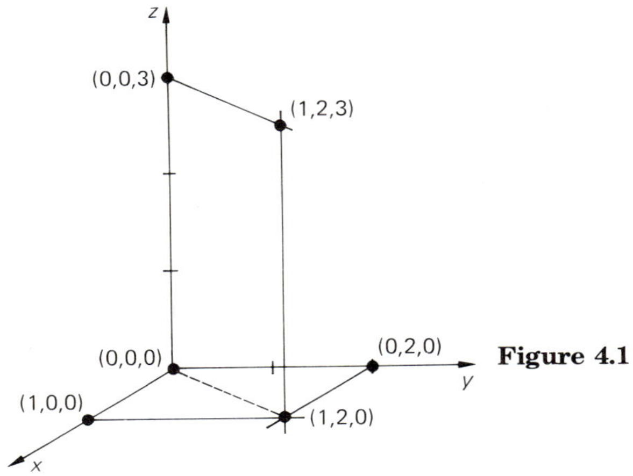


**Figure 4.1**

The rectangular array of numbers

$$\begin{bmatrix} 1 & 12 & 3.1 \\ 4 & -5 & 6.3 \end{bmatrix}$$

is a matrix consisting of *two* rows and *three* columns. We can construct matrices in any size; for example, the rectangular array

$$\begin{bmatrix} 3 & 1/2 & 0 \\ 4 & 2/3 & 0 \\ 0 & -5.1 & 2 \\ 2 & 173 & 6 \end{bmatrix}$$

is a matrix with four rows and three columns. These examples lead to the basic definition.

> A matrix is a rectangular array of numbers.

When describing a matrix, we usually specify its size by the number of rows and columns that it contains. If we say that a given rectangular array is a $4 \times 5$ matrix, we mean that it has four rows and five columns. Thus, the rectangular arrays

$$(1, 2, 3, 4), \quad \begin{bmatrix} 1 \\ 2 \\ 3 \end{bmatrix}, \quad \begin{bmatrix} 0 & 2 & 5 \\ 0 & 3 & 6 \\ 1 & 4 & 9 \end{bmatrix}$$

are all matrices; the first is a $1 \times 4$ matrix, the second is a $3 \times 1$ matrix, and the last is a $3 \times 3$ matrix.

> An $m \times n$ matrix has $m$ rows and $n$ columns.

The term *vector* is reserved for matrices having either one row or one column. A $1 \times n$ matrix is often called a *row vector*, while an $m \times 1$ matrix is said to be a *column vector*. The $1 \times 4$ and $3 \times 1$ matrices shown above are both vectors.

A matrix that has the same number of rows and columns is said to be *square*. The $3 \times 3$ matrix shown above is a square matrix.

The use of matrices arises naturally in organizing data. The next four examples illustrate some typical situations.

***Example 1*** When reporting National Hockey League standings, sportswriters often use a matrix to show the number of games a team has won, lost, or tied.

For example, the final standings in the East Division for the 1969–1970 season were

$$\begin{array}{l} \\ \text{Chicago} \\ \text{Boston} \\ \text{Detroit} \\ \text{New York} \\ \text{Montreal} \\ \text{Toronto} \end{array} \begin{array}{c} \begin{array}{ccc} \text{W} & \text{L} & \text{T} \end{array} \\ \begin{bmatrix} 45 & 22 & 9 \\ 40 & 17 & 19 \\ 40 & 21 & 15 \\ 38 & 22 & 16 \\ 38 & 22 & 16 \\ 29 & 34 & 13 \end{bmatrix} \end{array}.$$

This is a $6 \times 3$ matrix.

***Example 2*** The daily reports of the New York stock market are another example of a matrix. Newspapers often report a stock's activity by giving its high, low, and closing prices. For example, a partial report might list

$$\begin{array}{l} \\ \text{American Airlines} \\ \text{American Motors} \\ \text{Atlantic Richfield} \\ \text{General Electric} \\ \text{General Motors} \end{array} \begin{array}{c} \begin{array}{ccc} \text{High} & \text{Low} & \text{Close} \end{array} \\ \begin{bmatrix} 13\frac{5}{8} & 13\frac{1}{4} & 13\frac{1}{4} \\ 4\frac{1}{8} & 4 & 4\frac{1}{8} \\ 60\frac{1}{8} & 58\frac{1}{2} & 58\frac{1}{2} \\ 53\frac{1}{4} & 52\frac{3}{8} & 52\frac{3}{8} \\ 75 & 73 & 74 \end{bmatrix} \end{array}.$$

This is a $5 \times 3$ matrix.

***Example 3*** A more complicated situation is given by the social interactions within small groups of acquaintances. Suppose that a sociologist is trying to determine the most dominant individual in a family by asking each person to state whose opinion he or she seeks in making a difficult decision, and tabulating their responses. Suppose that Table 4.1 lists the responses of this family. The sociologist can use this table to construct a

**Table 4.1**

| Member | Person whose opinion is sought |
|---|---|
| Father | Mother |
| Mother | Father or daughter |
| Oldest son | His own |
| Daughter | Mother or father |
| Youngest son | Mother or brother |

5 × 5 matrix of zeros and ones. The rows of the matrix represent the family member and the columns represent the person whose opinion is sought. The sociologist enters a one in the row representing the person whose opinion is sought, and a zero in all the others. The following matrix, whose use we will investigate in Section 4.5, is obtained.

Person whose opinion is sought

| | Father | Mother | Oldest son | Daughter | Youngest son |
|---|---|---|---|---|---|
| Father | 0 | 1 | 0 | 0 | 0 |
| Mother | 1 | 0 | 0 | 1 | 0 |
| Oldest son | 0 | 0 | 1 | 0 | 0 |
| Daughter | 1 | 1 | 0 | 0 | 0 |
| Youngest son | 0 | 1 | 1 | 0 | 0 |

***Example 4*** Frontier Airlines services a number of cities in the central and western United States. Below we list a number of the cities that they service and indicate with a one those cities connected by nonstop flights; all other connections are denoted by a zero (as of 1 June 1976).

To

| From | | Bozeman | Denver | Missoula | Salt Lake | St. Louis |
|---|---|---|---|---|---|---|
| | Bozeman | 0 | 0 | 1 | 1 | 0 |
| | Denver | 0 | 0 | 0 | 1 | 1 |
| From | Missoula | 1 | 0 | 0 | 1 | 0 |
| | Salt Lake | 1 | 1 | 0 | 0 | 0 |
| | St. Louis | 0 | 1 | 0 | 0 | 0 |

We will discuss an application for this matrix in Section 4.5.

---

We now consider three arithmetic operations involving matrices. The first two are quite straightforward, but the last one will surprise you.

**Addition of Matrices** *A set of matrices can be added provided they all have the same size. The sum is obtained by adding all corresponding elements.* For example,

$$\begin{bmatrix} 2 & 3 & 1 \\ -1 & 2 & 0 \end{bmatrix} + \begin{bmatrix} -3 & -14 & 6 \\ 10 & 0.5 & 0 \end{bmatrix}$$

$$= \begin{bmatrix} 2+(-3) & 3+(-14) & 1-6 \\ (-1)+10 & 2+(0.5) & 0-0 \end{bmatrix} = \begin{bmatrix} -1 & -11 & 7 \\ 9 & 2.5 & 0 \end{bmatrix},$$

and

$$\begin{bmatrix} 1 & 2 \\ 3 & 4 \end{bmatrix} + \begin{bmatrix} 5 & 6 \\ 7 & 8 \end{bmatrix} + \begin{bmatrix} 0 & -1 \\ -2 & 3 \end{bmatrix} = \begin{bmatrix} 1+5+0 & 2+6-1 \\ 3+7-2 & 4+8+3 \end{bmatrix} = \begin{bmatrix} 6 & 7 \\ 8 & 15 \end{bmatrix},$$

but

$$\begin{bmatrix} 2 & 3 \\ 4 & 5 \end{bmatrix} + \begin{bmatrix} 6 & 7 & 8 \\ 9 & 0 & 1 \end{bmatrix}$$

cannot be added, since the first is a $2 \times 2$ matrix, while the second is a $2 \times 3$ matrix.

Note that if we add a matrix of zeros to any matrix, we obtain the original matrix:

$$\begin{bmatrix} 1 & 2 & 3 \\ 4 & 5 & 6 \end{bmatrix} + \begin{bmatrix} 0 & 0 & 0 \\ 0 & 0 & 0 \end{bmatrix} = \begin{bmatrix} 1 & 2 & 3 \\ 4 & 5 & 6 \end{bmatrix}.$$

Such zero matrices are often called *additive identities*, and behave very much like the number 0 does in ordinary addition.

**Scalar Multiplication of a Matrix**

In algebra, when four objects $x$ are added together, we represent their sum as

$$x + x + x + x = 4x.$$

To apply this concept to matrices, let's see what happens if we replace the object $x$ with a matrix. Then

$$4\begin{bmatrix} 1 & 2 \\ 3 & 4 \end{bmatrix} = \begin{bmatrix} 1 & 2 \\ 3 & 4 \end{bmatrix} + \begin{bmatrix} 1 & 2 \\ 3 & 4 \end{bmatrix} + \begin{bmatrix} 1 & 2 \\ 3 & 4 \end{bmatrix} + \begin{bmatrix} 1 & 2 \\ 3 & 4 \end{bmatrix}$$
$$= \begin{bmatrix} 1+1+1+1 & 2+2+2+2 \\ 3+3+3+3 & 4+4+4+4 \end{bmatrix} = \begin{bmatrix} 4 & 8 \\ 12 & 16 \end{bmatrix}.$$

Observe that in effect we have merely multiplied each entry in the matrix by the number 4. [In mathematics, the term *scalar* is often used to denote a number without specifying what kind (integer, real, complex, etc.) it is.] These observations lead to the following definition.

*Any matrix may be multiplied by a scalar simply by multiplying each entry in the matrix by that scalar.*

The scalar need not be a positive integer. For example,

$$-2\begin{bmatrix} 3 & -5 \\ 0 & 2 \end{bmatrix} = \begin{bmatrix} -6 & 10 \\ 0 & -4 \end{bmatrix},$$

$$1/3\begin{bmatrix} 4 & 6 \\ 3 & -9 \end{bmatrix} = \begin{bmatrix} \frac{4}{3} & 2 \\ 1 & -3 \end{bmatrix},$$

and

$$\sqrt{2}\begin{bmatrix} 2 & \sqrt{8} \\ 0 & \sqrt{2} \end{bmatrix} = \begin{bmatrix} 2\sqrt{2} & \sqrt{8}\sqrt{2} \\ 0 & \sqrt{2}\sqrt{2} \end{bmatrix} = \begin{bmatrix} 2\sqrt{2} & 4 \\ 0 & 2 \end{bmatrix}.$$

If we combine the operations of addition and scalar multiplication by $(-1)$, we achieve the *subtraction* of the two matrices. For example,

$$\begin{bmatrix} 7 & 12 \\ 9 & 4 \end{bmatrix} - \begin{bmatrix} 6 & -8 \\ 11 & 7 \end{bmatrix} = \begin{bmatrix} 7 & 12 \\ 9 & 4 \end{bmatrix} + (-1)\begin{bmatrix} 6 & -8 \\ 11 & 7 \end{bmatrix}$$

$$= \begin{bmatrix} 7 & 12 \\ 9 & 4 \end{bmatrix} + \begin{bmatrix} -6 & 8 \\ -11 & -7 \end{bmatrix}$$

$$= \begin{bmatrix} 7-6 & 12+8 \\ 9-11 & 4-7 \end{bmatrix} = \begin{bmatrix} 1 & 20 \\ -2 & -3 \end{bmatrix}.$$

In practice, it is unnecessary to go through all the steps we went through in this example. Instead, you should *merely subtract corresponding entries*:

$$\begin{bmatrix} 7 & 12 \\ 9 & 4 \end{bmatrix} - \begin{bmatrix} 6 & -8 \\ 11 & 7 \end{bmatrix} = \begin{bmatrix} 7-6 & 12-(-8) \\ 9-11 & 4-7 \end{bmatrix} = \begin{bmatrix} 1 & 20 \\ -2 & -3 \end{bmatrix}.$$

Of course, only matrices of the same size can be subtracted.

We now turn to the operation of multiplication.

**Matrix Multiplication** One of the most surprising facts about matrices is that *matrix multiplication is not done by multiplying corresponding entries.* This is certainly startling, since multiplying corresponding entries seems to be the most natural procedure. There is, however, a very good reason *not* to multiply matrices in this way. The full explanation for the unusual method of multiplying two matrices is presented in Section 4.2. Here we will learn only how to multiply two matrices.

Rather than formally state the procedure, we will go through the steps of three such multiplications, explaining each stage of the process.

***Example 5*** We multiply the following matrices:

$$\begin{bmatrix} 1 & 2 \\ 3 & 4 \end{bmatrix} \cdot \begin{bmatrix} 5 & 6 \\ 7 & 8 \end{bmatrix}. \tag{4.1}$$

Matrix multiplication, like driving, requires the use of both hands! The product of the two matrices in (4.1) is also a $2 \times 2$ matrix (more will be said about this later). Thus, we must find four entries for the product matrix:

$$\begin{bmatrix} - & - \\ - & - \end{bmatrix}.$$

To find the top left entry we will consider the top *row* of the left matrix and the first *column* of the right matrix only:

$$\begin{bmatrix} 1 & 2 \\ \bullet & \bullet \end{bmatrix} \cdot \begin{bmatrix} 5 & \bullet \\ 7 & \bullet \end{bmatrix} = \begin{bmatrix} - & \bullet \\ \bullet & \bullet \end{bmatrix}. \tag{4.2}$$

We multiply the first number in the row by the first number in the column and add the result to the product of the second entry in the row and the second entry in the column:

$$(1 \cdot 5) + (2 \cdot 7) = 19. \tag{4.3}$$

The answer, 19, is placed in the top left entry of the product matrix in (4.2):

$$\begin{bmatrix} 1 & 2 \\ \bullet & \bullet \end{bmatrix} \cdot \begin{bmatrix} 5 & \bullet \\ 7 & \bullet \end{bmatrix} = \begin{bmatrix} 19 & \bullet \\ \bullet & \bullet \end{bmatrix}.$$

This completes the procedure for finding this one entry. We should note two points in the procedure.

**1.** If we place a finger of our left hand on the leftmost entry of the top row of the left matrix, and a finger of our right hand on the top entry of the first column of the right matrix, we have the first two numbers that are multiplied in (4.3). Moving the left finger one entry to the right and the right finger one entry down, we have the next pair of numbers that we need to multiply. (This is what we meant when we said that two hands are required to multiply matrices!)

**2.** The location (row and column) of the answer is determined by the *row* that our left finger is on and the *column* that our right finger is on. In this case, it is the first row and first column (we number rows from the top and columns from the left).

Let's now find the top right entry of the product. Since this entry is in the first row and second column, we must use the first row of the left matrix and the second column of the right matrix:

$$\begin{bmatrix} 1 & 2 \\ \bullet & \bullet \end{bmatrix} \cdot \begin{bmatrix} \bullet & 6 \\ \bullet & 8 \end{bmatrix} = \begin{bmatrix} \bullet & (1 \cdot 6) + (2 \cdot 8) \\ \bullet & \bullet \end{bmatrix}$$

$$= \begin{bmatrix} \bullet & 22 \\ \bullet & \bullet \end{bmatrix}.$$

Similarly,

$$\begin{bmatrix} \bullet & \bullet \\ 3 & 4 \end{bmatrix} \cdot \begin{bmatrix} 5 & \bullet \\ 7 & \bullet \end{bmatrix} = \begin{bmatrix} \bullet & \bullet \\ (3 \cdot 5) + (4 \cdot 7) & \bullet \end{bmatrix}$$

$$= \begin{bmatrix} \bullet & \bullet \\ 43 & \bullet \end{bmatrix},$$

and

$$\begin{bmatrix} \bullet & \bullet \\ 3 & 4 \end{bmatrix} \cdot \begin{bmatrix} \bullet & 6 \\ \bullet & 8 \end{bmatrix} = \begin{bmatrix} \bullet & \bullet \\ \bullet & (3 \cdot 6) + (4 \cdot 8) \end{bmatrix}$$

$$= \begin{bmatrix} \bullet & \bullet \\ \bullet & 50 \end{bmatrix}.$$

Therefore, the final answer is

$$\begin{bmatrix} 1 & 2 \\ 3 & 4 \end{bmatrix} \cdot \begin{bmatrix} 5 & 6 \\ 7 & 8 \end{bmatrix} = \begin{bmatrix} 19 & 22 \\ 43 & 50 \end{bmatrix}. \tag{4.4}$$

***Example 6*** Multiply the following two matrices:

$$\begin{bmatrix} 0 & 1 \\ 2 & 3 \end{bmatrix} \cdot \begin{bmatrix} 4 & 5 & 6 \\ 7 & 8 & 9 \end{bmatrix}.$$

Although these two matrices have *different* sizes, they *can* be multiplied. First we multiply the top row of the left matrix with first column of the right matrix, obtaining

$$(0 \cdot 4) + (1 \cdot 7) = 7$$

so that we have

$$\begin{bmatrix} 0 & 1 \\ \bullet & \bullet \end{bmatrix} \cdot \begin{bmatrix} 4 & \bullet & \bullet \\ 7 & \bullet & \bullet \end{bmatrix} = \begin{bmatrix} 7 & & \\ & & \end{bmatrix}.$$

We repeat the procedure using the top row of the left matrix and the second column of the right matrix, obtaining

$$\begin{bmatrix} 0 & 1 \\ \bullet & \bullet \end{bmatrix} \cdot \begin{bmatrix} 4 & 5 & \bullet \\ 7 & 8 & \bullet \end{bmatrix} = \begin{bmatrix} 7 & 8 & \\ & & \end{bmatrix}.$$

Finally, we can multiply the top row of the left matrix by the third column of the right matrix, which gives us

$$\begin{bmatrix} 0 & 1 \\ \bullet & \bullet \end{bmatrix} \cdot \begin{bmatrix} 4 & 5 & 6 \\ 7 & 8 & 9 \end{bmatrix} = \begin{bmatrix} 7 & 8 & 9 \\ & & \end{bmatrix}.$$

This completes the first row of the product matrix. Repeating this procedure using the second row of the left matrix and multiplying its entries with the entries in the columns of the right matrix, we have

$$\begin{bmatrix} 0 & 1 \\ 2 & 3 \end{bmatrix} \cdot \begin{bmatrix} 4 & 5 & 6 \\ 7 & 8 & 9 \end{bmatrix} = \begin{bmatrix} 7 & 8 & 9 \\ 29 & 34 & 39 \end{bmatrix},$$

since

$$(2 \cdot 4) + (3 \cdot 7) = 29,$$

$$(2 \cdot 5) + (3 \cdot 8) = 34,$$

and

$$(2 \cdot 6) + (3 \cdot 9) = 39.$$

Thus, the product is a $2 \times 3$ matrix.

***Example 7*** Is the multiplication

$$\begin{bmatrix} 4 & 5 & 6 \\ 7 & 8 & 9 \end{bmatrix} \cdot \begin{bmatrix} 0 & 1 \\ 2 & 3 \end{bmatrix}$$

possible? To determine this, we consider the top row of the left matrix and the first column of the right matrix:

$$\begin{bmatrix} 4 & 5 & 6 \\ \bullet & \bullet & \bullet \end{bmatrix} \cdot \begin{bmatrix} 0 & \bullet \\ 2 & \bullet \end{bmatrix}.$$

We multiply the 4 by 0 and the 5 by 2, but there is no entry with which to multiply the 6, since the columns in the right matrix have only two entries. In this case, *multiplication of the matrices is impossible.*

---

This example illustrates the basic limitation of matrix multiplication: Matrix multiplication is possible only if each row of the left matrix has exactly as many entries as each column of the right matrix. Since the number of entries in each row of the left matrix equals the number of columns of the left matrix, while the number of entries in each column of the right matrix equals the number of rows of the right matrix, it follows that *matrix multiplication is possible only if the number of columns in the left matrix equals the number of rows in the right matrix.* This is why

$$\begin{bmatrix} 0 & 1 \\ 2 & 3 \end{bmatrix} \cdot \begin{bmatrix} 4 & 5 & 6 \\ 7 & 8 & 9 \end{bmatrix} = \begin{bmatrix} 7 & 8 & 9 \\ 29 & 34 & 39 \end{bmatrix}, \quad \textbf{(4.5)}$$

while

$$\begin{bmatrix} 4 & 5 & 6 \\ 7 & 8 & 9 \end{bmatrix} \cdot \begin{bmatrix} 0 & 1 \\ 2 & 3 \end{bmatrix} \tag{4.6}$$

has no product. It is useful to write the sizes of the two matrices in question. In Eq. (4.5) we have

$$(2 \times 2) \cdot (2 \times 3),$$

same (the inner 2 and 2)

while in Eq. (4.6) we have

$$(2 \times 3) \cdot (2 \times 2).$$

different (the inner 3 and 2)

One further observation can be made from these examples: *The number of rows in the product matrix equals the number of rows in the left matrix, while the number of columns in the product matrix equals the number of columns in the right matrix.* This fact follows when we note that the location of each entry in the product matrix depends on the row in the left matrix and the column in the right matrix that are multiplied to obtain the entry. Thus, the product of two $2 \times 2$ matrices is a $2 \times 2$ matrix, while

$$(2 \times 2) \cdot (2 \times 3)$$

rows (the first 2) columns (the 3)

yields a $2 \times 3$ matrix (see Examples 5 and 6).

We may summarize these facts in the following formal definition of matrix multiplication.

**DEFINITION 4.1** **The product $AB$ of the $m \times n$ matrix $A$ and the $n \times p$ matrix $B$ is an $m \times p$ matrix. The entry in the $i$th row and $j$th column of $AB$ is found by summing the products of each consecutive entry in the $i$th row of $A$ by each consecutive entry in the $j$th column of $B$.**

**If we denote the $i$th row and $j$th column entry of the matrix $A$ by $a_{ij}$ and of the matrix $B$ by $b_{ij}$, then the entry in the $i$th row and the $j$th column of $AB$ is given by the formula**

$$a_{i1}b_{1j} + a_{i2}b_{2j} + a_{i3}b_{3j} + \cdots + a_{in}b_{nj}. \tag{4.7}$$

***Example 8*** Multiplying, we have

$$\begin{bmatrix} 2 & -3 & 4 & 1 \\ 5 & 0 & 2 & 6 \\ 1 & 3 & -4 & 2 \end{bmatrix} \cdot \begin{bmatrix} -7 & 6 \\ 2 & -1 \\ 0 & 3 \\ 4 & 5 \end{bmatrix} = \begin{bmatrix} -16 & 32 \\ -11 & 66 \\ 7 & 1 \end{bmatrix}.$$

***Example 9*** Multiplying, we have

$$\begin{bmatrix} 5 & 6 \\ 7 & 8 \end{bmatrix} \cdot \begin{bmatrix} 1 & 2 \\ 3 & 4 \end{bmatrix} = \begin{bmatrix} 23 & 34 \\ 31 & 46 \end{bmatrix}. \tag{4.8}$$

Observe that this answer is quite different from Eq. (4.4), in which the same matrices are multiplied in reverse order. Thus, *the order in which matrices are multiplied may affect the answer.* Indeed, as we have seen, it may even be impossible to multiply two matrices if their order is reversed (see Examples 6 and 7).

---

Mathematicians express the fact that $AB \neq BA$ by saying that matrix multiplication is *not commutative.* This general statement does not rule out those occasional instances in which the same product is obtained. Clearly, if $A$ and $B$ are both the same square matrix, we have

$$AB = AA = BA,$$

so that every square matrix commutes with itself. The following example defines a very special square matrix which commutes with every matrix having the same size.

***Example 10*** We shall denote by $I$ any $n \times n$ matrix having ones down the diagonal starting at the upper left corner and zeros everywhere else.

a) If $n = 2$, then $I$ is the matrix

$$\begin{bmatrix} 1 & 0 \\ 0 & 1 \end{bmatrix}.$$

Note that

$$\begin{bmatrix} 1 & 0 \\ 0 & 1 \end{bmatrix} \cdot \begin{bmatrix} a & b \\ c & d \end{bmatrix} = \begin{bmatrix} a & b \\ c & d \end{bmatrix},$$

regardless of what the entries $a$, $b$, $c$, and $d$ of the second matrix might be, and that

$$\begin{bmatrix} a & b \\ c & d \end{bmatrix} \cdot \begin{bmatrix} 1 & 0 \\ 0 & 1 \end{bmatrix} = \begin{bmatrix} a & b \\ c & d \end{bmatrix}.$$

Thus $IA = A = AI$, so that $I$ commutes with *any* $2 \times 2$ matrix $A$. Furthermore, multiplication by $I$ does *not* change $A$, so $I$ behaves very much like the number 1 does in ordinary multiplication. For this reason, $I$ is called the *multiplicative identity.*

b) If $n = 3$, then $I$ is the matrix

$$\begin{bmatrix} 1 & 0 & 0 \\ 0 & 1 & 0 \\ 0 & 0 & 1 \end{bmatrix}.$$

Note that

$$\begin{bmatrix} 1 & 0 & 0 \\ 0 & 1 & 0 \\ 0 & 0 & 1 \end{bmatrix} \cdot \begin{bmatrix} a & b & c \\ d & e & f \\ g & h & i \end{bmatrix} = \begin{bmatrix} a & b & c \\ d & e & f \\ g & h & i \end{bmatrix}$$

and that

$$\begin{bmatrix} a & b & c \\ d & e & f \\ g & h & i \end{bmatrix} \cdot \begin{bmatrix} 1 & 0 & 0 \\ 0 & 1 & 0 \\ 0 & 0 & 1 \end{bmatrix} = \begin{bmatrix} a & b & c \\ d & e & f \\ g & h & i \end{bmatrix}.$$

It should be apparent that $I$ will have the same properties for any value of the integer $n$.

---

## EXERCISES 4.1

*In Exercises 1–20, find $A + B$, $A + 3B$, $A - B$, $AB$, and $BA$ wherever possible. Do the matrices $A$ and $B$ commute?*

**1.** $A = \begin{bmatrix} 5 & 2 \\ -1 & 3 \end{bmatrix}$, $B = \begin{bmatrix} 4 & 6 \\ 0 & 7 \end{bmatrix}$

**2.** $A = \begin{bmatrix} 1 & 2 \\ 3 & 4 \end{bmatrix}$, $B = \begin{bmatrix} 4 & -2 \\ -3 & 1 \end{bmatrix}$

**3.** $A = \begin{bmatrix} 7 & 2 \\ 1 & 6 \end{bmatrix}$, $B = \begin{bmatrix} 6 & -2 \\ -1 & 7 \end{bmatrix}$

**4.** $A = \begin{bmatrix} 4 & 7 \\ -1 & 6 \end{bmatrix}$, $B = \begin{bmatrix} 0 & 2 \\ 9 & 8 \end{bmatrix}$

**5.** $A = \begin{bmatrix} 2 & 4 \\ 6 & 8 \end{bmatrix}$, $B = \begin{bmatrix} 1 & 3 & 5 \\ 7 & 0 & -1 \end{bmatrix}$

**6.** $A = (1 \quad 2 \quad 3)$, $B = (4 \quad 5 \quad 6)$

**7.** $A = (0 \quad 2 \quad 4)$, $B = (-1 \quad -3 \quad -5)$

**8.** $A = (1 \quad 5 \quad 9)$, $B = (2 \quad 4)$

**9.** $A = \begin{bmatrix} 1 & 2 & 3 \\ 0 & 4 & 6 \end{bmatrix}$, $B = (5 \; 8)$

**10.** $A = (1 \quad 2)$, $B = \begin{bmatrix} 3 \\ 4 \end{bmatrix}$

**11.** $A = (1 \quad 2 \quad 3)$, $B = \begin{bmatrix} 0 \\ -2 \\ 6 \end{bmatrix}$

**12.** $A = \begin{bmatrix} 1 & 2 & 3 \\ 0 & 4 & 6 \end{bmatrix}$, $B = \begin{bmatrix} 0 \\ 2 \\ -6 \end{bmatrix}$

**13.** $A = \begin{bmatrix} 5 & 4 & 8 \\ 1 & 2 & 3 \\ 0 & 9 & 6 \end{bmatrix}$, $B = \begin{bmatrix} 1 \\ 5 \\ 9 \end{bmatrix}$

**14.** $A = \begin{bmatrix} 3 & 7 & 9 \\ 2 & 5 & -6 \\ 1 & 0 & -4 \end{bmatrix}$, $B = \begin{bmatrix} 1 & 3 \\ 4 & -2 \\ 7 & 1 \end{bmatrix}$

**15.** $A = \begin{bmatrix} 1 & 4 & 7 \\ 2 & -6 & 5 \end{bmatrix}$, $B = \begin{bmatrix} 1 & -9 \\ 2 & 7 \\ -3 & 3 \end{bmatrix}$

**16.** $A = \begin{bmatrix} 1 & 3 \\ 4 & -2 \\ 7 & 1 \end{bmatrix}$, $B = \begin{bmatrix} 1 & -9 \\ 2 & 7 \\ -3 & 3 \end{bmatrix}$

**17.** $A = \begin{bmatrix} 1 & 2 & 3 \\ 4 & 5 & 6 \\ 7 & 8 & 9 \end{bmatrix}$, $B = \begin{bmatrix} 1 & 0 & 1 \\ 0 & 1 & 0 \\ 1 & 0 & 1 \end{bmatrix}$

**18.** $A = \begin{bmatrix} 1 & 7 & 4 \\ 2 & 8 & 5 \\ 3 & 9 & 6 \end{bmatrix}$, $B = \begin{bmatrix} 0 & 0 & -1 \\ 0 & -1 & 0 \\ -1 & 0 & 0 \end{bmatrix}$

**19.** $A = \begin{bmatrix} 1 & 2 & 3 \\ 0 & 1 & 0 \\ 4 & 5 & 6 \end{bmatrix}$, $B = \begin{bmatrix} 1 & 7 & 0 \\ 0 & 8 & 0 \\ 0 & 9 & 1 \end{bmatrix}$

**20.** $A = \begin{bmatrix} 1 & 2 & 3 \\ 4 & 5 & 6 \\ 7 & 9 & 8 \end{bmatrix}$, $B = \begin{bmatrix} -14 & 11 & -3 \\ 10 & -13 & 6 \\ 1 & 5 & -3 \end{bmatrix}$

## 4.2 WRITING A LINEAR SYSTEM IN MATRIX NOTATION

In this section we will reveal why matrices are multiplied in the peculiar fashion that we described in Section 4.1.

Consider the following linear system of the two equations in two unknowns $x$ and $y$:

$$\begin{aligned} 3x + 2y &= 6 \\ 4x - y &= 2. \end{aligned} \tag{4.9}$$

We can rewrite this system using matrices. We begin by forming the $2 \times 2$ matrix of coefficients of the variables on the left side of the linear system (4.9)

$$\begin{bmatrix} 3 & 2 \\ 4 & -1 \end{bmatrix}. \tag{4.10}$$

[Notice that the coefficient of the $y$-term in the second equation is $-1$, since $(-1)y = -y$.] Now if we multiply this matrix (4.10) by the $2 \times 1$ matrix (or vector) of variables

$$\begin{bmatrix} x \\ y \end{bmatrix}$$

we obtain

$$\begin{bmatrix} 3 & 2 \\ 4 & -1 \end{bmatrix} \cdot \begin{bmatrix} x \\ y \end{bmatrix} = \begin{bmatrix} 3x + 2y \\ 4x - y \end{bmatrix}. \tag{4.11}$$

The product matrix in Eq. (4.11) is a $2 \times 1$ matrix whose entries are the left sides of the two equations in (4.9). Equating the left sides of (4.9) with their right sides we have the *matrix equation*

$$\begin{bmatrix} 3 & 2 \\ 4 & -1 \end{bmatrix} \cdot \begin{bmatrix} x \\ y \end{bmatrix} = \begin{bmatrix} 6 \\ 2 \end{bmatrix}. \tag{4.12}$$

Observe that this matrix equation is equivalent to the linear system (4.9), since the left side is the $2 \times 1$ matrix consisting of the left sides of (4.9), while the right side of (4.12) is the $2 \times 1$ matrix formed from the right sides of (4.9).

***Example 1*** Write the system

$$\begin{aligned} x - \tfrac{5}{3}y &= 19/7 \\ -2x + y &= 0 \end{aligned} \tag{4.13}$$

in matrix notation.

Again we form the matrix of coefficients of the variables on the left side of (4.13), recalling that $x = (1)x$, so that we have

$$\begin{bmatrix} 1 & -\frac{5}{3} \\ -2 & 1 \end{bmatrix}. \tag{4.14}$$

Multiplying the matrix (4.14) by the vector of variables and equating the resulting $2 \times 1$ matrix with the right sides of (4.13), we have

$$\begin{bmatrix} 1 & -\frac{5}{3} \\ -2 & 1 \end{bmatrix} \cdot \begin{bmatrix} x \\ y \end{bmatrix} = \begin{bmatrix} 19/7 \\ 0 \end{bmatrix}. \tag{4.15}$$

The matrix equation (4.15) is equivalent to the system (4.13).

---

We may also express larger systems in matrix form.

***Example 2*** Write the system

$$\begin{aligned} 4x - 2y + z &= 9 \\ 3x + y - 2z &= 9 \\ x - 4y &= 2 \end{aligned} \tag{4.16}$$

in matrix notation.

Observe that the third equation in (4.16) has no $z$-term. We can insert a $z$-term since $0 \cdot z = 0$, so that the matrix of coefficients of the variables is

$$\begin{bmatrix} 4 & -2 & 1 \\ 3 & 1 & -2 \\ 1 & -4 & 0 \end{bmatrix}. \tag{4.17}$$

We now multiply the matrix (4.17) by the column vector of variables, obtaining

$$\begin{bmatrix} 4 & -2 & 1 \\ 3 & 1 & -2 \\ 1 & -4 & 0 \end{bmatrix} \cdot \begin{bmatrix} x \\ y \\ z \end{bmatrix} = \begin{bmatrix} 4x - 2y + z \\ 3x + y - 2z \\ x - 4y + 0z \end{bmatrix} = \begin{bmatrix} 9 \\ 9 \\ 2 \end{bmatrix}$$

when we equate the left and right sides of the system (4.16). Thus, (4.16) becomes

$$\begin{bmatrix} 4 & -2 & 1 \\ 3 & 1 & -2 \\ 1 & -4 & 0 \end{bmatrix} \cdot \begin{bmatrix} x \\ y \\ z \end{bmatrix} = \begin{bmatrix} 9 \\ 9 \\ 2 \end{bmatrix}.$$

---

In general, suppose we are given a system of $m$ equations in $n$ variables $x_1, x_2, \ldots, x_n$:

$$\begin{aligned} a_{11}x_1 + a_{12}x_2 + \cdots + a_{1n}x_n &= b_1, \\ a_{21}x_1 + a_{22}x_2 + \cdots + a_{2n}x_n &= b_2, \\ \vdots \qquad \vdots \qquad\qquad \vdots \quad &\quad \vdots \\ a_{m1}x_1 + a_{m2}x_2 + \cdots + a_{mn}x_n &= b_m, \end{aligned} \tag{4.18}$$

where the $a_{ij}$ and $b_j$ are constants. If we multiply the $m \times n$ matrix of coefficients $a_{ij}$ by the column vector of variables

$$\begin{bmatrix} a_{11} & a_{12} & \cdots & a_{1n} \\ a_{21} & a_{22} & \cdots & a_{2n} \\ \vdots & \vdots & & \vdots \\ a_{m1} & a_{m2} & \cdots & a_{mn} \end{bmatrix} \cdot \begin{bmatrix} x_1 \\ x_2 \\ \vdots \\ x_n \end{bmatrix},$$

we obtain the left side of the linear system (4.18). Thus (4.18) is equivalent to the matrix equation

$$\begin{bmatrix} a_{11} & a_{12} & \cdots & a_{1n} \\ a_{21} & a_{22} & \cdots & a_{2n} \\ \vdots & \vdots & & \vdots \\ a_{m1} & a_{m2} & \cdots & a_{mn} \end{bmatrix} \cdot \begin{bmatrix} x_1 \\ x_2 \\ \vdots \\ x_n \end{bmatrix} = \begin{bmatrix} b_1 \\ b_2 \\ \vdots \\ b_m \end{bmatrix}. \tag{4.19}$$

Note that it is not necessary for $m$ to equal $n$, since the number of equations need not equal the number of variables. *In what follows, however, we shall assume that* $m = n$.

If we denote the $n \times n$ matrix of coefficients by $A = (a_{ij})$, the column vector of variables by $X = (x_i)$, and the column vector of constants by $B = (b_j)$, we can abbreviate Eq. (4.19) by the equation

$$AX = B. \tag{4.20}$$

This matrix equation is very similar in appearance to the single variable equation

$$ax = b, \tag{4.21}$$

where $a$ and $b$ are constants and $x$ is an unknown. Equation (4.21) can easily be solved for $x$, if $a \neq 0$, by multiplying both sides by $a^{-1}$, since $a^{-1} \cdot a = 1$, so that

$$a^{-1}ax = a^{-1}b,$$

or

$$x = a^{-1}b.$$

In Sections 4.3 and 4.4, we will extend this idea to the matrix equation (4.20). We will determine conditions under which the matrix $A$ has a "reciprocal" matrix $A^{-1}$, called an *inverse*, satisfying the identity $AA^{-1} = A^{-1}A = I$, where $I$ is the $n \times n$ multiplicative identity described in Example 10 in Section 4.1. Should such an inverse exist, we can multiply both sides of the matrix equation (4.20) on the left by $A^{-1}$, obtaining

$$A^{-1}AX = A^{-1}B,$$

or

$$IX = A^{-1}B. \tag{4.22}$$

Since

$$\begin{bmatrix} 1 & 0 & 0 & \cdots & 0 \\ 0 & 1 & 0 & \cdots & 0 \\ 0 & 0 & 1 & \cdots & 0 \\ \cdot & \cdot & \cdot & & \cdot \\ \cdot & \cdot & \cdot & & \cdot \\ \cdot & \cdot & \cdot & & \cdot \\ 0 & 0 & 0 & \cdots & 1 \end{bmatrix} \cdot \begin{bmatrix} x_1 \\ x_2 \\ x_3 \\ \cdot \\ \cdot \\ \cdot \\ x_n \end{bmatrix} = \begin{bmatrix} x_1 \\ x_2 \\ x_3 \\ \cdot \\ \cdot \\ \cdot \\ x_n \end{bmatrix} = X,$$

Eq. (4.22) becomes

$$X = A^{-1}B. \tag{4.23}$$

Thus, the vector of unknowns is equal to the vector of constants obtained by multiplying $A^{-1}$ by $B$. *This is the solution of the original system of linear equations* (4.19). Clearly, this procedure gives us a powerful tool for solving systems of linear equations.

### EXERCISES 4.2

*In Exercises 1–10, write the given system of equations in matrix notation.*

1. $\begin{aligned} 3x - y &= 1 \\ 2x + 3y &= 8 \end{aligned}$

2. $\begin{aligned} 4x + y &= -5 \\ -x + 3y &= 11 \end{aligned}$

3. $\begin{aligned} 3x - 5y &= 12 \\ x + 4y &= 4 \end{aligned}$

4. $\begin{aligned} 3x - y &= 2 \\ 2x + 4y &= -8 \end{aligned}$

5. $\begin{aligned} x + y &= 2 \\ x - y &= 1 \end{aligned}$

6. $\begin{aligned} 2x - 3y &= 1 \\ 6x + 6y &= -7 \end{aligned}$

7. $\begin{aligned} x + y + z &= 6 \\ 2x + y - z &= 3 \\ 3x - y + 2z &= 4 \end{aligned}$

8. $\begin{aligned} x + y + z &= 6 \\ 2x + y &= 3 \\ -y + 2z &= 4 \end{aligned}$

9. $\begin{aligned} x + y + z &= 0 \\ 2x - z &= 3 \\ 3x - y &= 3 \end{aligned}$

10. $\begin{aligned} 2x + y + 3z &= 8 \\ 4x - 8y + 3z &= 9 \\ -6x + 7y - 6z &= 1 \end{aligned}$

*In Exercises 11–20, find the linear system of equations associated with the given matrix equation. (Hint: Multiply.)*

11. $\begin{bmatrix} 4 & -2 \\ -2 & 1 \end{bmatrix}\begin{bmatrix} x \\ y \end{bmatrix} = \begin{bmatrix} 8 \\ 4 \end{bmatrix}$

12. $\begin{bmatrix} -3 & 2 \\ 6 & -4 \end{bmatrix}\begin{bmatrix} x \\ y \end{bmatrix} = \begin{bmatrix} 1 \\ -2 \end{bmatrix}$

13. $\begin{bmatrix} 6 & -15 \\ -2 & 5 \end{bmatrix}\begin{bmatrix} x \\ y \end{bmatrix} = \begin{bmatrix} 1 \\ 2 \end{bmatrix}$

14. $\begin{bmatrix} 0 & 1 \\ -1 & 3 \end{bmatrix}\begin{bmatrix} x \\ y \end{bmatrix} = \begin{bmatrix} 4 \\ 3 \end{bmatrix}$

15. $\begin{bmatrix} 2 & 7 \\ -3 & 0 \end{bmatrix}\begin{bmatrix} x \\ y \end{bmatrix} = \begin{bmatrix} 0 \\ 0 \end{bmatrix}$

16. $\begin{bmatrix} 9 & -8 \\ -7 & 6 \end{bmatrix}\begin{bmatrix} x \\ y \end{bmatrix} = \begin{bmatrix} 5 \\ 4 \end{bmatrix}$

17. $\begin{bmatrix} 1 & 1 & 1 \\ 2 & 1 & -1 \\ -4 & -2 & 2 \end{bmatrix}\begin{bmatrix} x \\ y \\ z \end{bmatrix} = \begin{bmatrix} 6 \\ 3 \\ 4 \end{bmatrix}$

18. $\begin{bmatrix} 2 & 1 & 3 \\ 4 & -2 & 3 \\ -6 & 7 & -6 \end{bmatrix}\begin{bmatrix} x \\ y \\ z \end{bmatrix} = \begin{bmatrix} 8 \\ 3 \\ 1 \end{bmatrix}$

19. $\begin{bmatrix} 2 & 1 & 3 \\ 0 & -8 & 3 \\ -6 & 7 & 0 \end{bmatrix}\begin{bmatrix} x \\ y \\ z \end{bmatrix} = \begin{bmatrix} 8 \\ -9 \\ 1 \end{bmatrix}$

20. $\begin{bmatrix} 1 & 0 & 1 & 0 \\ 0 & -1 & 2 & -3 \\ 1 & 0 & 0 & -4 \\ 0 & 7 & 0 & 6 \end{bmatrix}\begin{bmatrix} x \\ y \\ z \\ w \end{bmatrix} = \begin{bmatrix} 1 \\ 2 \\ -3 \\ -4 \end{bmatrix}$

21. Show that the matrices

$$\begin{bmatrix} 4 & 5 \\ 3 & 4 \end{bmatrix} \text{ and } \begin{bmatrix} 4 & -5 \\ -3 & 4 \end{bmatrix}$$

are inverses, and use this fact to solve the system of equations

$$4x + 5y = 1$$
$$3x + 4y = 2.$$

**22.** Show that the matrices

$$\begin{bmatrix} 3 & -7 \\ 4 & -9 \end{bmatrix} \quad \text{and} \quad \begin{bmatrix} -9 & 7 \\ -4 & 3 \end{bmatrix}$$

are inverses, and use this fact to solve the system of equations

$$\begin{aligned} 3x - 7y &= 4 \\ 4x - 9y &= 6. \end{aligned}$$

---

## 4.3 THE AUGMENTED MATRIX PROCEDURE

The augmented matrix procedure is used to keep track of the steps involved in simultaneously solving a system of linear equations. We will show in Section 4.4 that this procedure can be extended into a tool for finding the inverse of a matrix if an inverse exists. Before describing this procedure we need to make these concepts more precise.

**DEFINITION 4.2** **A square matrix $A$ is *invertible* if we can find a square matrix $B$ such that**

$$AB = I = BA. \tag{4.24}$$

Now suppose that the matrix $A$ is invertible. Is it possible to find two different matrices $B_1$ and $B_2$ *both satisfying* Eq. (4.24)? To see that this is impossible, observe that by (4.24)

$$B_1 = B_1I = B_1(AB_2) = (B_1A)B_2 = IB_2 = B_2,$$

implying that the two matrices $B_1$ and $B_2$ are identical. Thus only one matrix can exist that satisfies Eq. (4.24). Such a matrix is called the *inverse* of $A$ and is denoted by $A^{-1}$. Hence, if $A$ is invertible we have

$$AA^{-1} = I = A^{-1}A. \tag{4.25}$$

We should raise the question as to whether all square matrices are invertible. Recall Example 3 in Section 2.3, where we saw that the system

$$\begin{aligned} 4x - 2y &= 6 \\ -6x + 3y &= 8 \end{aligned} \tag{4.26}$$

has no solution. Writing this system in matrix notation

$$AX = \begin{bmatrix} 4 & -2 \\ -6 & 3 \end{bmatrix}\begin{bmatrix} x \\ y \end{bmatrix} = \begin{bmatrix} 6 \\ 8 \end{bmatrix} = B, \tag{4.27}$$

we see that the matrix of coefficients

$$A = \begin{bmatrix} 4 & -2 \\ -6 & 3 \end{bmatrix}$$

cannot have an inverse, because if it did we could multiply both sides of Eq. (4.27) by that inverse and obtain a solution for the system (4.27):

$$X = A^{-1}AX = A^{-1}B. \tag{4.28}$$

Since the system (4.27) has *no* solution, $A$ cannot be invertible. Thus, *not all square matrices are invertible.*

In order to motivate and explain the augmented matrix procedure, we shall review the method of solving a system of linear equations simultaneously, which we used in Section 2.3. As we solve the system of equations simultaneously, we make a point of observing the effect of each step on the system *when it is written in matrix notation.* Recall the system given in Example 1 in Section 2.3:

$$\begin{aligned} 3x + 2y &= 6 \\ 4x - \phantom{2}y &= 2 \end{aligned} \quad \text{or} \quad \begin{bmatrix} 3 & 2 \\ 4 & -1 \end{bmatrix} \begin{bmatrix} x \\ y \end{bmatrix} = \begin{bmatrix} 6 \\ 2 \end{bmatrix}. \tag{4.29}$$

In that example we multiplied each term in the second equation by 2, obtaining

$$\begin{aligned} 3x + 2y &= 6 \\ 8x - 2y &= 4 \end{aligned} \quad \text{or} \quad \begin{bmatrix} 3 & 2 \\ 8 & -2 \end{bmatrix} \begin{bmatrix} x \\ y \end{bmatrix} = \begin{bmatrix} 6 \\ 4 \end{bmatrix}.$$

Note that this corresponds to multiplying the bottom row of the matrix of coefficients $A$ and of the vector of constants $B$ by 2, but leaving the vector of unknowns alone. We then added the top equation to the bottom equation, getting $11x = 10$. We will also do this here, but we will keep the second equation:

$$\begin{aligned} 11x \phantom{- 2y} &= 10 \\ 8x - 2y &= 4 \end{aligned} \quad \text{or} \quad \begin{bmatrix} 11 & 0 \\ 8 & -2 \end{bmatrix} \begin{bmatrix} x \\ y \end{bmatrix} = \begin{bmatrix} 10 \\ 4 \end{bmatrix}.$$

Observe that we could have obtained the new matrix of coefficients $A$ and vector of constants $B$ by adding the bottom rows to the top rows of the previous matrix of coefficients $A$ and vector of constants $B$. Again the vector of unknowns is unchanged. The next step in the process is to divide the first equation by 11, obtaining the value of $x$:

$$\begin{aligned} x \phantom{- 2y} &= 10/11 \\ 8x - 2y &= 4 \end{aligned} \quad \text{or} \quad \begin{bmatrix} 1 & 0 \\ 8 & -2 \end{bmatrix} \begin{bmatrix} x \\ y \end{bmatrix} = \begin{bmatrix} 10/11 \\ 4 \end{bmatrix},$$

which amounts to dividing all entries in the top row of the $A$ and $B$ matrices by 11. Again the vector of unknowns $X$ is not modified. We next subtract eight times the top equation from the bottom, obtaining

$$\begin{aligned} x &= 10/11 \\ -2y &= -36/11 \end{aligned} \quad \text{or} \quad \begin{bmatrix} 1 & 0 \\ 0 & -2 \end{bmatrix} \begin{bmatrix} x \\ y \end{bmatrix} = \begin{bmatrix} 10/11 \\ -36/11 \end{bmatrix},$$

which corresponds to subtracting eight times the top rows of $A$ and $B$ from the bottom rows, and leaving $X$ alone. Finally, we divide the bottom equation by $-2$ to get the solution of Eq. (4.29)

$$\begin{aligned} x &= 10/11 \\ y &= 18/11 \end{aligned} \quad \text{or} \quad \begin{bmatrix} x \\ y \end{bmatrix} = \begin{bmatrix} 1 & 0 \\ 0 & 1 \end{bmatrix} \begin{bmatrix} x \\ y \end{bmatrix} = \begin{bmatrix} 10/11 \\ 18/11 \end{bmatrix},$$

which corresponds to dividing the bottom entries of $A$ and $B$ by $-2$.

From what we have done above, it is clear that we could keep track of every step in the process by ignoring the vector $X$ and merely keeping track of the coefficients in the $A$ and $B$ matrices. To do this we construct the *augmented matrix* $A \mid B$ obtained by adjoining the initial vector of constants $B$ on the right of the original matrix $A$:

$$A \mid B = \left[\begin{array}{cc|c} 3 & 2 & 6 \\ 4 & -1 & 2 \end{array}\right].$$

Add twice the bottom row to the top row to obtain

$$\left[\begin{array}{cc|c} 11 & 0 & 10 \\ 4 & -1 & 2 \end{array}\right].$$

Next, divide the top row by 11

$$\left[\begin{array}{cc|c} 1 & 0 & 10/11 \\ 4 & -1 & 2 \end{array}\right],$$

and subtract four times the top row from the bottom row

$$\left[\begin{array}{cc|c} 1 & 0 & 10/11 \\ 0 & -1 & -18/11 \end{array}\right].$$

Finally, multiply the bottom row by $-1$ to obtain

$$\left[\begin{array}{cc|c} 1 & 0 & 10/11 \\ 0 & 1 & 18/11 \end{array}\right].$$

The final augmented matrix is equivalent to the matrix equation

$$\begin{bmatrix} x \\ y \end{bmatrix} = \begin{bmatrix} 1 & 0 \\ 0 & 1 \end{bmatrix} \begin{bmatrix} x \\ y \end{bmatrix} = \begin{bmatrix} 10/11 \\ 18/11 \end{bmatrix},$$

implying that $x = 10/11$, $y = 18/11$ is the solution to the system (4.29).

The method we have used above is called the *augmented matrix procedure.* It consists of changing the invertible matrix $A$, of the augmented matrix $A \mid B$, into an identity matrix $I$ by using the following three operations.

1. Interchanging two rows of the augmented matrix.
2. Multiplying a row of $A \mid B$ by a nonzero constant.
3. Adding a multiple of one row of $A \mid B$ to another.

To understand why the procedure works, note that if we knew the inverse and multiplied the augmented matrix $A \mid B$ on the left by $A^{-1}$, we would obtain

$$\underbrace{A^{-1}}_{n\times n}(\underbrace{A}_{n\times n} \mid \underbrace{B}_{n\times 1}) = \underbrace{A^{-1}A}_{n\times n} \mid \underbrace{A^{-1}B}_{n\times 1} = \underbrace{I}_{n\times n} \mid \underbrace{A^{-1}B}_{n\times 1},$$

with the last column of the resulting augmented matrix equal to the solution vector $X$ (see Eq. 4.28). Matrix multiplication on the left is equivalent to the three row operations mentioned above (see Exercise 17 on p. 169). If the matrix $A$ is not invertible it will be impossible to convert $A$ into an identity matrix by using the three row operations. The following three examples illustrate the use of the procedure. As the size of the system increases, so do the savings in time and effort.

***Example 1*** Solve the system of equations (see Example 4 in Section 2.3.)

$$\begin{aligned} 4x - 2y + z &= 9 \\ 3x + y - 2z &= 9 \\ x - 4y + 5z &= 0. \end{aligned} \tag{4.30}$$

The augmented matrix is

$$\left[\begin{array}{rrr|r} 4 & -2 & 1 & 9 \\ 3 & 1 & -2 & 9 \\ 1 & -4 & 5 & 0 \end{array}\right].$$

Rather than divide the first row by 4 to get a 1 in the top left corner, we can interchange the first and third rows (by the first row operation). This does not affect the outcome, since it is equivalent to writing the system of Eqs. (4.30) in reverse order:

$$\left[\begin{array}{rrr|r} 1 & -4 & 5 & 0 \\ 3 & 1 & -2 & 9 \\ 4 & -2 & 1 & 9 \end{array}\right].$$

Now subtract three times the first row from the second and four times the first from the third, eliminating the other entries in the first column:

$$\left[\begin{array}{rrr|r} 1 & -4 & 5 & 0 \\ 0 & 13 & -17 & 9 \\ 0 & 14 & -19 & 9 \end{array}\right].$$

Dividing the second row by 13 and eliminating the other entries in the second column by adding appropriate multiples of the new second row, we obtain

$$\left[\begin{array}{rrr|r} 1 & 0 & -3/13 & 36/13 \\ 0 & 1 & -17/13 & 9/13 \\ 0 & 0 & -9/13 & -9/13 \end{array}\right].$$

Multiplying the third row by $(-13/9)$ and eliminating the remaining entries in the third column yields

$$\left[\begin{array}{rrr|r} 1 & 0 & 0 & 3 \\ 0 & 1 & 0 & 2 \\ 0 & 0 & 1 & 1 \end{array}\right].$$

This augmented matrix is equivalent to the matrix equation

$$\begin{bmatrix} x \\ y \\ z \end{bmatrix} = \begin{bmatrix} 1 & 0 & 0 \\ 0 & 1 & 0 \\ 0 & 0 & 1 \end{bmatrix} \begin{bmatrix} x \\ y \\ z \end{bmatrix} = \begin{bmatrix} 3 \\ 2 \\ 1 \end{bmatrix},$$

implying that $x = 3$, $y = 2$, $z = 1$ is the solution of the system (4.30).

---

What happens if the system has no solution or infinitely many solutions, as in Examples 2 and 3 in Section 2.3?

***Example 2*** The system in Example 2 in Section 2.3 reduces to the augmented matrix

$$\left[\begin{array}{rr|r} 3 & -9 & 6 \\ -1 & 3 & -2 \end{array}\right].$$

Dividing the first row by 3 and adding the first row to the second row, we get

$$\left[\begin{array}{rr|r} 1 & -3 & 2 \\ 0 & 0 & 0 \end{array}\right].$$

The equivalent matrix equation and system of equations are

$$\begin{bmatrix} 1 & -3 \\ 0 & 0 \end{bmatrix}\begin{bmatrix} x \\ y \end{bmatrix} = \begin{bmatrix} 2 \\ 0 \end{bmatrix} \quad \text{and} \quad \begin{aligned} x - 3y &= 2 \\ 0 &= 0 \end{aligned},$$

indicating that the equations in the original system were multiples of each other. Hence this system has every point on the line $x - 3y = 2$ as a solution.

***Example 3*** The system of equations in Example 3 in Section 2.3 is equivalent to the augmented matrix

$$\left[\begin{array}{rr|r} 4 & -2 & 6 \\ -6 & 3 & 8 \end{array}\right].$$

Multiplying the first line by $(-3/2)$ yields

$$\left[\begin{array}{rr|r} -6 & 3 & -9 \\ -6 & 3 & 8 \end{array}\right]$$

so that when the first row is subtracted from the second, we have the augmented matrix

$$\left[\begin{array}{rr|r} -6 & 3 & -9 \\ 0 & 0 & -17 \end{array}\right].$$

But the equivalent matrix equation and system of equations are

$$\begin{bmatrix} -6 & 3 \\ 0 & 0 \end{bmatrix}\begin{bmatrix} x \\ y \end{bmatrix} = \begin{bmatrix} -9 \\ 17 \end{bmatrix} \quad \text{and} \quad \begin{aligned} -6x + 3y &= -9 \\ 0 &= 17 \end{aligned},$$

and the second equation is false! Thus the two lines in this problem are parallel, and no solution exists.

---

In neither of these two examples were we able to transform the matrix of coefficients $A$ into an identity matrix $I$ by adding products of the rows together. The inability to convert $A \mid B$ into the augmented matrix $I \mid A^{-1}B$ signals the following two facts: (1) The matrix $A$ is not invertible; and (2) the system does not have a unique solution. Indeed, there is either no solution or infinitely many solutions. It is easy to distinguish between the two cases of no solution and infinitely many solutions by using the following rule.

> If any row of zeros on the left of the vertical bar has a nonzero constant on the right, there are no solutions.

## EXERCISES 4.3

*In Exercises 1–16, solve the given system of equations by using the augmented matrix procedure. If the procedure fails, determine whether there are infinitely many solutions or no solution.*

**1.** $\begin{aligned} x + 2y &= 5 \\ 5x - y &= 3 \end{aligned}$

**2.** $\begin{aligned} x + 2y &= 8 \\ 5x - y &= 7 \end{aligned}$

**3.** $\begin{aligned} 2x + 3y &= 7 \\ 4x + y &= 19 \end{aligned}$

**4.** $\begin{aligned} 2x + 7y &= -6 \\ x + 3y &= -2 \end{aligned}$

**5.** $\begin{aligned} 4x + 3y &= 3 \\ -x + 2y &= -9 \end{aligned}$

**6.** $\begin{aligned} -2x - 3y &= 4 \\ 3x + 2y &= 9 \end{aligned}$

**7.** $\begin{aligned} 7x - 3y &= -9 \\ 6x + y &= 3 \end{aligned}$

**8.** $\begin{aligned} -5x + y &= 3 \\ 17x - 2y &= 8 \end{aligned}$

**9.** $\begin{aligned} 2x + 3y + z &= 1 \\ x - y - 2z &= 3 \\ 3x - 2y - 3z &= 2 \end{aligned}$

**10.** $\begin{aligned} x + 3y + 2z &= 2 \\ 2x + 6y - 4z &= 2 \\ 5x - 3y - 2z &= 1 \end{aligned}$

**11.** $\begin{aligned} 2x - 3y + z &= 14 \\ 4x - 2y - 2z &= 4 \\ -3x - y + 3z &= -1 \end{aligned}$

**12.** $\begin{aligned} 4x - 2y - 2z &= 20 \\ -x + 3y - 2z &= 0 \\ 3x - 5y + 2z &= 8 \end{aligned}$

**13.** $\begin{aligned} 6x - 5y + 4z &= 8 \\ 3x + 10y - 2z &= 1/2 \\ x + y + z &= 137/60 \end{aligned}$

**14.** $\begin{aligned} x - y - 2z &= 3 \\ -x + 3y - 2z &= 1 \\ 2x + y - 10z &= 12 \end{aligned}$

**15.** $\begin{aligned} x - y - z &= 12 \\ -x + 2y - z &= 0 \\ 2x - 5y + 4z &= -12 \end{aligned}$

**16.** $\begin{aligned} w + x - 3y + 5z &= 7 \\ 2w - x + 4y - 2z &= -4 \\ 5w - 3x - 7y + 3z &= 6 \\ 2w - 7x - 20y + 16z &= 6 \end{aligned}$

**17.** Let $A$ be the $2 \times 2$ matrix

$$A = \begin{bmatrix} a & b \\ c & d \end{bmatrix}.$$

a) Find a $2 \times 2$ matrix $B$ such that the two rows of $A$ are interchanged when $A$ is multiplied by $B$ *on the left*.

b) Find a $2 \times 2$ matrix $C$ such that the entries in the first row of $CA$ are twice the corresponding entries in $A$, while the entries of the second row are one-third the corresponding entries in $A$.

c) Find a $2 \times 2$ matrix $D$ such that

$$DA = \begin{bmatrix} a & b \\ c + 2a & d + 2b \end{bmatrix},$$

that is, it adds a multiple of the first row of $A$ to the second.

**18.** Solve the system below simultaneously and using the augmented matrix procedure. Compare the two methods for speed and accuracy.

$$\begin{aligned} 3u + 2v + 4w - 3x + 7y - 2z &= 6 \\ 4u + 5v - w + 3x + 6y + z &= -10 \\ -2u + 3v - 5w + 2x - 9y + 3z &= 3 \\ 3u + 5v + 3w - 6x + 8y - 3z &= 12 \\ u - v + w + 2x - 4y + 2z &= -2 \\ 5u + 7v - w - 6x + 2y + z &= 0 \end{aligned}$$

---

## 4.4 MATRIX INVERSION BY GAUSSIAN ELIMINATION

In addition to being a most effective method for solving systems of linear equations, the augmented matrix procedure can be extended into an algorithm for determining inverses of invertible matrices. This algorithm is called *Gaussian elimination.* For simplicity, we will use $2 \times 2$ matrices, although the comments we make can be generalized to $n \times n$ matrices. Let

$$A = \begin{bmatrix} a & c \\ b & d \end{bmatrix}$$

be a matrix with known coefficients, for which we are seeking an inverse. If $A$ is invertible the inverse

$$A^{-1} = \begin{bmatrix} x & u \\ y & v \end{bmatrix}$$

will satisfy the equation $AA^{-1} = I$, or

$$\begin{bmatrix} a & c \\ b & d \end{bmatrix}\begin{bmatrix} x & u \\ y & v \end{bmatrix} = \begin{bmatrix} 1 & 0 \\ 0 & 1 \end{bmatrix}. \qquad \textbf{(4.31)}$$

This equation is equivalent to *two* systems of simultaneous equations

$$\begin{bmatrix} a & c \\ b & d \end{bmatrix}\begin{bmatrix} x \\ y \end{bmatrix} = \begin{bmatrix} 1 \\ 0 \end{bmatrix} \quad \text{and} \quad \begin{bmatrix} a & c \\ b & d \end{bmatrix}\begin{bmatrix} u \\ v \end{bmatrix} = \begin{bmatrix} 0 \\ 1 \end{bmatrix}, \qquad \textbf{(4.32)}$$

since the first column of the identity matrix in Eq. (4.31) is obtained by multiplying the matrix $A$ by the first column of $A^{-1}$ and the second column of $I$ is the product of $A$ and the second column of $A^{-1}$. Applying the augmented matrix procedure to each system,

$$\left[\begin{array}{cc|c} a & c & 1 \\ b & d & 0 \end{array}\right] \quad \text{and} \quad \left[\begin{array}{cc|c} a & c & 0 \\ b & d & 1 \end{array}\right],$$

we attempt to transform the matrix $A$ into an identity matrix $I$ by adding products of the rows together. If we succeed, the value we obtain to the right of the vertical bars must be the columns of the inverse $A^{-1}$, since they are the solutions of the two systems (4.32):

$$\left[\begin{array}{cc|c} 1 & 0 & x \\ 0 & 1 & y \end{array}\right] \quad \text{and} \quad \left[\begin{array}{cc|c} 1 & 0 & u \\ 0 & 1 & v \end{array}\right].$$

The steps used to transform $A$ into $I$ by operating on the rows *will be identical for both augmented matrices*, so we will save steps if we do both procedures *at the same time*. This can be accomplished by considering only *one* augmented matrix at the start

$$A \mid I = \left[\begin{array}{cc|cc} a & c & 1 & 0 \\ b & d & 0 & 1 \end{array}\right]. \tag{4.33}$$

Operating on the rows of this augmented matrix, we attempt to transform the matrix $A$ into an identity matrix $I$. *If we succeed*, the entries in the matrix to the right of the vertical bar will be the inverse $A^{-1}$. Summarizing, we have the following general procedure.

**Gaussian Elimination**

To invert the $n \times n$ matrix $A$, construct the augmented matrix

$$A \mid I.$$

**1.** If it is possible to transform the entries left of the vertical line into an identity matrix by adding products of one row of $A \mid I$ to another, the resulting square matrix on the right of the vertical line is the inverse of $A$.

**2.** If at any step we obtain a row of zeros to the left of the vertical line, the matrix $A$ has no inverse.

Since the row operations correspond to multiplying $A \mid I$ by $A^{-1}$ on the left, we can express the Gaussian elimination procedure as

$$A \mid I \xrightarrow{\text{row operations}} I \mid A^{-1}.$$

The following three examples illustrate the process of Gaussian elimination.

***Example 1*** Invert the matrix

$$A = \begin{bmatrix} 4 & -2 & 1 \\ 3 & 1 & -2 \\ 1 & -4 & 5 \end{bmatrix}$$

using Gaussian elimination.

We begin by considering the augmented matrix

$$A \mid I = \left[\begin{array}{rrr|rrr} 4 & -2 & 1 & 1 & 0 & 0 \\ 3 & 1 & -2 & 0 & 1 & 0 \\ 1 & -4 & 5 & 0 & 0 & 1 \end{array}\right]. \tag{4.34}$$

We can now convert the first matrix into an identity matrix by adding and subtracting multiples of one row with another. *Whenever we make one of these row transformations on the first matrix, we make the same row transformation on the second matrix.* To begin, since we want the upper left entry to be a one, we divide the top rows by 4, obtaining

$$\left[\begin{array}{rrr|rrr} 1 & -1/2 & 1/4 & 1/4 & 0 & 0 \\ 3 & 1 & -2 & 0 & 1 & 0 \\ 1 & -4 & 5 & 0 & 0 & 1 \end{array}\right].$$

Now we subtract three times row 1 from row 2 to obtain a zero entry in the second row, first column:

$$\left[\begin{array}{rrr|rrr} 1 & -1/2 & 1/4 & 1/4 & 0 & 0 \\ 0 & 5/2 & -11/4 & -3/4 & 1 & 0 \\ 1 & -4 & 5 & 0 & 0 & 1 \end{array}\right].$$

We subtract row 1 from row 3, getting

$$\left[\begin{array}{rrr|rrr} 1 & -1/2 & 1/4 & 1/4 & 0 & 0 \\ 0 & 5/2 & -11/4 & -3/4 & 1 & 0 \\ 0 & -7/2 & 19/4 & -1/4 & 0 & 1 \end{array}\right].$$

Since the entry in the second row, second column must be a one, we multiply row 2 by 2/5 to get

$$\left[\begin{array}{rrr|rrr} 1 & -1/2 & 1/4 & 1/4 & 0 & 0 \\ 0 & 1 & -11/10 & -3/10 & 2/5 & 0 \\ 0 & -7/2 & 19/4 & -1/4 & 0 & 1 \end{array}\right].$$

The other two entries in the second column must be zero, so we add 1/2 of row 2 to row 1 and then add 7/2 of row 2 to row 3, getting

$$\left[\begin{array}{ccc|ccc} 1 & 0 & -3/10 & 1/10 & 1/5 & 0 \\ 0 & 1 & -11/10 & -3/10 & 2/5 & 0 \\ 0 & 0 & 9/10 & -13/10 & 7/5 & 1 \end{array}\right].$$

Multiplying row 3 by 10/9 yields an entry of one in the lower right corner:

$$\left[\begin{array}{ccc|ccc} 1 & 0 & -3/10 & 1/10 & 1/5 & 0 \\ 0 & 1 & -11/10 & -3/10 & 2/5 & 0 \\ 0 & 0 & 1 & -13/9 & 14/9 & 10/9 \end{array}\right].$$

Finally we add 3/10 of row 3 to row 1 and 11/10 of row 3 to row 2, to get zeros in the rest of column 3; we now have

$$\left[\begin{array}{ccc|ccc} 1 & 0 & 0 & -1/3 & 2/3 & 1/3 \\ 0 & 1 & 0 & -17/9 & 19/9 & 11/9 \\ 0 & 0 & 1 & -13/9 & 14/9 & 10/9 \end{array}\right].$$

The second of these matrices is $A^{-1}$. To check, multiply $A$ by $A^{-1}$.

***Example 2*** Find the inverse of

$$A = \begin{bmatrix} 1 & 3 & -2 & 4 \\ 7 & 6 & 0 & 5 \\ 3 & -2 & 0 & 7 \\ -1 & 2 & 5 & 1 \end{bmatrix}.$$

Arranging $A$ and $I$ side by side, we change the first-column entries of rows 2, 3, and 4 to zero by adding appropriate multiples of row 1 to them, obtaining

$$\left[\begin{array}{cccc|cccc} 1 & 3 & -2 & 4 & 1 & 0 & 0 & 0 \\ 0 & -15 & 14 & -23 & -7 & 1 & 0 & 0 \\ 0 & -11 & 6 & -5 & -3 & 0 & 1 & 0 \\ 0 & 5 & 3 & 5 & 1 & 0 & 0 & 1 \end{array}\right].$$

Now we divide row 2 by $-15$ and set the remaining entries in column 2 equal to zero by adding appropriate multiples of row 2:

$$\left[\begin{array}{cccc|cccc} 1 & 0 & 4/5 & -3/5 & -2/5 & 1/5 & 0 & 0 \\ 0 & 1 & -14/15 & 23/15 & 7/15 & -1/15 & 0 & 0 \\ 0 & 0 & -64/15 & 178/15 & 32/15 & -11/15 & 1 & 0 \\ 0 & 0 & 23/3 & -8/3 & -4/3 & 1/3 & 0 & 1 \end{array}\right].$$

We multiply row 3 by −15/64 and eliminate the remaining entries in column 3, obtaining

$$\left[\begin{array}{cccc|cccc} 1 & 0 & 0 & 13/8 & 0 & 1/16 & 3/16 & 0 \\ 0 & 1 & 0 & -17/16 & 0 & 3/32 & -7/32 & 0 \\ 0 & 0 & 1 & -89/32 & -1/2 & 11/64 & -15/64 & 0 \\ 0 & 0 & 0 & 597/32 & 5/2 & -63/64 & 115/64 & 1 \end{array}\right].$$

Finally, we multiply row 4 by 32/597 and eliminate the remaining entries in column 4 to get

$$\left[\begin{array}{cccc|cccc} 1 & 0 & 0 & 0 & -130/597 & 59/398 & 37/1194 & -52/597 \\ 0 & 1 & 0 & 0 & 85/597 & 15/398 & -139/1194 & 34/597 \\ 0 & 0 & 1 & 0 & -76/597 & 5/199 & 20/597 & 89/597 \\ 0 & 0 & 0 & 1 & 80/597 & -21/398 & 115/1194 & 32/597 \end{array}\right].$$

Since 1194 is a common multiple of all the denominators in $A^{-1}$, we can rewrite $A^{-1}$ as

$$A^{-1} = (1/1194)\begin{bmatrix} -260 & 177 & 37 & -104 \\ 170 & 45 & -139 & 68 \\ -152 & 30 & 40 & 178 \\ 160 & -63 & 115 & 64 \end{bmatrix}.$$

---

We might wonder what happens when we apply Gaussian elimination to a matrix $A$ that does not have an inverse. The following example shows that in this case we are unable to reduce the first matrix in $A \mid I$ to the identity matrix $I$ by row transformations. Thus, *the technique will fail when no inverse exists.*

***Example 3*** Consider the matrix

$$A = \begin{bmatrix} 1 & 2 & 3 \\ 4 & 5 & 6 \\ 9 & 8 & 7 \end{bmatrix}. \tag{4.35}$$

We will apply the Gaussian elimination technique to $A \mid I$. First we eliminate the entries in the bottom two rows of the first column:

$$\left[\begin{array}{ccc|ccc} 1 & 2 & 3 & 1 & 0 & 0 \\ 0 & -3 & -6 & -4 & 1 & 0 \\ 0 & -10 & -20 & -9 & 0 & 1 \end{array}\right].$$

Dividing row 2 by −3 yields

$$\left[\begin{array}{rrr|rrr} 1 & 2 & 3 & 1 & 0 & 0 \\ 0 & 1 & 2 & 4/5 & -1/3 & 0 \\ 0 & -10 & -20 & -9 & 0 & 1 \end{array}\right].$$

Subtracting twice row 2 from row 1 and adding 10 times row 2 to row 3, we have

$$\left[\begin{array}{rrr|rrr} 1 & 0 & -1 & -5/3 & 2/3 & 0 \\ 0 & 1 & 2 & 4/3 & -1/3 & 0 \\ 0 & 0 & 0 & 13/3 & -10/3 & 1 \end{array}\right].$$

But now all the entries in row 3 of the first matrix are zero, so we cannot continue the process of Gaussian elimination. Since the Gaussian elimination procedure fails, *no inverse exists.*

---

### EXERCISES 4.4

*In Exercises 1–12, find the inverse (if any) of the given matrix by Gaussian elimination.*

**1.** $\begin{bmatrix} 5 & 4 \\ 4 & 3 \end{bmatrix}$

**2.** $\begin{bmatrix} 7 & 6 \\ 2 & 8 \end{bmatrix}$

**3.** $\begin{bmatrix} 1 & 0 & 2 \\ 0 & 3 & 4 \\ 5 & 6 & 0 \end{bmatrix}$

**4.** $\begin{bmatrix} 1 & 2 & 3 \\ 4 & 5 & 6 \\ 7 & 8 & 9 \end{bmatrix}$

**5.** $\begin{bmatrix} 3 & 4 & 5 \\ 2 & 1 & 6 \\ 9 & 8 & 7 \end{bmatrix}$

**6.** $\begin{bmatrix} 3 & 4 & -2 \\ 2 & -1 & 5 \\ 1 & 0 & 6 \end{bmatrix}$

**7.** $\begin{bmatrix} 7 & 0 & 3 \\ 2 & 6 & 1 \\ 4 & 9 & 5 \end{bmatrix}$

**8.** $\begin{bmatrix} -1 & -8 & 5 \\ 6 & -2 & 7 \\ 4 & -9 & -3 \end{bmatrix}$

**9.** $\begin{bmatrix} 3 & -1 & 3 & 1 \\ 3 & 1 & -1 & 3 \\ 1 & 1 & 3 & 3 \\ 0 & 3 & 1 & -1 \end{bmatrix}$

**10.** $\begin{bmatrix} -1 & -2 & 3 & 1 \\ -2 & 1 & 0 & 2 \\ 1 & -2 & 2 & 3 \\ 2 & -1 & -1 & 4 \end{bmatrix}$

**11.** $\begin{bmatrix} 1 & 5 & 5 & 4 \\ 6 & 1 & 4 & 8 \\ 6 & 3 & 2 & 8 \\ 3 & 7 & 7 & 2 \end{bmatrix}$

**12.** $\begin{bmatrix} 4 & 0 & -1 & -2 \\ 0 & 5 & 0 & -3 \\ 1 & 0 & 6 & 0 \\ 3 & 2 & 0 & 7 \end{bmatrix}$

*Solve Exercises 13–16 by using Gaussian elimination to invert the matrix of coefficients A.*

**13.**
$$\begin{aligned} x + y + z &= 1 \\ x + 2y - z &= 0 \\ -x + y - 2z &= 2 \end{aligned}$$

**14.**
$$\begin{aligned} x + y + z &= 6 \\ 2x + y - z &= 3 \\ 3x - y + 2z &= 4 \end{aligned}$$

**15.**
$$\begin{aligned} 4x \qquad - 2z \qquad &= 2 \\ -x + y + z \qquad &= 1 \\ - 2y - z + w &= 0 \\ x - 3y - 2z + 2w &= 2 \end{aligned}$$

**16.**
$$\begin{aligned} 3w - x + 3y + z &= 1 \\ 3w + x - y + 3z &= 2 \\ w + x + 3y + 3z &= 3 \\ 3x + y - z &= 4 \end{aligned}$$

**17.** Consider the matrix equations $AX = B_k$, where

$$A = \begin{bmatrix} 1 & 2 & 3 \\ 2 & 1 & 3 \\ 3 & 3 & 5 \end{bmatrix}$$

and

$$B_1 = \begin{bmatrix} 1 \\ 2 \\ 3 \end{bmatrix}, \quad B_2 = \begin{bmatrix} 4 \\ 5 \\ 6 \end{bmatrix}, \quad B_3 = \begin{bmatrix} 7 \\ 8 \\ 9 \end{bmatrix}, \quad B_4 = \begin{bmatrix} -1 \\ 0 \\ -2 \end{bmatrix}.$$

Use Gaussian elimination to find $A^{-1}$ and find the four sets of solutions corresponding to the vectors $B_1$, $B_2$, $B_3$, and $B_4$.

**18.** Repeat Exercise 17 with the matrix

$$A = \begin{bmatrix} 1 & 4 & 2 \\ 7 & 2 & 1 \\ 6 & -2 & 2 \end{bmatrix}$$

but use the same vectors $B_1$, $B_2$, $B_3$, and $B_4$.

**19.** Let $A$ be the $2 \times 2$ matrix

$$A = \begin{bmatrix} a & b \\ c & d \end{bmatrix}.$$

Define the *determinant* of $A$ by $\det A = ad - bc$. Prove that if the $\det A \neq 0$, then the inverse of $A$ is

$$A^{-1} = \frac{1}{\det A} \begin{bmatrix} d & -b \\ -c & a \end{bmatrix}.$$

**20.** Show that the matrix $A$ in Exercise 19 is not invertible when $\det A = 0$. (*Hint:* Multiply the top row of $A \mid I$ by $d$ and the bottom row by $b$.)

*Use the results of Exercises* 19 *and* 20 *to find the inverse (if any) of the following matrices.*

**21.** $\begin{bmatrix} 5 & 3 \\ 3 & 2 \end{bmatrix}$

**22.** $\begin{bmatrix} 2 & 1 \\ 7 & 4 \end{bmatrix}$

**23.** $\begin{bmatrix} 4 & -2 \\ -2 & 1 \end{bmatrix}$

**24.** $\begin{bmatrix} 2 & 3 \\ 3 & 4 \end{bmatrix}$

**25.** $\begin{bmatrix} 2 & -3 \\ 3 & 4 \end{bmatrix}$

**26.** $\begin{bmatrix} 6 & -2 \\ -9 & 3 \end{bmatrix}$

**27.** $\begin{bmatrix} 4 & 8 \\ -3 & -6 \end{bmatrix}$

**28.** $\begin{bmatrix} 15 & 3 \\ 4 & 1 \end{bmatrix}$

**29.** $\begin{bmatrix} -7 & 5 \\ 8 & 6 \end{bmatrix}$

**30.** $\begin{bmatrix} 5 & -6 \\ -1 & 1.2 \end{bmatrix}$

**31.** $\begin{bmatrix} 0 & 2 \\ -1 & 3 \end{bmatrix}$

**32.** $\begin{bmatrix} 1.2 & 0.4 \\ 0.6 & 0.2 \end{bmatrix}$

**33.** $\begin{bmatrix} 2/3 & 1/12 \\ 6 & 3/4 \end{bmatrix}$

**34.** $\begin{bmatrix} 2/3 & 1/12 \\ 6/5 & 3/4 \end{bmatrix}$

**35.** $\begin{bmatrix} 1.2 & 1.0 \\ 2.1 & 0.4 \end{bmatrix}$

**36.** $\begin{bmatrix} 3.7 & 4.2 \\ -1.0 & 1.6 \end{bmatrix}$

**37.** Show that if $A$ and $B$ are invertible matrices of the same size, then $AB$ is invertible. (*Hint:* $(AB)^{-1} = B^{-1}A^{-1}$.)

**38.** Define the determinant of the $3 \times 3$ matrix

$$A = \begin{bmatrix} a & b & c \\ d & e & f \\ g & h & i \end{bmatrix}$$

by $\det A = (aei + bfg + cdh) - (gec + hfa - idb)$. Show that $A$ is not invertible if $\det A = 0$. (*Hint:* Eliminate the entries in the $d$, $g$, and $h$ positions.)

---

## 4.5 SOME APPLICATIONS OF MATRICES

We will present in this section some typical applications of matrices in other disciplines and in business. The examples we will consider are highly simplified, since real-world situations typically involve dozens of conditions and produce large matrices that can be efficiently handled only on a computer. Although our models are simplistic, they indicate the power and versatility of matrix techniques.

The first two examples illustrate the usefulness of matrix multiplication.

***Example 1*** In Example 4 in Section 4.1 we formed a matrix listing nonstop air service by Frontier Airlines to five cities in the central and western United States. Nonstop service between two cities was indicated by a 1, while all other service was denoted by a 0. In that example we obtained the matrix $A$ given by

| | | To | | | | |
|---|---|---|---|---|---|---|
| | | Bozeman | Denver | Missoula | Salt Lake | St. Louis |
| | Bozeman | 0 | 0 | 1 | 1 | 0 |
| | Denver | 0 | 0 | 0 | 1 | 1 |
| From | Missoula | 1 | 0 | 0 | 1 | 0 |
| | Salt Lake | 1 | 1 | 0 | 0 | 0 |
| | St. Louis | 0 | 1 | 0 | 0 | 0 |

Multiplying $A$ by itself we have

$$A^2 = \begin{bmatrix} 0 & 0 & 1 & 1 & 0 \\ 0 & 0 & 0 & 1 & 1 \\ 1 & 0 & 0 & 1 & 0 \\ 1 & 1 & 0 & 0 & 0 \\ 0 & 1 & 0 & 0 & 0 \end{bmatrix} \begin{bmatrix} 0 & 0 & 1 & 1 & 0 \\ 0 & 0 & 0 & 1 & 1 \\ 1 & 0 & 0 & 1 & 0 \\ 1 & 1 & 0 & 0 & 0 \\ 0 & 1 & 0 & 0 & 0 \end{bmatrix},$$

or

$$A^2 = \begin{bmatrix} 2 & 1 & 0 & 1 & 0 \\ 1 & 2 & 0 & 0 & 0 \\ 1 & 1 & 1 & 1 & 0 \\ 0 & 0 & 1 & 2 & 1 \\ 0 & 0 & 0 & 1 & 1 \end{bmatrix}.$$

We now wish to determine the significance of the entries in the matrix $A^2$. Again we associate the names of the cities of departure and arrival and consider the routes shown in Fig. 4.2.

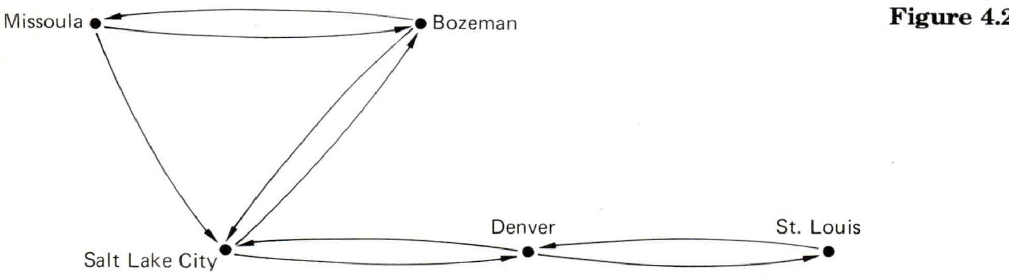


**Figure 4.2**

| | | To | | | | |
|---|---|---|---|---|---|---|
| | | Bozeman | Denver | Missoula | Salt Lake | St. Louis |
| | Bozeman | 2 | 1 | 0 | 1 | 0 |
| | Denver | 1 | 2 | 0 | 0 | 0 |
| From | Missoula | 1 | 1 | 1 | 1 | 0 |
| | Salt Lake | 0 | 0 | 1 | 2 | 1 |
| | St. Louis | 0 | 0 | 0 | 1 | 1 |

If we compare the routes shown in Fig. 4.2 with the entries in $A^2$, we soon discover that the entries in $A^2$ are simply the *number of one-stop flights between the two cities.* For example, there is only one one-stop flight from Bozeman to Denver (via Salt Lake), no one-stop flight from Bozeman to St. Louis, and two one-stop flights from Bozeman to itself! These latter flights are the round trips to Missoula and Salt Lake. The reader should check the remaining entries in the matrix $A^2$.

Multiplying $A^2$ by $A$, we obtain the matrix

$$A^3 = \begin{bmatrix} 0 & 0 & 1 & 1 & 0 \\ 0 & 0 & 0 & 1 & 1 \\ 1 & 0 & 0 & 1 & 0 \\ 1 & 1 & 0 & 0 & 0 \\ 0 & 1 & 0 & 0 & 0 \end{bmatrix} \begin{bmatrix} 2 & 1 & 0 & 1 & 0 \\ 1 & 2 & 0 & 0 & 0 \\ 1 & 1 & 1 & 1 & 0 \\ 0 & 0 & 1 & 2 & 1 \\ 0 & 0 & 0 & 1 & 1 \end{bmatrix}$$

$$= \begin{bmatrix} 1 & 1 & 2 & 3 & 1 \\ 0 & 0 & 1 & 3 & 2 \\ 2 & 1 & 1 & 3 & 1 \\ 3 & 3 & 0 & 1 & 0 \\ 1 & 2 & 0 & 0 & 0 \end{bmatrix},$$

whose entries are the number of two-stop flights connecting the two cities. For example, there is only one two-stop flight from Denver to Missoula, but there are three two-stop flights from Denver to Salt Lake: via St. Louis and Denver, via Salt Lake and Bozeman, and via Salt Lake and Denver.

While it is clear that many of these routes are ridiculous, it is also evident that matrix multiplication can be used to find the most direct route (the least number of stops) between two cities. Of course, in this simple example there is no need to use matrices, since a quick glance at Fig. 4.2 instantly reveals the most direct route. However, if we consider all the cities in the United States serviced by the major commercial airlines, it is evident that the matrix technique is useful in determining the most desirable routes.

---

Our next example illustrates a similar use of matrix multiplication in sociology.

***Example 2*** Consider the matrix that we found in Example 3 in Section 4.1 indicating the social interactions within a family.

| | Persons whose opinion is sought | | | | |
|---|---|---|---|---|---|
| | Father | Mother | Oldest son | Daughter | Youngest son |
| Father | 0 | 1 | 0 | 0 | 0 |
| Mother | 1 | 0 | 0 | 1 | 0 |
| Oldest son | 0 | 0 | 1 | 0 | 0 |
| Daughter | 1 | 1 | 0 | 0 | 0 |
| Youngest son | 0 | 1 | 1 | 0 | 0 |

The entries in this matrix correspond to *direct influences*. If we multiply the matrix by itself, we will obtain a matrix whose entries are those *indirect influences* involving one intermediary. This situation is exactly the same as the one-stop flights we considered in the previous example. Here,

$$A^2 = \begin{bmatrix} 0 & 1 & 0 & 0 & 0 \\ 1 & 0 & 0 & 1 & 0 \\ 0 & 0 & 1 & 0 & 0 \\ 1 & 1 & 0 & 0 & 0 \\ 0 & 1 & 1 & 0 & 0 \end{bmatrix} \begin{bmatrix} 0 & 1 & 0 & 0 & 0 \\ 1 & 0 & 0 & 1 & 0 \\ 0 & 0 & 1 & 0 & 0 \\ 1 & 1 & 0 & 0 & 0 \\ 0 & 1 & 1 & 0 & 0 \end{bmatrix} = \begin{bmatrix} 1 & 0 & 0 & 1 & 0 \\ 1 & 2 & 0 & 0 & 0 \\ 0 & 0 & 1 & 0 & 0 \\ 1 & 1 & 0 & 1 & 0 \\ 1 & 0 & 1 & 0 & 0 \end{bmatrix}.$$

Thus, the father in this family exerts a considerable indirect influence through the mother, even though only the mother and daughter consult him. Chains involving two intermediaries are obtained by multiplying $A^2$ by $A$, and so forth. We can assume that as the number of intermediaries increases, the indirect influence exerted diminishes. Thus, the sociologist may wish to assign different weights to the entries in each matrix before deciding which member exerts the most influence in the family.

---

These two examples of the use of matrix multiplication in transportation problems and group interactions are by no means the only uses for this technique. For example, we can also analyze family trees in trying to determine which heir has the strongest claim or the relative importance of various animal species in an ecosystem. Some examples of these situations are given in the exercises (see pp. 189–193).

***Example 3*** A cafeteria manager wishes to purchase the basic ingredients for the three specials of the day: chiliburgers, soft-shelled tacos, and sloppy joes. He generally serves 6 oz of the meat mixture on each bun or tortilla. The main ingredients in each mixture are the same, although the proportions differ. His recipe for chiliburgers requires 4 oz of cooked beans with 1 oz of meat and 1 oz of fresh tomato, the recipe for sloppy joes requires 4 oz of meat with 1 oz of cooked beans and 1 oz of fresh tomatoes, while tacos require 2 oz of each of the three ingredients. He estimates that he will sell 150 chiliburgers, 500 tacos, and 200 sloppy joes. Assuming there is no weight loss in the preparation of the mixtures, how much of each ingredient should he buy?

We can easily set this problem up in the notation of a matrix equation. First we form a matrix $A$ of the ingredients in each sandwich

| | Chiliburgers | Tacos | Sloppy joes |
|---|---|---|---|
| Beans | 4 | 2 | 1 |
| Meat | 1 | 2 | 4 |
| Tomatoes | 1 | 2 | 1 |

,

and we multiply this matrix by a column vector of the expected number of sales of each sandwich

$$\begin{matrix}\text{Chiliburgers} \\ \text{Tacos} \\ \text{Sloppy joes}\end{matrix} \quad \begin{bmatrix} 150 \\ 500 \\ 200 \end{bmatrix}.$$

Note that we have ordered the entries in this vector so that the *row* of each sandwich corresponds to the *column* of its ingredients. When we multiply we have

$$\begin{bmatrix} 4 & 2 & 1 \\ 1 & 2 & 4 \\ 1 & 2 & 1 \end{bmatrix}\begin{bmatrix} 150 \\ 500 \\ 200 \end{bmatrix} = \begin{bmatrix} 1800 \\ 1950 \\ 1350 \end{bmatrix}.$$

Thus, he must purchase 1800 oz (112.5 lb) of cooked beans, 1950 oz (121.88 lb) of meat, and 1350 oz (84.38 lb) of fresh tomatoes.

Suppose, on the other hand, that the cafeteria manager returns from vacation and discovers that his assistant has forgotten to make plans for the next day. The manager has 80 lb of cooked beans, 110 lb of meat, and 75 lb of fresh tomatoes on hand. How much of each of the three specials should he prepare to use up these ingredients?

In this situation we don't know the number of sandwiches, but the matrix of ingredients $A$ and the vector of total amounts of each ingredient are both known. We may describe the problem by the matrix equation

$$\begin{bmatrix} 4 & 2 & 1 \\ 1 & 2 & 4 \\ 1 & 2 & 1 \end{bmatrix}\begin{bmatrix} x \\ y \\ z \end{bmatrix} = \begin{bmatrix} 1280 \\ 1760 \\ 1200 \end{bmatrix}, \tag{4.36}$$

since $80 \times 16 = 1280$, $110 \times 16 = 1760$, and $75 \times 16 = 1200$. Our problem now is to find the unknown numbers $x$, $y$, and $z$. We may do so by inverting the matrix $A$ and multiplying each side of Eq. (4.36) on the left by $A^{-1}$. We use Gaussian elimination to obtain $A^{-1}$.

Dividing the first row of

$$\left[\begin{array}{ccc|ccc} 4 & 2 & 1 & 1 & 0 & 0 \\ 1 & 2 & 4 & 0 & 1 & 0 \\ 1 & 2 & 1 & 0 & 0 & 1 \end{array}\right]$$

by 4 and subtracting the result from the next two rows yields

$$\left[\begin{array}{ccc|ccc} 1 & 1/2 & 1/4 & 1/4 & 0 & 0 \\ 0 & 3/2 & 15/4 & -1/4 & 1 & 0 \\ 0 & 3/2 & 3/4 & -1/4 & 0 & 1 \end{array}\right].$$

We now multiply row 2 by 2/3 and eliminate the other entries in column 2 to obtain

$$\left[\begin{array}{ccc|ccc} 1 & 0 & -1 & 1/3 & -1/3 & 0 \\ 0 & 1 & 5/2 & -1/6 & 2/3 & 0 \\ 0 & 0 & -3 & 0 & -1 & 1 \end{array}\right].$$

Finally, we divide row 3 by $-3$ and eliminate the remaining entries in column 3 to obtain

$$\left[\begin{array}{ccc|ccc} 1 & 0 & 0 & 1/3 & 0 & -1/3 \\ 0 & 1 & 0 & -1/6 & -1/6 & 5/6 \\ 0 & 0 & 1 & 0 & 1/3 & -1/3 \end{array}\right].$$

This last matrix is $A^{-1}$, so we now multiply each side of Eq. (4.36) on the left by $A^{-1}$ and obtain

$$\begin{bmatrix} x \\ y \\ z \end{bmatrix} = \begin{bmatrix} 1/3 & 0 & -1/3 \\ -1/6 & -1/6 & 5/6 \\ 0 & 1/3 & -1/3 \end{bmatrix} \begin{bmatrix} 1280 \\ 1760 \\ 1200 \end{bmatrix} = \begin{bmatrix} 26\frac{2}{3} \\ 493\frac{1}{3} \\ 186\frac{2}{3} \end{bmatrix}.$$

Thus the manager should prepare enough for approximately 27 chiliburgers, 493 tacos, and 187 sloppy joes.

□ ***Example 4*** A fish-and-game commission is planning to stock four small lakes that have been completely fished out. They plan to introduce three species of fish in each lake: black bass, bullheads, and rainbow trout. The daily eating habits (in kilograms per kilogram of each species of fish) is given in the matrix $A$ below:

| | | Species of fish (per kilogram) | | |
|---|---|---|---|---|
| | | Black bass | Bullheads | Rainbows |
| Types of food (in kilograms) | Mosquito larvae | 0.064 | 0.082 | 0.314 |
| | Freshwater crustaceans | 0.141 | 0.163 | 0.082 |
| | Minnows and worms | 0.555 | 0.493 | 0.406 |

$= A$.

The daily production of such food (in grams) in each of the four lakes has been determined to be in excess of the following vectors:

$$L_1 = \begin{bmatrix} 3{,}000 \\ 800 \\ 3{,}950 \end{bmatrix}, \quad L_2 = \begin{bmatrix} 6{,}300 \\ 14{,}000 \\ 55{,}000 \end{bmatrix},$$

$$L_3 = \begin{bmatrix} 460 \\ 390 \\ 1{,}450 \end{bmatrix}, \quad L_4 = \begin{bmatrix} 1{,}000 \\ 500 \\ 5{,}000 \end{bmatrix}.$$

How many grams of each species of fish should they place in each lake?

Letting $x_k$, $y_k$, and $z_k$ denote the number of grams of black bass, bullheads, and rainbows in lake $k$ ($k = 1, 2, 3, 4$, respectively), we see that this problem reduces to solving the four matrix equations

$$AX_k = L_k, \quad k = 1, 2, 3, 4,$$

where

$$X_k = \begin{bmatrix} x_k \\ y_k \\ z_k \end{bmatrix}.$$

It is obvious that finding $A^{-1}$ will be of great help, since we can then solve each problem by simply multiplying $A^{-1}L_k$. We find $A^{-1}$ by Gaussian elimination. Dividing the first row of

$$\left[\begin{array}{ccc|ccc} 0.064 & 0.082 & 0.314 & 1 & 0 & 0 \\ 0.141 & 0.168 & 0.082 & 0 & 1 & 0 \\ 0.555 & 0.493 & 0.406 & 0 & 0 & 1 \end{array}\right].$$

by 0.064 and eliminating the remaining entries of column 1 yields

$$\left[\begin{array}{ccc|ccc} 1 & 1.281 & 4.906 & 15.625 & 0 & 0 \\ 0 & -0.013 & -0.610 & -2.203 & 1 & 0 \\ 0 & -0.218 & -2.317 & -8.672 & 0 & 1 \end{array}\right].$$

Now we divide row 2 by $-0.013$ and eliminate the rest of column 2, getting

$$\left[\begin{array}{ccc|ccc} 1 & 0 & -55.202 & -201.456 & 98.538 & 0 \\ 0 & 1 & 46.923 & 169.462 & -76.923 & 0 \\ 0 & 0 & 7.912 & 28.271 & -16.769 & 1 \end{array}\right].$$

Finally, we divide row 3 by 7.912 and clear the other entries in column 3 to get

$$\left[\begin{array}{ccc|ccc} 1 & 0 & 0 & -4.219 & -18.435 & 6.955 \\ 0 & 1 & 0 & 1.806 & 22.507 & -5.912 \\ 0 & 0 & 1 & 3.573 & -2.119 & 0.126 \end{array}\right].$$

Although round-off error prevents this last matrix from exactly equaling $A^{-1}$, it is still reasonably close, since

$$A^{-1}A = \begin{bmatrix} 0.99067 & -0.01422 & -0.01271 \\ 0.00791 & 1.01465 & 0.01239 \\ -0.00018 & -0.00089 & 0.99932 \end{bmatrix}.$$

Additional accuracy could have been obtained by carrying more significant digits in each calculation.

Finally, we multiply each of the vectors $L_k$ on the left by $A^{-1}$ and obtain

$$X_1 = A^{-1}L_1 = \begin{bmatrix} 67.25 \\ 71.2 \\ 9{,}521.5 \end{bmatrix}, \qquad X_2 = A^{-1}L_2 = \begin{bmatrix} 97{,}855.3 \\ 1{,}315.8 \\ 1{,}089.7 \end{bmatrix},$$

$$X_3 = A^{-1}L_3 = \begin{bmatrix} 954.36 \\ 1{,}036.09 \\ 999.87 \end{bmatrix}, \qquad X_4 = A^{-1}L_4 = \begin{bmatrix} 21{,}338.5 \\ -16{,}500.0 \\ 3{,}143.5 \end{bmatrix}.$$

These entries indicate the number of grams of live fish of the three species that the lakes can support. Note that lake 1 is ideal for rainbow trout, lake 2 is excellent for black bass but can also support small populations of bullheads and rainbows, while lake 3 can support approximately equal populations of the three species. The middle entry for lake 4 is rather mysterious, since we can't have a negative number of fish in the lake. Actually, this indicates that *the resources of the lake cannot be used in an optimal fashion*. Evidently lake 4 lacks enough of one of the favorite foods of bullheads. Most probably, lake 4 does not produce enough freshwater crustaceans. This being the case, stocking the lake with bullheads might have serious consequences: The bullheads will seriously deplete the lake's crustacean population, and will die if they cannot substitute some other food for the crustaceans. At any rate, the ecological balance in the lake will be seriously affected. In stocking lake 4, the best policy is to avoid bullheads entirely and determine a mix of black bass and rainbows that stays within the bounds of the daily food production. We will discuss this procedure in Section 5.3.

***Example 5*** Figure 4.3 illustrates a typical traffic situation common to many cities in the United States: Alternating streets are one-way in opposite directions.

The numbers at the ends of each street shown in Fig. 4.3 represent the average number of vehicles per hour entering or leaving these intersections during the peak traffic hours of 4:00 to 6:00 P.M.

Suppose that the water mains under Jefferson Street between Broadway and Main must be repaired. This will require closing the street to traffic. The city engineer would like to make these repairs with a minimum of resulting traffic congestion. She can reset the traffic signals at the six intersections to regulate the flow of traffic, and has determined that each street $x_1, \ldots, x_7$ can handle 800 VPH (vehicles per

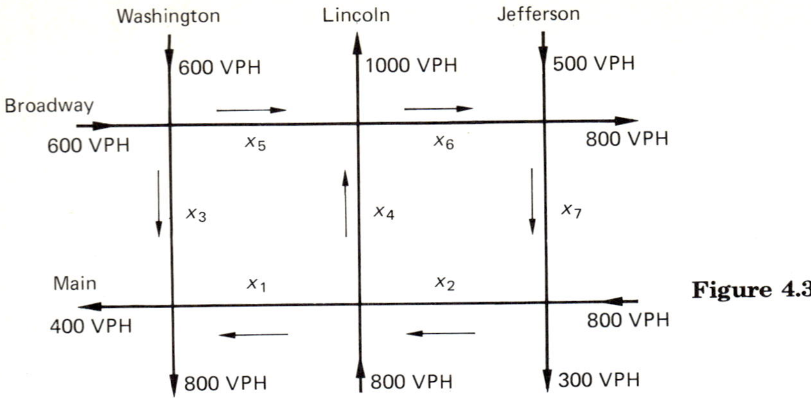


**Figure 4.3**

hour) without causing congestion. What traffic patterns will cause the least amount of vehicular delay?

Let $x_1, x_2, \ldots, x_7$ denote the number of vehicles per hour that will be handled on each of the seven block-long streets. The arrows indicate the direction of flow of the traffic. If there is no congestion, *all traffic entering an intersection must also leave that intersection.* This conservation of mass principle leads to six linear equations, one at each intersection:

$$\begin{aligned}
\textit{Traffic in} &= \textit{Traffic out} \\
x_1 + x_3 &= 400 + 800 \\
x_2 + 800 &= x_1 + x_4 \\
x_7 + 800 &= x_2 + 300 \\
600 + 600 &= x_3 + x_5 \\
x_4 + x_5 &= x_6 + 1000 \\
x_6 + 500 &= x_7 + 800 .
\end{aligned} \tag{4.37}$$

We may rewrite this system of equations (4.37) in a more convenient form by keeping all the unknowns on one side:

$$\begin{aligned}
x_1 \quad + x_3 \qquad\qquad\qquad &= 1200 \\
x_1 - x_2 \quad + x_4 \qquad\qquad &= 800 \\
x_2 \qquad\qquad\qquad - x_7 &= 500 \\
x_3 \quad + x_5 \qquad\quad &= 1200 \\
x_4 + x_5 - x_6 \quad &= 1000 \\
x_6 - x_7 &= 300.
\end{aligned} \tag{4.38}$$

The problem now is to determine values for the unknowns $x_1, \ldots, x_6$ so that when we set $x_7$ equal to 0 (closing the street so that no traffic may pass), none of the unknowns $x_1, \ldots, x_6$ exceeds 800.

We can rewrite the system (4.38) in the form $AX = B$, where $A$ is a $6 \times 7$ matrix of coefficients and $B$ is a $1 \times 6$ vector of numbers. Although $A$ is not square, we may apply the augmented matrix procedure to the arrays $A \mid B$:

$$\left[\begin{array}{rrrrrrr|r} 1 & 0 & 1 & 0 & 0 & 0 & 0 & 1200 \\ 1 & -1 & 0 & 1 & 0 & 0 & 0 & 800 \\ 0 & 1 & 0 & 0 & 0 & 0 & -1 & 500 \\ 0 & 0 & 1 & 0 & 1 & 0 & 0 & 1200 \\ 0 & 0 & 0 & 1 & 1 & -1 & 0 & 1000 \\ 0 & 0 & 0 & 0 & 0 & 1 & -1 & 300 \end{array}\right]. \tag{4.39}$$

We will try to get a diagonal of ones in the matrix. Subtracting row 1 from row 2 and adding the result to row 3 yields

$$\left[\begin{array}{rrrrrrr|r} 1 & 0 & 1 & 0 & 0 & 0 & 0 & 1200 \\ 0 & -1 & -1 & 1 & 0 & 0 & 0 & -400 \\ 0 & 0 & -1 & 1 & 0 & 0 & -1 & 100 \\ 0 & 0 & 1 & 0 & 1 & 0 & 0 & 1200 \\ 0 & 0 & 0 & 1 & 1 & -1 & 0 & 1000 \\ 0 & 0 & 0 & 0 & 0 & 1 & -1 & 300 \end{array}\right].$$

Now, we change sign in row 2 and add row 3 to row 4 to obtain

$$\left[\begin{array}{rrrrrrr|r} 1 & 0 & 1 & 0 & 0 & 0 & 0 & 1200 \\ 0 & 1 & 1 & -1 & 0 & 0 & 0 & 400 \\ 0 & 0 & -1 & 1 & 0 & 0 & -1 & 100 \\ 0 & 0 & 0 & 1 & 1 & 0 & -1 & 1300 \\ 0 & 0 & 0 & 1 & 1 & -1 & 0 & 1000 \\ 0 & 0 & 0 & 0 & 0 & 1 & -1 & 300 \end{array}\right].$$

At this point we change sign in row 3, subtract the result from row 2, and subtract row 4 from row 5, getting

$$\left[\begin{array}{rrrrrrr|r} 1 & 0 & 1 & 0 & 0 & 0 & 0 & 1200 \\ 0 & 1 & 0 & 0 & 0 & 0 & -1 & 500 \\ 0 & 0 & 1 & -1 & 0 & 0 & 1 & -100 \\ 0 & 0 & 0 & 1 & 1 & 0 & -1 & 1300 \\ 0 & 0 & 0 & 0 & 0 & -1 & 1 & -300 \\ 0 & 0 & 0 & 0 & 0 & 1 & -1 & 300 \end{array}\right].$$

Finally, we add row 4 to row 3, change the sign of row 5, and subtract row 5 from row 6, to get

$$\left[\begin{array}{ccccccc|c} 1 & 0 & 1 & 0 & 0 & 0 & 0 & 1200 \\ 0 & 1 & 0 & 0 & 0 & 0 & -1 & 500 \\ 0 & 0 & 1 & 0 & 1 & 0 & 0 & 1200 \\ 0 & 0 & 0 & 1 & 1 & 0 & -1 & 1300 \\ 0 & 0 & 0 & 0 & 0 & 1 & -1 & 300 \\ 0 & 0 & 0 & 0 & 0 & 0 & 0 & 0 \end{array}\right]. \tag{4.40}$$

The array in (4.40) is equivalent to the system of equations

$$\begin{aligned} x_1 + x_3 \qquad\quad &= 1200 \\ x_2 - x_7 \qquad\quad &= \phantom{0}500 \\ x_3 + x_5 \qquad\quad &= 1200 \\ x_4 + x_5 - x_7 &= 1300 \\ x_6 - x_7 \qquad\quad &= \phantom{0}300. \end{aligned} \tag{4.41}$$

Since we wish to set $x_7 = 0$, (4.41) implies immediately that $x_2 = 500$ and $x_6 = 300$. Three equations remain:

$$\begin{aligned} x_1 + x_3 &= 1200 \\ x_3 + x_5 &= 1200 \\ x_4 + x_5 &= 1300. \end{aligned} \tag{4.42}$$

The system (4.42) is underdetermined since there are more unknowns than equations. This means that we can arbitrarily pick one of the variables. Since we want each of the unknowns to be no more than 800, we can pick $x_5 = 600$. Then $x_3 = 600$, $x_4 = 700$, and $x_1 = 600$. Hence, the city engineer could time the lights so that $x_1 = 600$, $x_2 = 500$, $x_3 = 600$, $x_4 = 700$, $x_5 = 600$, and $x_6 = 300$, when she closes $x_7 = 0$.

Although this example is a highly simplified situation, it suggests how more complicated traffic flows can easily be handled. Some cities have installed computerized traffic systems in their downtown areas. These systems automatically adjust the timing of traffic lights as the traffic flow along certain key streets changes, minimizing the possibility of traffic congestion. The models that the computers use in adjusting the timing are based, in part, on the principles explained above.

## EXERCISES 4.5

1. Northwest Airlines has nonstop service between Chicago and the cities of Atlanta, Tampa, Ft. Lauderdale, and Miami. There is nonstop service between Atlanta and the cities of Tampa and Miami and between Miami and the cities of Tampa and Ft. Lauderdale. Use matrices to find the number of one-stop and two-stop flights between any of these cities.
2. Northwest Airlines has nonstop service between Spokane and the cities of Butte, Great Falls, and Missoula. It also has nonstop service between Missoula and the cities of Butte, Helena, and Great Falls; between Butte and the cities of Helena and Bozeman; between Great Falls and the cities of Helena and Billings; between Billings and the cities of Helena and Bozeman. Use the matrix method to find the number of one-stop and two-stop flights among these seven cities.
3. Figure 4.4 shows five southwest United States cities that are serviced by Frontier Airlines. The lines joining two cities indicate nonstop jet service between them. Use the matrix method to find the number of one-stop and two-stop flights among these five cities.

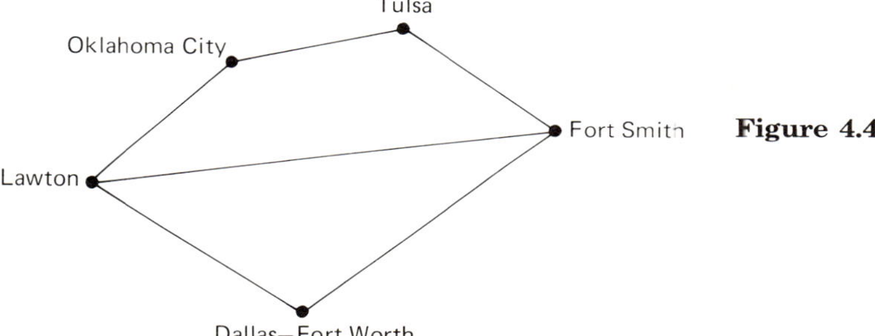


Figure 4.4

4. Figure 4.5 shows eight cities serviced by nonstop jet service by Frontier Airlines. Use the matrix method to find the number of one-stop, two-stop, and three-stop flights among these cities.

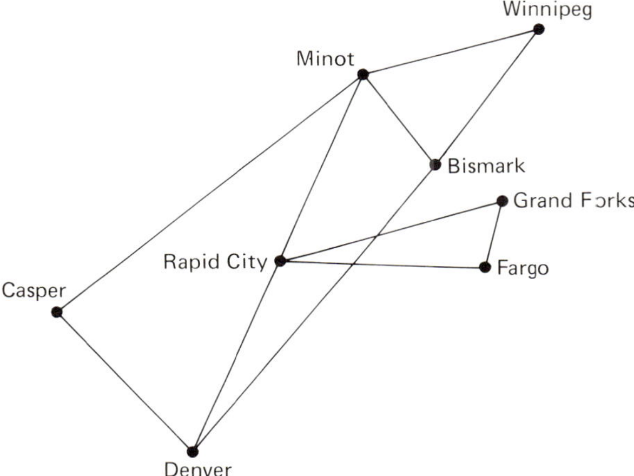


Figure 4.5

*In Exercises 5–9, form a matrix A of zeros and ones from the given genealogical table by numbering each person in the table and entering a one for the entry in the jth row and the kth column if the kth person is a son or daughter of the jth person. For example, if the genealogical table is the one for the Danish kings (see Fig. 4.6), then the matrix A is*

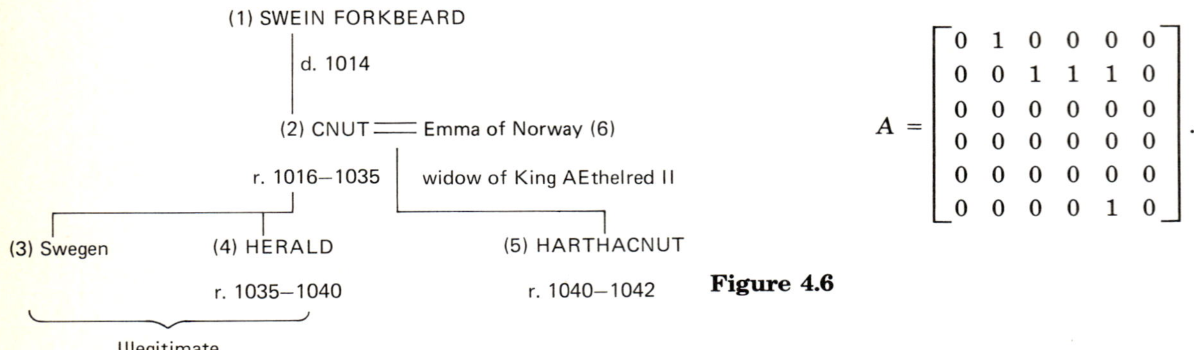


$$A = \begin{bmatrix} 0 & 1 & 0 & 0 & 0 & 0 \\ 0 & 0 & 1 & 1 & 1 & 0 \\ 0 & 0 & 0 & 0 & 0 & 0 \\ 0 & 0 & 0 & 0 & 0 & 0 \\ 0 & 0 & 0 & 0 & 0 & 0 \\ 0 & 0 & 0 & 0 & 1 & 0 \end{bmatrix}.$$

**Figure 4.6**

*Multiplying A by itself repeatedly will yield the number of links in the chain joining the two people. Generally, in determing royal succession, the shorter the chain, the stronger the claim.*

**5.** Use the matrix $A$ in the genealogical table for the Danish kings to find all two-link chains.

**6.** The kings in the House of Cerdic, from Eadward the Elder, include the abbreviated genealogical chart shown in Fig. 4.7. Using the seven numbered persons, construct a matrix $A$ and determine whether Eadward the Confessor or Harthacnut had the stronger claim.

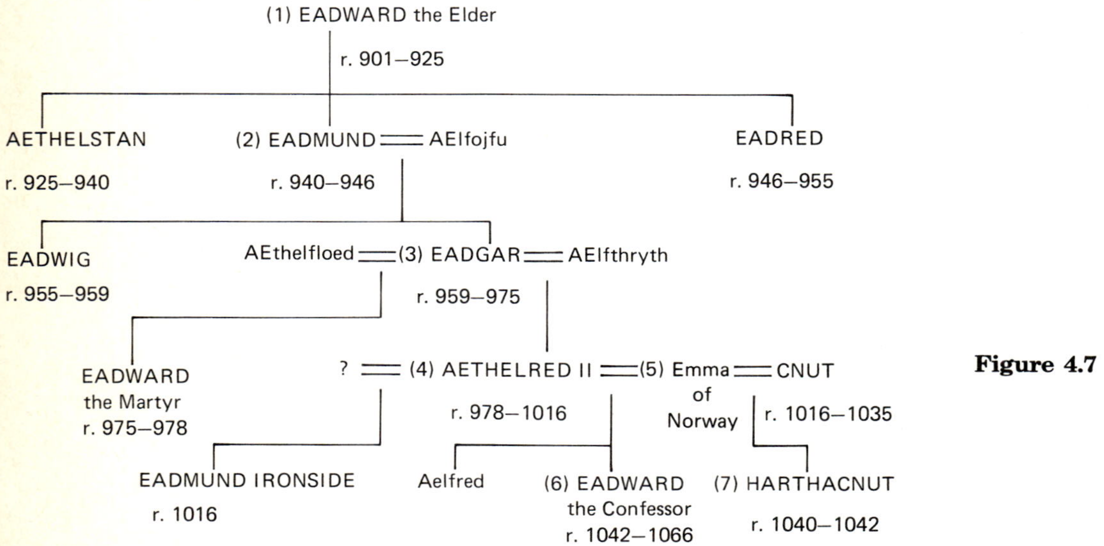


**Figure 4.7**

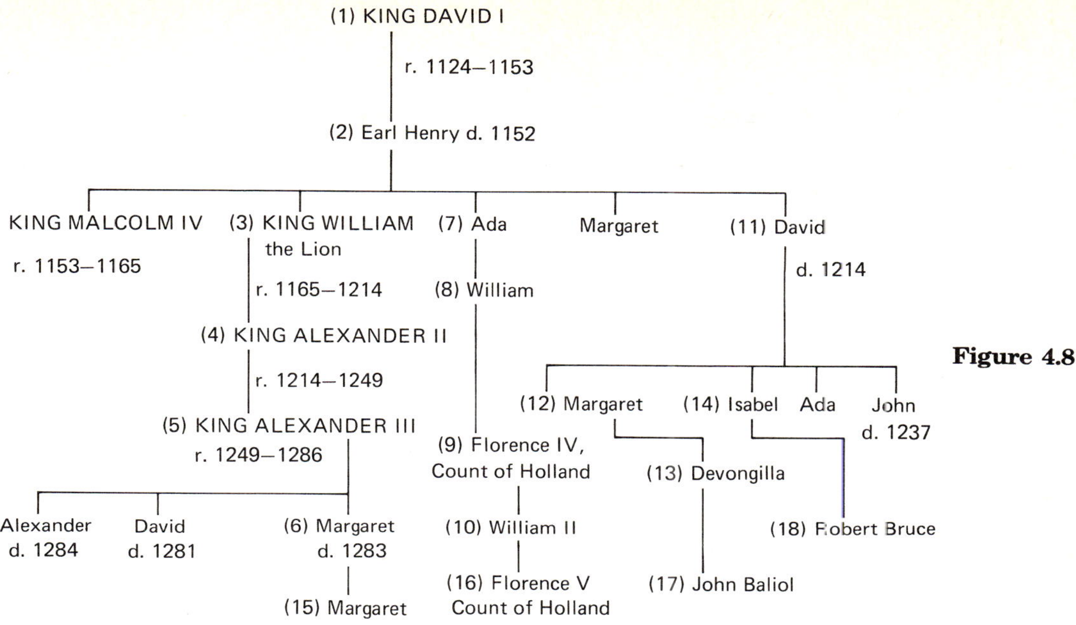


Figure 4.8

**7.** The genealogical table of Scots succession is given in Fig. 4.8. Determine which of the four competitors (numbered 15–18) had the strongest claim. King Edward I of England appointed John Baliol King of the Scots, which led to the wars of Scottish independence and the emergence of the hero King Robert the Bruce (the grandson of the Robert Bruce in Fig. 4.8).

**8.** The fifteenth century was marked by considerable turmoil concerning the rights to the crown of England. Both the houses of York and Lancaster had strong claims to the throne. The genealogical table is given in Fig. 4.9. The

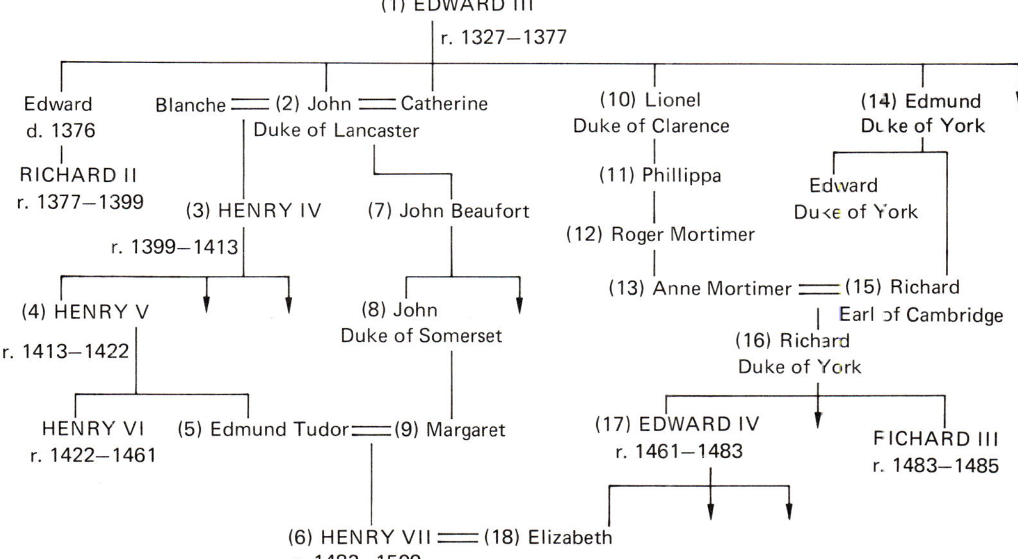


Figure 4.9

so-called War of the Roses was resolved with the marriage of Henry VII of the House of Lancaster to Elizabeth of York, which united both claims to the throne.

Using the matrix method, decide which of the following had the stronger claim to the crown.

a) Edmund Tudor or Richard, Duke of York
b) Henry VII or Edward IV
c) John, Duke of Somerset or Richard, Earl of Cambridge

**9.** The Carolingian Dynasty, founded by the Emperor Charlemagne, is partially illustrated in Fig. 4.10. Who had the stronger claim to the empire: Louis IV, King of Germany, or Louis IV d'Outremer? Use the matrix method to decide.

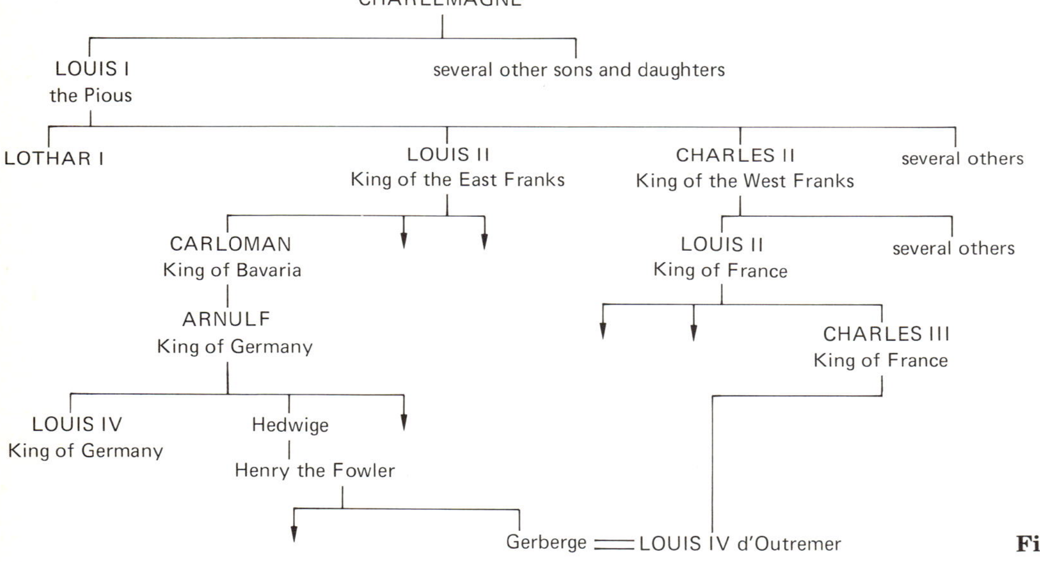


**Figure 4.10**

**10.** Indo-European is the root language of many of the world's languages. Figure 4.11 shows some of the languages that are derived from Indo-European. Construct a matrix $A$ of zeros and ones in which each row (and corresponding column) denotes a language derived from Indo-European. Enter a one whenever that column's language is derived from a given row's language. Calculate $A^2$ and $A^3$.

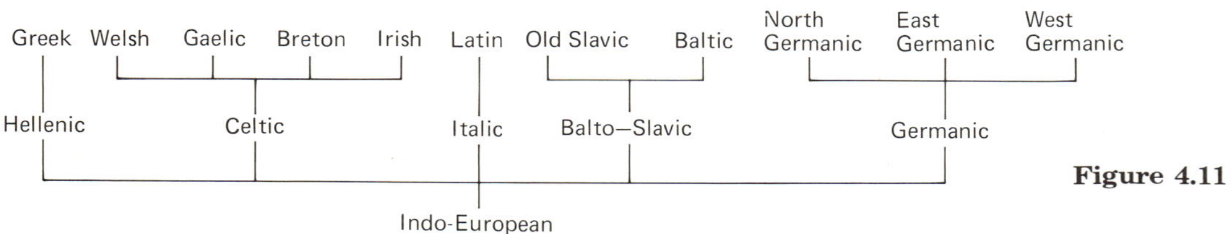


**Figure 4.11**

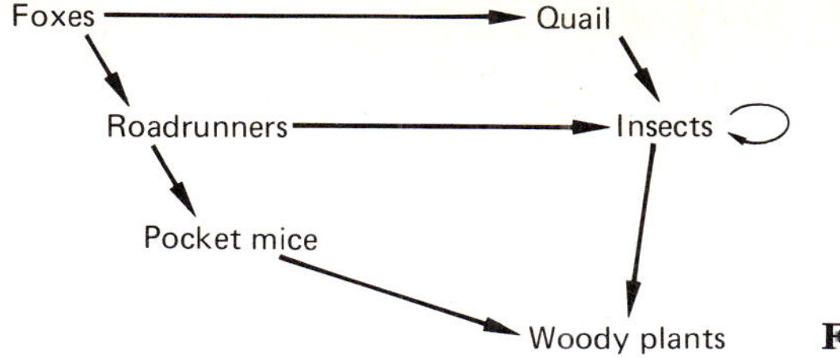


**Figure 4.12**

**11.** Figure 4.12 illustrates a food web. The food sources of each population are shown by arrows (for example, quail feed on insects). Construct a matrix $A$ of zeros and ones in which each row and corresponding column denotes an object in the food chain. Enter a one whenever the object corresponding to a given row feeds on the object in the given column. Calculate and interpret $A^2$, $A^3$, and $A^4$.

**12.** Figure 4.13 illustrates the carbon cycle in nature. Construct a matrix $A$ of zeros and ones in which a one indicates that the object corresponding to that row produces the object in the given column. Calculate and interpret $A^2$, $A^3$, and $A^4$. Find the longest chain of distinct objects, and the shortest and longest cycles.

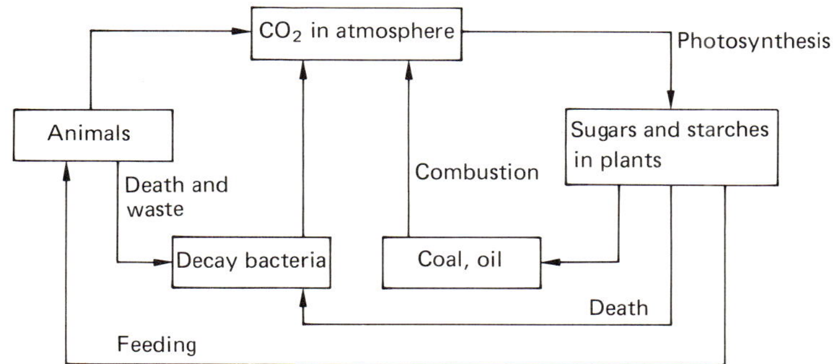


**Figure 4.13**
The carbon cycle.

**13.** Figure 4.14 illustrates the nitrogen cycle in nature. Construct a matrix $A$ as in Exercise 12, and calculate and interpret $A^2$, $A^3$, $A^4$, and $A^5$. Find the longest chain of distinct objects, and the shortest and longest cycles.

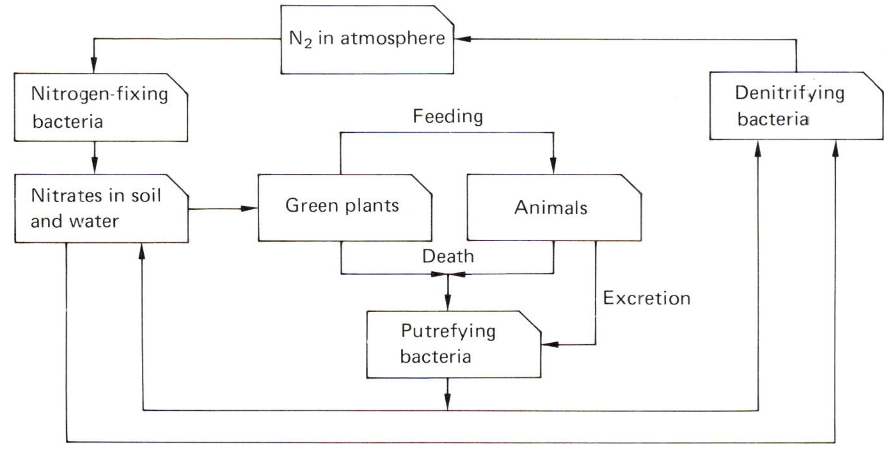


**Figure 4.14**
The nitrogen cycle.

*The following exercises assume the data concerning the recipes for chiliburgers, soft-shelled tacos, and sloppy joes given in Example* 3 *on p.* 181.

**14.** How much of each of the three basic ingredients should the cafeteria manager order if he plans to produce 200 chiliburgers, 600 tacos, and 400 sloppy joes?

**15.** How much of each ingredient is required for 300 chiliburgers, 200 tacos, and 500 sloppy joes?

**16.** What can he produce if he has 130 lb of cooked beans, 175 lb of meat, and 85 lb of fresh tomatoes on hand?

**17.** What can he produce if he has 117.5 lb of cooked beans, 125 lb of meat, and 57.5 lb of fresh tomatoes?

**18.** What can he produce to use up 112 lb of cooked beans, 160 lb of meat, and 64 lb of fresh tomatoes?

*The cafeteria manager in Example* 3 *replaces chiliburgers with bean burritos. The recipe for bean burritos requires* 3 oz *of beans,* 1 oz *of meat, and* 2 oz *of fresh tomatoes per serving. In Exercises* 19–22, *indicate the quantities of each item that he will produce if the supplies on hand are as given.*

**19.** 80 lb of beans, 112 lb of meat, and 72 lb of fresh tomatoes.

**20.** As in Exercise 16.

**21.** As in Exercise 17.

**22.** As in Exercise 18.

☐ *Exercises* 23–26 *assume the same matrix as that given in Example* 4 *on p.* 183. *Find the number of grams of fish with which they should stock lake* 5 *if it can produce the following.*

**23.** 3200 g of mosquito larvae, 840 g of crustaceans, and 4200 g of minnows and worms.

**24.** 400 g of mosquito larvae, 250 g of crustaceans, and 900 g of minnows and worms.

**25.** 150 g of mosquito larvae, 300 g of crustaceans, and 1000 g of minnows and worms.

**26.** 350 g of mosquito larvae, 400 g of crustaceans, and 1500 g of minnows and worms.

*What should the city engineer in Example* 5 (*see p.* 185) *do if, instead of closing Jefferson between Broadway and Main, she must do each of the following?*

**27.** Close Broadway between Lincoln and Jefferson.

**28.** Reduce the traffic on Washington between Broadway and Main as much as possible.

**29.** Handle a sudden surge of traffic entering the grid on Jefferson of 800 VPH if at most 600 VPH can leave on Jefferson.

**30.** Close Main between Washington and Lincoln.

**31.** Suppose you are the city manager and must close the street labeled $x_{12}$ in the traffic grid shown in Fig. 4.15. Assume that no street can handle more than 1000 VPH without causing congestion. Find, using Gaussian elimination, a traffic pattern that requires only one of the streets $x_1, \ldots, x_{11}$ to carry 1000 vehicles.

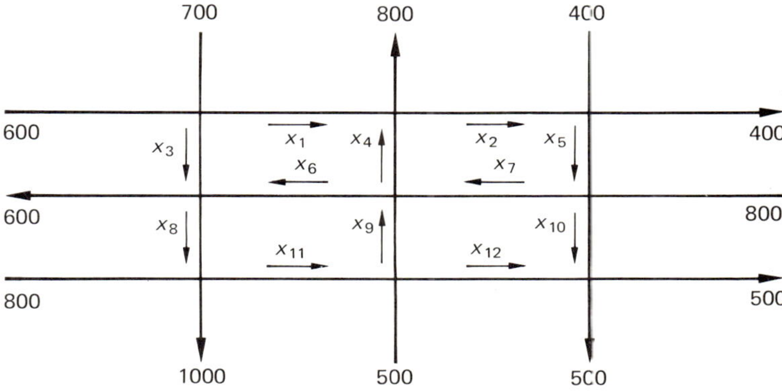


**Figure 4.15**

## CHAPTER VOCABULARY

**additive identity** 0

**augmented matrix** $A \mid B$

**augmented matrix procedure** $A \mid B \rightarrow I \mid A^{-1}B$

**column vector**

**columns of a matrix**

**commutative multiplication** $AB = BA$

**determinant** $\det A$

**direct influence**

**Gaussian elimination** $A \mid I \rightarrow I \mid A^{-1}$

**genealogical table**

**indirect influence**

**inverse of a matrix** $A^{-1}$

**invertible matrix**

**matrix** $A$

**matrix addition**

**matrix equation** $AX = B$

**matrix multiplication**

**multiplicative identity** $I$

**row operation**

**row transformation**

**row vector**

**row of a matrix**

**square matrix**

**traffic problem**

**vector**

# CHAPTER 5 LINEAR PROGRAMMING

In this chapter we discuss one of the most important methods used in decision making. Linear programming is widely used in industry and government as a tool for making management decisions about such diverse topics as the use of raw materials, natural resources, goods inventories, personnel assignments, and transportation.

Linear programming originated primarily during World War II as a procedure for determining the shipping routes that would minimize the distance that the Allied shipping fleet would have to travel.* Other uses were quickly discovered: optimizing the allocation of supplies, deployment of troops, and distribution of petroleum products. The Air Force's Project SCOOP developed most of the techniques that now form the theory of linear programming. George B. Dantzig,† one of the leaders of this project, is credited with developing the simplex algorithm, which we will describe in Section 5.3.

In the first two sections of this chapter we will examine the method of linear programming from a geometrical point of view. Once the basic principles are evident, we will describe the simplex algorithm in Sections 5.3 and 5.4, and show how this matrix method may be used to solve very large and complicated problems.

---

## 5.1 LINEAR INEQUALITIES

In this section we will discuss some useful properties of *linear inequalities*. Consider, for example, the inequality

$$y < x. \tag{5.1}$$

Clearly, every point $(x, y)$ in the Cartesian plane that satisfies this inequality must have a $y$-coordinate that is less than its $x$-coordinate. In particular, no point on the line $y = x$ satisfies Eq. (5.1). If we consider the graph shown in Fig. 5.1, it is apparent that only those points in the shaded half-plane satisfy Eq. (5.1), since each has an $x$-coordinate greater than its $y$-coordinate. The points in the unshaded half-plane satisfy the reverse inequality

$$y \geq x.$$

The properties of this simple example can be applied to more general situations as well. The linear equation

$$ax + by = c$$

---

* F. L. Hitchcock, "The Distribution of a Product from Several Sources to Numerous Localities." *Journal of Mathematical Physics* **20** (1941): 224–230.

† G. B. Dantzig, *Activity Analysis of Production and Allocation*, Ch. 21. New York: John Wiley, 1951.

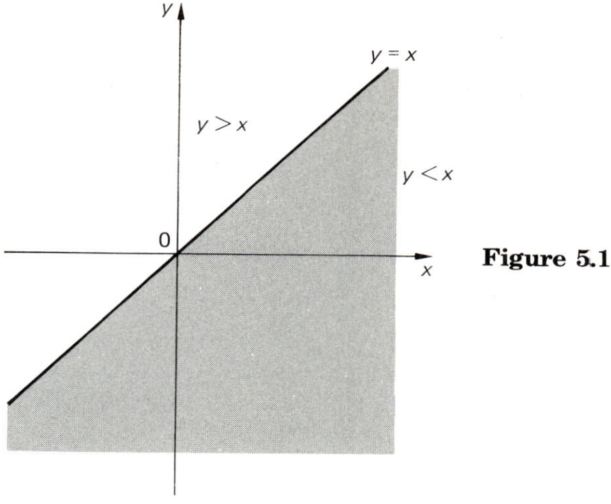


Figure 5.1

separates the Cartesian plane into two half-planes. One of these half-planes satisfies the inequality

$$ax + by < c,$$

while the other satisfies the inequality

$$ax + by > c.$$

In particular, *all we need to do to determine which inequality corresponds to a given half-plane is to test one of its points.* We illustrate this procedure in the following example.

***Example 1*** Determine the half-plane of points that satisfy the linear inequality

$$2x - y < 5. \tag{5.2}$$

Adding $y - 5$ to both sides of the inequality yields the equivalent inequality

$$2x - 5 < y.$$

If we graph the linear equation

$$y = 2x - 5,$$

we obtain a straight line with slope 2 and $y$-intercept $(-5)$ (see Fig. 5.2). We now pick *any* point that is not on the line $y = 2x - 5$; say, for example, the origin. Substituting the origin's coordinates into inequal-

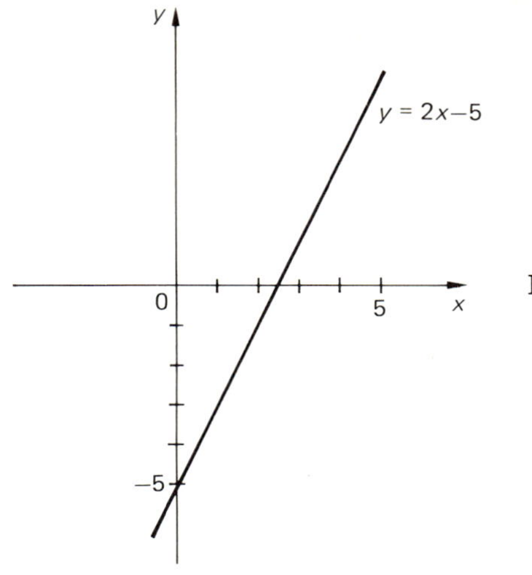


Figure 5.2

ity (5.2), we see that

$$0 = 2(0) - 0 < 5,$$

so that the origin $(0, 0)$ satisfies (5.2). Thus, every point on the origin's side of the line $y = 2x - 5$ satisfies the inequality (5.2).

Observe that the inequality

$$2x - y \leq 5$$

consists of all points in the half-plane *on and above* the line

$$y = 2x - 5.$$

---

Just as we study systems of linear equations, we can also consider systems of linear inequalities.

***Example 2*** Determine the points that satisfy both of the inequalities

$$\begin{aligned} 3x - y &\leq 2, \\ -x + 4y &> 5. \end{aligned} \tag{5.3}$$

First we graph the lines

$$3x - y = 2$$

$$-x + 4y = 5$$

by plotting two points on each line (see Fig. 5.3). We note that the origin satisfies the first inequality of (5.3) but not the second, since the left-hand side of each inequality will be zero at the origin. Thus, the

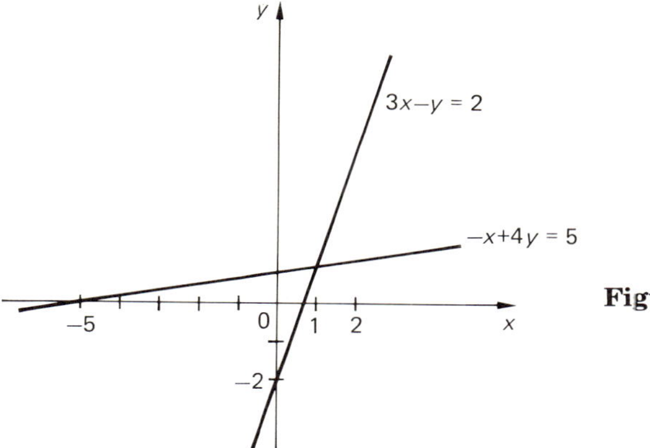


**Figure 5.3**

half-plane on and above the line $3x - y = 2$ satisfies the first inequality of the system (5.3), and the half-plane above the line $-x + 4y = 5$ satisfies the second inequality. Shading in these two half-planes, we see that the points in the darkest region of Fig. 5.4 satisfy both inequalities. Observe that the region includes the edge bounded by the line $3x - y = 2$, but excludes the edge bounded by the line $-x + 4y = 5$.

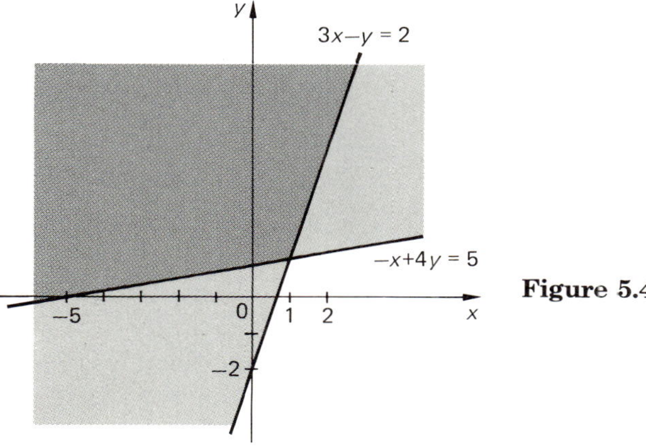


**Figure 5.4**

***Example 3*** Find all points satisfying the system of inequalities

$$\begin{aligned} y - x &< 3, \\ y + 2x &< 4, \\ x &> 0, \\ y &> 0. \end{aligned} \tag{5.4}$$

Graphing the lines

$$\begin{aligned} y - x &= 3, \\ y + 2x &= 4, \\ x &= 0, \\ y &= 0, \end{aligned}$$

in Fig. 5.5, we obtain the half-planes of points below the lines $y = x + 3$, and $y = -2x + 4$. The graphs of the inequalities $x > 0$ and $y > 0$ give the half-plane to the right of the $y$-axis and the half-plane above the $x$-axis, respectively. The shaded region in Fig. 5.5 satisfies all four inequalities.

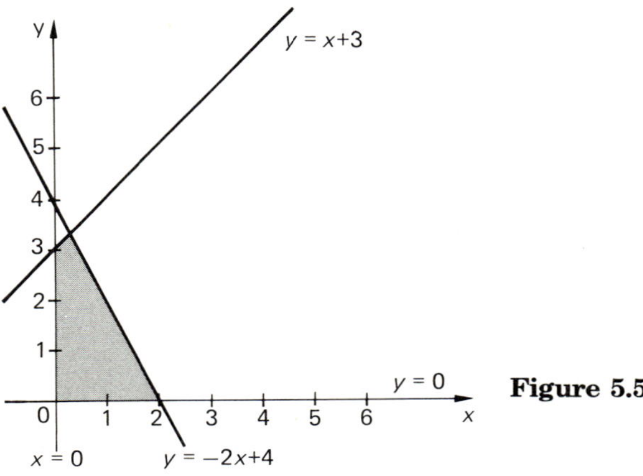


**Figure 5.5**

---

Just as systems of linear equations need not have a solution, it is possible to have a system of linear inequalities which is not satisfied by any point.

***Example 4*** Consider the system of linear inequalities

$$\begin{aligned} y - x &< 2 \\ y - x &> 4. \end{aligned} \tag{5.5}$$

The lines $y = x + 2$ and $y = x + 4$ are shown in Fig. 5.6. Since the two lines are parallel (the slopes are equal), and since the first inequality in (5.5) is satisfied by points below the line $y = x - 2$, while the second inequality is satisfied by points above the line $y = x + 4$, the two

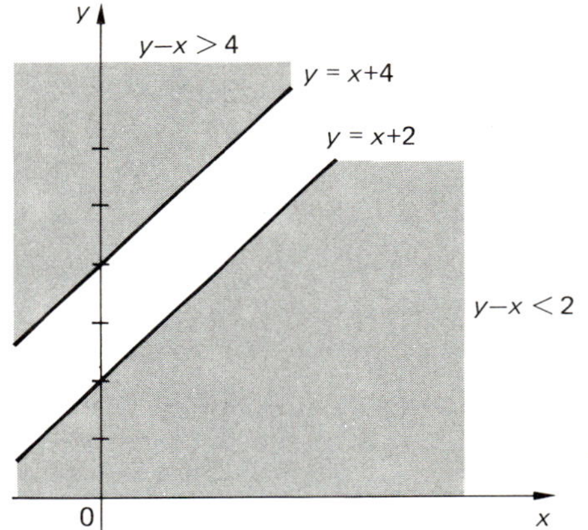


**Fig. 5.6**
No points in common.

half-planes have no points in common. Thus, no point in the $xy$-plane satisfies the system (5.5).

---

## EXERCISES 5.1

*In Exercises* 1–16, *shade the region of all points that satisfy the given inequality or system of inequalities.*

**1.** $x \le 2y$

**2.** $y \le x + 1$

**3.** $2x - 3y \le 6$

**4.** $5y - 4x \ge 8$

**5.** $\begin{aligned} 7x + 2y &\le 14 \\ x &\le y \end{aligned}$

**6.** $\begin{aligned} 4x - y &\le 2 \\ x + y &\le 1 \end{aligned}$

**7.** $\begin{aligned} 3x + y &\ge 4 \\ 2x + 3y &\le 8 \end{aligned}$

**8.** $\begin{aligned} 2x - 3y &\ge 6 \\ 6y - 4x &\le 6 \end{aligned}$

**9.** $\begin{aligned} 5x + 2y &\le 10 \\ x - y &\ge 3 \\ x &\ge 0 \end{aligned}$

**10.** $\begin{aligned} 5x + 4y &\le 20 \\ -2x + 3y &\le 6 \\ y &\ge 0 \end{aligned}$

**11.** $5x + 3y \leq 15$
$x + 2y \leq 6$
$x \geq 0$
$y \geq 0$

**12.** $x + y \leq 3$
$-x + y \leq 1$
$y \geq 0$
$x \geq 4$

**13.** $x + y \leq 1$
$2x + 2y \geq 3$

**14.** $-x + y \geq 1$
$x > 0$
$y \leq 0$

**15.** $4x - y \leq 2$
$x + y \geq 1$
$y \leq 0$

**16.** $2x - 3y \geq 6$
$6y - 4x \geq 6$
$0 \leq y \leq 4$

## 5.2 LINEAR PROGRAMMING: GEOMETRIC METHOD

Systems of linear inequalities play a vital role in decision making. The following example illustrates the method of linear programming.

***Example 1*** The university cafeteria manager has 300 pounds of ground beef, which she can use to make hamburgers or chili. Each hamburger contains 1/4 pound of ground beef, while each bowl of chili contains 1/6

pound of ground beef. It requires an average of 40 seconds either to prepare a hamburger (several are cooking at the same time) or to serve three bowls of chili. The cafeteria is open for 10 hours, and makes 30 cents profit on each hamburger and 18 cents profit on each bowl of chili. If we can assume that everything the cooks prepare is eaten, find the number of hamburgers and bowls of chili that should be produced to maximize profits.

Although this problem appears very complicated, we are able to write it as a simple system of linear inequalities. Our main task is to translate each sentence into mathematical symbols. Since we don't know how many hamburgers or bowls of chili should be produced, we will let $x$ represent the number of hamburgers and $y$ the number of bowls of chili. Then, two linear inequalities are obvious:

$$x \geq 0, \qquad y \geq 0 \tag{5.6}$$

since the number of hamburgers or bowls of chili cannot be negative.

We get another linear inequality by determining how many hamburgers and bowls of chili can be made from 300 pounds of ground beef. Either four hamburgers or six bowls of chili require one pound of ground beef. Thus, if we divide the number of hamburgers $x$ by four *and* the number of bowls of chili $y$ by six, we get the total number of pounds of ground beef used in making them. That is,

$$\frac{x}{4} + \frac{y}{6} \leq 300, \tag{5.7}$$

since *at most* 300 pounds of ground beef are available.

Our final linear inequality is obtained by determining how many hamburgers or bowls of chili can be served in the available 10 hours of operating time. For simplicity, we will do all our calculations in minutes. Thus $x$ hamburgers require $(2/3)x$ minutes for their production. since 40 seconds equals 2/3 of a minute. Similarly, 9 bowls of chili require 2 minutes of serving time, so that each bowl requires 2/9 of a minute and $(2/9)y$ is the number of minutes required to serve $y$ bowls of chili. Since the number of minutes required to serve the hamburgers and chili cannot exceed 600 (10 hr $\times$ 60 min/hr), we obtain the inequality

$$\tfrac{2}{3}x + \tfrac{2}{9}y \leq 600. \tag{5.8}$$

Thus, we have the system of four linear inequalities (5.6)–(5.8), or

$$\begin{aligned} &x \geq 0, \qquad y \geq 0, \\ &\frac{x}{4} + \frac{y}{6} \leq 300, \\ &\tfrac{2}{3}x + \tfrac{2}{9}y \leq 600. \end{aligned} \tag{5.9}$$

Observe that we have *not* tried to obtain an equation concerning profit, which is the item we want to maximize. Instead we have concentrated solely on determining what can be produced within the limits of the time and quantity of ground beef available. The collection of points $(x, y)$ that satisfy the system of inequalities (5.9) is called the *feasible set*, and *it is from among this set of points that we must choose the one that maximizes our profit.*

We now graph the four half-planes given by the inequalities (5.9). The points they have in common, the feasible set, are represented by the shaded region in Fig. 5.7.

*After* we have determined the feasible set, we consider the profit (in cents). This is obtained by multiplying the number of hamburgers $x$ by the profit per hamburger and adding this to the number of bowls of chili times the profit per bowl. Thus

$$30x + 18y = P = \text{profit}. \tag{5.10}$$

Now, regardless of what number $P$ is, Eq. (5.10) determines a straight line, since subtracting $30x$ from both sides yields

$$18y = -30x + P,$$

so that

$$y = \frac{-30}{18}x + \frac{P}{18}$$

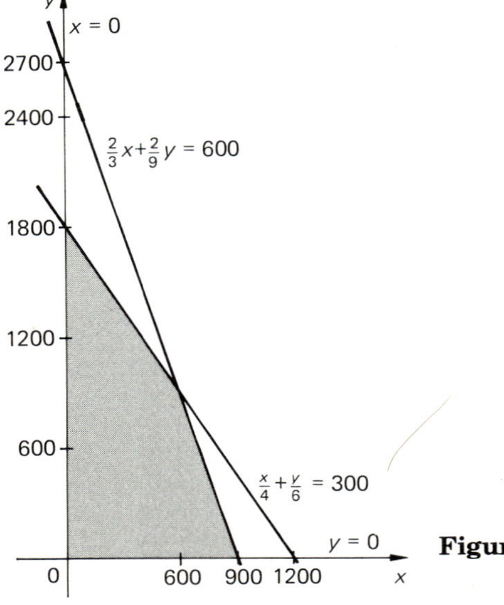


**Figure 5.7**

or

$$y = \frac{-5}{3}x + \frac{P}{18}. \tag{5.11}$$

Thus, the profit equation (5.10) yields a straight line having slope $-5/3$ and $y$-intercept $P/18$. As we increase the number $P$, the straight line (5.11) rises (see Fig. 5.8). Since we want to *maximize* profit, we need to find *the maximal number P such that the line* (5.11) *still intersects the feasible set.* It is obvious that as we slide the line (5.11) upward we will finally intersect the feasible set only at one point. The coordinates of this point of intersection are the number of hamburgers $x$ and bowls of chili $y$ that should be produced to maximize the profit. It is evident (see Fig. 5.8) that this point will be *a corner* (or *vertex*) *of the feasible set*—in this case, the point at which the lines

$$\frac{x}{4} + \frac{y}{6} = 300 \tag{5.12}$$

and

$$\tfrac{2}{3}x + \tfrac{2}{9}y = 600 \tag{5.13}$$

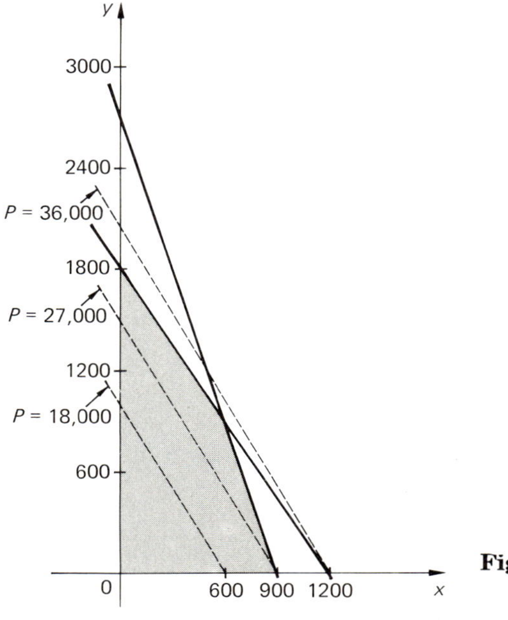


Figure 5.8

meet. Hence, we need to solve the two equations (5.12) and (5.13) simultaneously to obtain the coordinates of the point at which the profit is maximized. Multiplying the first equation by 12 and the second by 9 gives us the system

$$3x + 2y = 3600$$

$$6x + 2y = 5400.$$

Subtracting the first equation from the second, we have $3x = 1800$, or $x = 600$ so that $y = 900$. Substituting the values $x = 600$ and $y = 900$ into Eq. (5.10) yields the maximum profit

$$P_{\text{max}} = 30(600) + 18(900) = 34{,}200,$$

or $P_{\text{max}} = \$342.00$. This profit occurs whenever the cafeteria produces and sells 600 hamburgers and 900 bowls of chili.

***Example 2*** Suppose all the information in Example 1 is unchanged except that only 2 cents profit is made on each bowl of chili. In this case, the profit equation (5.10) must be changed to

$$P = 30x + 2y \tag{5.14}$$

so that

$$y = -15x + \frac{P}{2}. \tag{5.15}$$

The vertices of the feasible set are the same as before: $(0, 0)$, $(900, 0)$, $(600, 900)$, and $(0, 1800)$. If we substitute the coordinates of the vertices into Eq. (5.14) and denote by $P(x, y)$ the profit made by selling $x$ hamburgers and $y$ bowls of chili, we get

$$P(0, 0) = 30(0) + 2(0) = 0,$$

$$P(900, 0) = 30(900) + 2(0) = 27{,}000,$$

$$P(600, 900) = 30(600) + 2(900) = 19{,}800,$$

and

$$P(0, 1800) = 30(0) + 2(1800) = 3600,$$

so that a maximum profit of $P_{\text{max}} = \$270.00$ occurs when we produce and sell 900 hamburgers and *no* chili. In this situation the manager might consider raising the price of chili to increase the profit per bowl or dropping the item from the menu.

The change in the optimal solution is not surprising when we consider the fact that the lines determined by Eq. (5.15) are almost vertical so that the rightmost point of the feasible set is the one that meets the line with maximal value $P$.

---

Examples 1 and 2 are illustrations of the technique of *linear programming*. Although at first glance the technique may seem complicated, the method of linear programming reduces to three steps:

1. Determining the *feasible set* of points that satisfy some given system of linear inequalities. These inequalities are the physical constraints of the problem we are investigating.
2. Finding all the *vertices* (or corners) of the feasible set. The equation we are optimizing will assume its optimal value at a vertex of the feasible set.
3. Substituting the coordinates of each vertex into the equation that we wish to optimize. The optimal value so obtained is the solution to the problem.

It may happen that two vertices yield the same optimal value. This occurs when the equation we are optimizing has the same slope as one of the edges of the feasible set. In this case, *any point on that edge will yield the optimal value.* The decision maker can now select whichever of these choices he or she prefers. Note, however, that the optimal value still occurs at some vertex.

Linear programming has wide applicability in business and in the social and biological sciences. The following example illustrates a possible use in environmental protection.

***Example 3*** A coal-fired generating plant burns at least 500 tons of coal per day to maintain its electric operating capacity. It can purchase 4% sulfur coal at \$8.00 per ton or 1% sulfur coal at \$12.00 per ton. Burning a ton of coal produces 38 pounds of $SO_2$ for each percent of sulfur in the coal.* To meet pollution emission standards, the plant cannot exceed a daily output of 22,800 pounds of $SO_2$. How much 1% sulfur coal should the plant burn to minimize cost, while complying with the emission standards?

Let $x$ and $y$ represent the number of tons of 4% sulfur coal and 1% sulfur coal, respectively, that the plant will consume each day. Obviously we have the inequalities $x \geq 0$ and $y \geq 0$ since these amounts

---

* See R. G. Bond and C. P. Straub, *Handbook of Environmental Control*, Vol. 1. Cleveland: CRC Press, 1972, p. 232.

cannot be negative. Adding these two quantities, we know that

$$x + y \geq 500,$$

since *at least* 500 tons of coal must be burned each day. To calculate the amount of $SO_2$ that will be produced, we multiply 38 pounds by 4 to obtain the amount that each ton of 4% sulfur coal will produce. Thus, the amount of $SO_2$ produced daily must satisfy the inequality

$$38(4)x + 38y \leq 22{,}800.$$

Gathering all these inequalities, we obtain the system

$$\begin{aligned} 152x + 38y &\leq 22{,}800, \\ x + \quad y &\geq 500, \\ x \geq 0, \quad y &\geq 0, \end{aligned} \tag{5.16}$$

which yields the feasible set shown by the shaded region in Fig. 5.9. Solving the system of equations

$$\begin{aligned} 152x + 38y &= 22{,}800 \\ x + \quad y &= \quad 500 \end{aligned} \tag{5.17}$$

simultaneously, we multiply the bottom equation by 38 and subtract from the top equation, obtaining $114x = 3800$, or $x = 33\frac{1}{3}$. Thus, $y = 466\frac{2}{3}$, so the vertices of the feasible set are $(0, 500)$, $(0, 600)$, and $(33\frac{1}{3}, 466\frac{2}{3})$.

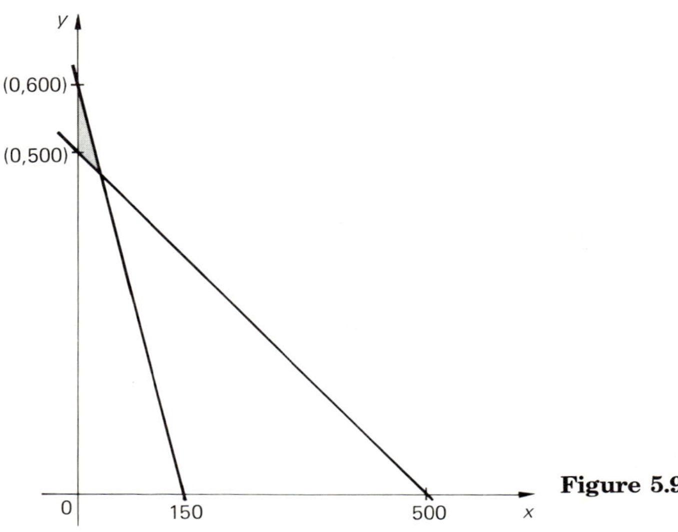


**Figure 5.9**

The cost equation is given by

$$\text{Cost} = C = 8x + 12y. \tag{5.18}$$

Substituting the vertices of the feasible set into Eq. (5.18), we get

$$C(0, 500) = 8(0) + 12(500) = \$6000,$$

$$C(0, 600) = 8(0) + 12(600) = \$7200,$$

$$C(33\tfrac{1}{3}, 466\tfrac{2}{3}) = 8(33\tfrac{1}{3}) + 12(466\tfrac{2}{3}) = \$5866.67.$$

Thus the minimum cost is attained using $33\frac{1}{3}$ tons of 4% sulfur coal and $466\frac{2}{3}$ tons of 1% sulfur coal.

***Example 4*** An investor wishes to build some duplexes and 8-plexes on a tract of land that he owns. Each building requires the same amount of ground space and enough land is available for nine buildings to be erected on the tract. Zoning laws require that there be at least as many duplexes as 8-plexes in a development of this type. The cost of building each duplex is \$60,000, while an 8-plex costs \$180,000. The maximum amount of capital that the investor has available for this project is \$1,000,000. If the total monthly rent from each duplex is \$600, while each 8-plex yields \$2000, how many buildings of each kind should he build?

Let $x$ be the number of duplexes and $y$ the number of 8-plexes. Again, $x \geq 0$ and $y \geq 0$, since these quantities cannot be negative. Since at most nine buildings can be constructed, we have

$$x + y \leq 9.$$

The zoning law yields the inequality

$$y \leq x,$$

while the available capital produces the inequality

$$60{,}000x + 180{,}000y \leq 1{,}000{,}000.$$

Gathering these five inequalities, we have the system

$$\begin{aligned} 60{,}000x + 180{,}000y &\leq 1{,}000{,}000, \\ x + \phantom{180{,}000}y &\leq 9, \\ y &\leq x, \\ x &\geq 0, \\ y &\geq 0. \end{aligned} \tag{5.19}$$

The feasible set is shown in Fig. 5.10, and has the vertices (0, 0), (9, 0), (31/6, 23/6), and (25/6, 25/6). The monthly rent equation is

$$\text{Rent} = R = 600x + 2000y. \tag{5.20}$$

Substituting the vertices of the feasible set into Eq. (5.20) yields

$$R(0,0) = \$0, \qquad R(9,0) = \$5400,$$

$$R(\tfrac{31}{6}, \tfrac{23}{6}) = \$10{,}766.67, \qquad R(\tfrac{25}{6}, \tfrac{25}{6}) = \$10{,}833.33.$$

Although the vertex (25/6, 25/6) yields the maximum rent, it is impossible to build fractions of a duplex or 8-plex. In this case we are limited to considering only those points of the feasible set that have integers for both coordinate values. This collection of points is called a *lattice*. The rent lines (5.20) are all parallel to the dashed line in Fig. 5.10. Sliding this line toward the lattice, we see that the first lattice point we meet is (4, 4), with a monthly rent of $R(4, 4)$ = \$10,400. This, then, is the final solution to this problem: The investor should build four duplexes and four 8-plexes.

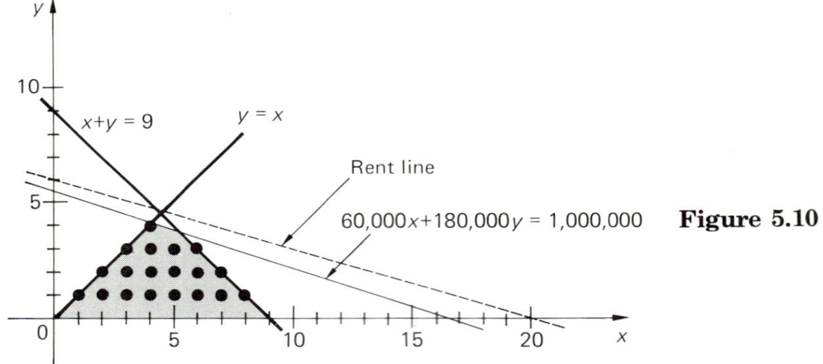


**Figure 5.10**

It is worth noting that the point (5, 4) lies just outside the feasible set and satisfies all but the first inequality of the system (5.19). Since

$$60{,}000(5) + 180{,}000(4) = 1{,}020{,}000,$$

the investor would notice that an additional \$20,000 would increase his monthly rent by

$$R(5,4) - R(4,4) = 11{,}000 - 10{,}400 = \$600.$$

It would then be very tempting to raise the additional capital needed for the fifth duplex.

Those linear programming problems in which only points with integer-valued coordinates are considered are frequently called *integer programming* problems.

As these four examples indicate, linear programming is a very useful tool for making decisions. However, in practice it is rare to find a case involving only two variables. Often dozens of variables enter a given problem. When more than three variables are involved, it is tiresome and difficult to find the vertices of the feasible set. For this reason various algorithms have been developed to reduce the work involved. We will study one of these techniques, called the *simplex method*, in Section 5.3.

## EXERCISES 5.2

*In Exercises 1–12, find the feasible set of the given inequality or system of inequalities.*

**1.** $x \le 2y$

**2.** $y \le x + 1$

**3.** $2x - 3y \le 6$

**4.** $5y - 4x \ge 8$

**5.** $7x + 2y \le 14$
$x \le y$

**6.** $4x - y \le 2$
$x + y \le 1$

**7.** $3x + y \ge 4$
$2x + 3y \le 8$

**8.** $2x - 3y \ge 6$
$6y - 4x \le 6$

**9.** $5x + 2y \le 10$
$x - y \ge 3$
$x \ge 0$

**10.** $5x - 4y \le 20$
$-2x + 3y \le 6$
$y \ge 0$

**11.** $5x + 3y \le 15$
$x + 2y \le 6$
$x \ge 0$
$y \ge 0$

**12.** $x + y \le 3$
$-x + y \le 1$
$y \ge 0$
$x \ge 4$

**13.** Suppose that all the information given in Example 1 (see p. 204) is unchanged except that 20 cents profit is made on each bowl of chili. Show that two vertices of the feasible set yield the maximum profit. Verify that the point (400, 1200) also yields the same profit. How much is the maximal profit?

**14.** If all the information in Example 1 is unchanged except that the cafeteria stays open for only eight hours daily, how many hamburgers and bowls of chili should be produced to maximize profits? What is the maximal profit?

**15.** Suppose in Example 1 that the cafeteria manager is required by the student government to prepare at least 1200 bowls of chili. What should she do to maximize profits?

**16.** Suppose that the power plant in Example 3 (see p. 209) is granted a variance in its emission standards allowing it to emit 45,600 pounds of $SO_2$ daily. What amounts of 4% and 1% sulfur coal should it buy to minimize cost?

**17.** Suppose that the power plant in Example 3 can no longer obtain 4% sulfur coal, but instead can purchase 2% sulfur coal at $10.00 per ton. If all other conditions remain unchanged, what amounts of 2% and 1% sulfur coal should be purchased to minimize cost?

**18.** A feedlot manager wishes to fatten his cattle as economically as possible. Each steer requires at least either a bushel of corn or three bales of hay (or a linear combination of these two items) to maintain an acceptable daily weight gain. In addition, each steer must consume at least one bale of hay daily for roughage. If hay costs $1.00 a bale and corn $2.70 a bushel, which combination of hay and corn should he use?

**19.** The fish-and-game agency wishes to stock a lake with large-mouth black bass and bullheads. In a study by A. S. Pearse [*Ecology* **5** (1924): 256], the following daily eating habits were discovered for each 1-kg fish of these two species:

| Food (in grams) | Black bass | Bullheads |
|---|---|---|
| Minnows | 344 | 148 |
| Insects | 64 | 82 |
| Crustaceans | 141 | 168 |
| Earthworms | 211 | 345 |

Insects and worms abound, but the lake can only sustain a daily production of 100 kg of minnows and 90 kg of crustaceans. Assuming that no fish presently feed on the minnows and crustaceans, and that the weight gained by the stocked fish will equal that lost through fishing and other causes, how many 1-kg fish of each species should the lake be stocked with to maximize the total number of fish?

**20.** A cement factory produces at least 3.5 million tons of cement annually. Two electrostatic precipitators are used to reduce the dust emissions from the plant. One of them includes a multicyclone and reduces emissions to 1.2 lb per ton. The other reduces emissions to only 4.2 lb per ton. The plant must comply with an annual emission standard of at most 6 million lb of dust. If it costs 20¢ per ton of cement produced to use the multicyclone electrostatic precipitator and 15¢ per ton to use the other electrostatic precipitator, how should they be used to minimize cost?

**21.** A credit union requires at least 40 minutes of computer time each month to handle its accounts. Daily balances are computed by batch programming and processed at night, but withdrawals and deposits must be handled on-line. At least 10 minutes of on-line time is required. The computer center insists on selling at least twice as much batch time as on-line time, and charges $120 per minute for on-line time and $80 per minute for batch time. How much of each time should the credit union buy to minimize its costs?

**22.** In Exercise 21 suppose that the computer center changes its sales policy and insists on selling at least as much on-line time as batch time. What time allocations will produce minimum costs for the credit union manager, all other conditions remaining the same?

**23.** A small accounting firm wishes to determine how many audits and tax returns it should handle each week in order to maximize revenue. There are 360 hours of staff time and 40 hours of review time available each week. Each audit contributes $500 in revenue but requires 40 hours of staff time and 4 hours of review time. A tax return contributes $60 in revenue and requires 5 hours of staff time and 1 hour of review time. What is the optimal mix of audits and tax returns?

**24.** The accounting firm in Exercise 23 raises the charge for preparing a tax return to $100. What is the optimal mix of audits and tax returns?

**25.** A manufacturer allocates a maximum of $400,000 for advertising his product on national media. Each minute of television time costs $40,000 and each one-page newspaper ad costs $10,000. Each television commercial is believed to reach 10,000,000 viewers, while the newspaper ads reach 2,000,000 readers. The advertising company advises the manufacturer to use at least four newspaper ads and six television commercials. How should the budget be divided to maximize exposure?

**26.** Suppose the investor in Example 4 (see p. 211) can only raise $900,000 for his development. How many buildings of each kind should he build?

---

## 5.3 LINEAR PROGRAMMING: THE SIMPLEX ALGORITHM

Since matrices are very useful in solving systems of linear equations, it is not surprising that they are also useful in linear programming. In this section we will describe the most useful technique in solving linear programming problems: the *simplex tableau*. The procedure is very similar to the augmented matrix procedure, particularly as it is applied to Example 5 in Section 4.5 on traffic congestion.

To see how the method works, it is best to illustrate the technique on a linear programming problem for which we already know the solution. For this reason, we will again consider the situation in Example 1 in Section 5.2, in which a university cafeteria manager is trying to decide how to use 300 lb of ground beef. She can use the meat to produce hamburgers or chili. Each hamburger contains 1/4 lb of ground beef, while each bowl of chili requires 1/6 lb of ground beef. Forty seconds of individual attention are required to cook each hamburger (several are cooking at the same time), or to serve three bowls of chili. The cafeteria is open for 10 hours and makes 30 cents profit on each hamburger and 18 cents profit on each bowl of chili. Assuming that the labor costs are fixed, what number of hamburgers and bowls of chili will maximize profits?

To determine a system of inequalities describing the above situation, we set $x$ equal to the number of hamburgers and $y$ to the number of bowls of chili that are to be produced. Since neither of these numbers can be negative, we obtain the two inequalities $x \geq 0$ and $y \geq 0$. Another inequality is obtained by observing that the 300 lb of ground beef on hand limit the total number of hamburgers and bowls of chili that can be produced. Since each hamburger takes 1/4 lb and each bowl of chili requires 1/6 lb, we have

$$\frac{x}{4} + \frac{y}{6} \leq 300.$$

Finally, the cafeteria is open a limited number of hours per day. The time required to cook the hamburgers and serve the bowls of chili can not exceed the time available. Since each hamburger takes 40 seconds (= 2/3 minute) and each bowl requires 40/3 seconds (= 2/9 minute) to serve, the time used in production is

$$\tfrac{2}{3}x + \tfrac{2}{9}y \leq 600,$$

since 10 hours = 600 minutes. Thus, we obtained the system of four inequalities:

$$\begin{aligned} \frac{x}{4} + \frac{y}{6} &\leq 300, \\ \tfrac{2}{3}x + \tfrac{2}{9}y &\leq 600, \\ x \geq 0, \qquad y &\geq 0. \end{aligned} \tag{5.21}$$

All possible points $(x, y)$ in the Cartesian plane satisfying these four inequalities form the *feasible set*, and we must choose that (those) point(s) of the feasible set that maximize the *profit function*

$$P = 30x + 18y, \tag{5.22}$$

which is the total number of cents profit that will be made by producing $x$ hamburgers and $y$ bowls of chili.

*Since inequalities are not as easy to work with as equalities, we will convert the system* (5.21) *into a system of two equations.* Let $t_1$ and $t_2$ be two variables that take up the slack in the inequalities, that is, such that

$$\frac{x}{4} + \frac{y}{6} + t_1 = 300,$$

and

$$\tfrac{2}{3}x + \tfrac{2}{9}y + t_2 = 600.$$

Then *all four of the variables* $x$, $y$, $t_1$, *and* $t_2$ *are nonnegative.* Since $t_1$ and $t_2$ make up the difference in the previous inequalities, they are called *slack variables*. We also rewrite Eq. (5.22) in the form

$$-30x - 18y + P = 0.$$

These three equations yield the system

$$\begin{aligned} \frac{x}{4} + \frac{y}{6} + t_1 \qquad\qquad &= 300, \\ \tfrac{2}{3}x + \tfrac{2}{9}y \qquad + t_2 \qquad &= 600, \\ -30x - 18y \qquad\qquad + P &= 0, \end{aligned} \tag{5.23}$$

where $x \geq 0$, $y \geq 0$, $t_1 \geq 0$, and $t_2 \geq 0$.

We will now determine the optimal solution by using matrix methods. We begin by rewriting system (5.23) in the form $A \mid B$, where $A$ is the matrix of coefficients and $B$ is the vector of constants on the right side of each equation. We call the augmented matrix $A \mid B$ a *simplex tableau*:

$$A \mid B = \begin{array}{c} \begin{array}{ccccc|c} x & y & t_1 & t_2 & P & \end{array} \\ \left[\begin{array}{ccccc|c} 1/4 & 1/6 & 1 & 0 & 0 & 300 \\ 2/3 & 2/9 & 0 & 1 & 0 & 600 \\ -30 & -18 & 0 & 0 & 1 & 0 \end{array}\right] \end{array}. \tag{5.24}$$

Instead of proceeding as we did using the augmented matrix procedure, we follow these steps for the *simplex algorithm*:

1. Locate the negative entry in the bottom row of $A$ whose absolute value is largest. If more than one satisfies this condition, select any of them. If all entries in the bottom row are positive, the simplex tableau is in *final form*.
2. Divide the entries in $B$ by the corresponding entries in the column above the entry selected in step 1.
3. Select the divisor that yields the smallest positive quotient.
4. Divide each entry in that row by that divisor, creating a 1 in that location, and use that 1 to eliminate the remaining entries in that column.
5. Return to step 1.

We will perform each of these steps on $A \mid B$. The largest negative entry in the bottom row of $A$ is the $-30$ in the first column. We divide 300 by 1/4 and 600 by 2/3, obtaining

$$\rightarrow \left[\begin{array}{ccccc|c} 1/4 & 1/6 & 1 & 0 & 0 & 300 \\ \textcircled{2/3} & 2/9 & 0 & 1 & 0 & 600 \\ -30 & -18 & 0 & 0 & 1 & 0 \end{array}\right], \quad \begin{aligned} 300/(1/4) &= 1200, \\ 600/(2/3) &= 900, \end{aligned}$$
$$\uparrow$$

so the entry circled above is the divisor satisfying step 3. Such an entry is called a *pivot*. Dividing each entry in the second row by 2/3, we obtain

$$\left[\begin{array}{ccccc|c} 1/4 & 1/6 & 1 & 0 & 0 & 300 \\ 1 & 1/3 & 0 & 3/2 & 0 & 900 \\ -30 & -18 & 0 & 0 & 1 & 0 \end{array}\right]. \tag{5.25}$$

Now we eliminate the other entries in column 1 by adding or subtracting multiples of row 2, getting the matrix

$$\rightarrow \left[\begin{array}{ccccc|c} 0 & \textcircled{1/12} & 1 & -3/8 & 0 & 75 \\ 1 & 1/3 & 0 & 3/2 & 0 & 900 \\ 0 & -8 & 0 & 45 & 1 & 27{,}000 \end{array}\right]. \tag{5.26}$$
$$\uparrow$$

Returning to step 1, we find that there is a negative entry $-8$ in the second column. Dividing 75 by 1/12 and 900 by 1/3, we have

$$75/(1/12) = 900, \qquad 900/(1/3) = 2700,$$

so the circled entry in (5.26) is the new pivot. Multiplying each entry in row 1 by 12 yields

$$\left[\begin{array}{ccccc|c} 0 & 1 & 12 & -9/2 & 0 & 900 \\ 1 & 1/3 & 0 & 3/2 & 0 & 900 \\ 0 & -8 & 0 & 45 & 1 & 27{,}000 \end{array}\right]. \tag{5.27}$$

Eliminating the other entries in column 2, we have

$$\left[\begin{array}{ccccc|c} 0 & 1 & 12 & -9/2 & 0 & 900 \\ 1 & 0 & -4 & 0 & 0 & 600 \\ 0 & 0 & 96 & 9 & 1 & 34{,}200 \end{array}\right]. \tag{5.28}$$

When we return to step 1, we see that the simplex tableau is in final form, since all entries in the bottom row of the matrix are positive. *We must now interpret the meaning of the matrix* (5.28), which is equivalent to the system of equations

$$\begin{aligned} y + 12t_1 - 9/2t_2 \quad &= \quad 900, \\ x \quad - \quad 4t_1 \quad &= \quad 600, \\ 96t_1 + \quad 9t_2 + P &= 34{,}200, \end{aligned} \tag{5.29}$$

where $x \geq 0$, $y \geq 0$, $t_1 \geq 0$, and $t_2 \geq 0$. If the slack variables $t_1$ and $t_2$ are zero, the profit $P$ will be a maximum. Setting $t_1 = t_2 = 0$ in the system (5.29) yields the solution

$$x = 600, \qquad y = 900, \qquad P = 34{,}200$$

of the linear programming problem. Hence, a maximum profit of $P =$ 34,200 cents = \$342.00 is attained by producing $x = 600$ hamburgers and $y = 900$ bowls of chili. This is exactly the same answer that we obtained in Example 1 in Section 5.2.

The addition of the slack variables to the equations defining the feasible set provides a convenient way of limiting consideration to only the points on the boundary of the feasible set. This is a useful stratagem since, as we proved in Section 5.2, only the corner points need be considered in determining the optimal value of the linear equation that we are optimizing. In (5.25) we are at the origin and $P = 0$. Steps 1 and 2 guarantee that the simplex algorithm moves in the direction in which each unit change increases $P$ the most. In this case, each unit in the $x$ direction increases $P$ by 30, while each unit in the $y$ direction increases $P$ by 18. Step 3 stops us at the corner (900, 0) of the feasible set and step 4 substitutes the coordinates of this corner point, yielding $P = 27{,}000$. We now return to step 1 where the negative entry in the bottom row informs us that this vertex is not optimal. Again steps 1 and 2 move us in the direction of the next vertex that increases $P$ the most (per unit change), and so on until we reach the optimal vertex.

In practice, the last column of the matrix $A$ is usually omitted since the entries in this column do not change or affect the outcome. However, to emphasize the connection with the associated system of inequalities, we will continue writing this column in each calculation.

***Example 1*** Maximize the function $P = 4x + 3y$ subject to the inequalities

$$\begin{aligned} 2x + 3y &\leq 12, \\ -3x + 2y &\leq 6, \\ x \geq 0, \qquad y &\geq 0. \end{aligned} \tag{5.30}$$

Let $t_1 \geq 0$ and $t_2 \geq 0$ be the slack variables that convert the first two inequalities in (5.30) into equalities. We then have the system

$$\begin{aligned} 2x + 3y + t_1 \qquad\qquad &= 12, \\ -3x + 2y \qquad + t_2 \qquad &= 6, \\ -4x - 3y \qquad\qquad + P &= 0, \end{aligned} \tag{5.31}$$

with associated augmented matrix

$$A \mid B = \begin{array}{c} \rightarrow \\ \\ \\ \end{array} \left[\begin{array}{ccccc|c} \textcircled{2} & 3 & 1 & 0 & 0 & 12 \\ -3 & 2 & 0 & 1 & 0 & 6 \\ -4 & -3 & 0 & 0 & 1 & 0 \end{array}\right]. \tag{5.32}$$
$$\uparrow$$

The first step in using the simplex algorithm is to find the largest negative entry in the bottom row; hence we select the first column of (5.32). Dividing the entries in $B$ by the corresponding entries in the first column, we obtain

$$(12/2) = 6 \quad \text{and} \quad (6/-3) = -2.$$

The smallest *positive* quotient is obtained using the divisor 2; thus the circled entry in (5.32) is the first pivot. Dividing each entry in the first row by 2 and eliminating the other entries in the first column, we have

$$\left[\begin{array}{ccccc|c} 1 & 3/2 & 1/2 & 0 & 0 & 6 \\ 0 & 13/2 & 3/2 & 1 & 0 & 24 \\ 0 & 3 & 2 & 0 & 1 & 24 \end{array}\right]. \tag{5.33}$$

Since all entries in the last row of (5.33) are positive, the simplex tableau is in final form. The augmented matrix is equivalent to the system

$$\begin{aligned} x + \tfrac{3}{2}y + \tfrac{1}{2}t_1 \qquad\qquad &= 6, \\ \tfrac{13}{2}y + \tfrac{3}{2}t_1 + t_2 \qquad &= 24, \\ 3y + 2t_1 \qquad + P &= 24, \end{aligned} \tag{5.34}$$

where $x \geq 0$, $y \geq 0$, $t_1 \geq 0$, and $t_2 \geq 0$. Hence,

$$P = 24 - 3y - 2t_1$$

will be maximized if $y = t_1 = 0$, yielding the solution

$$x = 6, \qquad y = 0, \qquad P = 24.$$

***Example 2*** Minimize the function $P = 3x + 5y$ subject to the inequalities

$$\begin{aligned} &2x + 5y \geq 48, \\ &3x + 4y \geq 58, \\ &x \geq 0, \qquad y \geq 0. \end{aligned} \tag{5.35}$$

We convert the system of inequalities (5.35) into a system of equations by introducing the slack variables $t_1$ and $t_2$. In this problem, however, the slack variables *are nonpositive*, since the left sides of the system (5.35) equal or exceed the right sides. Thus we have the system

$$\begin{aligned} 2x + 5y + t_1 \qquad\qquad &= 48, \\ 3x + 4y \qquad + t_2 \qquad &= 58, \\ -3x - 5y \qquad\qquad + P &= 0, \end{aligned} \tag{5.36}$$

where $x \geq 0$, $y \geq 0$, $t_1 \leq 0$, and $t_2 \leq 0$. The system (5.36) is equivalent to the matrix

$$\begin{matrix}\rightarrow \\ \\ \\ \end{matrix}\left[\begin{array}{rrrrr|r} 2 & \textcircled{5} & 1 & 0 & 0 & 48 \\ 3 & 4 & 0 & 1 & 0 & 58 \\ -3 & -5 & 0 & 0 & 1 & 0 \end{array}\right]. \tag{5.37}$$
$$\qquad\quad\uparrow$$

Using the simplex algorithm, we select the second column and, since $48/5 = 9.6$ and $58/4 = 14.5$, we begin by pivoting on the entry 5. Dividing row 1 by 5 and eliminating the remaining entries in column 2, we obtain

$$\begin{matrix} \\ \rightarrow \\ \\ \end{matrix}\left[\begin{array}{rrrrr|l} 2/5 & 1 & 1/5 & 0 & 0 & 9.6 \\ \textcircled{7/5} & 0 & -4/5 & 1 & 0 & 19.6 \\ -1 & 0 & 1 & 0 & 1 & 48 \end{array}\right]. \tag{5.38}$$
$$\uparrow$$

Dividing 19.6 by 7/5 and 9.6 by 2/5 yields

$$(9.6) \div (2/5) = 24, \qquad (19.6) \div (7/5) = 14,$$

so the 7/5 entry is the pivot. Multiplying the second row by 5/7 and eliminating the remaining entries in column 1, we have

$$\left[\begin{array}{rrrrr|r} 0 & 1 & 3/7 & -2/7 & 0 & 4 \\ 1 & 0 & -4/7 & 5/7 & 0 & 14 \\ 0 & 0 & 3/7 & 5/7 & 1 & 62 \end{array}\right]. \tag{5.39}$$

This simplex tableau is in final form. The matrix (5.39) is equivalent to the system

$$\begin{aligned} y + \tfrac{3}{7}t_1 - \tfrac{2}{7}t_2 \qquad &= 4, \\ x \qquad - \tfrac{4}{7}t_1 + \tfrac{5}{7}t_2 \qquad &= 14, \\ \tfrac{3}{7}t_1 + \tfrac{5}{7}t_2 + P &= 62, \end{aligned} \tag{5.40}$$

where $x \geq 0$, $y \geq 0$, $t_1 \leq 0$, and $t_2 \leq 0$. Since $t_1$ and $t_2$ are nonpositive, we will minimize

$$P = 62 - \tfrac{3}{7}t_1 - \tfrac{5}{7}t_2$$

if we set $t_1 = t_2 = 0$. Then, substituting these values in (5.40) we obtain $x = 14$, $y = 4$, and $P = 62$. This is the solution to our linear programming problem.

---

So far we have not solved any problem that could not be handled by the geometrical method. The great advantage of the simplex algorithm is that it may also be used on problems involving more than two variables. The following two examples involve three or more variables. Solving the linear programming problem by the geometrical method would be either extremely difficult or impossible.

***Example 3*** In Example 4 of Section 4.5, we consider the problem of stocking several lakes with three species of fish. The fourth lake required stocking with a *negative* number of bullheads to completely utilize the daily food production of the lake. Since this is impossible, the fish-and-game commissioners would realize that no positive combination of the three species of fish would completely utilize the lake's resources. Instead, they might wish to maximize the fish population in the lake while not exceeding the lake's resources. If $x$, $y$, and $z$ are the number of grams of live black bass, bullheads, and rainbows that we wish to put in the lake, we have the matrix inequality $AX \leq L_4$, or

$$\begin{aligned} 0.064x + 0.082y + 0.314z &\leq 1000, \\ 0.141x + 0.168y + 0.082z &\leq \phantom{0}500, \\ 0.555x + 0.493y + 0.406z &\leq 5000, \end{aligned} \tag{5.41}$$

where $x \geq 0$, $y \geq 0$, and $z \geq 0$. These six inequalities determine the feasible set in three-dimensional space. Subject to these inequalities, we wish to maximize the total mass of fish in the pond, that is, to maximize

$$M = x + y + z.$$

Introducing the three nonnegative slack variables $t_1$, $t_2$, and $t_3$ into the inequalities (5.41), we obtain the matrix $A \mid L_4$:

$$\begin{array}{c} \begin{matrix} x & \quad y & \quad z & \quad t_1 & t_2 & t_3 & M & \end{matrix} \\ \left[\begin{array}{ccccccc|c} 0.064 & 0.082 & 0.314 & 1 & 0 & 0 & 0 & 1000 \\ 0.141 & 0.168 & 0.082 & 0 & 1 & 0 & 0 & 500 \\ 0.555 & 0.493 & 0.406 & 0 & 0 & 1 & 0 & 5000 \\ -1 & -1 & -1 & 0 & 0 & 0 & 1 & 0 \end{array}\right]. \end{array}$$

Since all three negative entries are equal we may as well pick the first column. Then, since

$$1000/0.064 = 15{,}625, \quad 500/0.141 = 3546.1,$$

and

$$5000/0.555 = 9009.01,$$

it follows that the pivot is 0.141. Dividing the second row by 0.141 and eliminating the remaining entries in column 1, we have

$$\left[\begin{array}{ccccccc|c} 0 & 0.006 & 0.277 & 1 & -0.454 & 0 & 0 & 773.05 \\ 1 & 1.191 & 0.582 & 0 & 7.092 & 0 & 0 & 3546.1 \\ 0 & -0.168 & 0.083 & 0 & -3.936 & 1 & 0 & 3031.915 \\ 0 & 0.191 & -0.418 & 0 & 7.092 & 0 & 1 & 3546.1 \end{array}\right].$$

Clearly, we must now work with column 3. Here we have

$$\frac{773.05}{0.277} = 2790.79, \quad \frac{3546.1}{0.582} = 6092.96, \quad \text{and} \quad \frac{3031.915}{0.083} = 36{,}529.1,$$

so that 0.277 is the pivot. Dividing row 1 by 0.277 and eliminating the other entries in column 3, we have

$$\left[\begin{array}{ccccccc|c} 0 & 0.022 & 1 & 3.61 & -1.639 & 0 & 0 & 2790.794 \\ 1 & 1.178 & 0 & -2.101 & 8.046 & 0 & 0 & 1921.858 \\ 0 & -0.170 & 0 & -0.3 & -3.8 & 1 & 0 & 2800.279 \\ 0 & 0.2 & 0 & 1.509 & 6.407 & 0 & 1 & 4712.652 \end{array}\right].$$

This simplex tableau is in final form, and is equivalent to the system

$$\begin{aligned} 0.022y + z + 3.61\,t_1 - 1.639t_2 \qquad &= 2790.794, \\ x + 1.178y \qquad - 2.101t_1 + 8.046t_2 \qquad &= 1921.858, \\ -\,0.170y \qquad - 0.3\,t_1 - 3.8\,t_2 + t_3 &= 2800.279, \end{aligned} \tag{5.42}$$

and

$$M = 4712.652 - 0.2y - 1.509t_1 - 6.407t_2.$$

Since $y \geq 0$, $t_1 \geq 0$, and $t_2 \geq 0$, $M$ is maximized by setting all three of these variables equal to zero. Then $M = 4712.652$ grams is the total mass of fish that the lake can support on an indefinite basis. Replacing these values in the system (5.42) yields the values $z = 2790.794$, $x = 1921.858$, and $t_3 = 2800.279$. Thus, the fish-and-game commission should stock the fourth lake with 1921.858 grams of black bass and 2790.794 grams of rainbow trout.

***Example 4*** A feed manufacturer wishes to market a brand of feed that is guaranteed to contain at least 10% fiber, 2% fat, and 30% protein (by weight). The manufacturer purchases three primary feeds: alfalfa meal, fish meal, and soybean meal. The percentage contents of the three basic nutrients for these feeds are given in Table 5.1. The net costs of alfalfa meal, fish meal, and soybean meal are \$70, \$12, and \$93 per ton, respectively. The manufacturer can add a nonfood ingredient, which provides none of the nutrients, to make up any difference in weight. This filler costs \$2 per ton. What is the minimum cost to the manufacturer for one ton of such a mix?

**Table 5.1**

| | Percent (by weight) of nutrient in feed | | |
|---|---|---|---|
| | Alfalfa meal | Fish meal | Soybean meal |
| Fiber | 25 | 3 | 6 |
| Fat | 2 | 5 | 1 |
| Protein | 17 | 25 | 45 |

Let $x_1$, $x_2$, $x_3$, and $x_4$ be the decimal parts of a ton that will be required of the alfalfa meal, fish meal, soybean meal, and filler. Then we have the following equations and inequalities:

$$x_1 + x_2 + x_3 + x_4 = 1$$

since the total weight of the ingredients is one ton;

$$25x_1 + 3x_2 + 6x_3 \geq 10,$$

since these ingredients must contain at least 10% fiber (by weight);

$$2x_1 + 5x_2 + x_3 \geq 2,$$

since the ingredients must contain at least 2% fat; and

$$17x_1 + 25x_2 + 45x_3 \geq 30,$$

since at least 30% protein is in the mix. The total cost of the mix is

$$C = 70x_1 + 112x_2 + 93x_3 + 2x_4.$$

Thus, we obtain the simplex tableau

$$\begin{array}{cccccccc|c}
x_1 & x_2 & x_3 & x_4 & t_1 & t_2 & t_3 & C & \\
\hline
1 & 1 & 1 & 1 & 0 & 0 & 0 & 0 & 1 \\
25 & 3 & 6 & 0 & 1 & 0 & 0 & 0 & 10 \\
2 & 5 & 1 & 0 & 0 & 1 & 0 & 0 & 2 \\
17 & 25 & 45 & 0 & 0 & 0 & 1 & 0 & 30 \\
-70 & -112 & -93 & -2 & 0 & 0 & 0 & 1 & 0
\end{array},$$

where the slack variables $t_1 \leq 0$, $t_2 \leq 0$, and $t_3 \leq 0$. The first pivot is around 5, so we obtain

$$\left[\begin{array}{rrrrrrrr|r} 0.6 & 0 & 0.8 & 1 & 0 & -0.2 & 0 & 0 & 0.6 \\ 23.8 & 0 & 5.4 & 0 & 1 & -0.6 & 0 & 0 & 8.8 \\ 0.4 & 1 & 0.2 & 0 & 0 & 0.2 & 0 & 0 & 0.4 \\ 7.0 & 0 & 40.0 & 0 & 0 & -5.0 & 1 & 0 & 20.0 \\ -25.2 & 0 & -70.6 & -2 & 0 & 22.4 & 0 & 1 & 44.8 \end{array}\right],$$

The next pivot point is 40, so we have

$$\left[\begin{array}{rrrrrrrr|r} 0.46 & 0 & 0 & 1 & 0 & -0.1 & -0.02 & 0 & 0.2 \\ 22.855 & 0 & 0 & 0 & 1 & 0.075 & -0.135 & 0 & 6.1 \\ 0.365 & 1 & 0 & 0 & 0 & 0.225 & -0.005 & 0 & 0.3 \\ 0.175 & 0 & 1 & 0 & 0 & -0.125 & 0.025 & 0 & 0.5 \\ -12.845 & 0 & 0 & -2 & 0 & 13.575 & 1.765 & 1 & 80.1 \end{array}\right].$$

Our next pivot is 22.855. Dividing the second row by 22.855 and clearing the rest of column 1 yields

$$\left[\begin{array}{rrrrrrrr|r} 0 & 0 & 0 & 1 & -0.02 & -1.02 & -0.017 & 0 & 0.077 \\ 1 & 0 & 0 & 0 & 0.044 & 0.003 & -0.006 & 0 & 0.267 \\ 0 & 1 & 0 & 0 & -0.016 & 0.224 & -0.003 & 0 & 0.203 \\ 0 & 0 & 1 & 0 & -0.008 & -0.126 & 0.026 & 0 & 0.453 \\ 0 & 0 & 0 & -2 & 0.562 & 13.617 & 1.689 & 1 & 83.528 \end{array}\right].$$

Finally, we clear the entries in the fourth column to get

$$\left[\begin{array}{rrrrrrrr|r} 0 & 0 & 0 & 1 & -0.02 & -1.02 & -0.017 & 0 & 0.077 \\ 1 & 0 & 0 & 0 & 0.044 & 0.003 & -0.006 & 0 & 0.267 \\ 0 & 1 & 0 & 0 & -0.016 & 0.224 & -0.003 & 0 & 0.203 \\ 0 & 0 & 1 & 0 & -0.008 & -0.126 & 0.026 & 0 & 0.453 \\ 0 & 0 & 0 & 0 & 0.522 & 11.577 & 1.655 & 1 & 83.682 \end{array}\right].$$

Thus, a ton of this mix costs the manufacturer \$83.68 and consists approximately of 0.077 tons of filler, 0.267 tons of alfalfa meal, 0.203 tons of fish meal, and 0.453 tons of soybean meal. Any inexactitudes in this answer are the result of round-off error caused by carrying only three decimal places.

## EXERCISES 5.3

*In Exercises 1–16, solve the given linear programming problem by the simplex algorithm.*

**1.** Maximize $P = 5x + 2y$
subject to
$3x + 4y \leq 5,$
$x + 2y \leq 4,$
$x \geq 0, \quad y \geq 0.$

**2.** Maximize $P = x + 3y$
subject to
$2x + 5y \leq 7,$
$3x + y \leq 5,$
$x \geq 0, \quad y \geq 0.$

**3.** Maximize $P = x + y$
subject to
$2x + 2y \leq 5,$
$x + 3y \leq 6,$
$x \geq 0, \quad y \geq 0.$

**4.** Maximize $P = 2x + 3y$
subject to
$4x + 6y \leq 7,$
$x + y \leq 5,$
$x \geq 0, \quad y \geq 0.$

**5.** Maximize $P = 4x + 3y$
subject to
$x - y \leq 5,$
$3x + 15y \leq 7,$
$x \geq 0, \quad y \geq 0.$

**6.** Minimize $P = x + 3y$
subject to
$3x + 4y \geq 5,$
$x + 2y \geq 4,$
$x \geq 0, \quad y \geq 0.$

**7.** Minimize $P = 4x + 3y$
subject to
$2x + 2y \geq 5,$
$-x + 3y \geq 6,$
$x \geq 0, \quad y \geq 0.$

**8.** Minimize $P = 2x + 3y$
subject to
$2x + 5y \geq 7,$
$3x + y \geq 5,$
$x \geq 0, \quad y \geq 0.$

**9.** Minimize $P = x + y$
subject to
$2x + 2y \geq 5,$
$x + 5y \geq 6,$
$x \geq 0, \quad y \geq 0.$

**10.** Minimize $P = 5x + 2y$
subject to
$10x + 4y \geq 3,$
$x - 2y \geq 1,$
$x \geq 0, \quad y \geq 0.$

**11.** Maximize $P = 2x + 2y + z$
subject to
$4x + 2y + 3z \leq 10,$
$6x + 3y + z \leq 12,$
$x + 2y - z \leq 5,$
$x \geq 0, \quad y \geq 0, \quad z \geq 0.$

**12.** Maximize $P = x + y + z$
subject to
$x + y - z \leq 2,$
$x - y + 2z \leq 5,$
$3x + y + 2z \leq 9,$
$x \geq 0, \quad y \geq 0, \quad z \geq 0.$

**13.** Minimize $P = x + y + z$
subject to
$x + y - z \geq 2,$
$-x + y + 2z \geq 5,$
$3x + y + 2z \geq 9,$
$x \geq 0, \quad y \geq 0, \quad z \geq 0.$

**14.** Minimize $P = x + 2y + 3z$
subject to
$4x + 2y + 3z \geq 8,$
$6x + 3y + z \geq 10,$
$x + 2y - z \geq 4,$
$x \geq 0, \quad y \geq 0, \quad z \geq 0.$

**15.** Maximize $P = x + y + 2z - w$ subject to

$$\begin{aligned} x + y + z + w &\le 5, \\ 2x + 3y + 4z &\le 7, \\ 3y + 5z + 7w &\le 14, \\ x + 2z + 3w &\le 6, \\ x \ge 0, \quad y \ge 0, \quad z \ge 0, \quad w &\ge 0. \end{aligned}$$

**16.** Minimize $P = x + 2y + 3z + 4w$ subject to

$$\begin{aligned} x + y + z + w &\ge 5, \\ 2x + 3y + 4z &\ge 7, \\ y + 3z + 5w &\ge 15, \\ x + 2z + 3w &\ge 6, \\ x \ge 0, \quad y \ge 0, \quad z \ge 0, \quad w &\ge 0. \end{aligned}$$

*Using the simplex algorithm, solve the following exercises in Section* 5.2 (*see pp.* 213–215).

**17.** Exercise 13

**18.** Exercise 14

**19.** Exercise 15

**20.** Exercise 18

**21.** Exercise 19

**22.** Exercise 20

**23.** Exercise 21

**24.** Exercise 22

**25.** Exercise 23

**26.** Exercise 24

**27.** Suppose that lake 4 in Example 3 (see p. 222) can sustain a daily yield of 3200 g of mosquito larvae, 840 g of crustaceans, and 4200 g of minnows and worms. How should the lake be stocked in order to maximize the weight of the fish in that lake?

**28.** Solve Exercise 27 if the lake can sustain a yield of 400 g of mosquito larvae, 250 g of crustaceans, and 900 g of minnows and worms.

**29.** Suppose that the feed manufacturer in Example 4 (see p. 224) wishes to produce a brand of feed guaranteeing a content of 5% fiber, 3% fat, and 20% protein (by weight). What is the minimum cost to the manufacturer per ton, if he plans to use the same nutrients and filler as those in Example 4?

**30.** Solve Exercise 29 if the contents of the feed exceed 7% fiber, 2% fat, and 28% protein.

**31.** A restaurant has been reserved for a Christmas party. A total of 510 persons are expected to attend. Three main courses will be served: prime rib, steak and lobster, and chicken Kiev. Experience indicates that the steak and lobster dinner is selected at least twice as often as chicken Kiev, while prime rib is selected at least three times as often as chicken Kiev. It costs the restaurant $3 to make the prime rib dinner, $4 for the steak dinner, and $2 for the chicken Kiev. How many dinners of each kind should be produced so that the total cost to the restaurant is minimized?

**32.** Solve Exercise 31 if instead of prime rib being chosen three times as often as chicken Kiev, we need at least 90 more orders of prime rib than chicken Kiev.

**33.** Solve Exercise 31 if instead of steak and lobster being chosen twice as often as chicken Kiev, we need at least 80 more orders of steak and lobster than chicken Kiev.

**34.** A Scotch whisky blender imports two kinds of straight Scotch whisky, Glenfish at $12.00 a fifth and Glencannon at $10.00 a fifth, which he mixes with neutral spirits at $1.00 a fifth. He sells three blended Scotch whiskies mixed according to the following specifications.

| Blend | Specification | Price per fifth |
|---|---|---|
| Old Satin | 40% Glenfish<br>30% Glencannon | 9.75 |
| Black Stallion | 25% Glenfish<br>40%Glencannon | 8.50 |
| Three Violets | 10% Glenfish<br>35% Glencannon | 6.75 |

His permits allow him to purchase at most 1500 fifths of Glenfish, 2000 fifths of Glencannon, and 3000 fifths of neutral spirits each day. Find the production policy that maximizes profits.

**35.** Mama Rosa's Spaghetti Sauce Company markets three types of spaghetti sauce: with meat, with mushrooms, and combined. The ingredients of each sauce are listed below.

| Ingredient | Sauce with meat | Sauce with mushrooms | Combined |
|---|---|---|---|
| Tomato sauce | 80% | 85% | 70% |
| Ground beef | 20% | 0% | 15% |
| Mushrooms | 0% | 15% | 15% |

The factory can produce 4000 lb of tomato sauce, 1000 lb of ground beef, and 500 lb of cooked mushrooms every hour. The cost of these ingredients is 40¢ per lb for tomato sauce, 90¢ per lb for ground beef, and $1.20 per lb for cooked mushrooms. Mama Rosa sells the sauce with meat and the sauce with mushrooms at 70¢ per lb, while the combined sauce sells for 82¢ per lb. Determine the proportions of the three sauces that will maximize Mama Rosa's profits.

## 5.4 NONSTANDARD LINEAR PROGRAMMING PROBLEMS

If we examine the problems that we have just solved using the simplex algorithm, we will discover that they are of two types:

1. A *maximum problem*, which consists of maximizing a function

   $P = CX,$

subject to the inequalities

$$AX \leq B, \qquad X \geq 0.$$

Here $A$ is a matrix of constants, $B$ is a column vector of constants, $C$ is a row vector of constants, and $X$ is the column vector of unknowns;

2. A *minimum problem*, which consists of minimizing a function

$$P = CX$$

subject to the inequalities

$$AX \geq B, \qquad X \geq 0.$$

In each case the vector $B$ has contained only nonnegative entries. *If we try to use the simplex algorithm in a different situation, it may not provide the solution.* Frequently, however, it is possible to modify the simplex algorithm and obtain an answer. The formulation of a set of rules that would always give us a solution is too complicated to be considered here. However, the following examples illustrate some of the difficulties that may arise.

***Example 1*** The pollution control problem that we solved in Example 3 in Section 5.2 requires that we minimize the cost function

$$C = 8x + 12y \quad \text{(dollars)}$$

subject to the inequalities

$$\begin{aligned} 152x + 38y &\leq 22{,}800, \\ x + \quad y &\geq \quad 500, \\ x \geq 0, \qquad y &\geq 0. \end{aligned} \tag{5.43}$$

The first inequality in the system (5.43) violates the requirements of the minimum problem. Hence, Example 3 in Section 5.2 is not of either type. Let's see what happens when we apply the simplex algorithm to the corresponding simplex tableau:

$$\begin{array}{ccccc} x & y & t_1 & t_2 & C \end{array}$$
$$\left[\begin{array}{rrrrr|r} 152 & 38 & 1 & 0 & 0 & 22{,}800 \\ 1 & 1 & 0 & 1 & 0 & 500 \\ -8 & -12 & 0 & 0 & 1 & 0 \end{array}\right],$$

where $t_1 \geq 0$ but $t_2 \leq 0$. The 1 in the second column is the pivot, so we get

$$\left[\begin{array}{ccccc|c} 114 & 0 & 1 & -38 & 0 & 3800 \\ 1 & 1 & 0 & 1 & 0 & 500 \\ 4 & 0 & 0 & 12 & 1 & 6000 \end{array}\right].$$

This simplex tableau is in final form. However, the corresponding cost function is

$$C = 6000 - 4x - 12t_2. \tag{5.44}$$

Since $t_2 \leq 0$, the cost will increase unless we choose $t_2 = 0$ in Eq. (5.44). Hence, we can set $t_2 = 0$. However, the cost will *decrease* if we choose $x > 0$. Thus, we cannot set $x = 0$. Instead, we need to eliminate this term in the last row of the simplex tableau, without introducing any $y$-terms into that row. We can do so by dividing the first row by 114 and eliminating the other entries in column 1. Thus we obtain

$$\left[\begin{array}{ccccc|c} 1 & 0 & \frac{1}{144} & -\frac{1}{3} & 0 & 33\frac{1}{3} \\ 0 & 1 & -\frac{1}{144} & 1 & 0 & 466\frac{2}{3} \\ 0 & 0 & -\frac{1}{36} & 13\frac{1}{3} & 1 & 5866\frac{2}{3} \end{array}\right].$$

Here the cost function is

$$C = 5866\tfrac{2}{3} + \tfrac{1}{36}t_1 - \tfrac{40}{3}t_2,$$

and since $t_1 \geq 0$ and $t_2 \leq 0$, the minimum is attained if $t_1 = t_2 = 0$. Thus, the minimum cost of \$5866.67 is obtained by buying $x = 33\frac{1}{3}$ tons of the 4% sulfur coal and $y = 466\frac{2}{3}$ tons of the 1% sulfur coal.

This example demonstrates the necessity of *not* proceeding mechanically with the simplex algorithm when attacking a nonstandard linear programming problem. Observe that we did not have the solution when the simplex tableau was in final form; instead the solution was reached with a simplex tableau not in final form. Hence, it is important to carefully examine the cost function to determine whether the solution to the problem has been attained.

***Example 2*** Vino Winery produces 90,000 liters of pure Cabernet grape wine and 300,000 liters of Gamay grape wine each year. Vino bottles two wines: a varietal labeled "Cabernet Sauvignon," which sells for \$5 a liter, and a generic labeled "Red Table Wine" selling for \$2 a liter. Varietal wines are required to contain at least 50% of that variety of wine. The blending specifications of the two types of wine follow.

| Bottle | Gamay wine | Cabernet wine |
|---|---|---|
| Cabernet Sauvignon | At most 50% | At least 50% |
| Red table wine | At most 80% | At least 20% |

Let $x$ represent the number of liter bottles of Cabernet Sauvignon and $y$ the number of liters of red table wine. There is at most 50% Gamay in the Cabernet Sauvignon and at most 80% Gamay in the red table wine. Hence

$$0.5x + 0.8y \geq 300{,}000,$$

since all the Gamay wine is used and the amount on the left side of the inequality is at least equal to the amount available. Similarly,

$$0.5x + 0.2y \leq 90{,}000,$$

since there is at least 50% Cabernet wine in the Cabernet Sauvignon and at least 20% in the red table wine. Clearly, $x \geq 0$ and $y \geq 0$, and we are interested in maximizing the revenue

$$R = 5x + 2y \quad \text{(dollars)}.$$

Thus we have the linear programming problem of maximizing

$$R = 5x + 2y \tag{5.45}$$

subject to the inequalities

$$\begin{aligned} 0.5x + 0.8y &\geq 300{,}000, \\ 0.5x + 0.2y &\leq \ \ 90{,}000, \\ x \geq 0, \qquad y &\geq 0. \end{aligned} \tag{5.46}$$

Notice that the first inequality of (5.46) violates the requirements for a maximum problem. Forming the simplex tableau, we have

$$\begin{array}{ccccc} x & y & t_1 & t_2 & R \end{array}$$
$$\left[\begin{array}{ccccc|r} 0.5 & 0.8 & 1 & 0 & 0 & 300{,}000 \\ 0.5 & 0.2 & 0 & 1 & 0 & 90{,}000 \\ -5 & -2 & 0 & 0 & 1 & 0 \end{array}\right], \tag{5.47}$$

where $t_1 \leq 0$ and $t_2 \geq 0$, since we must subtract from the left side of the first inequality of (5.46) to get an equality. The 0.5 in the second row is the pivot, yielding

$$\left[\begin{array}{ccccc|c} 0 & 0.6 & 1 & -1 & 0 & 210{,}000 \\ 1 & 0.4 & 0 & 2 & 0 & 180{,}000 \\ 0 & 0 & 0 & 10 & 1 & 900{,}000 \end{array}\right], \tag{5.48}$$

which is in final form. To simplify (5.48), multiply the top row by 5/3 and use it to clear the remaining entry in the second column:

$$\left[\begin{array}{ccccc|c} 0 & 1 & 5/3 & -5/3 & 0 & 350{,}000 \\ 1 & 0 & -2/3 & 8/3 & 0 & 40{,}000 \\ 0 & 0 & 0 & 10 & 1 & 900{,}000 \end{array}\right]. \tag{5.49}$$

The simplex tableaux (5.48) and (5.49) are both in final form. The revenue function is

$$R = 900{,}000 - 10t_2,$$

and since $t_2 \geq 0$, we can maximize the revenue by choosing $t_2 = 0$. The problem now is to determine the production levels $x$ and $y$. From (5.49) we obtain (with $t_2 = 0$)

$$y = 350{,}000 - \frac{5}{3}t_1,$$

$$x = 40{,}000 + \frac{2}{3}t_1,$$

where $t_1 \leq 0$. We must find $t_1$. To do so, we see how much of each unblended wine is required to produce $40{,}000 + (2/3)t_1$ bottles of Cabernet Sauvignon and $350{,}000 - (5/3)t_1$ bottles of red table wine. For the Cabernet wine we have

$$0.5[40{,}000 + (2/3)t_1] + 0.2[350{,}000 - (5/3)t_1] = 90{,}000 \text{ liters},$$

while for the Gamay wine we have

$$0.5[40{,}000 + (2/3)t_1] + 0.8[350{,}000 - (5/3)t_1] = 300{,}000 - t_1 \text{ liters}.$$

Since 90,000 liters of Cabernet and 300,000 liters of Gamay are available we must set $t_1 = 0$; otherwise, since $t_1 \leq 0$, more than the available supply of Gamay will be needed.

Thus, Vino Winery should produce 40,000 bottles of Cabernet Sauvignon and 350,000 bottles of red table wine.

The following example illustrates how small changes in one of the parameters may cause large changes in the solution. When this happens we say that the linear programming problem is *sensitive* to changes in the parameters. *Sensitivity analysis* is an important tool for the manager or scientist, since price fluctuations or measurement inaccuracies are often present in most mathematical models of real situations. A sensitive model serves as an indication that caution should be exercised by the user. For example, it may indicate that a certain product is reaching market saturation and that an immediate change in production is essential.

***Example 3*** Suppose that both wines in Example 2 sold at \$2 per liter. Total revenue for producing 350,000 bottles of red table wine and 40,000 bottles of Cabernet Sauvignon is \$780,000, and all the unblended wine is used.

Now assume that because of an overabundance, varietal wines drop their price at bottling time to \$1.75 per liter. Rather than lose \$10,000 [= 40,000 · (25¢)] by bottling Cabernet Sauvignon, Vino Winery should decide to bottle only red table wine. Mixing all their wines together, we see that the percentage of Gamay in the mixture is 300/390 ≈ 77%, while 90/390 ≈ 23% is Cabernet, well within the specifications of red table wine. Indeed, *any* decrease in the \$2 price of varietal wines should cause Vino Winery to eliminate the production of Cabernet Sauvignon. Thus, this maximum problem is sensitive to changes in the parameters.

Although this problem is easy to solve geometrically (see Fig. 5.11), the simplex algorithm must be modified substantially. Students interested in how such difficulties are resolved should consult more advanced books on linear programming, such as *Introduction to Linear Programming*, by W. W. Garvin (New York: McGraw-Hill, 1960) and *Linear Programming Methods*, by E. O. Heady and W. Chandler (Ames, Iowa: Iowa State College Press, 1958).

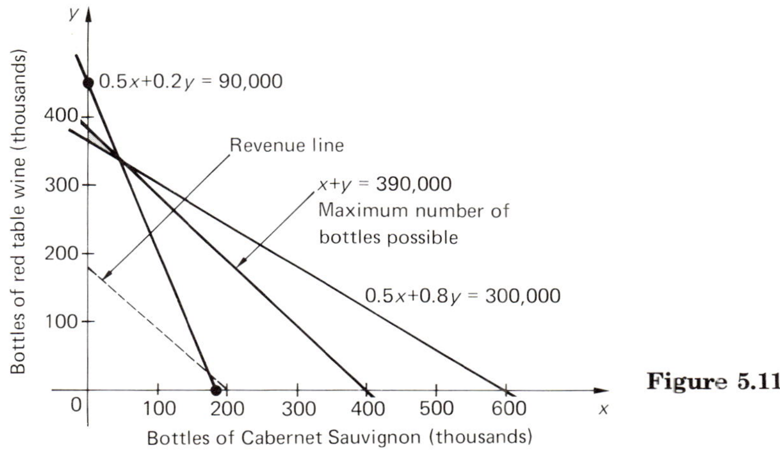


**Figure 5.11**

## EXERCISES 5.4

*In Exercises 1–8, solve the given linear programming problem by modifying the simplex algorithm. Check your solution using the geometric method.*

**1.** Maximize $P = 5x + 2y$
subject to
$3x + 4y \leq 5,$
$4x + 2y \geq 4,$
$x \geq 0, \quad y \geq 0.$

**2.** Maximize $P = x + 3y$
subject to
$2x + 5y \geq 7,$
$3x + y \leq 5,$
$x \geq 0, \quad y \geq 0.$

**3.** Minimize $P = x + y$
subject to
$2x + 2y \leq 5,$
$x + 3y \geq 6,$
$x \geq 0, \quad y \geq 0.$

**4.** Minimize $P = 2x + 3y$
subject to
$4x + 6y \leq 7,$
$5x + y \geq 5,$
$x \geq 0, \quad y \geq 0.$

**5.** Minimize $P = 4x + 3y$
subject to
$x + y \geq 5,$
$3x + 15y \leq 7,$
$x \geq 0, \quad y \geq 0.$

**6.** Minimize $P = x + 3y$
subject to
$3x + 4y \leq 5,$
$x + 4y \geq 4,$
$x \geq 0, \quad y \geq 0.$

**7.** Minimize $P = 4x + 3y$
subject to
$2x + 2y \geq 5,$
$x + 3y \geq -6,$
$x \geq 0, \quad y \geq 0.$
(*Hint*: Change the signs of the second equation.)

**8.** Minimize $P = 2x + 3y$
subject to
$-2x + 5y \geq -7,$
$3x + y \geq 5,$
$x \geq 0, \quad y \geq 0.$

*Modify the simplex algorithm to solve the following exercises from Section 5.2* (*see* p. 214).

**9.** Exercise 16

**10.** Exercise 17

*Determine whether the following exercises are sensitive by using the geometric method.*

**11.** Maximize and minimize $P = x + y$
subject to
$2x + 2y \leq 5,$
$x + 3y \geq 6,$
$x \geq 0, \quad y \geq 0.$

**12.** Maximize and minimize $P = 2x + 3y$
subject to
$2x + y \leq 10,$
$4x + 6y \geq 30,$
$x \geq 0, \quad y \geq 0.$

**13.** Wally's Nut Company processes 1000 lb of cashews, 700 lb of peanuts, and 300 lb of walnuts each day. It prepares two mixes whose ingredients are listed below.

| Mix | Cashews | Peanuts | Walnuts |
|---|---|---|---|
| Mixed nuts | At least 50% | At most 30% | At least 20% |
| Cashew mix | At least 50% | At most 50% | None |

a) If both mixes sell at \$2 per lb, what should the daily production of each mix be to maximize revenue?
b) Suppose that the mixed nuts sell at \$2.25 per lb, while cashews sell at \$2.00 per lb. What production will maximize revenue?
c) Suppose that the mixed nuts sell for \$2.00 per lb, while the cashew mix sells at \$2.25 per lb. Maximize the revenue.
d) Suppose that cashews cost \$1.60 per lb, peanuts cost 60¢ per lb, and walnuts cost \$1.50 per lb. If the mixed nuts sell for \$2.25 per lb and the cashew mix sells at \$2.00 per lb, what production will maximize profits?
e) Answer part (d) if the cashew mix sells at \$2.25, while the mixed nuts sell at \$2.00 per lb.

## CHAPTER VOCABULARY

**augmented matrix** $A \mid B$

**corner (vertex) of feasible set**

**feasible set**

**final form of tableau**

**half-plane**

**integer programming**

**lattice**

**linear inequality** $ax + by > c$

**linear programming** geometric method; simplex algorithm

**maximum problem** $P = CX, AX \leq B, X \geq 0$

**minimum problem** $P = CX, AX \geq B, X \geq 0$

**optimal value**

**profit function** $P = CX$

**sensitivity analysis**

**set** collection

**simplex algorithm**

**simplex method**

**simplex tableau**

**slack variables** $t_i$

# PART II PROBABILITY and STATISTICS

# CHAPTER 6 PROBABILITY

## 6.1 INTRODUCTION

It would be most unusual for any of us to have a day in which we faced no situation involving uncertainties. Will the car ahead of us flashing a left-turn signal really turn left? Will the events scheduled for the weekend occur as planned? Have you studied the right material for the test? Usually we must make our decisions instinctively since it is impossible to quantify the choices. However, in many situations it is possible to assign a numerical value to each choice indicating how likely it is to happen. Often we say that the "chances are 1 in 5" or the "odds are 4 to 1 against" a certain event happening. These numerical values are associated with a mathematical model that only approximates the real-world situation. For example, we might assume that in flipping an ordinary coin each side is equally likely to turn up, although no effort has been made by the United States Mint to mint "fair" coins.

The techniques used to obtain these numerical values and their use in decision making is called *probability theory.*

Before we begin a systematic development of the study of probability, it would be interesting to see how well you do instinctively. Here is a true–false test involving a variety of situations. Take your time; some of the questions are tricky!

*True False* **1.** The doctor tells you that you only have a 50% chance of surviving any heart attack. That means that even if you survive the first attack, you will certainly die the second time you have a heart attack.

*True False* **2.** In tossing a coin, we find that heads has come up five times in a row. The chances that heads will come up again on the next toss are less than 1 in 10.

*True False* **3.** A man has two coins in his pocket, a fair one and one that has heads on both sides. Suppose he reaches into his pocket, pulls out a coin, and sees that the top side shows heads. Then the odds are even that he has selected the fair coin.

*True False* **4.** Two equally good handball players play a four-game match. The odds are even that the final score will be two games each.

*True False* **5.** There are 25 people at a party. There is less than 1 chance in 10 that two of them will have the same birthday.

*True False* **6.** At least one of these questions is true.

We shall now provide answers to these questions in a highly intuitive manner. Later sections will provide justification for each solution. The main purpose of this test is to demonstrate that a systematic approach is necessary.

**Question 1** If you answered true to this question, you need to read it more carefully a second time. Actually, it states that you have a 50% chance of surviving *each* heart attack. Granted, only 50% of the people experiencing the first attack survive, but 50% of these survivors will also survive the second heart attack. Since 50% of 50% is 25%, you have one chance in four of surviving two heart attacks. Thus, the answer to question 1 is false. The probabilities of surviving any number of heart attacks can easily be determined by using the geometrical model, called a *tree*, shown in Fig. 6.1.

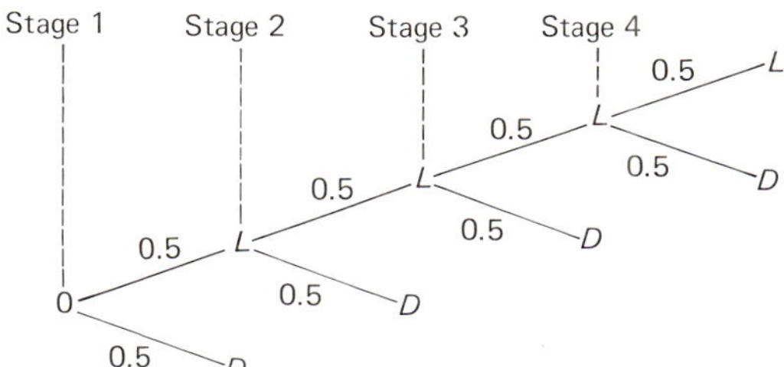


**Figure 6.1**

Any situation that advances in stages can be illustrated by a tree, constructed one stage at a time from an initial vertex 0. Starting at this vertex we draw as many lines as there are possible outcomes to the first stage. In this case, the first stage is the occurrence of the first heart attack, from which there are only two possible outcomes: living, which we denote by the letter L, and dying, by D. We label each line with the probability with which the outcome occurs, in this case, 0.5 (= 50%). If the patient has died, no further heart attacks are possible, so the second stage (the occurrence of the second heart attack) is only associated with people who survived the first attack. Using that outcome as a new vertex, we again draw two lines indicating the possible outcomes of the second attack and label their ends L and D. Each line is again labeled by the probability of that outcome occurring. We repeat this process again for the third attack, and so on, obtaining the tree shown in Fig. 6.1.

Now, to calculate the probability of surviving three heart attacks but not the fourth, we multiply the probabilities along the path OLLLD together and obtain

$$(0.5)^4 = \left(\frac{1}{2}\right)^4 = \frac{1}{2^4} = \frac{1}{16},$$

so the chances of this happening are 1 in 16.

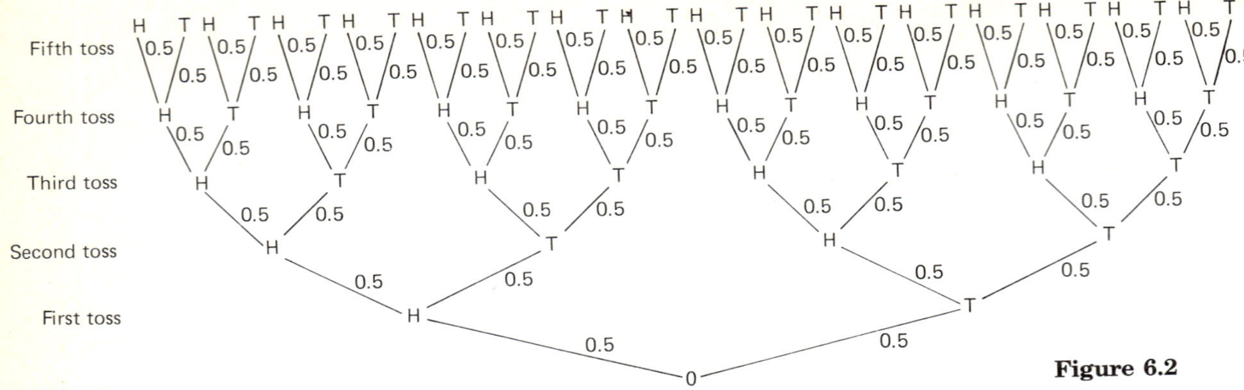


**Figure 6.2**

**Question 2** Here again the answer is false, since each toss of a coin is independent of all previous outcomes. Thus, the probability that heads will come up again is $\frac{1}{2}$, if the coin is fair.

If you answered this question incorrectly, it is probably because you misinterpreted the question as asking what the odds are on heads coming up five times in a row. Had that been the question, we could have found the exact probability by drawing a tree, as shown in Fig. 6.2. Each stage in this tree is another flip of the coin, which must be associated with all the outcomes of the previous stage.

Multiplying the probabilities along the leftmost "branch" of the tree OHHHHH, we obtain

$$(0.5)^5 = (\tfrac{1}{2})^5 = \tfrac{1}{32},$$

which is the probability that five heads will be thrown in succession. Notice that each path from the "root" O to the top of the tree is one of $2^5 = 32$ possible ways of flipping a coin five times. Of course, the total number of heads and tails may be the same along two paths, but the order in which they are obtained will be different.

**Question 3** You may have deceived yourself into thinking that since it could be either coin and since the probability of selecting each coin is 0.5, then the answer is true. Actually, the answer is false, because you must take into account the fact that heads can come up in *three* ways: two ways on the two-headed coin and one way on the fair coin. Thus, the two-headed coin is *twice* as likely to have been selected as the fair coin is when the top side shows heads.

The important point here is the need to count the total number of ways in which an event can happen. If we had two fair coins and the two-headed coin, four of the six sides would show heads. Were we then to draw a coin and notice that its top side was heads, we would know that this could happen in two ways with the trick coin, and in two ways with the fair coins. In this case, the probability of the coin being fair is 0.5.

Returning to the situation in which we had one fair coin and one two-headed coin, we might ask what the probability is of flipping heads five times in a row, if after each flip the coin is replaced in the pocket and a coin is again arbitrarily selected for the next toss. Since three of the four sides are heads and each coin is equally likely to be selected, heads will appear 75% of the time. The probabilities in the tree of Fig. 6.2 must be changed to reflect these odds: Each line rising to an H must be labeled 0.75, while each line rising to a T is labeled 0.25 (see Fig. 6.3). Thus, the branch OHHHHH has the probability

$$(0.75)^5 = \left(\frac{3}{4}\right)^5 = \frac{243}{1024} \approx 0.2373$$

of occurring.

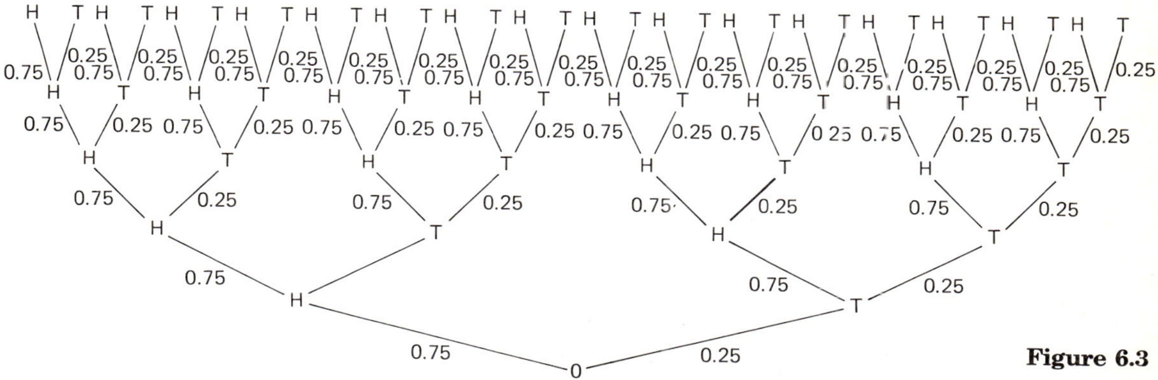


Figure 6.3

**Question 4** This question can also be answered by drawing a tree. First, we label the players A and B. Starting at the vertex O, we observe there are two outcomes to the first game: It is won by either A or B. We indicate who won the game by drawing a line labeled with the letter of the winning player. Since each player has an equal chance of winning, the probability associated to each line is 0.5 (see Fig. 6.4).

We observe that there are 16 possible ways in which the match could have been played, each with a probability of $(0.5)^4 = (\frac{1}{2})^4 = \frac{1}{16}$ of occurring. The winning order of each possible match is shown to the right of each branch. Only the *six* ways indicated by an asterisk lead to a drawn match. Thus, the probability that the final score will be two

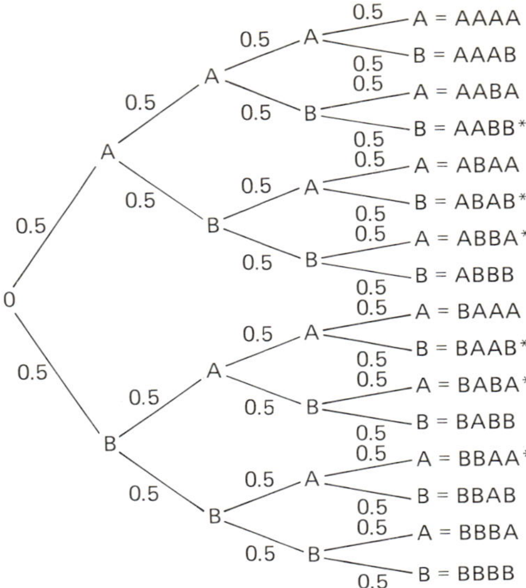


**Figure 6.4**

games apiece is

$$6\left(\frac{1}{16}\right) = \frac{3}{8} = 0.375,$$

since each of the six ways in which the match can be drawn has probability $\frac{1}{16}$ of occurring. Hence, question 4 is false.

**Question 5** Surprisingly, the answer to this question is also false! In fact we will see that there is a better than 50–50 chance that two of the 25 people will have the same birthday.

We can consider this problem as a tree with 25 stages, each stage (corresponding to a given person's birthday) having 365 possible outcomes (the number of days in a year if we ignore leap years). However, it quickly becomes evident that drawing this tree is impossible in practice, since each of the 365 outcomes of each stage splits into 365 ways at the next stage, and so on. Thus, we would have to draw $(365)^{25} \approx 1.141 \times 10^{64}$ branches! Therefore, we must find a different approach to this problem.

Instead of trying to find the probability that two people at the party have the same birthday, we can find the probability that everyone at the party has a different birthdate. Once we have the probability $p$ that no two birthdays are the same, the difference $1 - p$ is the probability that at least two people have the same birthday.

To find $p$ we must make an assumption that is probably not statistically correct, namely, that all 365 birthdays are equally likely to occur. Statistical evidence suggests that because of a variety of conditions, such as climate, geographical location, and habits, more births occur at certain times of the year than at others. However, *because it leads to a simpler mathematical model*, we will assume that each day has probability 1/365 of occurring.

As we ask each person at the party to tell us his or her birthdate, we know that there are 365 possible answers that the first person can give (discounting February 29), but only 364 answers that the second person can give without matching the first person's birthday. Thus, the probability that the first two answers are different is 364/365. Of the 365 possible answers, only 363 are allowed for the third person so that his or her birthday does not match the previous two. Thus, the probability that the first three persons have different birthdays is

$$\frac{364}{365}\cdot\frac{363}{365}.$$

Continuing in this fashion, we find that there are 362 remaining days that the fourth person can give without matching the previous answers, so the probability that four people have different birthdays is

$$\frac{364}{365}\cdot\frac{363}{365}\cdot\frac{362}{365}$$

and so on; thus, the probability that 25 people have different birthdays is

$$p = \frac{364}{365}\cdot\frac{363}{365}\cdot\frac{362}{365}\cdot \quad \cdots \quad \cdot\frac{341}{365} \approx 0.401. \tag{6.1}$$

Hence, the probability that at least two people will have the same birthday is

$$1 - p \approx 1 - 0.401 = 0.569.$$

(The calculation in Eq. 6.1 can be checked quickly on a calculator.)

The procedure above can easily be adapted to groups of other sizes. Table 6.1 gives the probability that at least two people in a group of $n$ persons have the same birthday, $2 \leq n \leq 30$.

**Question 6** Your answer to this question is correct regardless of how you answered, since the other five questions are false. This question is a good illustration of the traps into which the unwary can stumble. In the study of probability, it is essential that we adequately define what is

**Table 6.1**
**Probability $p(A)$ that at least two persons in a group of $n$ have the same birthday, $2 \leq n \leq 30$.**

| $n$ | $p(A)$ | $n$ | $p(A)$ | $n$ | $p(A)$ |
|---|---|---|---|---|---|
| 2 | 0.003 | 11 | 0.141 | 21 | 0.444 |
| 3 | 0.008 | 12 | 0.167 | 22 | 0.476 |
| 4 | 0.016 | 13 | 0.194 | 23 | 0.507 |
| 5 | 0.027 | 14 | 0.223 | 24 | 0.538 |
| 6 | 0.040 | 15 | 0.253 | 25 | 0.569 |
| 7 | 0.056 | 16 | 0.284 | 26 | 0.598 |
| 8 | 0.074 | 17 | 0.315 | 27 | 0.627 |
| 9 | 0.095 | 18 | 0.347 | 28 | 0.654 |
| 10 | 0.117 | 19 | 0.379 | 29 | 0.681 |
| | | 20 | 0.411 | 30 | 0.706 |

being investigated. Care must be exercised in reading (or writing) each problem, so that each outcome can be properly classified in a *single* way.

If you answered all six questions correctly, you probably don't need the following advice: *Make sure that you thoroughly understand what each problem asks before you try to solve it.*

The test above was designed to illustrate some of the pitfalls encountered in dealing with uncertainty. Two points are worth reemphasizing. First, in applying mathematics to a real phenomenon, we construct a mathematical model of the phenomenon. In doing so, we simplify the real situation by eliminating those factors that we deem inessential. The mathematical model, together with the probabilities that we assign to each outcome, is only an approximation of reality. If the model has been poorly designed it is unlikely to produce worthwhile results. Second, it is essential that we develop adequate mathematical methods to deal with these models, since intuition will often lead us astray. Do not be concerned if you did not understand every detail in the solutions above; the following sections provide a systematic development of the techniques and concepts of probability theory.

## 6.2 SETS AND VENN DIAGRAMS

Modern mathematics is based to a great extent on the concept of sets and the transformation of one set into another. This universality is one of the main reasons for the introduction of this concept in most grade school and high school mathematics courses. We will review briefly some of the fundamental ideas in elementary set theory, since these ideas will help in the explanation of certain concepts that we will encounter in this text.

A *set* is simply a collection of objects. For example, we can consider the set of the three primary colors: red, blue, and yellow. To write this set symbolically, we enclose the collection of objects in the set by a pair of braces:

{red, blue, yellow}.

Similarly, the set of even integers between 5 and 13 is the set

$\{6, 8, 10, 12\}$.

An example of a set of unrelated objects is the list for a scavenger hunt;

{bottle cap, maple leaf, chicken feather, chopstick, lighter fluid}.

Frequently a set contains so many elements that it is not possible to specify each one individually. If the elements of the set can be specified by a short sentence or formula, we can include this description following a vertical bar between the braces, as in the set

$\{x \mid x$ is a state in the United States$\}$,

which contains the names of the 50 states in the Union, or the set

$$\{x \mid x = n^2 + 1, \quad n = 2, 4, 6, 8, \cdots\},$$

which consists of the integers 5, 17, 37, 65, 101, $\cdots$, and so forth.

The objects in a set are called the *elements*, or *members*, of the set. We can indicate that the element 6 *belongs* to the set $\{6, 8, 10, 12\}$ by writing

$$6 \in \{6, 8, 10, 12\}.$$

If an element does not belong to a given set, we indicate this by drawing a vertical bar through the $\in$. For example,

$$7 \notin \{6, 8, 10, 12\}$$

and

purple $\notin$ {red, blue, yellow}.

One of the difficulties that we quickly encounter in real situations is deciding what elements are in the set. This problem is often the result of not being sufficiently precise. For example, suppose the state of California decides to collect a head tax of one thousand dollars for each person in California. The set in question consists of all persons in California. However, since a particular time has not been stated, it is difficult to determine exactly what is meant by "all persons in California", since people constantly move in and out of the state. If we then added the condition "all persons in California at noon on 1 January 1980", there might still be ambiguities: Does a passenger in an airline over California at that time belong to the set? How about a passenger on a ship within 3 miles, 10 miles, or 200 miles of the California coast? What about someone who dies or is born at noon? You can easily see that such imprecise wording would lead to a vast number of court actions by persons unwilling to pay this tax. Thus, *a set is well defined only when it is clear whether an element belongs or does not belong to the set.*

A set may contain a *finite* or an *infinite* number of elements. For example, the set of primary colors has only three elements, but the set of all positive integers

$$\{1, 2, 3, 4, 5, \ldots\}$$

is infinite. (The three dots after the number 5 mean "and so forth.") If there are no elements in the set we denote this by using the symbol $\varnothing$ for the set, and calling it the *empty set*. The empty set is a very useful concept if we want to show that something *doesn't* happen. We will see many applications of this concept in this book.

Two sets are *equal* if they contain exactly the same elements. For example,

$$\{6, 8, 10, 12\} = \{8, 12, 6, 10\}$$

but

$$\{6, 8, 10, 12\} \neq \{6, 8, 10\},$$

because 12 is not an element of the set on the right. However, since each element in the set on the right is also an element in the set on the left, we say that the set $\{6, 8, 10, 12\}$ *contains* the set $\{6, 8, 10\}$ and write

$$\{6, 8, 10, 12\} \supset \{6, 8, 10\}.$$

We can also say that $\{6, 8, 10\}$ is a *subset* of $\{6, 8, 10, 12\}$.

***Example 1*** The set of all students present in this class is a subset of the set of all students on campus.

---

Since each set contains all its own elements it is a subset of itself, or, symbolically,

$$A \subset A.$$

Observe that the empty set is a subset of any set $A$,

$$\varnothing \subset A,$$

since $A$ contains all the elements (none) that are in $\varnothing$. Finally, suppose that $A$ contains $B$ *and* $B$ contains $A$:

$$A \subset B \quad \text{and} \quad B \subset A.$$

Then every element in $A$ belongs to $B$ and every element in $B$ belongs to $A$. Thus, $A$ and $B$ must contain precisely the same elements. Therefore,

$$A = B$$

whenever

$$A \subset B \quad \text{and} \quad B \subset A, \tag{6.2}$$

and conversely. In other words, to show that two sets are equal we must show that each contains all the elements of the other.

In discussing a particular situation it is often possible to define a set containing all the elements that can arise. Such a set is called a *universal set* and is represented by the symbol $\mathcal{U}$. For example, suppose we are interested in determining the total number of runs scored by both teams in a baseball game. In this case we could let

$$\mathcal{U} = \{0, 1, 2, 3, \ldots\},$$

the set of all nonnegative integers. Obviously such numbers as 99,999 are extremely unlikely, but every nonnegative integer is possible.

The *complement* of a subset $A$ of the universal set $\mathcal{U}$ is the set $A'$ consisting of all elements of $\mathcal{U}$ that are not in $A$.

It is easy to visualize these concepts by using *Venn diagrams*. Here we let the points on and within a rectangular box represent the universal set $\mathcal{U}$. If $A$ is a subset of $\mathcal{U}$, we represent $A$ by the points on and inside a circular region in the box. The unshaded region in Fig. 6.5(a)

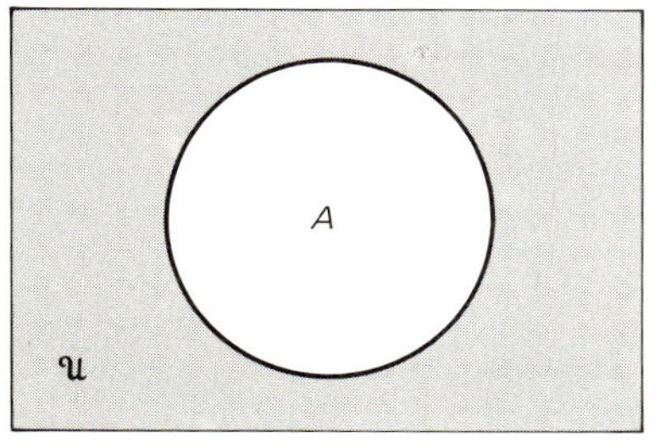


(a)

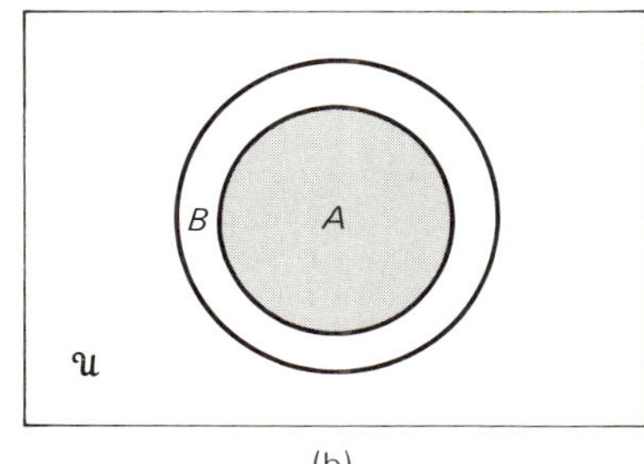


(b)

**Fig. 6.5**
(a) The shaded region is $A'$; (b) $A \subset B$.

represents the set $A$, while the shaded region in the box $\mathcal{U}$ is $A'$, the complement of $A$. In Fig. 6.5(b) we illustrate the meaning of the statement "$A$ is contained in $B$."

Suppose that $A$ and $B$ are both subsets of a universal set $\mathcal{U}$. Then the set of all elements belonging to both sets is called the *intersection* of $A$ and $B$. We denote the intersection of $A$ and $B$ by the symbol

$$A \cap B$$

and illustrate its meaning in Fig. 6.6. For example, if $A = \{1, 4, 7, 10, 13\}$ and $B = \{2, 4, 6, 8, 10\}$, then

$$A \cap B = \{1, 4, 7, 10, 13\} \cap \{2, 4, 6, 8, 10\} = \{4, 10\}.$$

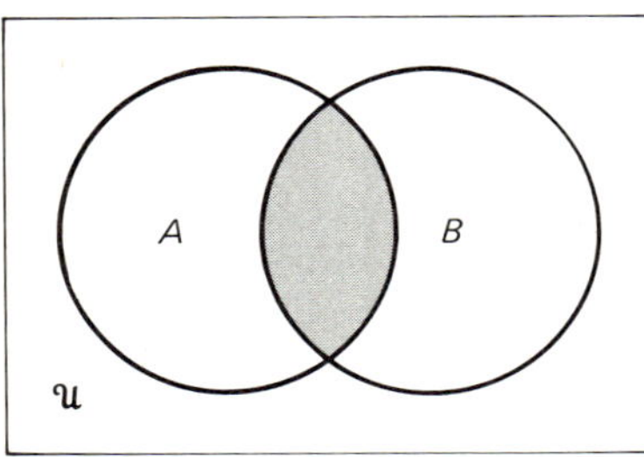


**Fig. 6.6**
$A \cap B$ is shaded.

The sets $A$ and $B$ are said to be *disjoint* if they have *no* elements in common. Clearly, the statement "$A$ and $B$ are disjoint" is equivalent to the equation

$$A \cap B = \varnothing. \qquad \textbf{(6.3)}$$

This situation is illustrated in Fig. 6.7.

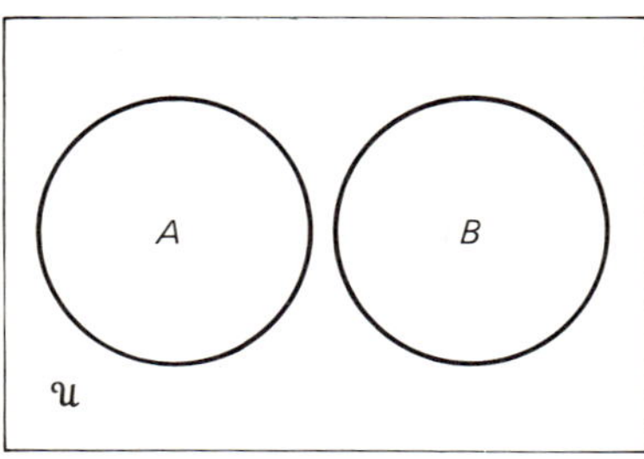


**Fig. 6.7**
$A$ and $B$ are disjoint.

***Example 2*** Let

$$A = \{x \mid x \text{ is a country touching the Indian Ocean}\}$$

and

$$B = \{x \mid x \text{ is a country in South America}\}.$$

Then certainly $A \cap B = \varnothing$ and the two sets are disjoint.

---

If $A$ and $B$ are two subsets of a universal set $\mathcal{U}$, then the set of all elements belonging to either or both of the sets $A$ and $B$ is called the *union* of $A$ and $B$, and is written

$$A \cup B.$$

If $A = \{1, 4, 7, 10, 13\}$ and $B = \{2, 4, 6, 8, 10\}$, then

$$A \cup B = \{1, 4, 7, 10, 13\} \cup \{2, 4, 6, 8, 10\}$$
$$= \{1, 2, 4, 6, 7, 8, 10, 13\}.$$

We illustrate this concept in Fig. 6.8.

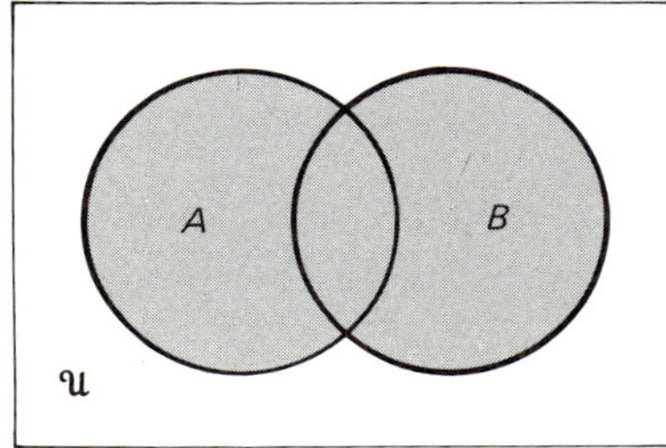


**Fig. 6.8**
$A \cup B$ is shaded.

It is possible to draw Venn diagrams representing more complicated situations. For example, the set $A \cup B'$ is shown in Fig. 6.9(a), while $A \cup (B \cap C')$ is the shaded region in Fig. 6.9(b). The set $A \cup B'$ consists of all elements that either belong to $A$ or do not belong to $B$. Similarly, $A \cup (B \cap C')$ consists of all elements belonging either to $A$ or to the set $B \cap C'$, which consists of all elements in $B$ *and* not in $C$.

Venn diagrams aid certain operations with sets. For example, suppose we wish to find the complement of the set $A \cup B'$, that is, $(A \cup B')'$. We see that this is the unshaded region in Fig. 6.9(a) consisting of those points in $B$ that are not in $A$. This last set is the

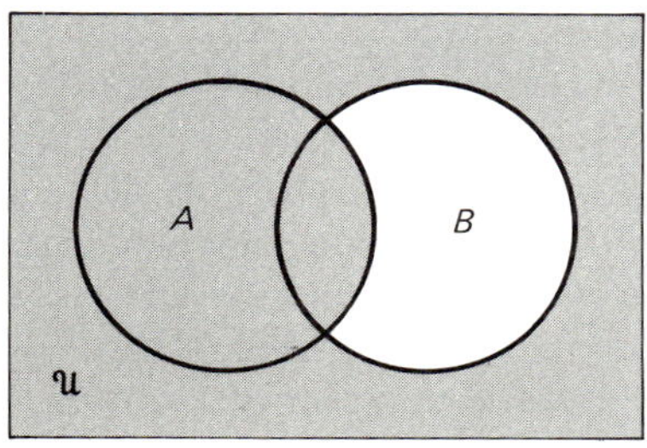


(a)

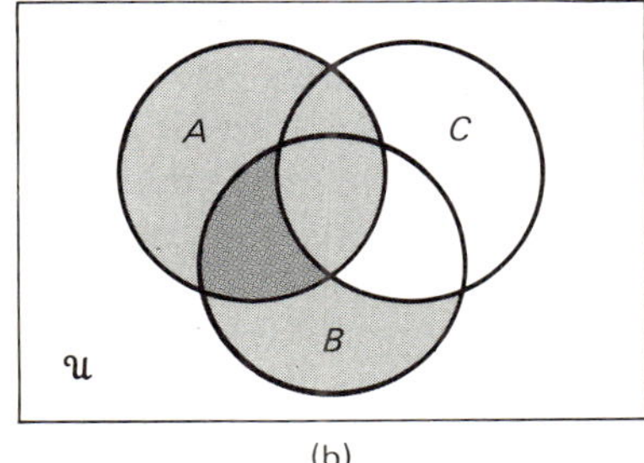


(b)

**Fig. 6.9**
(a) $A \cup B'$ is shaded; (b) $A \cup (B \cap C')$ is shaded.

intersection of $B$ and $A'$, so

$$(A \cup B')' = A' \cap B. \tag{6.4}$$

Venn diagrams can also be used to obtain quantitative information, as we show in the following examples.

***Example 3*** A survey of 1000 randomly chosen families produced the following data concerning the automakers of their family cars.

523 families owned a GM product.
407 families owned a Ford product.
290 families owned a Chrysler product.
123 families owned both a GM and a Ford product.
107 families owned both a GM and a Chrysler product
94 families owned both a Ford and a Chrysler product.
15 families owned cars made by all three automakers.

On the basis of this data we wish to determine the following.

a) How many families owned *no* cars made by these automakers?
b) How many families only owned GM products?
c) How many families only owned Ford products?
d) How many families only owned Chrysler products?

Many of the families are listed more than once in the data above. For example, if a family owned both a Chevrolet and a Dodge, it would be counted three times: as a family owning a GM product (the Chevrolet), as a family owning a Chrysler product (the Dodge), and as a family owning both a GM and a Chrysler product. We can use Venn diagrams to clarify this overlapping. Define each of the three circular regions in Fig. 6.10(a) as the set of families owning cars made by each of the three automakers. The shaded region in Fig. 6.10(a) is the set of families owning cars made by all three automakers, since

$$\{\text{GM}\} \cap \{\text{Ford}\} \cap \{\text{Chrysler}\}$$

is the overlap of all three sets. From our data we know that this set consists of 15 families, so we mark 15 inside this region (see Fig. 6.10(b). Now $\{\text{GM}\} \cap \{\text{Ford}\}$ contains 123 families, of which 15 are already accounted for; thus, 108 families are in the other part of $\{\text{GM}\} \cap \{\text{Ford}\}$. Similarly, the second part of $\{\text{GM}\} \cap \{\text{Chrysler}\}$ contains $107 - 15 = 92$ families, and the other region of $\{\text{Ford}\} \cap \{\text{Chrysler}\}$ contains $94 - 15 = 79$ families. We can now find the number of families owning only GM cars since the entire set $\{\text{GM}\}$ consists of 523 families,

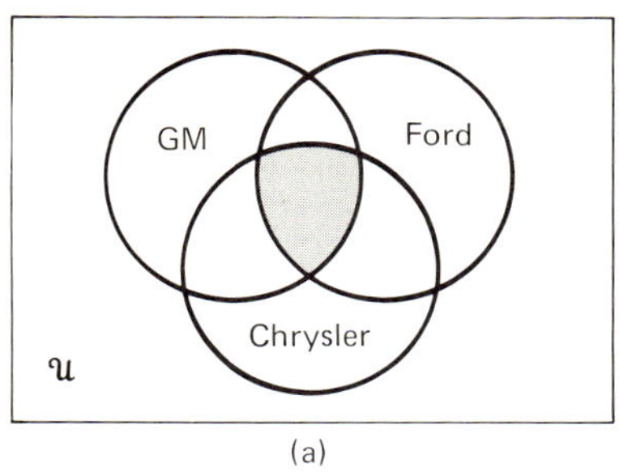


(a)

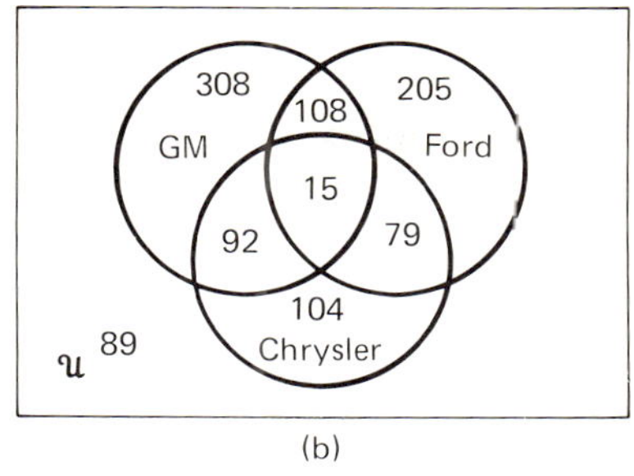


(b)

**Figure 6.10**

of which

$$108 + 15 + 92 = 215$$

have already been counted. Thus, 308 families own only GM cars. Similarly, the number of all families owning only Fords is

$$407 - (108 + 15 + 79) = 205,$$

while the number of all families owning only Chryslers is

$$290 - (92 + 15 + 79) = 104.$$

This provides an answer to parts (b), (c), and (d). To answer part (a) we add all the families in the seven regions of

$$\{\text{GM}\} \cup \{\text{Ford}\} \cup \{\text{Chrysler}\}$$

and subtract this quantity from 1000:

$$1000 - [308 + 205 + 104 + 108 + 92 + 79 + 15] = 89,$$

so 89 families own no cars made by the three automakers.

***Example 4*** Human blood is often classified according to whether or not it contains three antigens A, B, and Rh. Whenever the antigen A or B is present it is listed, but if both of these antigens are absent the blood is said to be of type O. If the Rh antigen is present, the blood is said to be positive; otherwise it is negative. Thus, the set of all main blood types consists of the members

$$\{O^+, O^-, A^+, A^-, B^+, B^-, AB^+, AB^-\}$$

(see Fig. 6.11a). During 1973, the Red Cross gathered data concerning the percentage of the general population having the three antigens.

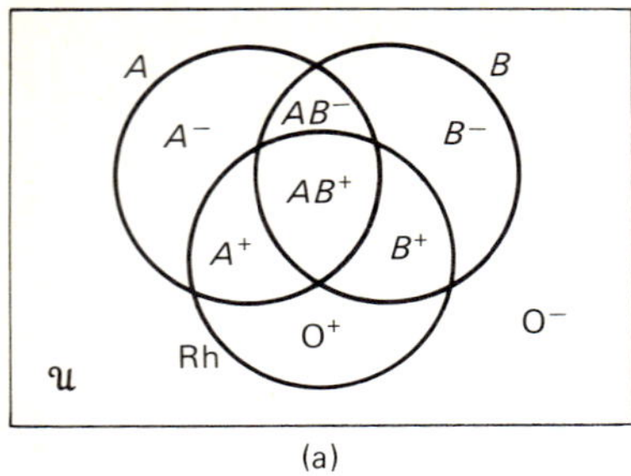


(a)

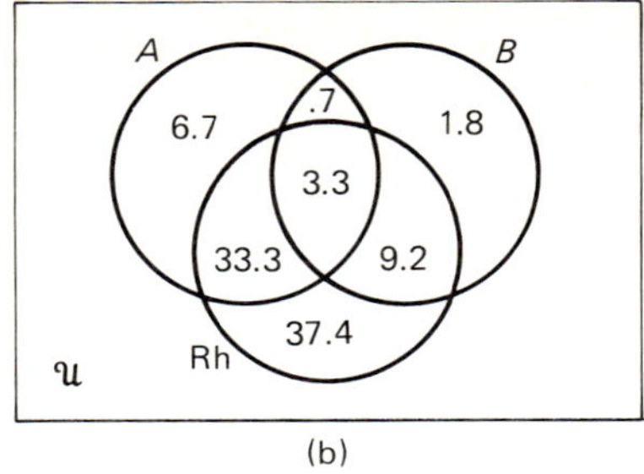


(b)

**Figure 6.11**

They found that

44.0% has antigen A,
16.1% has antigen B,
83.2% has antigen Rh,
36.6% has both antigens A and Rh,
12.5% has both antigens B and Rh,
4.0% has both antigens A and B, and
3.3% has all three antigens.

What percentage has blood type $AB^-$? What percentage has type $O^-$?

Again we begin by assigning the number 3.3 to the region $AB^+$. Since 4.0% have both antigens A and B, and $AB^+$ accounts for 3.3%, $AB^-$ will be 0.7%. Similarly, $A^+$ is 36.6% − 3.3% = 33.3% and $B^+$ is 12.5% − 3.3% = 9.2%. Then, the percentage of $A^-$ is 44% − (33.3% + 3.3% + 0.7%) = 6.7%, while $B^-$ is 16.1% − (9.2% + 3.3% + 0.7%) = 1.8% and $O^+$ is 83.2% − (33.3% + 3.3% + 9.2%) = 37.4%. Subtracting the values we now have in Fig. 6.11(b) from 100% yields the information that $O^-$ is 7.6%. Clearly, the rarest blood type is $AB^-$.

---

## EXERCISES 6.2

*In Exercises 1–12, let* $A = \{1, 2, 3, 4, 5, 6\}$, $B = \{2, 4, 6, 8, 10\}$, *and* $\mathcal{U} = \{1, 2, 3, 4, 5, 6, 7, 8, 9, 10\}$. *Find each of the given sets.*

**1.** $A \cup B$

**2.** $A \cap B$

**3.** $A'$

**4.** $B'$

**5.** $A \cup B'$

**6.** $A' \cap B$

**7.** $A' \cup B$

**8.** $A \cap B'$

**9.** $A' \cap B'$

**10.** $A' \cup B'$

**11.** $A \cup (A' \cap B)$

**12.** $B \cap (A' \cup B')$

*In Exercises 13–20, suppose that $M$ = {all math majors}, $A$ = {all students taking an accounting course}, and $B$ = {all students taking a botany course}. Describe the following sets in words.*

**13.** $A \cap M$
**14.** $B \cup M$
**15.** $A \cap B$
**16.** $M \cap B$
**17.** $A'$
**18.** $M' \cup A$
**19.** $M' \cap B$
**20.** $B' \cup M$

*In Exercises 21–28, verify the given identity using Venn diagrams with two sets A and B.*

**21.** $(A \cup B)' = A' \cap B'$
**22.** $(A' \cup B)' = A \cap B'$
**23.** $(A \cap B')' = A' \cup B$
**24.** $A \cap A' = \varnothing$
**25.** $A \cup (A' \cap B) = A \cup B$
**26.** $A \cap (A' \cup B) = A \cap B$
**27.** $(A \cup B) \cap B' = A \cap B'$
**28.** $(A \cap B) \cap B' = \varnothing$

*Shade a Venn diagram to illustrate each of the following sets.*

**29.** $(A \cap B) \cup C$
**30.** $(A \cup B) \cap C$
**31.** $(A \cup B) \cup C'$
**32.** $(A \cup B)' \cup C$
**33.** $(A \cap B) \cap C'$
**34.** $A \cap (B \cap C)'$
**35.** $(A \cap B)' \cap C$
**36.** $(A \cup B)' \cap C$

**37.** In a genetics experiment 100 pea plants were classified according to certain characteristics, with the following results:

25 plants had white flowers,
60 plants had smooth peas, and
35 plants had purple flowers and wrinkled peas.

a) How many plants had white flowers and smooth peas?
b) How many plants had purple flowers and smooth peas?

**38.** In a survey of 600 families it was found that 450 subscribe to *TV Guide*, 250 subscribe to *Time*, and 100 families receive neither magazine.

a) How many families receive both magazines?
b) How many families only receive *TV Guide*?

**39.** Course schedules in a freshman class of 37 students were compared and it was found that

28 students were taking English composition,
17 were taking chemistry,
21 were taking sociology,
8 were taking both chemistry and sociology,
14 were taking both English and sociology,
12 were taking both English and chemistry, and
4 were taking all three.

a) How many students were not taking any of these courses?
b) How many students were taking only English?
c) How many students were taking only English and sociology?

**40.** A brewery conducts a survey to determine the most popular beer. It asks 700 randomly chosen people to list their three favorite brands of beer, and obtains the following data:

175 like Pabst,
232 like Budweiser,
205 like Schlitz,
61 like both Pabst and Schlitz,
78 like both Budweiser and Schlitz,
32 like all three, and
312 prefer other beers.

a) How many like both Budweiser and Pabst but not Schlitz?
b) How many like Budweiser but do not like Pabst or Schlitz?

---

## 6.3 EVENTS AND THEIR PROBABILITIES

In Section 6.1 we examined a number of situations in which elementary counting arguments were used to assign probabilities to various possible outcomes. In this section we will formalize some of these notions and lay the foundations for many other concepts in probability theory.

The set of all possible outcomes of a particular experiment is called the *sample space* of that experiment, and is denoted by the letter $S$. Hence, the set $S$ consists of the collection of all possible outcomes of the experiment. For example, if the experiment consists of tossing a fair coin three times, the sample space

$$S = \{\text{HHH, HHT, HTH, THH, HTT, THT, TTH, TTT}\}$$

is the set of all eight possible outcomes of the experiment. Note that each of these eight outcomes is a path from the root to the top of the tree in Fig. 6.12. If the experiment consists of finding which of two teams A and B wins two games first, the sample space is the set of six outcomes

$$S = \{\text{AA, ABA, ABB, BAA, BAB, BB}\},$$

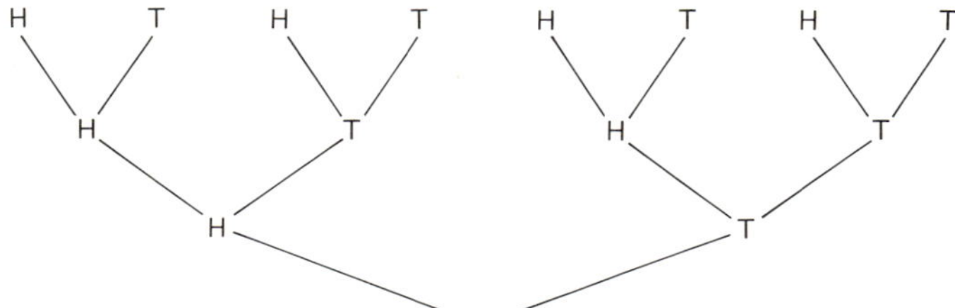


**Figure 6.12**

where each letter represents the winner of that game (see Fig. 6.13). The sample space for the experiment of rolling a single die once is the set of the six values that can be obtained:

$$S = \{1, 2, 3, 4, 5, 6\}.$$

Each outcome in the sample space is called a *sample point* of that experiment. For example, HTH is a sample point of the coin tossing experiment, while 4 is a sample point of the die rolling experiment.

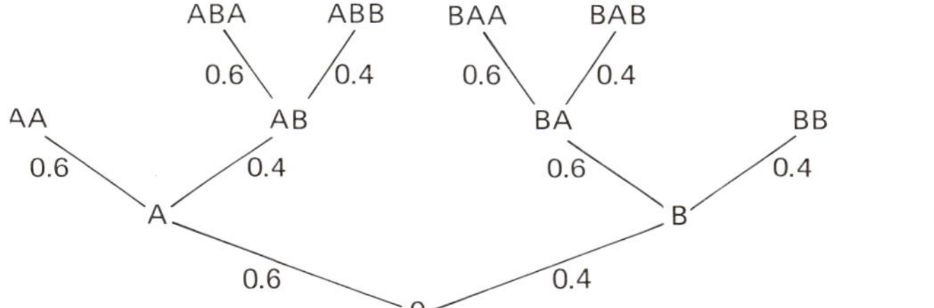


**Figure 6.13**

An *event* $E$ is any collection of sample points. If we prefer to use the language of sets, we say that an event $E$ is any subset of the sample space $S$. We write $E \subset S$, and use the Venn diagram shown in Fig. 6.14 to illustrate the relationship between $E$ and $S$. For example, if $E$ is the event that the second coin toss is heads, then

$$E = \{\text{HTH}, \text{HTT}, \text{TTH}, \text{TTT}\}.$$

We say that the event $E$ *occurs* whenever the outcome $x$ of a particular experiment is a sample point that belongs to the set $E$, that is, $x \in E$ (see Fig. 6.14). Hence, if the result of flipping the coin three times is HTT, then the event $E$ occurred, since HTT $\in E$. On the other hand, if the outcome had been THH, the event $E$ did not occur, since THH $\notin E$. In Fig. 6.14, the sample point $y$ does not belong to $E$, so $E$ does not occur if the result of the experiment is the sample point $y$.

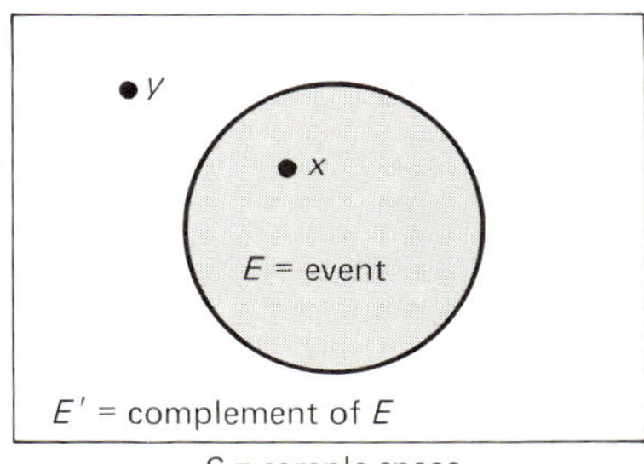


**Figure 6.14**

An event $E$ can contain all the points in the sample space, in which case we write $E = S$, or it may contain no sample points, so that $E = \varnothing$ (the empty set). Obviously, the event $E = S$ always occurs and the event $E = \varnothing$ never occurs. A single sample point $x$ can be considered as an event $E = \{x\}$.

The set of all sample points that do not belong to the event $E$ belong to the event $E'$, called the *complement* of $E$ (see Fig. 6.14). For example, in the experiment of finding which team wins two games first, the complement of the event $E$ that $A$ wins the third game,

$$E = \{\text{ABA, BAA}\},$$

is the event

$$E' = \{\text{AA, ABB, BAB, BB}\}.$$

Using the notation of set theory, we see that

$$E \cup E' = S \quad \text{and} \quad E \cap E' = \varnothing,$$

since every sample point is in either $E$ or $E'$ but not in both (see Fig. 6.14). Two events having no common sample points are called *mutually exclusive events* (see Fig. 6.15a). Thus, an event $E$ and its complement $E'$ are mutually exclusive events, but events need *not* be complementary to be mutually exclusive (see Fig. 6.15a). For example, the events

$$E = \{\text{AA, ABA}\} \quad \text{and} \quad F = \{\text{BAB, ABB}\}$$

are mutually exclusive events, but $F \neq E'$. Three or more events are mutually exclusive if each pair of them is mutually exclusive (see Fig. 6.15b).

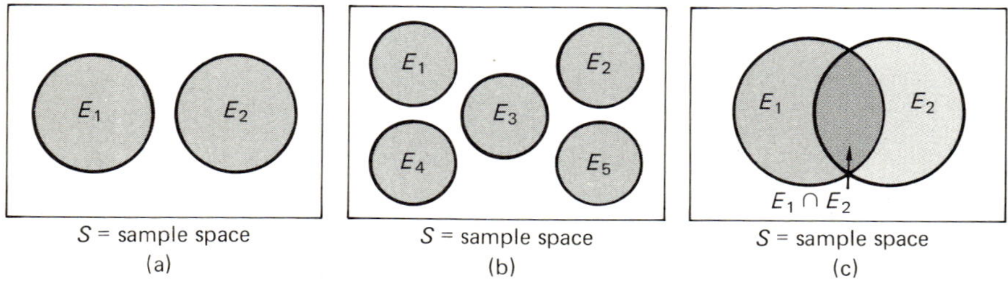


**Fig. 6.15**
(a) Mutually exclusive events; (b) several mutually exclusive events; (c) events that are not mutually exclusive.

***Example 1*** The three events

$$E = \{AA, ABA\}, \qquad F = \{BAB, ABB\}, \qquad G = \{BB, BAA\}$$

are mutually exclusive.

---

Of course, two events $E_1$ and $E_2$ need not be mutually exclusive. The darkest shaded area in Fig. 6.15(c) indicates the nonempty collection of sample points $E_1 \cap E_2$ belonging to both events $E_1$ and $E_2$. Hence these two events are not mutually exclusive.

**DEFINITION 6.1** **Suppose that $x_1, x_2, \ldots, x_n$ are different sample points in a sample space $S$. Let $E$ be the event $E = \{x_1, x_2, \ldots, x_n\}$. Then the probability that the event $E$ occurs is defined by**

$$p(E) = p(\{x_1\}) + p(\{x_2\}) + \cdots + p(\{x_n\}). \tag{6.5}$$

Definition 6.1 states that the probability that event $E$ occurs, written $p(E)$, equals the sum of the probabilities of occurrence of each sample point in $E$. A consequence of this definition is that

$$p(S) = 1(= 100\%), \tag{6.6}$$

since the event $S$ must always occur, and

$$p(\varnothing) = 0(= 0\%), \tag{6.7}$$

since the event $\varnothing$ never occurs.

If we use the Venn diagram in Fig. 6.15(a) and Definition 6.1, the following is clear.

**THEOREM 6.1** **If $E_1$ and $E_2$ are mutually exclusive events in a sample space $S$, the probability that either $E_1$ or $E_2$ occurs is**

$$p(E_1 \text{ or } E_2) = p(E_1 \cup E_2) = p(E_1) + p(E_2). \tag{6.8}$$

**Similarly, if three (or more) events $E_1$, $E_2$, and $E_3$ in a sample space $S$ are mutually exclusive, the probability that any one of these events occurs equals the sum of their individual probabilities of occurrence:**

$$p(E_1 \cup E_2 \cup E_3) = p(E_1) + p(E_2) + p(E_3).$$

The situation is more complicated if we have two (or more) events that are not mutually exclusive (see Fig. 6.15c).

**THEOREM 6.2** **If $E_1$ and $E_2$ are any two events, the probability that either $E_1$ or $E_2$ occurs is**

$$p(E_1 \cup E_2) = p(E_1) + p(E_2) - p(E_1 \cap E_2). \tag{6.9}$$

***Proof*** The proof of this theorem is quite easy. It amounts to noting that each sample point $x$ in $E_1 \cap E_2$ has been counted twice in the sum of the first two terms on the right side of Eq. (6.9), once as a point in $E_1$ and once as a point in $E_2$. Thus, we must subtract the probabilities of occurrence of all sample points in the overlap $E_1 \cap E_2$, obtaining Eq. (6.9). Note that Eq. (6.9) reduces to Eq. (6.8) when the events $E_1$ and $E_2$ are mutually exclusive, since $p(E_1 \cap E_2) = p(\varnothing) = 0$. ∎

So far we have discussed only events in a sample space *without indicating how to assign a probability to each sample point.* The difficulty here is that the only restrictions in assigning the probabilities are that

$$p(x) \geq 0, \tag{6.10}$$

that is, each sample point must be assigned a nonnegative probability, and that

$$p(S) = 1, \tag{6.11}$$

that is, the sum of the probabilities of all sample points in $S$ must equal 1. This freedom allows us to make the assignment in a variety of ways. For example, we could repeat the experiment a great number of times and count the number of times each sample point occurs. We could then assign a probability to the sample point $x$ by dividing the number of times $x$ occurred by the total number of times the experiment was performed:

$$p(x) = \frac{\text{number of times } x \text{ occurred}}{\text{total number of times}}. \tag{6.12}$$

This technique of assigning probabilities is called the method of *experimental relative frequencies.* The main problem with this method is that it depends entirely on one specific sequence of experiments and is very unlikely to be reproducible. It can, however, be used as a *guide* in developing the probability of each sample point. A more effective method of assigning probabilities to the sample points is in terms of some intrinsic property of the model. That is, the assigned probabilities are incorporated in the mathematical model of the real situation. The following three examples illustrate these two methods of assigning probabilities.

***Example 2*** Consider the experiment in which a coin is tossed three times. The sample space consists of the eight possible outcomes of the experiment:

$$S = \{\text{HHH, HHT, HTH, HTT, THH, THT, TTH, TTT}\}.$$

Each outcome is a branch of the tree in Fig. 6.12. Suppose we assume that the coin is fair, that is, that heads and tails are equally likely to

occur. Then we would not have any reason to select one outcome over any of the others. In this situation we would say that each outcome is *equally probable.* Since there are eight outcomes, we would assign each one probability $p = \frac{1}{8}$. Sample spaces in which each outcome is equally probable are called ***uniform sample spaces.***

Alternatively, we could assign each line in the tree a probability of $\frac{1}{2}$, reflecting the fact that on each toss heads is just as likely to occur as tails. If we multiply the probabilities along each line of a branch, we obtain

$$\tfrac{1}{2} \cdot \tfrac{1}{2} \cdot \tfrac{1}{2} = \tfrac{1}{8},$$

which is the probability associated with that branch or outcome.

We shall call this result the *tree multiplication property*.

**TREE MULTIPLICATION PROPERTY** Suppose that each sample point in some sample space corresponds to a branch in a given tree and vice versa. If we associate a probability to each line in the tree, then the probability of the sample point is obtained by multiplying the probabilities associated to each line in the corresponding branch.

***Example 3*** Suppose two teams A and B are playing a match in which the first team to win two games is proclaimed the victor. The sample space consists of six outcomes

$$S = \{\text{AA, ABA, ABB, BAA, BAB, BB}\},$$

where each letter represents the winner of that game. Suppose that the probability that A will win each game is 0.6; then 0.4 is the probability that B will win a given game. Labeling each line accordingly (see Fig. 6.13), we obtain the following probabilities for each of our sample points:

$$\begin{aligned}
p(\{\text{AA}\}) &= (0.6)(0.6) = 0.360,\\
p(\{\text{ABA}\}) &= (0.6)(0.4)(0.6) = 0.144,\\
p(\{\text{ABB}\}) &= (0.6)(0.4)(0.4) = 0.096,\\
p(\{\text{BAA}\}) &= (0.4)(0.6)(0.6) = 0.144,\\
p(\{\text{BAB}\}) &= (0.4)(0.6)(0.4) = 0.096,\\
p(\{\text{BB}\}) &= (0.4)(0.4) = 0.160.
\end{aligned}$$

This sample space is clearly not uniform, since the sample points are not equally probable. If the event in which we are interested is that "A wins," then $E = \{\text{AA, ABA, BAA}\}$ and

$$\begin{aligned}
p(E) &= p(\{\text{AA}\}) + p(\{\text{ABA}\}) + p(\{\text{BAA}\})\\
&= 0.36 + 0.144 + 0.144 = 0.648.
\end{aligned}$$

---

In the following example, experimental relative frequency has been used to assign the probabilities. Here, there is no *mathematical* reason why any particular antigen should be more common than any other. However, statistical evidence amply demonstrates that this sample space is not uniform.

***Example 4*** Human blood can contain a number of different antigens. In testing blood types, we commonly classify the blood according to whether it has the A, B, and Rh (Rhesus) antigens. If antigens A or B are present in the blood they are listed, but if they are both absent the blood is said to be type O. If the Rh antigen is present the blood is said to be positive; otherwise, it is negative. Thus an individual with type $A^-$ blood has only the antigen A, one with type $O^+$ has only the Rh antigen, and one with $AB^+$ has all three antigens.

If we let A, B, and $^+$ denote the events that a blood sample contains the antigens A, B, and Rh, respectively, we can use the Venn diagram shown in Fig. 6.16 to describe the sample space $S$ of all possible blood types. We see that $S$ consists of eight different blood types:

$$S = \{O^+, O^-, A^+, A^-, B^+, B^-, AB^+, AB^-\}.$$

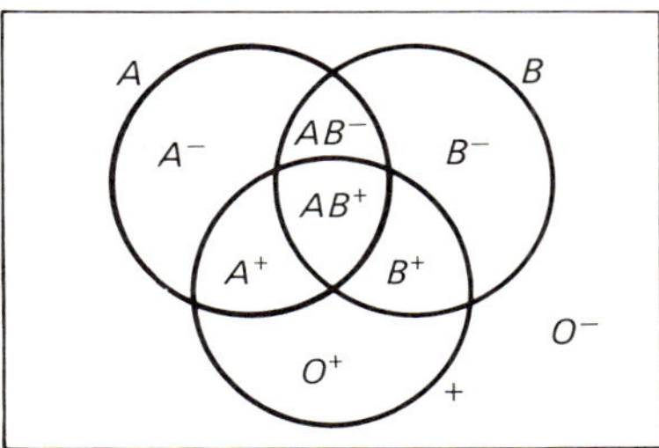


**Fig. 6.16**
Main blood types.

To get an idea of the relative frequency of the various blood types, we can use the percentages that were obtained by the Red Cross based on almost four million drawings during the year 1973 to 1974 (see Table 6.2).

**Table 6.2**
**Relative frequency of main blood types, 1973–1974**

| | |
|---|---|
| $O^+$ = 37.4% | $O^-$ = 7.6% |
| $A^+$ = 33.3% | $A^-$ = 6.7% |
| $B^+$ = 9.2% | $B^-$ = 1.8% |
| $AB^+$ = 3.3% | $AB^-$ = 0.7% |

When an individual receives a blood transfusion, it is important that he or she not be given blood containing an antigen that is absent from his or her blood. The result of being given a transfusion of blood containing an antigen absent from an individual's blood can be fatal, since the antigen triggers massive clotting of the blood. Thus, when seeking blood donors, one must find people lacking the same antigens as the recipient. For example, a person with $A^+$ blood can receive transfusions of types $O^-$, $O^+$, $A^-$, and $A^+$, but not types $B^+$, $B^-$, $AB^+$, and $AB^-$. A person with type $O^-$ blood can donate blood to all individuals, and is called a *universal donor.* On the other hand, an individual with type $AB^+$ blood may receive transfusions of all eight blood types, and is called a *universal recipient.*

Suppose that an individual has type $A^-$ blood. What is the probability that some acquaintance can give her blood? In this case, the event $E$ consists of the donor's blood containing either no antigens or only the A antigen. Thus, $E = \{O^-, A^-\}$, and

$$p(E) = p(\{O^-\}) + p(\{A^-\}) = 0.076 + 0.067 = 0.143$$

or 14.3%. Thus, approximately one in every seven acquaintances will be able to donate blood to this individual. A person with $A^+$ blood can receive transfusions from anyone having blood types in the set

$$E = \{O^-, O^+, A^-, A^+\},$$

so the probability that this person can find a donor is

$$\begin{aligned} p(E) &= p(\{O^-\}) + p(\{O^+\}) + p(\{A^-\}) + p(\{A^+\}) \\ &= 0.076 + 0.374 + 0.067 + 0.333 = 0.85. \end{aligned}$$

In other words, 17 of every 20 persons will be able to donate blood for this individual.

Note that although $O^+$ is the most common blood type, people with $O^+$ blood can only receive types $O^+$ and $O^-$ [$p(\{O^+, O^-\}) = 0.076 + 0.374 = 0.45$], while the much rarer $AB^+$ blood type can receive blood from anyone.

---

## EXERCISES 6.3

1. Suppose we toss a fair coin three times and count only the total number of heads and tails that appear. Describe a sample space $S$ for this experiment. Is it uniform?

2. Suppose we toss a pair of dice once. Describe two sample spaces that can be associated with the experiment. Is there a uniform sample space associated with this experiment?

*In Exercises 3–5, determine whether or not the events $E_1$ and $E_2$ are mutually exclusive.*

3. In throwing a pair of dice, let $E_1$ be the event that the outcome is odd and $E_2$ the event that one of the dice is a 6.

4. In flipping a coin four times, let $E_1$ be the event that at least two heads appear and $E_2$ the event that only the third toss is heads.
5. Two teams play until one of them wins three games. Let $E_1$ be the event that the last game is won by team A, and let $E_2$ be the event that team B wins the series.

*In Exercises 6–8 find the complement of the given event.*

6. In Exercise 5, let $E$ be the event that team A wins in at least a four-game series.
7. In Exercise 5, let $E$ be the event that at most four games are played.
8. In Exercise 5, let $E$ be the event that team B wins at least two games.

*Draw a tree and determine the probability of each sample point in Exercises 9–12 using the tree multiplication property.*

9. Teams A and B play until one has won three games, each team being equally likely to win each game.
10. Suppose in Exercise 9 that team A has probability 0.7 of winning each game.
11. Suppose in Exercise 9 that A is twice as likely as B to win the *last* game.
12. Suppose in Exercise 9 that each team is equally likely to win the first game, but that each time a team wins a game, the probability that it will win the next game is increased by 0.1.
13. Two patients need a blood transfusion. Patient $P_1$ has blood type $B^+$, while patient $P_2$ has blood type $AB^-$. Which patient will have the easiest time finding a donor?
14. A credit-union manager observes at random times the number of persons waiting in line at a teller's window. After many months of record keeping, she determines the following table of probabilities:

| People waiting | 0 | 1 | 2 | 3 | 4 or more |
|---|---|---|---|---|---|
| Probability | 0.40 | 0.25 | 0.20 | 0.10 | 0.05 |

   a) What is the probability that someone is waiting in line?
   b) What is the probability that fewer than two people are waiting in line?
   c) What is the probability that no more than three people are waiting in line?

15. Before a bill can reach the president of the United States for his signature it must pass the House and the Senate. A lobbyist assesses the bill's probability of passing the House as 0.8 and the probability of passing the Senate as 0.6. Additionally, she estimates that the probability that it will be passed by either the House or the Senate is 0.9.

a) What is the probability that the bill will reach the president?
b) Suppose a massive demonstration and letter writing campaign decreases the probability that it will pass the House to 0.5. If all other probabilities remain the same, what is the probability that the bill will clear both chambers?

**16.** Two customers arrive at a fast-food stand at the same time. Six double-burgers are already prepared, but the cook forgot to put a pickle on two of them. If each customer buys two double-burgers:

a) What is the probability that neither of them gets one of the pickle-less burgers?
b) What is the probability that at least one of them gets a pickle-less burger?
c) What is the probability that both of them get a pickle-less burger?
d) What is the probability that one of them gets both pickle-less burgers?

**17.** One hundred doctors are interviewed concerning their leisure activity. If 71 of them jog, 74 of them play tennis, 79 of them swim, and 84 of them play golf, how many at least do all four?

**18.** In Exercise 17, suppose we know that 50 doctors jog and play tennis. How many at least do all four?

**19.** A four-question true-false test is given. The instructor assigns grades by adding the number answered correctly minus half the number answered incorrectly. (Questions not answered are not counted.) If a score of 2 is passing, what is the probability of passing the test purely by guessing?

**20.** Suppose the instructor in Exercise 19 wants to discourage guessing and assigns grades by adding the number answered correctly minus twice the number answered incorrectly. What is the probability that guessing will give you a lower score than not answering at all?

---

## 6.4 UNIFORM SAMPLE SPACES

A particularly simple sample space is one in which each sample point is equally likely to occur. Such sample spaces are called *uniform sample spaces*. The sample space in Example 2 in Section 6.3,

$$S = \{\text{HHH}, \text{HHT}, \text{HTH}, \text{THH}, \text{HTT}, \text{THT}, \text{TTH}, \text{TTT}\},$$

associated with the experiment of tossing a fair coin three times, is an example of a uniform sample space. On the other hand, Example 3 in Section 6.3 is not a uniform sample space.

There is an easy way of determining the probability that an event $E$ occurs in a uniform sample space. Since each sample point is equally likely to occur, we merely divide the number of sample points in $E$ by the number of sample points in $S$.

**DEFINITION 6.2** **Let $S$ be a uniform sample space. Denote the number of sample points in $E$ by $n(E)$, for any event $E \subset S$. Then the probability that the event $E$**

occurs is

$$p(E) = \frac{n(E)}{n(S)}. \quad \textbf{(6.13)}$$

***Example 1*** In the experiment of tossing a fair coin three times, let $E$ be the event that at least one head turns up. Then

$$E = \{\text{HHH, HHT, HTH, THH, HTT, THT, TTH}\},$$

so that $n(E) = 7$. Since $n(S) = 8$, Eq. (6.13) yields

$$p(E) = \frac{n(E)}{n(S)} = \frac{7}{8}.$$

---

Theorems 6.1 and 6.2 are particularly easy to prove in a uniform sample space. If we count all the sample points in $E_1$ and add that number to the number of sample points in $E_2$, we have counted the number of points in the overlap $E_1 \cap E_2$ twice. Thus, we must subtract the number of points that have been twice counted, so that

$$n(E_1 \cup E_2) = n(E_1) + n(E_2) - n(E_1 \cap E_2). \quad \textbf{(6.14)}$$

If we now divide both sides of Eq. (6.14) by $n(S)$ and use Eq. (6.13), we obtain Theorem 6.2:

$$\begin{aligned} p(E_1 \cup E_2) &= \frac{n(E_1 \cup E_2)}{n(S)} = \frac{n(E_1)}{n(S)} + \frac{n(E_2)}{n(S)} - \frac{n(E_1 \cap E_2)}{n(S)} \\ &= p(E_1) + p(E_2) - p(E_1 \cap E_2). \end{aligned}$$

***Example 2*** In the experiment in which a fair coin is tossed three times, let $E_1$ be the event that heads comes up at least twice, and let $E_2$ be the event that the second toss is heads. Then

$$E_1 = \{\text{HHH, HHT, HTH, THH}\}$$

and

$$E_2 = \{\text{HHH, HHT, THH, THT}\},$$

so the probability that heads comes up twice or on the second toss is

$$\begin{aligned} p(E_1 \cup E_2) &= p(E_1) + p(E_2) - p(E_1 \cap E_2) \\ &= \frac{n(E_1)}{n(S)} + \frac{n(E_2)}{n(S)} - \frac{n(E_1 \cap E_2)}{n(S)} \\ &= \frac{4}{8} + \frac{4}{8} - \frac{3}{8} = \frac{5}{8}, \end{aligned}$$

since

$$E_1 \cap E_2 = \{\text{HHH}, \text{HHT}, \text{THH}\}.$$

***Example 3*** Consider the experiment of selecting one card from an ordinary deck of 52 playing cards. Let $E_1$ be the event that the card selected is a jack, and $E_2$ the event that the card selected is a heart. Each sample point in this situation is one of the cards in the deck, and since each card is equally likely to be selected, the sample space is uniform. Assigning a probability of 1/52 to each sample point, we have

$$p(E_1) = p(\{\text{J}\spadesuit\} \cup \{\text{J}\heartsuit\} \cup \{\text{J}\diamondsuit\} \cup \{\text{J}\clubsuit\}) = \frac{4}{52}.$$

Similarly, $p(E_2) = 13/52$ and

$$\begin{aligned} p(E_1 \cup E_2) &= p(E_1) + p(E_2) - p(E_1 \cap E_2) \\ &= \frac{4}{52} + \frac{13}{52} - \frac{1}{52} = \frac{16}{52}, \end{aligned}$$

since $E_1 \cap E_2 = \{\text{J}\heartsuit\}$.

---

## EXERCISES 6.4

1. Suppose we toss three different colored dice once and keep track of the outcome of each die. Is the sample space we obtain uniform?
2. Suppose we toss a pair of dice and add the total number of spots. Is this sample space uniform?
3. Describe a sample space for two teams A and B playing a series in which the first team to win three games is the victor. Is this sample space uniform?

*In Exercises 4–8, assume that a pair of fair dice is thrown once. Use the associated uniform sample space with 36 sample points (six for each die) to answer the questions.*

4. What is the probability that the sum of the faces is 7?
5. What is the probability that the sum of the faces is 6 or 8?
6. What is the probability of rolling craps (2, 3, or 12)?
7. What is the probability of rolling a 7 or an 11?
8. What is the probability of rolling a 6 or an 8 the hard way (that is, two threes or two fours)?

*In Exercises* 9–12, *assume that you are betting at a* 38-*slot roulette wheel numbered* 00, 0, *and* 1 *to* 36. *The house wins if the ball lands in the* 00 *or* 0 *slots.*

**9.** What is the probability that the ball lands in an odd slot?

**10.** What is the probability that the ball lands in a slot numbered 1 to 12?

**11.** What is the probability that the ball lands in slot 18?

**12.** What is the probability that the ball lands in slot 17 or 18?

*In Exercises* 13–16, *assume that you are playing with an ordinary deck of* 52 *cards.*

**13.** What is the probability of drawing a black queen?

**14.** What is the probability of drawing either a heart or the queen of spades?

**15.** What is the probability of drawing a face card?

**16.** What is the probability of drawing a card numbered 2 through 10?

**17.** A church raffle sells 1630 tickets. If you buy 8 tickets, what is the probability that you will win the raffle?

**18.** A paper mill operates 365 days per year. On an average, the mill experiences one shutdown per year due to electrical malfunction. If the malfunction is equally likely to happen any day of the year, what is the probability that it will happen in July?

---

## 6.5 CONDITIONAL PROBABILITY

In Sections 6.3 and 6.4, we dealt with experiments in which we were simply trying to determine the probability that a given event $E$ occurs. In this section we will see that if additional information is provided concerning an experiment, the probability that event $E$ occurs will change. For example, suppose the Dodgers win 60% of their home games. It is the end of the eighth inning in Los Angeles, and the Dodgers have a ten-run lead. Clearly, the chances that they will win this game are greater than 60%. Thus, we now wish to find the probability of occurrence of the event $E$ that the Dodgers will win, given the *condition* on event $F$ that they are ten runs ahead at the end of the eighth inning; this is written

$$p(E \mid F).$$

The following two examples will generate a mathematically precise definition of conditional probability.

***Example 1*** Suppose we select a card at random from a deck of 52. Find the probability of drawing a queen, given the condition that a red face card was drawn.

Let $E$ be the event that a queen is selected. Then $E$ consists of the four sample points

$$E = \{Q\spadesuit, Q\heartsuit, Q\diamondsuit, Q\clubsuit\}.$$

Let $F$ be the event that the card drawn is a red face card. Then $F$ consists of six sample points: two kings, two queens, and two jacks. Since every card is just as likely to be drawn in this experiment, the sample space $S$ is uniform so that

$$p(E) = \frac{n(E)}{n(S)} = \frac{4}{52} = \frac{1}{13}$$

and

$$p(F) = \frac{n(F)}{n(S)} = \frac{6}{52} = \frac{3}{26}.$$

But if event $F$ is known to have occurred, only six cards are involved and each is equally likely to have been selected. Clearly, we need only consider the sample points in $E$ that *also belong to* $F$, that is, the set

$$E \cap F = \{Q\heartsuit, Q\diamondsuit\}.$$

Thus, two of the six ways for $F$ to have occurred belong to $E$, so that the probability that $E$ occurs given that $F$ has occurred is

$$p(E \mid F) = \frac{n(E \cap F)}{n(F)} = \frac{2}{6} = \frac{1}{3}.$$

Observe that if we divide both the numerator and the denominator of the fraction

$$\frac{n(E \cap F)}{n(F)}$$

by $n(S)$, we obtain

$$p(E \mid F) = \frac{n(E \cap F)}{n(F)} = \frac{n(E \cap F)/n(S)}{n(F)/n(S)}$$
$$= \frac{p(E \cap F)}{p(F)}.$$

We can use this last expression to find $p(E \mid F)$ by first obtaining

$$p(E \cap F) = p(\{Q\heartsuit, Q\diamondsuit\})$$
$$= \frac{n(\{Q\heartsuit, Q\diamondsuit\})}{n(S)} = \frac{2}{52}.$$

Then we have

$$p(E \mid F) = \frac{p(E \cap F)}{p(F)} = \frac{2/52}{6/52} = \frac{2}{6} = \frac{1}{3}.$$

***Example 2*** Two equally matched teams A and B are playing a contest in which the first team to win two games is the champion. Find the probability that A wins the contest, given that B wins the second game.

Since the teams are equally matched, the probability that A wins the championship is usually equal to 0.5. However, the fact that B has won the second game substantially changes these odds.

Consider the tree shown in Fig. 6.17, and let $E$ be the event that team A wins the contest. Then $E$ contains the sample points

$$E = \{\text{AA, ABA, BAA}\}.$$

Let $F$ be the event that the second game is won by team B. Then $F$ has the sample points

$$F = \{\text{ABA, ABB, BB}\}.$$

Since $F$ has occurred, only the sample points ABA, ABB, and BB are possible outcomes of the experiment. Only one of these sample points, ABA, lies in both $E$ *and* $F$. So we need to find the relative frequency of this outcome in comparison to the possible outcomes (the sample points in $F$). Thus,

$$p(E \mid F) = \frac{p(\{\text{ABA}\})}{p(\{\text{ABA}\}) + p(\{\text{ABB}\}) + p(\{\text{BB}\})}$$

$$= \frac{(1/8)}{(1/8) + (1/8) + (1/4)} = \frac{(1/8)}{1/2} = \frac{1}{4}.$$

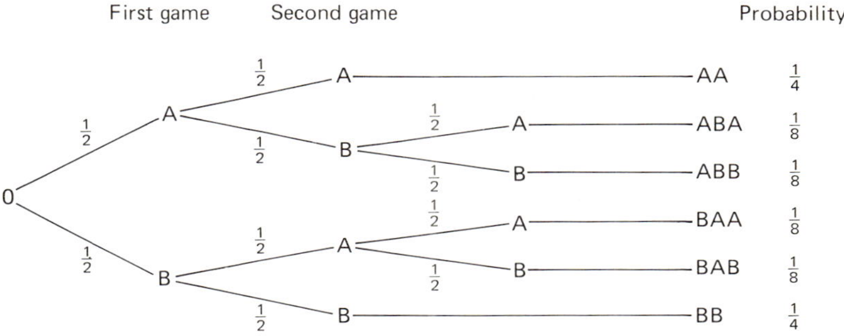


**Fig. 6.17**
First two out of three game series.

Again, we may observe that

$$p(E \mid F) = \frac{p(\{\text{ABA}\})}{p(\{\text{ABA}\}) + p(\{\text{ABB}\}) + p(\{\text{BB}\})}$$

$$= \frac{p(\{\text{ABA}\})}{p(\{\text{ABA, ABB, BB}\})}$$

$$= \frac{p(E \cap F)}{p(F)}.$$

---

In each of these examples, we have seen that the conditional probability can be obtained as the relative frequency of those sample points of $E$ that are in $F$ compared to the sample points in $F$. This leads to the following definition.

**DEFINITION 6.3** **The conditional probability of an event $E$, given that an event $F$ has occurred $[p(F) > 0]$, is given by**

$$p(E \mid F) = \frac{p(E \cap F)}{p(F)}. \tag{6.15}$$

A Venn diagram is useful in visualizing the meaning of conditional probability. Consider Fig. 6.18; the conditional probability $p(E \mid F)$ is obtained by reducing the sample space to the set $F$ and determining how much of this new sample space belongs to $E$. Thus only the relationship that $E$ bears to $F$ is important in finding the conditional probability $p(E \mid F)$.

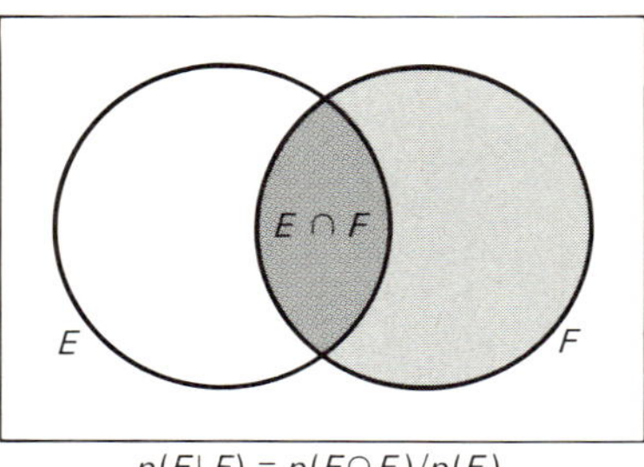

$p(E \mid F) = p(E \cap F)/p(F)$

**Figure 6.18**

***Example 3*** Red-green color blindness is an inherited trait that is linked with the gene that determines the sex of an individual. For this reason, red-green color blindness occurs much more frequently in males than in females.

Suppose that 100,000 individuals from a large population are tested, giving the percentages shown in Table 6.3; we see that color

**Table 6.3**

| | Male | Female | Total |
|---|---|---|---|
| Color-blind | 3.93% | 0.15% | 4.08% |
| Normal | 45.17% | 50.75% | 95.92% |
| Total | 49.1% | 50.9% | 100% |

blindness affects approximately 4% of that population. However, it might be more meaningful to ask what the probability is that an individual is color-blind if that individual is male (or female). Let $E$ be the event that an individual is red-green color-blind and let $F$ be the event that he is male. By Definition 6.3,

$$p(E \mid F) = \frac{p(E \cap F)}{p(F)},$$

and Table 6.3 indicates that the event $E \cap F$ (color-blind male) occurs with a frequency $p(E \cap F) = 0.0393$ (= 3.93%), while the event $F$ (male) occurs $p(F) = 0.491$, or 49.1% of the time. Hence,

$$p(E \mid F) = \frac{0.0393}{0.491} = 0.08;$$

that is, the probability that a male will be red-green color-blind is 8%. Similarly, if $G$ is the event that the individual is female, we have

$$p(E \mid G) = \frac{p(E \cap G)}{p(G)} = \frac{0.0015}{0.509} = 0.00295,$$

or 0.295%.

Two special situations are worth considering. If the event $E \supset F$, then

$$p(E \mid F) = \frac{p(E \cap F)}{p(F)} = \frac{p(F)}{p(F)} = 1.$$

On the other hand, if $E$ and $F$ have no sample points in common, $E \cap F = \varnothing$, and

$$p(E \mid F) = \frac{p(E \cap F)}{p(F)} = \frac{p(\varnothing)}{p(F)} = 0.$$

These two situations are illustrated in Fig. 6.19(a) and (b).

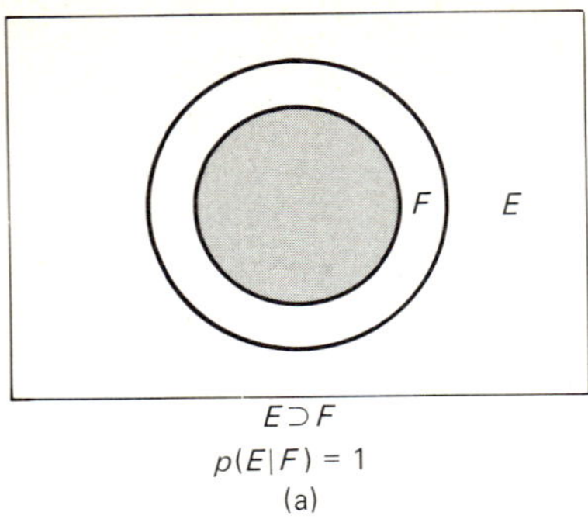


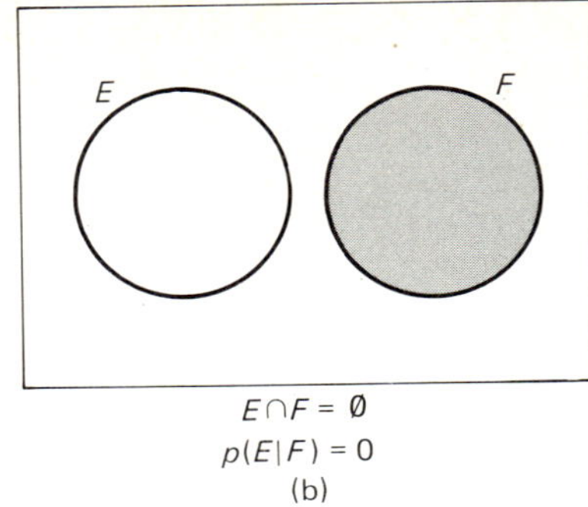


**Figure 6.19**

***Example 4*** A pair of dice is rolled and the number of dots on the two upper faces are totaled. Let $E_1$ be the event that the total is at most 8, let $E_2$ be the event that the total is exactly 9, and let $F$ be the event that at least one face shows a 1. We wish to find $p(E_1 | F)$ and $p(E_2 | F)$.

Figure 6.20 lists all the points in the sample space and describes the events $E_1$, $E_2$, and $F$. Clearly, $E_1 \supset F$ and $E_2 \cap F = \varnothing$, so that $p(E_1 | F) = 1$ and $p(E_2 | F) = 0$.

---

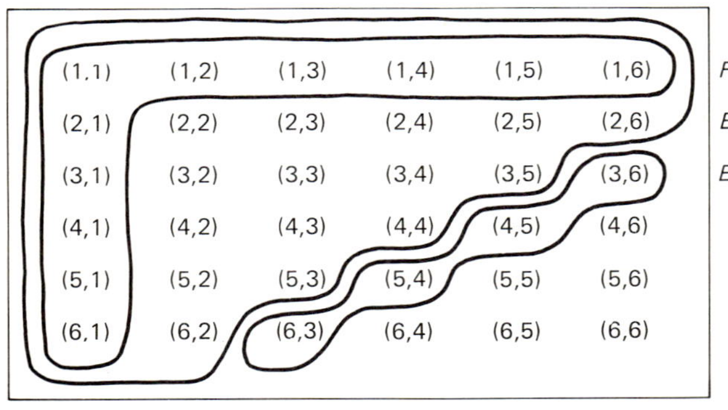


**Fig. 6.20** Sample space for a throw of two dice.

Equation (6.15) may be multiplied on both sides by $p(F)$ to obtain the *multiplication principle* of probability:

$$p(E \cap F) = p(E | F)p(F). \tag{6.16}$$

Equation (6.16) is often useful in finding the probability that two events $E$ and $F$ occur simultaneously.

***Example 5*** A merchant receives a shipment of 100 transistor radios, 12 of which are defective. He decides to place one in a display case and take one home to his family. What is the probability that the one he places in the display case is satisfactory but the one he takes home is defective?

Let $F$ be the event that the one placed in the display case is satisfactory and let $E$ be the event that the one taken home is defective. Since 88 of the radios are satisfactory, we have

$$p(F) = \frac{88}{100}.$$

After this radio has been placed in the display case, only 99 radios remain, 12 of which are defective. Thus, the probability of selecting a defective radio to take home, once a satisfactory radio has been placed in the display case, is

$$p(E \mid F) = \frac{12}{99}.$$

Hence both events will occur with probability

$$p(E \cap F) = p(E \mid F)p(F) = \frac{12}{99} \cdot \frac{88}{100} = 0.1067.$$

---

The definition of conditional probability,

$$p(E \mid F) = \frac{p(E \cap F)}{p(F)},$$

indicates that the probability that event $E$ occurs can be affected by knowing that $F$ has occurred. In some situations, however, knowing that $F$ has occurred causes *no change* in the probability of $E$, that is,

$$p(E \mid F) = p(E). \tag{6.17}$$

When Eq. (6.17) holds, the events $E$ and $F$ are said to be *independent.* For independent events, the multiplication principle of Eq. (6.16) becomes

$$p(E \cap F) = p(E)p(F), \tag{6.18}$$

that is, *the probability that two independent events will both occur equals the product of their probabilities.* The reader will recall that we have made frequent use of this concept in working with tree diagrams, where the outcome of each stage does not depend on the results of the previous stages (see Examples 2 and 3 in Section 6.3 or the coin tossing experiments in Section 6.1).

Events that are not independent are called *dependent.* We have seen several examples of dependent events in this section. For example, event $E$ in Example 5 depends on $F$ because $p(E) = 12/100$, while $p(E \mid F) = 12/99$.

Although it is easy to see that the outcome of flipping a fair coin is independent of the result of all previous tosses, it is often quite difficult to tell in a more complicated situation whether two events are independent. This becomes especially difficult when we must rely on experimental data, because the inaccuracies in the experimental results may show dependence where there is none. The following examples illustrate some of the difficulties.

***Example 6*** A pair of dice is thrown and the number of dots shown on the top faces are added. One die is smaller than the other. Let $F$ be the event that the smaller die shows one dot, and $E$ the event that the sum of the dots equals 7. If we let the first entry of each point in the sample space illustrated in Fig. 6.20 indicate the number of dots showing on the smaller die, we see that

$$p(E) = \frac{6}{36} = \frac{1}{6}, \qquad p(F) = \frac{6}{36} = \frac{1}{6},$$

and

$$p(E \mid F) = \frac{p(E \cap F)}{p(F)} = \frac{p(\{1, 6\})}{p(F)} = \frac{1/36}{1/6} = \frac{1}{6}.$$

Thus, $p(E \mid F) = p(E)$, so that $E$ and $F$ are independent events.

However, suppose we had to rely only on data that had been obtained experimentally. To obtain the following table of values, a pair of mismatched dice was thrown repeatedly and only those rolls in which the smaller die showed 1 were counted:

| | Sum of dots shown when smaller die was a 1 | | | | | | |
|---|---|---|---|---|---|---|---|
| | 2 | 3 | 4 | 5 | 6 | 7 | Total |
| Number of rolls | 4 | 9 | 7 | 5 | 8 | 9 | 42 |

A seven occurred in 9 of the 42 rolls, leading us to think that

$$p(E \mid F) = \frac{9}{42} = \frac{3}{14},$$

which is quite different from 1/6. It is clear from this experiment that we do not have enough data. If we continue rolling the dice, say, until the smaller die has shown a one 10,000 times, the number of times each given combination appears should be much closer to 1/6. Thus an adequate number of trials is essential if we wish to detect independence.

***Example 7*** A researcher wishes to find out whether there is any dependence between red-green color blindness and baldness in men. Suppose she collects the data shown in Table 6.4 after testing 1000 randomly chosen individuals. From her experimental data, she infers that the probability of being color-blind is

**Table 6.4**

| | Bald | Not bald | Total |
|---|---|---|---|
| Color-blind | 30 | 52 | 82 |
| Not color-blind | 393 | 525 | 918 |
| Total | 423 | 577 | 1000 |

$$p(C) = \frac{82}{1000} = 0.082 \text{ (or } 8.2\%\text{)},$$

and the probability of being bald is

$$p(B) = \frac{423}{1000} = 0.423 \text{ (or } 42.3\%\text{)}.$$

To see whether there is any dependence between baldness and color blindness, she must compute the conditional probability $p(C \mid B)$. Since $C \cap B$ is the event that the individual is both bald and color-blind, we have

$$p(C \cap B) = \frac{30}{1000} = 0.03 \text{ (or } 3\%\text{)},$$

so that

$$p(C \mid B) = \frac{p(C \cap B)}{p(B)} = \frac{0.030}{0.423} = 0.071 \text{ (or } 7.1\%\text{)}.$$

Since a difference of only one percentage point exists between $p(C \mid B)$ and $p(C)$, the researcher should be very hesitant about claiming that events $B$ and $C$ are dependent. Indeed, it could well be that they are independent, and the difference is a result of either insufficient data or an inadequate procedure in selecting the individuals for the experiment (for example, some individuals might be related).

The reader may wonder why it is necessary to show only one of the identities

$$p(E \mid F) = p(E)$$

or

$$p(F \mid E) = p(F)$$

to prove independence between the two events $E$ and $F$. The following theorem provides the answer to this question.

**THEOREM 6.3** **If $p(E \mid F) = p(E)$, then $p(F \mid E) = p(F)$.**

***Proof*** By Definition 6.3, we have

$$p(E) = p(E \mid F) = \frac{p(E \cap F)}{p(F)}$$

Cross multiplying, we have $p(E \cap F) = p(E)p(F)$, so that

$$p(F \mid E) = \frac{p(E \cap F)}{p(E)} = \frac{p(E)p(F)}{p(E)} = p(F). \blacksquare$$

## EXERCISES 6.5

1. Find the probability of drawing a red queen, given that a heart was drawn.
2. Find the probability of drawing an eight, given the condition that a face card was not drawn.
3. Two cards are drawn from a deck of 52. What is the probability that
   a) Both cards are spades?
   b) Both cards are red?
   c) The first card is a diamond and the second a spade?
   d) The first card is a spade and the second is red?
4. A couple plans to have children until they have either two boys or four children. If either sex is equally likely, what is the probability that
   a) They will have exactly three children, if the second child is a girl?
   b) They will have exactly three children, if the second child is a boy?
   c) They will have at least three children, if the second child is a girl?
   d) They will have more boys than girls, if the second child is a boy?
5. Five people are vying for two positions, president and vice-president, in a club. Mr. Martin and Ms. Smith are among the five. What is the probability that Mr. Martin is elected president, given that Ms. Smith has not been elected vice-president? What is the probability that Mr. Martin got one of the two positions?

**6.** A scientist gathers data trying to find a link between color blindness and race. Sampling 1000 random male individuals, he obtains the table below.

| | Black | Other | Total |
|---|---|---|---|
| Color-blind | 46 | 58 | 104 |
| Not color-blind | 572 | 324 | 896 |
| Total | 618 | 382 | 1000 |

Let $B$ be the event that an individual is black and $C$ the event that he is color-blind.

a) Find $p(B \mid C)$.

b) Find $p(C \mid B)$.

c) Are the events $B$ and $C$ independent?

**7.** Four people called North, East, South, and West, are playing contract bridge. Each is dealt 13 cards from an ordinary deck of 52.

a) What is the probability that North has two aces if South has one?

b) What is the probability that North has five spades if South has six spades?

c) If North and South together have two aces, what is the probability that East has the other two aces?

*In Exercises* 8–10, *use the mortality data provided in Table* 6.5.

**Table 6.5**

| Age (years) | Probability of death (%) |
|---|---|
| 0–9 | 2.94 |
| 10–19 | 0.73 |
| 20–29 | 1.37 |
| 30–39 | 2.07 |
| 40–49 | 4.78 |
| 50–59 | 11.53 |
| 60–69 | 26.63 |
| 70 and over | 49.95 |

*Source:* Adapted from "Expectation of Life and Mortality Rates." *Information Please Almanac,* 1969.

**8.** Find the probability that an individual 40 years of age will die before reaching the age of 50.

**9.** Find the probability that an individual 50 years old will die before reaching the age of 70.

**10.** Find the probability that an individual 20 years old will die before reaching the age of 60. [Hint: Let $F$ be the probability of surviving the first 20 years. Then $p(F) = 1 - p(F')$.]

**11.** A city council studies the traffic accident reports for the previous year. The following table shows the cause and age of the driver at fault in the 950 reported accidents.

| | Intoxication | Mechanical failure | Other |
|---|---|---|---|
| Under 26 | 248 | 120 | 182 |
| 26 and over | 217 | 136 | 47 |

a) Given that the accident was due to intoxication, what is the probability that the driver was under 26 years of age?
b) Given that the driver at fault is aged 26 years or over, what is the probability that he or she was intoxicated?
c) Given that the accident was due to mechanical failure, what is the probability that the driver is 26 years of age or older?

**12.** A company produces a lot consisting of 2500 integrated circuits, of which 185 are defective. A quality control specialist selects three circuits for testing.
a) What is the probability that at least one of these circuits is defective?
b) What is the probability that all the circuits selected are defective?

**13.** A power company finds that 3% of the mail payments it receives are drawn on insufficient funds. These checks are always postdated. Of the checks drawn with sufficient funds, 2% are postdated. If a check is postdated, what is the probability that it is drawn on insufficient funds? (*Hint:* Draw a tree diagram.)

**14.** Answer Exercise 13 if only 50% of the checks drawn on insufficient funds are postdated.

**15.** Weather studies in Phoenix, Arizona, suggest that a knowledge of today's weather can be used to forecast tomorrow's weather. Assuming that $p(\text{dry tomorrow} \mid \text{dry today}) = 4/5$ and $p(\text{dry tomorrow} \mid \text{rain today}) = 2/3$, find the probability that it will be dry two days from now if it is dry today. If it is raining today, what is the probability that it will be dry two days from now? (*Hint:* Use a tree diagram.)

**16.** In Exercise 15 suppose that it is dry today.
a) What is the probability that it will rain in the next three days?
b) What is the probability that it will rain three days from now?
c) What is the probability that at least two of the next three days are dry?
d) What is the probability that the next three days are dry?

---

## 6.6 BAYES'S FORMULA

In this section we will study more complicated situations involving conditional probability. We will make frequent use of the *multiplication principle*

$$p(E \cap F) = p(E \mid F)p(F),$$

which we obtained by multiplying both sides of the definition of conditional probability by $p(F)$. First of all, we should observe that we can reverse the roles of the events $E$ and $F$ in this equation, obtaining

$$p(F \cap E) = p(F \mid E)p(E).$$

But $E \cap F = F \cap E$, so we have

$$p(E \mid F)p(F) = p(F \mid E)p(E),$$

and dividing both sides by $p(F)$ yields

$$p(E \mid F) = \frac{p(F \mid E)p(E)}{p(F)}. \tag{6.19}$$

A quick glance at Fig. 6.21 shows that

$$F = (E \cap F) \cup (E' \cap F),$$

and the events $E \cap F$ and $E' \cap F$ are mutually exclusive, since they are disjoint sets. Hence, the probability that either of these mutually exclusive events occurs is the sum of their individual probabilities (see Theorem 6.1), so that

$$p(F) = p[(E \cap F) \cup (E' \cap F)] = p(E \cap F) + p(E' \cap F). \tag{6.20}$$

Applying the multiplication principle to each term on the right-hand side of Eq. (6.20) yields

$$p(F) = p(F \mid E)p(E) + p(F \mid E')p(E'). \tag{6.21}$$

Replacing the denominator of Eq. (6.19) with the right-hand side of Eq. (6.21) leads to *Bayes's formula:*

$$p(E \mid F) = \frac{p(F \mid E)p(E)}{p(F \mid E)p(E) + p(F \mid E')p(E')}. \tag{6.22}$$

The following examples illustrate the power and use of this formula.

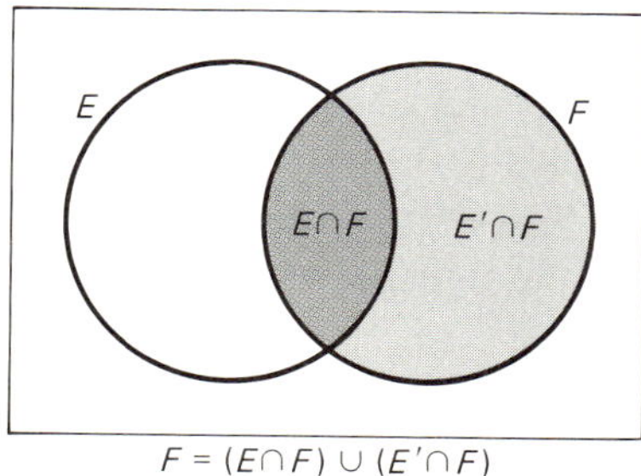


**Figure 6.21**

***Example 1*** Joe, the lonesome Maytag washer repairman, decides to improve his social life by repairing Frigidaire washers as a sideline. Maytag and Frigidaire produce 40% and 60% of the washers sold in his area, respectively. Approximately 10% of the Maytag washers and 35% of the Frigidaire washers require service every year. What is the probability that the next call Joe answers will be to repair a Maytag washer?

Let $E$ be the event that the washer is a Maytag and note that $E'$ is the event that the washer is a Frigidaire, since these are the only washers sold in the area. Then $p(E) = 0.4$ and $p(E') = 0.6$. Let $F$ be the event that the washer is defective. Then

$$p(F \mid E) = 0.1 \text{ (or 10\%)} \quad \text{and} \quad p(F \mid E') = 0.35 \text{ (or 35\%)}.$$

By Bayes's formula, the probability that the washer is a Maytag, given that it is defective, is

$$p(E \mid F) = \frac{p(F|E)p(E)}{p(F \mid E)p(E) + p(F \mid E')p(E')}$$

$$= \frac{(0.1)(0.4)}{(0.1)(0.4) + (0.35)(0.6)} = 0.16,$$

or 16% of the time. Thus, Joe should plan accordingly and carry

$$\frac{0.84}{0.16} = 5.25$$

times as many Frigidaire repair parts as Maytag repair parts.

***Example 2*** The probability that an individual with certain symptoms has tuberculosis is 60%. A TB tine test will confirm this diagnosis for 95% of the cases in which the disease is present, but will also give a positive diagnosis for 10% of the cases in which the individual is healthy. What is the probability that a person exhibiting these symptoms and reacting positively to the TB tine test actually has TB?

Let $E$ be the event that a person exhibiting the symptoms actually has TB, and let $F$ be the event that the patient's TB tine test is positive. We know that the patient has reacted positively to the TB tine test and we wish to know what the probability is that the patient has TB; that is, we wish to find

$$p(E \mid F).$$

By Bayes's formula, we have

$$p(E \mid F) = \frac{p(F \mid E)p(E)}{p(F \mid E)p(E) + p(F \mid E')p(E')}$$

$$= \frac{(0.95)(0.6)}{(0.95)(0.6) + (0.1)(0.4)} = \frac{0.57}{0.61} = 0.934,$$

so there is a 93.4% probability that the patient has TB.

---

Bayes's formula (Eq. 6.22) can be generalized to allow for more than two possible outcomes. Suppose $E_1, E_2, \ldots, E_n$ are mutually exclusive events whose union equals the sample space

$$S = \bigcup_{k=1}^{n} E_k.$$

Let $F$ be any event, and suppose that $F$ has occurred. Then, using Eq. (6.19), we obtain

$$p(E_k \mid F) = \frac{p(F \mid E_k)p(E_k)}{p(F)}. \tag{6.23}$$

From Fig. 6.22 we clearly see that

$$F = (E_1 \cap F) \cup (E_2 \cap F) \cup \cdots \cup (E_n \cap F),$$

and the events $E_1 \cap F, E_2 \cap F, \ldots, E_n \cap F$ are mutually exclusive. Thus, the probability of $F$ occurring equals the sum of their individual probabilities:

$$p(F) = p(E_1 \cap F) + p(E_2 \cap F) + \cdots + p(E_n \cap F). \tag{6.24}$$

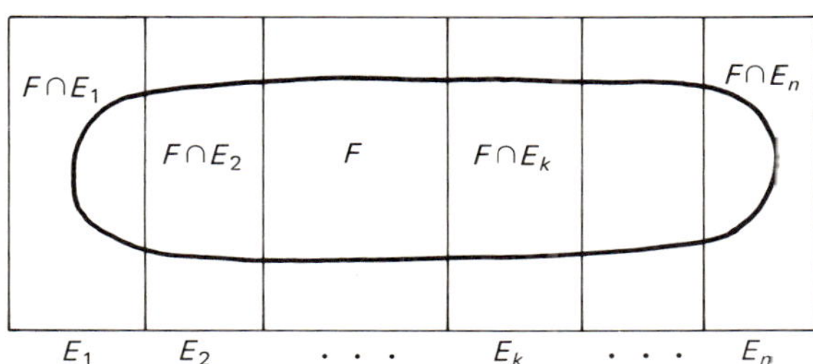


**Figure 6.22**

Applying the multiplication principle to each term on the right-hand side of Eq. (6.24) gives us

$$p(F) = p(F \mid E_1)p(E_1) + p(F \mid E_2)p(E_2) + \cdots + p(F \mid E_n)p(E_n). \tag{6.25}$$

Replacing $p(F)$ in Eq. (6.23) with the right-hand side of Eq. (6.25) yields the generalization of Bayes's formula:

$$p(E_k \mid F) = \frac{p(F \mid E_k)p(E_k)}{p(F \mid E_1)p(E_1) + p(F \mid E_2)p(E_2) + \cdots + p(F \mid E_n)p(E_n)}, \quad 1 \le k \le n. \tag{6.26}$$

***Example 3*** In a certain county 40% of the voters are Democrats, 25% are Republicans, and 35% are Independents. In a survey it is established that 60% of the Democrats, 10% of the Republicans, and 50% of the Independents are willing to vote for a well-qualified woman for county commissioner. Given that a voter has just cast a ballot for the woman, what is the probability that the voter is a Democrat?

Let $D$, $R$, and $I$ be the events that the voter is a Democrat, Republican, and Independent, respectively. Let $W$ be the event that the person votes for the woman. Then

$$p(D) = 0.40, \qquad p(R) = 0.25, \qquad p(I) = 0.35,$$

and

$$p(W \mid D) = 0.60, \qquad p(W \mid R) = 0.10, \qquad p(W \mid I) = 0.50.$$

Using Bayes's formula (Eq. 6.26), we have

$$p(D \mid W) = \frac{p(W \mid D)p(D)}{p(W \mid D)p(D) + p(W \mid R)p(R) + p(W \mid I)p(I)}$$

$$= \frac{(0.6)(0.4)}{(0.6)(0.4) + (0.1)(0.25) + (0.5)(0.35)} = 0.545,$$

or 54.5%. Similarly, $p(R \mid W) = 0.057$ and $p(I \mid W) = 0.398$.

---

**EXERCISES 6.6**

1. A store sells Hewlett-Packard and Texas Instruments calculators. Almost 70% of the calculators that it sells are Texas Instruments. However, 25% of the Texas Instrument calculators are returned because of defects, while only 5% of the Hewlett-Packard calculators are defective. What is the probability that the next calculator returned is a Hewlett-Packard? Texas Instruments?

2. A man has a fair coin and a two-headed coin in his pocket. He selects one at random, flips it, and observes that heads came up. What is the probability that the coin is the two-headed one?

3. Suppose in Exercise 2 that the man has five fair coins and a two-headed coin. What is the probability that the coin is a fair coin if heads came up?

4. A certain judge is very harsh in judging traffic violations. If the motorist committed the violation, he or she will be convicted 99% of the time. Even if the motorist did not commit the violation, there is a 60% probability that he or she will be convicted. Suppose that 80% of those arrested actually committed the traffic violation.
   a) What is the probability that you are innocent if you are convicted?
   b) What is the probability that you are innocent if you are acquitted?

5. High fever and yellowing of the white of the eye are early symptoms of hepatitis; 70% of patients having these symptoms have the disease. A blood test will confirm hepatitis in 98% of the patients having the disease. However, 10% of the patients not having hepatitis will also have positive test results.
   a) What is the probability that a patient with the early symptoms has hepatitis if his or her blood test result is positive?
   b) What is the probability that a patient with the early symptoms has hepatitis if his or her blood test is negative?

6. GM, Ford, and Chrysler sell 40%, 35%, and 25% of the pickup trucks in a certain town, respectively. The respective probabilities that each brand of truck requires engine repair are 10%, 12%, and 14%, respectively. What is the probability that the next pickup requiring engine repair is a GM? a Ford? a Chrysler?

7. Use Bayes's formula to answer Exercise 13 on p. 280.

8. Use Bayes's formula to answer Exercise 15 on p. 280.

9. In a federal study on employment and education of 7000 people, the following information was obtained:

| | Highest level of education | | |
|---|---|---|---|
| | No high school diploma | High school diploma | Bachelor's degree |
| Employed | 1612 | 2120 | 2610 |
| Unemployed | 206 | 240 | 212 |

   Using Bayes's formula:
   a) Find the probability that an individual has at most a high school diploma if he or she is employed.
   b) Find the probability that an individual has at least a high school diploma if he or she is unemployed.

**10.** An auto insurance company has determined that a teenager who has taken a course in driver education has a probability of 0.92 of completing the first year of driving without an accident, while one who has not taken the course has a probability of only 0.68. If 80% of all new teenage drivers have taken driver education, what is the probability that a teenage driver involved in an accident in his or her first year of driving did not take the course?

**11.** Answer Exercise 10 for straight-A students if the insurance company determines that they have a probability of 0.95 of completing the first year of driving without an accident, regardless of whether or not they took driver education. Assume that all other probabilities remain the same as those in Exercise 10.

**12.** A tomato shipper sends his produce to a supermarket chain by three means of transportation: refrigerated truck, rail, or air express. The probabilities that the produce will reach the supermarket within 24 hours by each means of transportation are 0.8, 0.5, and 0.98, respectively. Each means of transportation is equally likely to be used.

a) If the produce arrives within 24 hours, what is the probability that it was sent by rail?

b) If the produce does not arrive within 24 hours, what is the probability that it was sent by air express?

**13.** A New York importer places orders equally using three types of mail: registered airmail, ordinary airmail, and seamail. The probabilities that the order will reach the European manufacturer within one week via each type of mail are 0.8, 0.85, and 0.25, respectively.

a) If the order arrives within a week, what is the probability that it was sent by registered airmail?

b) If the order does not arrive within a week, what is the probability that it was sent by ordinary airmail?

---

## 6.7 COUNTING PROCEDURES

In this section we will present three different methods for counting the number of outcomes of a multistage experiment. An experiment is said to have *k-stages* if the final outcome of the experiment depends on the state of the experiment at $k$ different times during its execution. For example, the experiment may consist of tossing a coin $k$ times and keeping track of the result of each toss. The outcomes of this $k$-stage experiment would then consist of a $k$-term sequence of the letters T and H.

The first counting technique is straightforward.

**SEQUENTIAL COUNTING PRINCIPLE**

Suppose we are performing a $k$-stage experiment in which the states of the experiment at each stage do not depend on one another. If there are $n_1$ states at the first stage, $n_2$ states at the second stage, ..., and $n_k$ states at the $k$th stage, then the total number of $k$-stage outcomes is

$$n_1 \cdot n_2 \cdot \cdots \cdot n_k.$$

For example, if we toss an ordinary coin ten times and record the result of each toss, we find that there are two states at each stage of the experiment. Since the state of the $j$th toss does not depend on any other toss, we can use the sequential counting principle. Thus, there are

$$2 \cdot 2 \cdot \cdots \cdot 2 = 2^{10} = 1024$$

outcomes to this experiment.

So far we have discussed only experiments in which the state of the experiment at any stage is independent of all other states at other stages of the experiment. The next two procedures will differ in that the results at each stage will affect one another.

*A permutation of $k$ elements from a set of $n$ elements is any ordered selection of $k$ different objects from a collection of $n$ objects.* It might appear, superficially, that there is no difference between a permutation and the $k$-term outcomes of a $k$-stage experiment, since the latter also yields ordered sets consisting of $k$ objects (the results of each stage). However, there is one fundamental difference: The objects in a permutation are all *different*, whereas they need not be different in a $k$-term outcome.

We may use the sequential counting principle to compute the number $P_{n,k}$ of permutations of $k$ elements in a set of $n$ elements. There are $n$ ways of choosing the first object, which is then removed from the set. Thus, the second object can only be chosen in $(n - 1)$ ways, since only $(n - 1)$ objects remain. Once it has been selected, there are $(n - 2)$ ways of selecting the third object, and so on. The $k$th object may be chosen in $(n - k + 1)$ ways. Thus, multiplying the ways in which each object can be chosen, we obtain

$$P_{n,k} = n(n - 1)(n - 2) \cdots (n - k + 1). \tag{6.27}$$

In particular, we should note that the total number of ways of rearranging $n$ objects is

$$P_{n,n} = n(n - 1)(n - 2) \cdots 2 \cdot 1,$$

since $(n - n + 1) = 1$. For convenience, we use the symbol $n!$ (read "$n$ factorial") to denote this product, that is,

$$n! = n(n - 1)(n - 2) \cdots 3 \cdot 2 \cdot 1. \tag{6.28}$$

With this notation we can rewrite Eq. (6.27) using factorials by multiplying $P_{n,k}$ by $(n - k)!/(n - k)!$ (or 1); we obtain

$$\begin{aligned} P_{n,k} &= n(n - 1) \cdots (n - k + 1) \\ &= \frac{n(n - 1) \cdots (n - k + 1)(n - k) \cdots 1}{(n - k) \cdots 1} \\ &= \frac{n!}{(n - k)!}. \end{aligned} \tag{6.29}$$

To make Eq. (6.29) consistent with Eq. (6.28) when $k = n$, we define

$$(n - n)! = 0! = 1. \tag{6.30}$$

***Example 1*** A basketball team has four guards (call them $a$, $b$, $c$, and $d$). In how many ways can two starting guards be announced at the beginning of the game?

In this problem the objects must be different and the *order* makes a difference. Thus, we seek the number of permutations of two objects from a set of four objects. Hence, there are

$$P_{4,2} = \frac{4!}{(4-2)!} = \frac{4!}{2!} = \frac{4 \cdot 3 \cdot 2 \cdot 1}{2 \cdot 1} = 12$$

ways of announcing two starting guards. The twelve selections are shown below:

| | | | |
|---|---|---|---|
| $ab$, | $ba$, | $ca$, | $da$, |
| $ac$, | $bc$, | $cb$, | $db$, |
| $ad$, | $bd$, | $cd$, | $dc$. |

***Example 2*** A saleswoman wishes to visit six cities in her assigned territory. In how many different ways can she plan her schedule?

Clearly, the objects (cities) are all different and the order in which she visits them is significant. The number of possible schedules is

$$P_{6,6} = \frac{6!}{(6-6)!} = \frac{6!}{0!} = 6 \cdot 5 \cdot 4 \cdot 3 \cdot 2 \cdot 1 = 720.$$

Suppose, instead, that the saleswoman decides to visit only four of the cities, and do her marketing in the other cities by telephone. Then there are

$$P_{6,4} = \frac{6!}{(6-4)!} = \frac{6!}{2!} = \frac{6 \cdot 5 \cdot 4 \cdot 3 \cdot 2 \cdot 1}{2 \cdot 1} = 360$$

schedules she can make.

***Example 3*** In how many different ways can a disk jockey play four of the top ten songs?

Again the order is important and the selections are all different. Thus, the radio program can be aired in

$$P_{10,4} = \frac{10!}{(10-4)!} = \frac{10!}{6!} = 10 \cdot 9 \cdot 8 \cdot 7 = 5040$$

ways.

---

A *combination of k elements from a set of n elements* is any selection of $k$ different objects *without regard to order* from a collection of $n$ objects. The only difference between permutations and combinations is that order is immaterial for combinations. Hence we can calculate the number of combinations $C_{n,k}$ of $k$ objects taken from a set of $n$ objects by dividing the number of permutations of these objects $P_{n,k}$ by the number of ways that the $k$ objects can be ordered $P_{k,k}$:

$$C_{n,k} = \frac{P_{n,k}}{P_{k,k}}, \tag{6.31}$$

or in terms of the factorial notation

$$C_{n,k} = \frac{n!/(n-k)!}{k!/(k-k)!} = \frac{n!}{(n-k)!k!}. \tag{6.32}$$

***Example 4*** How many ways are there of picking two starting guards from the four guards $a$, $b$, $c$, and $d$ on the basketball team mentioned earlier?

In this problem the order does not matter; yet the objects are all different. Hence, we seek the number of combinations of two objects from a set of four objects, that is

$$C_{4,2} = \frac{4!}{(4-2)!2!} = \frac{4 \cdot 3 \cdot 2 \cdot 1}{2 \cdot 1 \cdot 2 \cdot 1} = 6.$$

The six selections are shown below, where the order indicated is immaterial:

$$\begin{array}{ll} \{a, b\}, & \{b, c\}, \\ \{a, c\}, & \{b, d\}, \\ \{a, d\}, & \{c, d\}. \end{array}$$

The reader should compare this result with Example 1.

***Example 5*** How many possible bridge hands are there?

A bridge hand consists of any 13 cards from a standard deck of 52 cards. Since the order of selection is immaterial, the number of possible bridge hands is

$$C_{52,13} = \frac{52!}{(52-13)!13!} = \frac{52!}{39!13!} = 635{,}013{,}559{,}600.$$

---

Occasionally we run across a problem in which two or more of these techniques must be applied. In the following example we must apply both the sequential counting principle and the method of combinations.

***Example 6*** A basketball team has two centers, four forwards, and four guards. In how many ways can the coach select a center, two forwards, and two guards?

From the discussion above we know that the order is immaterial, so the center can be picked in $C_{2,1}$ ways, the two forwards in $C_{4,2}$ ways, and the two guards in $C_{4,2}$ ways. Each selection can be regarded as a different stage, so that the choice of lineup becomes a three-stage experiment. By the sequential counting principle, we know that the total number of ways of obtaining a lineup is the product of the number of ways of making each selection, or

$$N = C_{2,1}C_{4,2}C_{4,2} = \frac{2!}{1!1!} \cdot \frac{4!}{2!2!} \cdot \frac{4!}{2!2!} = 72.$$

***Example 7*** A joint committee is selected from Congress, and consists of three representatives and three senators. There are 435 representatives and 100 senators. In how many ways can the joint committee be formed?

This is a two-stage experiment, one being the selection of the representatives, the other the selection of the senators. The first can be done in $C_{435,3}$ ways, while the second allows $C_{100,3}$ possible selections. Thus, the total number of ways of forming the committee is obtained by multiplying these two numbers:

$$\begin{aligned} N &= C_{435,3} \cdot C_{100,3} \\ &= \frac{435!}{432!3!} \cdot \frac{100!}{97!3!} = 2{,}203{,}056{,}586{,}500. \end{aligned}$$

---

**EXERCISES 6.7**

**1.** Calculate the following numbers by using Eq. (6.29).

a) $P_{5,3}$ b) $P_{7,4}$
c) $P_{9,2}$ d) $P_{9,7}$
e) $P_{8,5}$ f) $P_{10,6}$
g) $P_{9,9}$ h) $P_{8,8}$
i) $P_{5,4}$ j) $P_{10,9}$

**2.** Calculate the following numbers by using Eq. (6.32).

a) $C_{5,3}$ b) $C_{7,4}$
c) $C_{9,2}$ d) $C_{9,7}$
e) $C_{8,5}$ f) $C_{10,6}$
g) $C_{9,9}$ h) $C_{8,8}$
i) $C_{5,4}$ j) $C_{10,9}$

**3.** How many permutations of 5 elements can be found from a set of 8 elements?

4. How many combinations of 6 elements can be found from a set of 11 elements?

5. In how many different ways can 3 books be chosen from a collection of 50?

*Use the formulas for permutations and combinations to solve the following exercises.*

6. A football coach has six tackles. In how many ways can he select his two starting tackles if

   a) The first one selected is the left tackle?
   b) He will decide later which one plays left tackle?

7. A padlock combination consists of three numbers from 0 to 39, no two of which are alike. How many different combinations are possible?

8. In how many ways can 12 books be arranged on a bookshelf?

9. A book company has five representatives covering the colleges and universities in five western states: Oregon, Washington, Idaho, Montana, and Utah. Each representative will be assigned one state as his or her territory. In how many ways can the territories be assigned?

10. In how many possible ways could the gold, silver, and bronze medals of an Olympic event be assigned to a field of 18 competitors?

11. Ten steers are to be selected from a herd of 25. In how many ways can this be done?

12. A city council is composed of four Democrats and three Republicans. A five-person delegation is chosen to attend an inauguration.

    a) How many different delegations are possible?
    b) How many delegations could have two Republicans?
    c) How many delegations could have two Democrats?

13. How many bridge hands are there containing

    a) Six hearts and seven clubs?
    b) Four spades, four hearts, and five diamonds?
    c) Two spades, three hearts, four diamonds, and four clubs?

14. A shipment of 30 calculators contains 5 defective ones. Suppose 10 calculators are selected at random from the 30.

    a) How many different selections can be made?
    b) How many different selections contain no defective calculators?
    c) How many different selections contain 5 defective calculators?
    d) How many different selections contain 2 defective calculators?

15. Social security numbers are nine digits long. How many different numbers are possible?

16. In how many ways can people be classified by sex, age, and political party? Assume that no one is over 120 years of age, and that the political affiliations are Democratic, Republican, and Independent.

17. A couple decides to have children until they have a boy or three children, whichever comes first. How many possible ways are there of having such a family? If the chances are even for either sex being born, what is the probability that the family has fewer than three children?

18. A coin purse is filled with nickels and dimes. In how many different ways can coins be taken from the purse, one at a time, until we have enough to purchase a 25¢ can of soda?

19. How many different ways are there of making 15 cents in change?

20. A coin is tossed four times or until two heads have appeared. Draw a tree showing all possible outcomes. How many of these required four tosses?

21. Two teams are playing in a hockey tournament. The first team to win two games wins the trophy. How many possible matches are there?

22. Two baseball teams are competing in their league's playoff. The first team to win three games represents the league in the World Series. In how many different ways can the playoff turn out?

*In Exercises* 23–28, *find the probability that at least two persons in a group of $n$ persons have the same birthday. Adapt Eq.* (6.1) *to handle each situation.*

**23.** $n = 35$ **24.** $n = 40$ **25.** $n = 45$

**26.** $n = 50$ **27.** $n = 55$ **28.** $n = 60$

29. What is the probability that in a group of six people, two will have a common horoscope sign? Assume that the twelve horoscope signs are all equally likely.

30. What is the probability that in a group of four people, two will have been born in the same month? Assume that the twelve months are equally likely.

31. Answer Exercise 30 when the group consists of five people.

32. A researcher asks each of four people to choose a number between 1 and 10. She repeats this experiment 100 times and notes that on 50 occasions two people chose the same number. Is there enough evidence to suggest the existence of extrasensory perception?

---

## 6.8 EXAMPLES AND APPLICATIONS

In this section we give some additional examples of the counting techniques that we developed in Section 7. These examples indicate the wide-ranging applicability of these methods.

***Example 1*** Each letter in the Morse code consists of a sequence of dots and dashes. Can each letter in the alphabet be described by a sequence of four or fewer dots and dashes?

Since each signal is either a dot or a dash, repetition of signals is allowed, and order is important, the best way to answer this problem concerning sequences of events is to construct a tree. Such a tree is shown in Fig. 6.23, in which each junction in the tree represents the letter shown below it. Note that the sequences ··––, ·–·–, –––·, and –––– are not needed, since the number of possible sequences is 2 one-signal letters, 4 two-signal letters, 8 three-signal letters, and 16 four-signal letters, so that we have 30 sequences consisting of at most four signals. Samuel F. B. Morse thoughtfully assigned the shorter sequences to the more commonly used letters in the English language; this substantially reduces the number of signals needed in transmitting most messages.

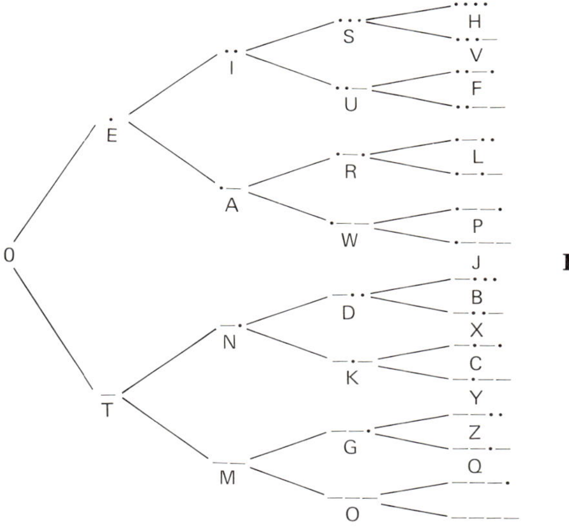


**Figure 6.23**

It is now very easy to answer two related questions: How many symbols can be described by sequences of exactly five dots and dashes, and how many can be described by five or fewer dots and dashes? Clearly, the next stage in the tree contains twice as many sequences as the fourth stage, since each sequence in the fourth stage branches two ways. Thus, there are 32 five-signal sequences and 62 sequences of five or fewer signals. This number of sequences is more than enough to include the digits 0 through 9, as well as most punctuation marks.

***Example 2*** The representatives of each league in baseball's World Series are decided by playoffs between the winners in each division. The first team to win three games is the representative. This means that each playoff series consists of 3, 4, or 5 games, since one team must win three games in any five-game series. In how many different ways can the playoff series be completed?

Since repetition is permitted and order of winning is important, we can best consider these sequences of events by constructing a tree. Suppose the White Sox were playing the Yankees. We indicate the winner of each game by the letter W for the White Sox or Y for the Yankees. Then the tree in Fig. 6.24 illustrates the possible outcomes of the series. We note that there are:

| | |
|---|---|
| 2 three-game series | WWW, YYY; |
| 6 four-game series | WWYW, WYWW, YWWW,<br>YYWY, YWYY, WYYY; |
| + 12 five-game series | WWYYW, WYWYW, WYYWW,<br>YWWYW, YWYWW, YYWWW,<br>WWYYY, WYWYY, WYYWY,<br>YWYWY, YWWYY, YYWWY. |
| = 20 possible playoff series. | |

Many factors, such as the advantage of playing in one's home park, the relative abilities of the starting pitchers, and the weather and condition of the field, have a significant effect on the outcome of each game. These conditions will change each team's chances of winning from game to game. Suppose the first game is played in Chicago, the next two in New York, and any other games are played in Chicago. If the White Sox have a 60% probability of winning in Chicago and a 30% probability of winning in New York, what is the probability that the White Sox will win the playoff? How likely is it that the playoff will finish in four or fewer games?

To answer these questions, we assign the probabilities given above to each branch of the tree (these values are underlined in Fig. 6.24). Observe that the Yankees have a 40% probability of winning in Chicago and a 70% probability of winning in New York. Each series ending in a W is won by the White Sox. Adding together the probabilities associated with the 10 ways in which the White Sox can win, we obtain

$$0.054 + 0.0756 + 0.03024 + 0.0756 + 0.03024 + 0.00864 + 0.03024 + 0.03024 = 0.46224,$$

which is the probability that the White Sox will win the playoff. Thus, the probability that the Yankees will win is 0.53776.

We can find the probability that the playoff will end in four or fewer games if we add the probabilities associated with each of the eight series of four or fewer games:

$$0.054 + 0.0756 + 0.0756 + 0.1176 + 0.0216 + 0.0336 + 0.0336 + 0.196 = 0.6176. \quad (6.33)$$

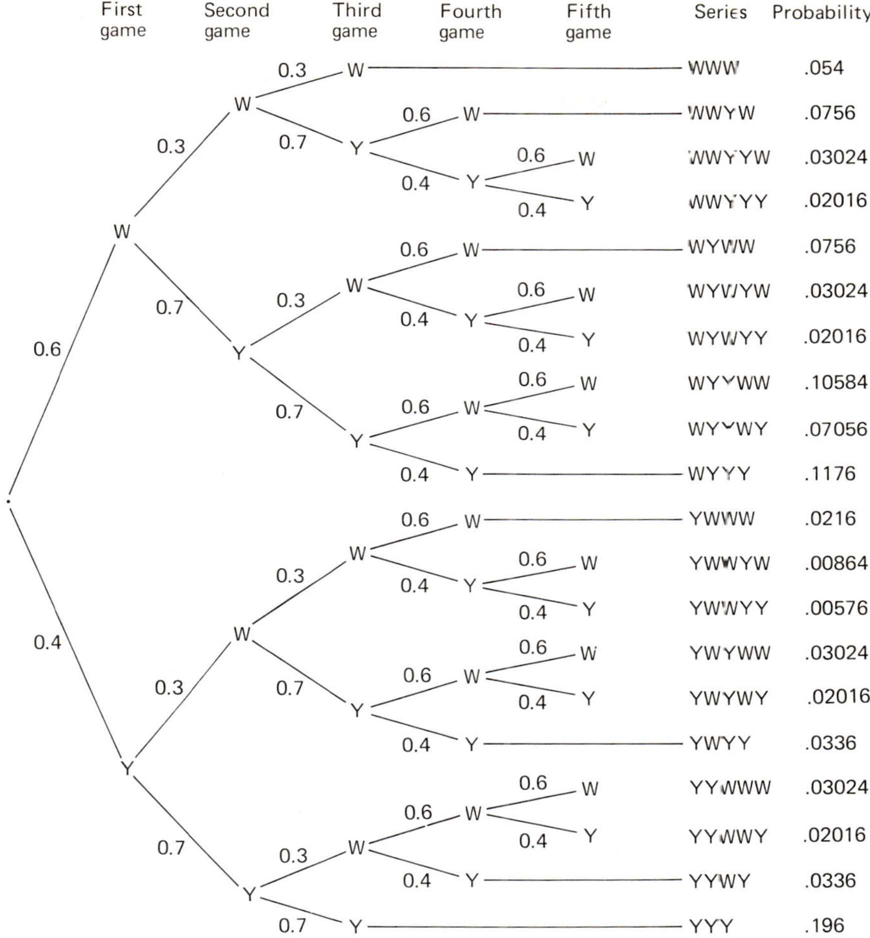


**Figure 6.24**

Note that the playoff has a 25% probability of ending in exactly three games, so if we subtract this amount from the left side of Eq. (6.33), we have

Probability of a three-game playoff = 0.25,
Probability of a four-game playoff = 0.3576,
Probability of a five-game playoff = 0.3924.

The chances are 3 in 4 that more than three games will be needed to decide the playoff with the probabilities we have assumed in this example.

***Example 3*** Probability theory is very useful in the study of genetics. Genetics is the branch of biology that deals with the processes by which certain characteristics are transmitted from parents to their offspring. The first systematic study of the laws of inheritance was published in 1865 by Mendel,* and was based on an eleven-year study of seven well-defined characteristics of the garden pea. Mendel discovered that certain traits, such as the color of the flowers, were determined by a pair of *genes*, one gene being inherited from each parent.

Mendel began his experiment with pure breeding lines of parents—one with purple flowers, the other with white. When the two lines were crossbred, the first generation (*hybrid*) plants all had purple flowers. But when these hybrids were crossbred, approximately 75% of their offspring had purple flowers and 25% had white flowers (705 plants had purple flowers, 224 had white flowers).

Mendel's explanation of this behavior can best be understood by designating the genes that determine purple and white flowers by the letters $A$ and $a$, respectively. There are then three possible gene pairs:

$$AA, \qquad Aa = aA, \qquad aa,$$

each of which is called a *genotype*. He assumed that the genotypes $AA$ and $Aa$ would both bear purple flowers, while the genotype $aa$ would yield white flowers. Since any genotype containing the $A$ gene yields purple flowers, he inferred that the $A$ gene was *dominant* and the $a$ gene *recessive*. If we begin by crossbreeding the two pure, or *homozygous*, strains $AA$ and $aa$, the first-generation offspring must all have *heterozygous* genotype $Aa$, since one gene is inherited from each parent (see Fig. 6.25). But when we crossbreed the first generation, all three genotypes $AA$, $Aa$, and $aa$ will arise, with the genotype $Aa$ $(= aA)$ twice as likely to occur as either of the homozygous pairs, as can be seen in the tree shown in Fig. 6.26. Thus, the dominant gene is likely to occur in three of every four cases, which accounts for the fact that approximately 75% of the second-generation plants had purple flowers.

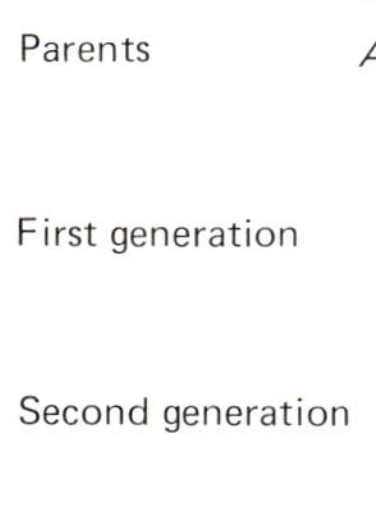


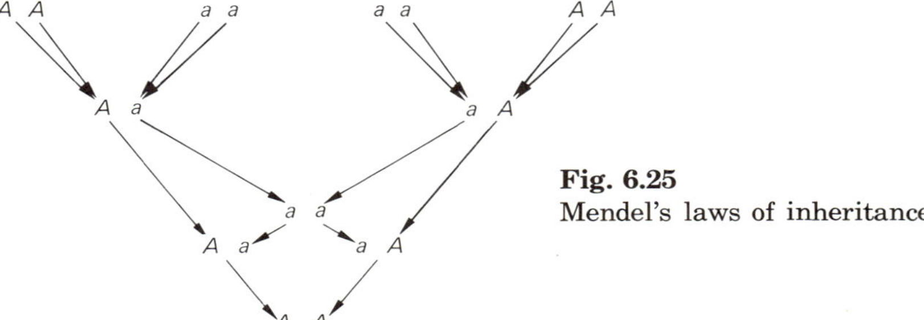


**Fig. 6.25**
Mendel's laws of inheritance.

---

* Johann Gregor Mendel (1822–1884), Abbot of the Augustinian monastery at Brno, Czechoslovakia.

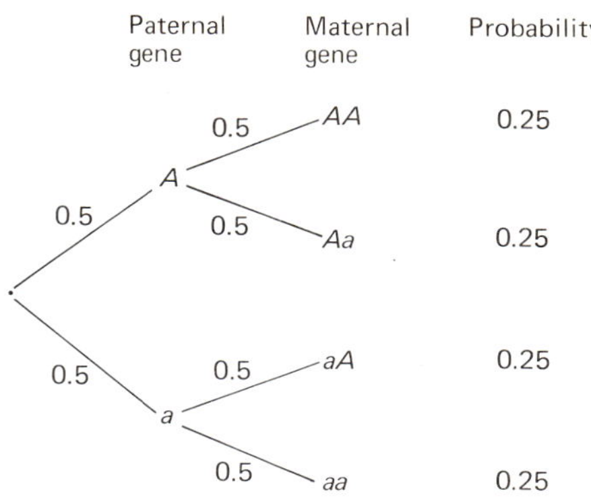


**Fig. 6.26**
Possible outcomes of the second generation, assuming equal probability for each event.

Mendel was able to discover the laws of inheritance because of three crucial steps that he performed in his experiment:

1. He examined only one specific trait at a time,
2. He used only purebreeding lines as parents, and
3. He carefully segregated the generations.

By focusing on a single characteristic he avoided the immense variety of differing traits exhibited in each generation, and was then able to note relative frequencies of occurrence in each generation.

Although many physical traits do not depend on a single pair of genes, the phenomenon is by no means isolated. Sickle-cell anemia in humans behaves in much the same way as the characteristics of Mendel's peas. The disease is due to the altering of one gene, causing the body to replace one amino acid residue by another during the synthesis of hemoglobin. This change reduces the ability of hemoglobin to combine with oxygen by a factor of 100.* Any reduction in the partial pressure of oxygen, which can occur while traveling due to the rarefaction of oxygen at higher altitudes, leads to a crystallization of the hemoglobin, producing the characteristic sickle shape of the red blood cells. Homozygous diseased individuals *aa* usually die in childhood, but heterozygous individuals *Aa* usually survive to adulthood and transmit their defective gene to some of their offspring. Interestingly, the abnormal hemoglobin in the red blood cells of heterozygous individuals protects them from malaria parasites, so the disadvantages of the disease are counterbalanced by the benefits of immunity to malaria. For this reason, sickle-cell anemia is more prevalent in areas in which malaria is endemic.†

---

* V. M. Ingram, *The Hemoglobins in Genetics and Evolution*. New York: Columbia University Press, 1963.

† A. C. Allison, "Protection afforded by sickle-cell trait against subtertian malarial infection." *British Medical Journal* **1** (1954): 290.

Although the hereditary factor involved in sickle-cell anemia is now understood, its discovery was hampered by the obvious impossibility of establishing a carefully controlled experiment such as the one Mendel was able to perform with his peas. Since homozygous diseased individuals do not reach maturity, it is not possible to begin the experiment with the *aa* genotype, and the experimenter has no control over the romantic attachments of human subjects. Indeed, even if we know the proportion of heterozygous individuals in the population, we may not be able to determine the probability of two of them mating, because social relationships are based on so many different conditions.

To examine what may occur in such situations, we make four assumptions:

1. No mutations occur; that is, each individual's genotype is determined strictly by inheritance.
2. Heterozygous individuals are twice as likely to mate with heterozygous individuals as with homozygous *AA* individuals.
3. Homozygous diseased individuals *aa* never reach maturity.
4. Each sex has the same proportion of genotypes, and is equally represented in the population.

Suppose 20% of the adult population has genotype *Aa* and 80% has genotype *AA* (normal individuals). We wish to determine the relative frequency of genotypes in the following generation.

To solve the problem we construct a tree, as shown in Fig. 6.27. The first stage consists of selecting a mature male from the population, the two branches being labeled by their respective probabilities. Now, the

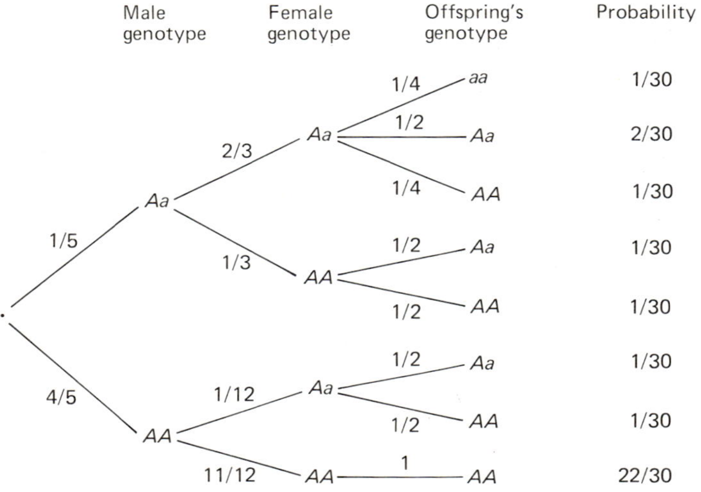

**Figure 6.27**

second assumption indicates that 2/3 of the heterozygous males will marry heterozygous females. Thus, only 1/3 of the heterozygous females will be available to marry $AA$ males. If there are 300 females in the population, 60 (= 20%) are heterozygous and 40 of these would wed heterozygous males. Thus, only 20 of the 240 females marrying normal individuals are heterozygous. Therefore, the probability that a normal male marries a heterozygous female is 20/240 = 1/12, while it is 11/12 that he marries a normal female. These probabilities are shown next to the second-stage branches in Fig. 6.27. The final stage consists of determining the possible genotypes of the offspring of each pair of parents. Obviously, if both parents are normal, so are the offspring, since no mutations are allowed. If only one parent is heterozygous, each child has a 50% chance of inheriting the recessive gene. If both parents are heterozygous, there are four equally likely gene pairs; these are shown in Fig. 6.28. But two of these are equivalent $Aa = aA$, so a heterozygous offspring occurs 50% of the time.

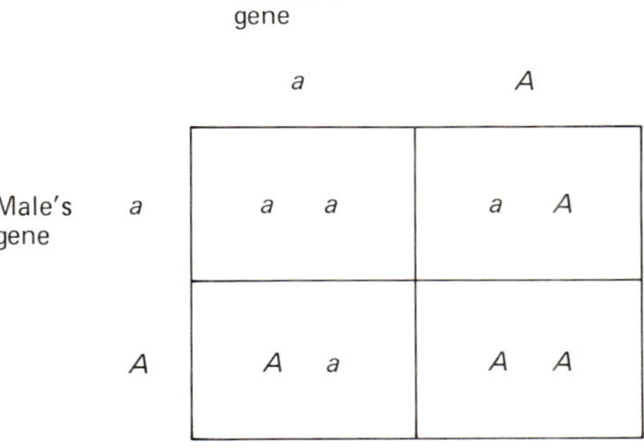


**Fig. 6.28**
Genotypes of the offspring of heterozygous individuals.

The probability of each outcome in the tree is obtained using the sequential counting principle: multiplying together the probabilities along the branch. These are shown in the last column of Fig. 6.27. To find the frequency of genotypes in the next generation, we add the probabilities for each genotype, getting

$$p(aa) = \frac{1}{30},$$

$$p(Aa) = \frac{2}{30} + \frac{1}{30} + \frac{1}{30} = \frac{4}{30},$$

$$p(AA) = \frac{1}{30} + \frac{1}{30} + \frac{1}{30} + \frac{22}{30} = \frac{25}{30}.$$

Thus, 5/6 of the first generation consists of normal individuals.

We may continue the process determining the relative frequency of each genotype for later generations in exactly the same fashion as above (see Exercises 14 and 15 on p. 304).

***Example 4*** Spaceship Alpha crashes on an "earth-type" planet of the star Rigel. Three men and twelve women survive the crash. To avoid conflicts the survivors decide to reinstitute the system of polygamy and allow each man to have four wives. In how many different ways can the liáisons be made?

If we assume that the order in which each wife is selected is not important, we know that there are $C_{12,4}$ ways in which we can select the four wives of the first man. Once these have been selected, there are $C_{8,4}$ ways of selecting the four wives for the second man. Finally, there are $C_{4,4}$ ways of providing for the last man. By the sequential counting principle, we see that there are

$$C_{12,4} \cdot C_{8,4} \cdot C_{4,4} = \frac{12!}{8!4!} \cdot \frac{8!}{4!4!} \cdot \frac{4!}{0!4!}$$
$$= 4950$$

ways in which the marriages can be arranged.

***Example 5*** In the game stud poker, one card is dealt face down, while the other four are dealt face up, one round at a time. After each round of cards has been dealt face up, the player with the best visible hand opens the betting. The best hand still in the game at the end of the betting following the deal of the last round of cards wins the pot. We can determine the probabilities of drawing various stud poker hands.

Suppose we wish to find the probability of drawing two pairs. Then, we must divide the number of ways of getting two pairs by the total number of stud poker hands:

$$p(\text{two pairs}) = \frac{n(\text{hands with two pairs})}{n(\text{hands})}.$$

Finding the total number of poker hands is easy. The order in which the cards are drawn is immaterial, and each hand consists of 5 cards from a deck of 52. Thus, there are

$$C_{52,5} = \frac{52!}{47!5!} = 2{,}598{,}960$$

poker hands. Finding total number of hands with two pairs is a little more complicated. The hand has the form

$$xxyyz.$$

The order of selecting the two denominations $x$ and $y$ from 13 denominations is not important, so there are

$$C_{13,2} = 78$$

ways of picking $x$ and $y$. Since there are four cards of each denomination and the order of their selection is immaterial, the two $x$'s (and the two $y$'s) can be chosen in

$$C_{4,2} = 6$$

ways. Finally, the card $z$ is any of the 44 cards not of the denominations $x$ and $y$. Thus $z$ can be chosen in

$$C_{44,1} = 44$$

ways. Hence, the number of hands with two pairs is

$$C_{13,2} \cdot C_{4,2} \cdot C_{44,1} = 123{,}552.$$

This yields the probability

$$p(\text{two pairs}) = 0.047539.$$

Finding the probability of drawing a flush is easy. All five cards are in the same suit, so there are

$$C_{13,5} = 1287$$

ways of selecting five cards without regard to order. Since there are four suits, the total number of flushes is $4C_{13,5}$. However, some of these flushes are also straights. These *royal* and *straight flushes* can happen in 10 ways in each suit. Removing the 40 royal and straight flushes, we are left with 5108 flushes. Thus the probability of drawing an ordinary flush is

$$p(\text{flush}) = \frac{5108}{C_{52,5}} = 0.001965.$$

Table 6.6 shows the number of ways and probabilities of drawing the various hands in stud poker.

**Table 6.6**

| Rank | $T$ = {Type of hand} | $n(T)$ | $p(T) = n(T)/C_{52,5}$ |
|---|---|---|---|
| 1 | Royal flush | 4 | 0.00000154 |
| 2 | Straight flush | 36 | 0.0000139 |
| 3 | Four of a kind | 624 | 0.00024 |
| 4 | Full house | 3,744 | 0.001441 |
| 5 | Flush* | 5,108 | 0.001965 |
| 6 | Straight* | 10,200 | 0.003925 |
| 7 | Three of a kind | 54,912 | 0.021128 |
| 8 | Two pairs | 123,552 | 0.047539 |
| 9 | One pair | 1,098,240 | 0.422569 |
| 10 | Bust hand | 1,302,540 | 0.501177 |

* Not including straight flushes or royal flushes.

***Example 6*** In contract bridge the declarer often must try to decide how many of the remaining cards in a suit are held by each of the two opponents. Suppose the declarer knows the whereabouts of all but 5 cards in spades. What is the probability that one opponent has three spades and the other the remaining two spades? What is the probability that there is a 4–1 split or a 5–0 split? Knowing these odds, the declarer can plan the best strategy for making the contract.

Suppose the opponent on the left has 3 spades. This can happen in $C_{5,3}$ ways. The remaining 10 cards in that hand can be chosen from the 21 nonspade cards in the two opponents' hands in $C_{21,10}$ ways. Thus, there are $C_{5,3} \cdot C_{21,10}$ ways in which the left opponent could have 3 spades. To find the number of ways in which one of the two opponents has three spades, we must also consider the situation in which the left opponent has two spades. This happens in $C_{5,2} \cdot C_{21,11}$ ways. Since there are $C_{26,13}$ ways of selecting the left opponent's cards, the probability that the spades are split 3–2 is

$$p(3\text{–}2) = \frac{C_{5,3} \cdot C_{21,10} + C_{5,2} \cdot C_{21,11}}{C_{26,13}} = 0.678.$$

Similarly,

$$p(4\text{–}1) = \frac{C_{5,4} \cdot C_{21,9} + C_{5,1} \cdot C_{21,12}}{C_{26,13}} = 0.283$$

and

$$p(5\text{–}0) = \frac{C_{5,5} \cdot C_{21,8} + C_{5,0} \cdot C_{21,13}}{C_{26,13}} = 0.039.$$

Other results of this type are shown in Table 6.7.

**Table 6.7**
**Distribution of cards in a suit.**

| Cards outstanding | Possible splits | Probability |
|---|---|---|
| 6 | 3–3 | 0.355 |
| | 4–2 | 0.484 |
| | 5–1 | 0.145 |
| | 6–0 | 0.015 |
| 5 | 3–2 | 0.678 |
| | 4–1 | 0.283 |
| | 5–0 | 0.039 |
| 4 | 2–2 | 0.407 |
| | 3–1 | 0.497 |
| | 4–0 | 0.096 |
| 3 | 2–1 | 0.780 |
| | 3–0 | 0.220 |
| 2 | 1–1 | 0.520 |
| | 2–0 | 0.480 |

## EXERCISES 6.8

1. Semaphore messages are sent letter by letter by holding two flags in any of the eight major compass directions: N, NE, E, SE, S, SW, W, NW. How many different letters can be made in this way?
2. How many different letters can be formed by raising at most three flags of three different colors?
3. How many words of four or fewer letters can be formed using a five-letter alphabet?
4. A firm has five identical positions available and a list of thirty-two qualified applicants. How many possible ways are there of filling these positions?
5. A basketball coach has three centers, five forwards, and six guards. In how many ways can she choose a starting lineup of one center, two forwards, and two guards?
6. A baseball manager has four right-handed pitchers and three southpaws. In how many ways can he select a starting right-hander followed by a southpaw relief pitcher?
7. Show that there are 1,098,240 stud poker hands with only one pair.
8. Show that there are 3744 ways of getting a full house in stud poker.
9. Show that there are 54,912 ways of drawing three of a kind in stud poker.

**10.** Find the number of possible straights in stud poker. How many of these are straight or royal flushes?

**11.** Show that in contract bridge the probability that six outstanding cards are split 3–3 is 0.355.

**12.** Show that 4–2 splits in contract bridge occur 48.4% of the time.

**13.** Show that a 3–1 split occurs almost 50% of the time. Find the exact probability.

**14.** Using the assumptions made in Example 3 on p. 298, find the relative frequencies of the genotypes *AA*, *Aa*, and *aa* in the second generation. Use the percentages we obtained for the first generation in assigning the probabilities in the tree.

**15.** If we change the probability in Example 3 that two heterozygous individuals will mate from $66\frac{2}{3}\%$ to 90%, what effect does this have on the first generation? on the second generation?

**16.** The first team to win four games in a World Series is the victor.

a) Find all possible ways of playing a World Series.
b) What is the probability that the series will end in five or fewer games, assuming each team is equally likely to win each game?*
c) Suppose the Yankees have a 60% chance of winning in New York, and the Dodgers have a $66\frac{2}{3}\%$ chance of winning in Los Angeles. The first two games are played in Los Angeles, the next three (if necessary) in New York, and any others in Los Angeles. What is the probability that the Dodgers will win the series?

---

## 6.9 A CASE STUDY: COUNTING AND THE GENETIC CODE

Biologists have been interested for many years in the nucleic acids, which, together with the proteins, are important molecular constituents of cells. There are two types of nucleic acid, ribonucleic acid (RNA) and deoxyribonucleic acid (DNA). Although RNA is distributed throughout the cell, DNA is concentrated in the nucleus. Since DNA is present in all cells, and the amount in any cell of a given species is essentially constant, it has long been associated with the chromosomes. In 1953, J. D. Watson and F. H. C. Crick were able to show, by means of a brilliant synthesis, that DNA is the actual genetic material of the cell and carries all the information necessary to direct every activity of the cell.

The Watson–Crick model of DNA is made up of two spiraling polynucleotide chains, which form a double helix (see Fig. 6.29). Each chain consists of alternating molecules of phosphate and deoxyribose

---

* In 63 such World Series between 1905 and 1970, there were 12 four-game, 15 five-game, 12 six-game, and 24 seven-game series. These statistics indicate the effects of the advantage of playing at home.

sugar, with either a purine or a pyrimidine base attached to each sugar molecule. The two chains are joined by hydrogen bonding between the bases. The effect is easily visualized if we imagine a flexible ladder that has been twisted into a double spiral, the rungs being the hydrogen-bonded bases.

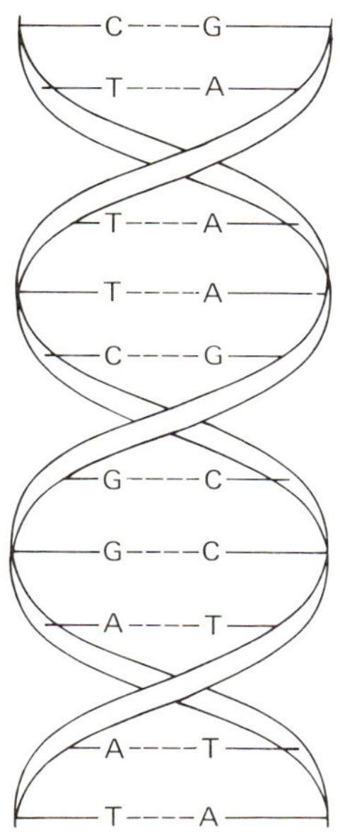


**Figure 6.29**

Four bases generally predominate, the two purines *adenine* (A) and *guanine* (G) and the two pyrimidines *thymine* (T) and *cytosine* (C). The crucial feature of the model is that it requires that adenine must bond with thymine (A-T) and guanine with cytosine (G-C). This requirement is based on the dimensions of the molecules compared to the diameter of the helix, and the positioning on the molecules of the hydrogen atoms that cause the bonding. Thus, the two chains are complementary, in the sense that if one chain is known, the other can be immediately determined by the required pairing. However, so far as a single chain is concerned, there is no theoretical restriction on the sequence of bases it can contain.

The DNA molecule replicates by unwinding from one end. As it unwinds in a pool of the four bases previously synthesized by the cell, the free bases form hydrogen bonds with their complementary bases on each of the chains. Once attached, they find themselves properly oriented so that the polymerization of a new chain can begin before the original chains are completely unwound (see Fig. 6.30).

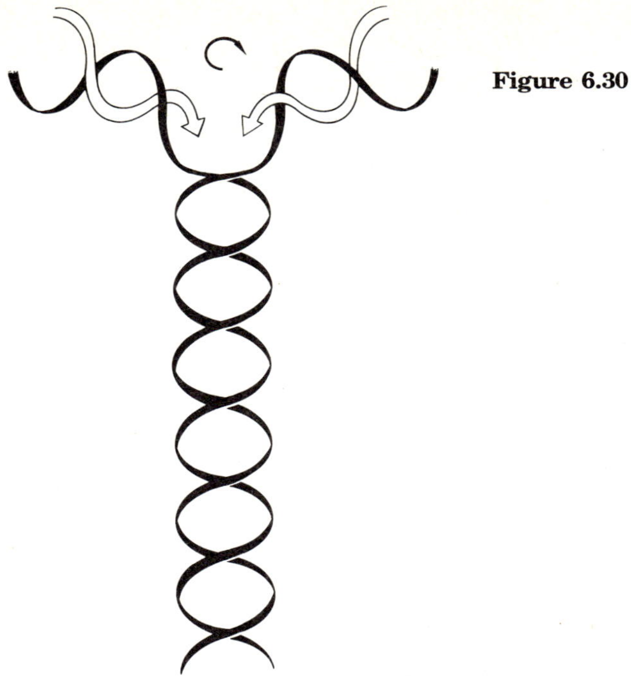

**Figure 6.30**

The main problem facing biologists is to determine the sequence of bases along each chain. At first glance this might seem impossible, but the difficulty is reduced when we realize that there are only 20 essential amino acids that must be synthesized or obtained from some outside source by the cell. There is strong evidence that each amino acid is coded by a sequence of three bases. Thus, the amino acids present may be used as a means of breaking the genetic code.

In addition to physical and chemical measurements of various amino acids, there is a mathematical reason for assuming that each amino acid is coded by three bases. Since 20 amino acids must be coded by sequences of the four bases A, U, C, and G, single- or double-letter codes do not suffice: There are only four single-letter codes; there are $4 \cdot 4 = 16$ double-letter codes. Of course, if we added the two sets together, we would get 20 one- and two-letter codes, but then some amino acids would have only half the molecular weight of the others. Furthermore, it has been found that certain amino acids can be specified by more than one sequence, so this *degeneracy* of the code implies that the sequences consist at least of triples, of which there are $4 \cdot 4 \cdot 4 = 64$ triple-letter codes. The tree in Fig. 6.31 shows the code words that have been confirmed.*

* F. H. C. Crick, "The genetic code: III." *Scientific American* **215,** No. 4, (1966): 55.

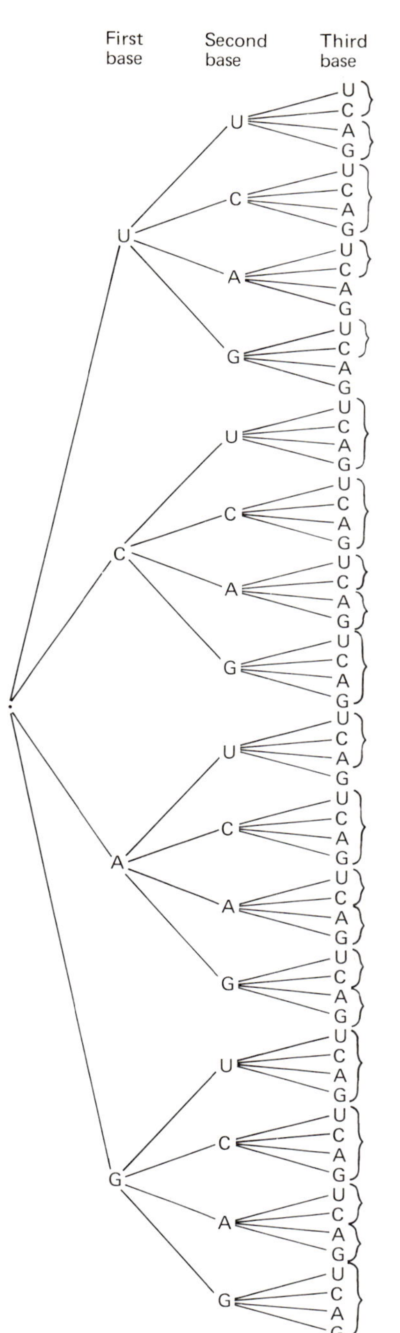


**Figure 6.31**

Each cell's DNA is preserved in the cell nucleus. The actual assembling of amino acids in the cytoplasm is done by the cell's RNA molecules, near replicas of the DNA. Each RNA molecule differs from DNA by replacement of a unit of *uracil* (U) for each unit of the DNA's thymine (T). The presence of uracil in RNA may be the key that allows RNA molecules to leave the cell nucleus.

It should now be apparent that we can determine the genetic code by finding the sequence of amino acids. The procedure requires a difficult chemical separation of a pure sample of RNA by stages into smaller polypeptide chains, each involving three or four amino acids. The order of the amino acids in these polypeptide chains is then determined. Fortunately not all RNA chains will separate at the same place, so the final stage is to fit them together by looking for overlapping amino acid sequences. For example, suppose we determine that a small part of the chain separates into five polypeptide chains:

| | |
|---|---|
| 1. MPM, | H = histidine |
| 2. HSSS, | L = lysine |
| 3. TLM, | M = methionine |
| 4. MVH, | P = proline |
| 5. SSLV. | S = serine |
| | T = tyrosine |
| | V = valine |

Attaching chain 5 to the end of chain 2 we get

HSSSLV.

The end of chain 4 can be attached to the beginning of the chain so far obtained, yielding

MVHSSSLV.

Although both chains 1 and 3 can be attached to the beginning of this last chain, we cannot attach chain 3 first. So the final result is

TLMPMVHSSSLV.

Of course, this is a gross simplification of the practical situation.

The counting techniques we have encountered in this chapter can be very useful in the task of determining possible orders of amino acids. For example, suppose we know that a chain contains 3M, 2L, 4V, and 3S, and we have identified the two subchains

MLVV and SMVS.

These two chains obviously do not overlap. We need to determine how many chains are possible. One M, L, and S are still unused. Letting $X$ = MLVV and $Y$ = SMVS, we see that the problem reduces to finding the number of permutations of the six letters M, L, V, S, X, and Y. Thus,

$$P_{6,6} = 6! = 720$$

chains are still possible. This number should encourage the researcher, since he or she has already reduced the possible number of chains by a factor of 385. To determine how many chains were originally possible, we assume first that each amino acid was coded by a different letter. Then the number of possible permutations is

$$P_{12,12} = 12! = 479{,}001{,}600. \tag{6.34}$$

However, since the 3M are indistinguishable, their order in the chain does not matter, so we must divide the result of Eq. (6.34) by $P_{3,3} = 3! = 6$. Similarly, the 2L, 4V, and 3S are indistinguishable. Dividing (6.34) by all their possible permutations, we arrive at the original number of possible chains with these 12 amino acids:

$$\frac{P_{12,12}}{P_{3,3} \cdot P_{2,2} \cdot P_{4,4} \cdot P_{3,3}} = \frac{12!}{3!2!4!3!} = 277{,}200.$$

## EXERCISES 6.9

1. How many different polypeptides can be formed using 4M and 5L amino acids?
2. How many different polypeptides can be formed using 1M, 2L, and 3S amino acids?
3. How many different polypeptides can be formed using 3M, 3L, and 3S amino acids?
4. How many different polypeptides can be formed using 4M, 3L, and 5S amino acids?
5. Suppose a polypeptide is known to contain 4L, 3V, and 3S amino acids, and the subchain SVLSL has been identified. How many polypeptide chains are still possible? How many were originally possible?
6. Suppose a polypeptide is known to consist of 2M, 2L, 4V, and 2S amino acids, and the chains VLM and LVM have been identified. How many polypeptide chains are still possible? Make sure you consider the situation with and without overlapping.
7. Suppose a polypeptide is known to contain 3M, 2V, 4L, and 5S amino acids. Three chains have been identified: MVVS, SLMV, and SSMS. How many polypeptide chains are possible? Make sure all overlapping possibilities are considered.
8. A polypeptide chain is known to contain 4L, 3M, 2V, and 1S amino acids. The following facts have been determined.

   a) The chain begins and ends with the same amino acid.
   b) The S amino acid is not adjacent to any M amino acids.
   c) Two LV and two ML chains are present.
   d) Each LV chain is adjacent to a different ML chain.
   e) At least one LL chain is present.

   How many polypeptide chains are possible?

## CHAPTER VOCABULARY

**Bayes's formula**

$$p(E \mid F) = \frac{p(F \mid E)p(E)}{p(F \mid E)p(E) + p(F \mid E')p(E')}$$

**belongs (to)**

**birthday problem**

**branch of a tree**

**chances**

**combination** $C_{n,k} = \dfrac{n!}{(n-k)!k!}$

**complement (of a set)** $A'$

**conditional probability** $p(E \mid F) = p(E \cap F)/p(F)$

**contained (in)** $A \subset B$

**dependent events**

**disjoint sets** $A \cap B = \varnothing$

**DNA** (= deoxyribonucleic acid)

**dominant gene**

**element** (= member)

**empty set** $\varnothing$

**equally probable**

**event** $E$

**experimental relative frequency**

**genes**

**genetic code**

**genotype**

**heterozygous**

**homozygous**

**hybrid**

**independent events** $p(E \cap F) = p(E)p(F)$

**intersection** $A \cap B$

**Mendel's laws of inheritance**

**mutually exclusive events**

**odds**

**permutation** $P_{n,k} = n!/(n-k)!$

**probability** $p(E)$

**probability multiplication principle** $p(E \cap F) = p(F)p(E \mid F)$

**recessive gene**

**RNA** (= ribonucleic acid)

**sample point**

**sample space** $S$

**sequential counting principle** $n_1 \cdot n_2 \cdots n_k$

**set** (= collection)

**set equality**

**subset**

**tree**

**tree multiplication property**

**uniform sample space**

**union** $A \cup B$

**universal donor**

**universal recipient**

**universal set** $\mathcal{U}$

**Venn diagram**

**Watson–Crick model**

# CHAPTER 7 STATISTICS AND DECISION THEORY

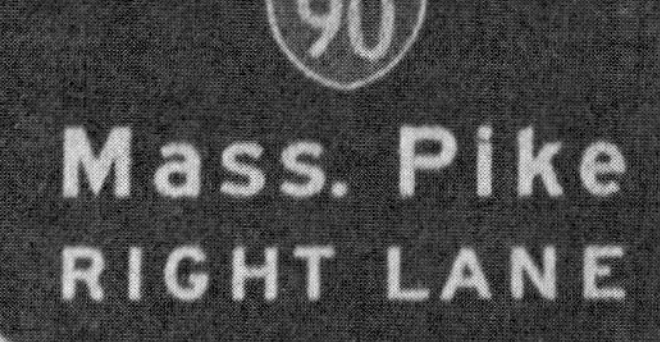

## 7.1 SAMPLING A POPULATION

*Statistics* is the science that deals with the methods of collecting, summarizing, and interpreting data. It has been described by some authors as the science of making decisions in the face of uncertainty. While this is certainly a valid description, too often it has been abused, either willfully or through ignorance, to promote invalid conclusions. Decisions based on statistics frequently affect everyone. For this reason it is important to have a basic understanding of the techniques and ideas that arise in statistics. A knowledge of these concepts will help you understand a diversity of everyday topics such as average wages, average family size, the cost of living, etc.

The data to be collected come from some universal set $U$, called a *sample space* or *population*. For example, if we are considering the ages of all females in the United States, the population $U$ consists of the set of ages, one for each American female.

If the population is very large it is often impossible or impractical to obtain all the data for that population. When this occurs we must examine a carefully selected subset of the population, called a *sample*. If this sample is a good copy of the entire population, we are able to draw accurate inferences about the population. This procedure is used by the Nielsen ratings for TV shows and by the various political pollsters. It is obvious that the accuracy of the conclusions we make from such a sample depends on how representative of the population the sample is. Selecting a representative sample is an important branch of statistics. We shall not dwell on this important topic other than to state that *random sampling* is usually desirable, that is, some method of selecting the sample in such a way that each member of the population has an equal probability of being selected for the sample. Most of the faulty conclusions in statistics are caused by bad sampling procedures.

Suppose now that we have found a method of selecting a representative sample of a population that we wish to study. The researcher then sets out to collect the raw data. It might appear at this point that there is nothing to worry about, and in many cases there is no problem in collecting the sample data. Often, however, this is a period of intense frustration for the researcher. For example, some of the people the researcher might have planned to interview may be gone, unwilling to cooperate, or unwilling to answer certain questions truthfully. Dropping the absent or uncooperative members will affect the sample and may cause it to be unrepresentative. Collecting false data obviously can lead to false conclusions. One remedy for the missing sample points is to design the selection procedure in such a way that a substitute is available. The false data can be avoided by the following scheme.

***Example 1*** Suppose the interviewer wishes to ask a sensitive question such as "have you ever smoked marijuana?" but knows that by asking that question, he or she is likely to get an evasive answer. To eliminate this problem, the following method was designed by S. L. Warner ("Randomized response: A survey technique for estimating evasive answer bias," *Journal of the American Statistical Association* **60** (1965): 63–69). After promising that no effort will be made to discover which question the person being interviewed answered, the interviewer gives a card containing two questions (one sensitive and one not) to the respondent. For example, the card might contain the questions:

(H) Have you ever smoked marijuana?

(T) Is the last digit of your social security number odd?

The respondent is then asked to flip a coin and, *hiding the outcome of the toss*, answer question (H) truthfully if heads occurs and question (T) truthfully if tails occurs.

How can we determine from such data the probability $p_0$ that a person in the United States population has smoked marijuana? We know that the probability of getting heads or of the last digit in the social security number being odd is 0.5, so we draw the tree shown in Fig. 7.1. Thus, the probability that we get a yes answer is

$$p(\text{yes}) = 0.5p_0 + 0.5(0.5) = \frac{p_0}{2} + \frac{1}{4}.$$

Let $m$ be the number of yes responses from the $n$ samples taken. If the sample is truly representative of the population, we obtain

$$\frac{m}{n} = p(\text{yes}) = \frac{p_0}{2} + \frac{1}{4},$$

or

$$p_0 = 2\left(\frac{m}{n} - \frac{1}{4}\right).$$

Thus, if 312 people in a random sample of 1000 answered yes, it would mean that

$$p_0 = 2\left(\frac{312}{1000} - \frac{1}{4}\right) = 0.124,$$

that is, 12.4% of the population has smoked marijuana.

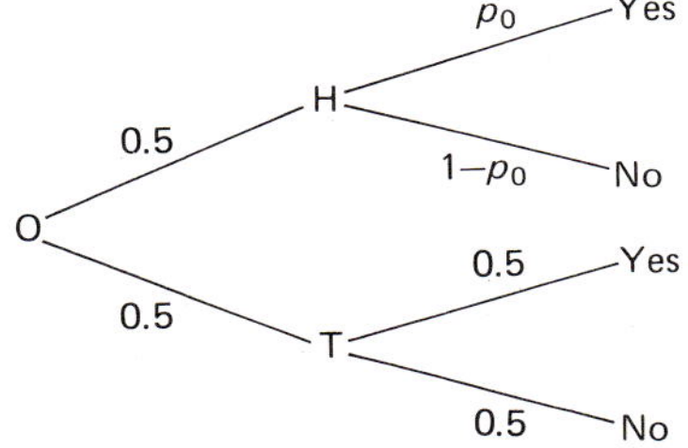


**Figure 7.1**

Once the data are collected, some organization is usually necessary before any patterns in the raw data can be detected. There are several ways of organizing data for analysis. We illustrate the most common ways of organizing real data by considering the data given in Table 7.1.

We can list each temperature from the highest to the lowest in a table and make a tally of the *frequency* with which each temperature occurred (see Table 7.2). The last column of Table 7.2 contains the frequency of all temperatures less than or equal to the one under

**Table 7.1**
**Daily high temperatures recorded at Stapleton International Airport, Denver, Colorado, in June 1977**

| | | | | | |
|---|---|---|---|---|---|
| 88 | 90 | 91 | 93 | 93 | 80 |
| 87 | 86 | 87 | 90 | 78 | 85 |
| 78 | 94 | 94 | 83 | 82 | 79 |
| 84 | 85 | 81 | 87 | 83 | 82 |
| 87 | 94 | 94 | 80 | 95 | 78 |

**Table 7.2**
**Frequency and cumulative frequency of high temperature data at Stapleton International Airport, June 1977**

| Temperature °F | Tally marks | Frequency | Cumulative frequency |
|---|---|---|---|
| 78 | ‖| | 3 | 3 |
| 79 | \| | 1 | 4 |
| 80 | ‖ | 2 | 6 |
| 81 | \| | 1 | 7 |
| 82 | ‖ | 2 | 9 |
| 83 | ‖ | 2 | 11 |
| 84 | \| | 1 | 12 |
| 85 | ‖ | 2 | 14 |
| 86 | \| | 1 | 15 |
| 87 | ‖‖ | 4 | 19 |
| 88 | \| | 1 | 20 |
| 89 | | 0 | 20 |
| 90 | ‖ | 2 | 22 |
| 91 | \| | 1 | 23 |
| 92 | | 0 | 23 |
| 93 | ‖ | 2 | 25 |
| 94 | ‖‖ | 4 | 29 |
| 95 | \| | 1 | 30 |

consideration, and is called the *cumulative frequency*. In addition to a frequency table such as Table 7.2, we can also display the frequencies in a *line chart* such as the one shown in Fig. 7.2.

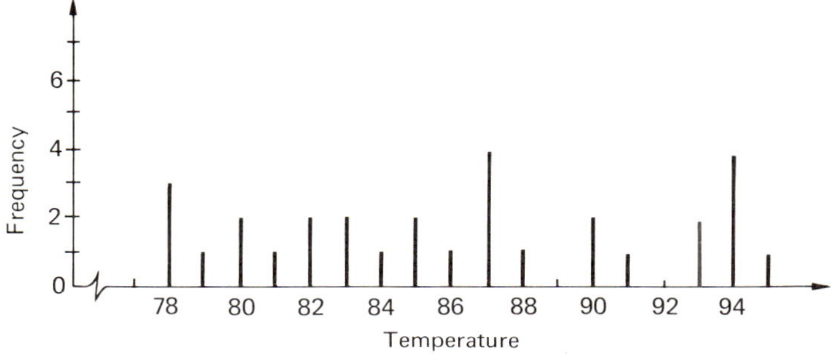


**Fig. 7.2**
Line chart for Table 7.2.

**Table 7.3**
**Grouped frequency and cumulative frequency**

<table>
<tr><th>Interval</th><th>Midpoint of interval</th><th>Tally marks</th><th>Frequency</th><th>Cumulative frequency</th></tr>
<tr><td>76.5–79.5</td><td>78</td><td>||||</td><td>4</td><td>4</td></tr>
<tr><td>79.5–82.5</td><td>81</td><td>𝍸</td><td>5</td><td>9</td></tr>
<tr><td>82.5–85.5</td><td>84</td><td>𝍸</td><td>5</td><td>14</td></tr>
<tr><td>85.5–88.5</td><td>87</td><td>𝍸|</td><td>6</td><td>20</td></tr>
<tr><td>88.5–91.5</td><td>90</td><td>|||</td><td>3</td><td>23</td></tr>
<tr><td>91.5–94.5</td><td>93</td><td>𝍸|</td><td>6</td><td>29</td></tr>
<tr><td>94.5–97.5</td><td>96</td><td>|</td><td>1</td><td>30</td></tr>
</table>

In many situations it is useful to group the data in intervals, so that each data point falls into one of six to fifteen abutting intervals of equal length. In this example, we could choose seven intervals of three degrees starting with 76.5°F. The frequencies and cumulative frequencies for these ten intervals are given in Table 7.3. In this case we refer to the information as *grouped data* and consider the *grouped frequency* and *grouped cumulative frequency*. We can display this information visually as a *histogram* (or bar graph), in which the height of the bar over each interval is determined by the grouped frequency for that interval (see Fig. 7.3). We can also display it as a *frequency polygon*, which joins the midpoints of the tops of adjacent histogram bars together with straight line segments (see Fig. 7.4). A graph that illustrates the grouped cumulative frequencies is the *cumulative frequency polygon* shown in Fig. 7.5. Here the grouped cumulative frequencies are plotted at the *right ends* (not the midpoints!) of the

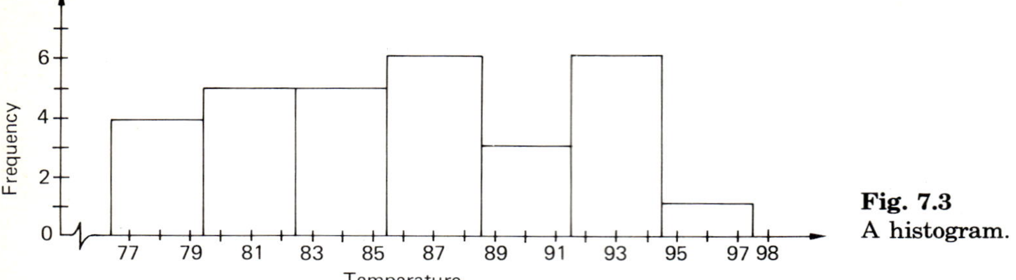


**Fig. 7.3**
A histogram.

**Fig. 7.4**
A frequency polygon.

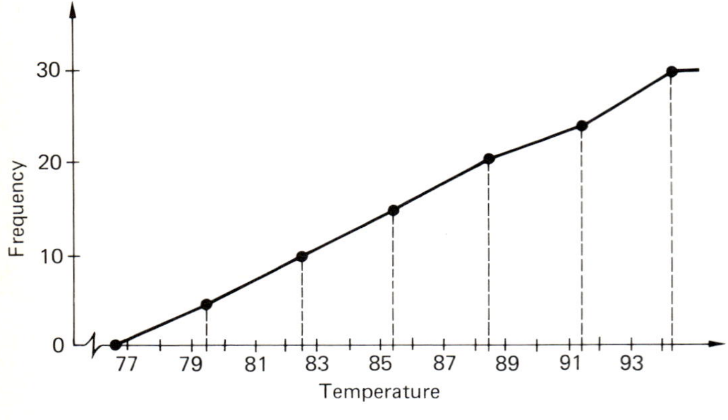


**Fig. 7.5**
A cumulative frequency polygon.

intervals, and the cumulative frequencies for adjacent intervals are joined by line segments. The ends are used rather than the midpoints because the grouped cumulative frequencies measure the frequency of all temperatures less than or equal to the highest temperature in the interval.

## EXERCISES 7.1

*In Exercises 1–4, group the data as indicated and prepare a grouped frequency and cumulative frequency table, histogram, and cumulative frequency polygon.*

1. The daily highs at Stapleton International Airport in May 1977 are given below. Group the data into ten intervals, using 56.5–59.5 as the first interval.

| | | | | | | |
|---|---|---|---|---|---|---|
| 70 | 77 | 80 | 74 | 73 | 74 | 71 |
| 81 | 79 | 79 | 74 | 80 | 76 | 65 |
| 73 | 78 | 80 | 79 | 68 | 57 | 67 |
| 76 | 85 | 84 | 78 | 75 | 75 | 59 |
| 73 | 79 | 84 | | | | |

2. The personal incomes per capita by state in 1973 are shown below. Group the data into ten intervals, using 3500–3749 as the first interval.

| | | | | | | |
|---|---|---|---|---|---|---|
| 4692 | 4054 | 4993 | 4571 | 4713 | 3931 | 4413 |
| 5154 | 3853 | 4340 | 5273 | 5029 | 3556 | 4682 |
| 5938 | 5778 | 5770 | 5304 | 3871 | 4282 | 4072 |
| 4082 | 5933 | 4987 | 5137 | 3952 | 3882 | 4695 |
| 5253 | 5489 | 5551 | 4841 | 4923 | 4095 | 5521 |
| 4694 | 5845 | 5076 | 5271 | 4395 | 4386 | 5745 |
| 4841 | 5705 | 4750 | 5695 | 4033 | 3961 | 4833 |
| 5541 | | | | | | |

3. The grant budgets for the 1977 SEED program of the National Science Foundation are listed below. Group the data into ten intervals, using 0–2600 as the first interval.

| | | | | |
|---|---|---|---|---|
| 8,900 | 16,100 | 22,900 | 1,250 | 1,150 |
| 22,900 | 20,900 | 25,300 | 1,400 | 2,100 |
| 19,600 | 22,700 | 20,300 | 800 | 1,400 |
| 22,800 | 8,800 | 17,200 | 4,650 | 1,900 |
| 7,900 | 21,400 | 1,850 | 2,050 | 200 |
| 2,000 | 1,400 | 1,200 | 2,900 | 1,000 |

4. Group the following 100 pairs of random digits into ten intervals beginning with 00–09.

| | | | | | | | | | |
|---|---|---|---|---|---|---|---|---|---|
| 65 | 82 | 66 | 44 | 80 | 66 | 85 | 86 | 42 | 38 |
| 05 | 31 | 75 | 15 | 67 | 53 | 09 | 89 | 97 | 36 |
| 09 | 21 | 69 | 99 | 42 | 20 | 74 | 09 | 10 | 37 |
| 27 | 48 | 49 | 87 | 74 | 30 | 88 | 10 | 05 | 22 |
| 66 | 29 | 78 | 27 | 48 | 28 | 83 | 53 | 60 | 80 |
| 58 | 49 | 75 | 76 | 26 | 90 | 06 | 94 | 05 | 33 |
| 99 | 69 | 56 | 38 | 30 | 82 | 90 | 58 | 61 | 22 |
| 11 | 77 | 20 | 74 | 01 | 60 | 10 | 80 | 93 | 12 |
| 19 | 38 | 26 | 53 | 35 | 97 | 95 | 06 | 72 | 89 |
| 05 | 31 | 60 | 55 | 24 | 05 | 17 | 48 | 11 | 55 |

**5.** Construct a frequency table for the digits in Exercise 4 and prepare a cumulative frequency polygon for these data. Which digits appear most frequently?

**6.** Suppose 524 respondents from a random sample of 1243 answer yes to the question in Example 1. What percentage of the population can be assumed to have smoked marijuana?

**7.** Suppose an interviewer presents the respondents with a card containing the following sensitive and nonsensitive questions:

(H) Have you ever engaged in premarital sex?
(T) Did the flip of the coin yield heads?

What does a yes answer signify? Would this card lead to evasive answers?

**8.** Suppose an interviewer presents the respondents with a card containing the questions:

(H) Have you ever engaged in premarital sex?
(T) Were you born on an even day of the month?

If 589 of 912 respondents answer yes, what can you infer about the percentage of the population that have engaged in premarital sex? (*Hint:* There are more odd days than even days in a year.)

**9.** Suppose a respondent is asked to secretly pick one of the integers 1, 2, or 3. After making a choice he or she is asked to respond truthfully to that question on the following card:

(1) Have you ever cheated on your income tax?
(2) Did you select the number 3?
(3) Is the last number of your social security number even?

Find the probability that the answer is yes, assuming that each number is equally likely to be selected. If 543 of 1800 respondents answer yes, what percentage of the population can we assume cheat on their income tax?

---

## 7.2 MEAN, VARIANCE, AND EXPECTED VALUE

In Section 7.1 we described some of the methods of collecting a sample and organizing or summarizing the raw data. Once those two steps have been taken, we need to assign a meaning to the information that we have obtained. Usually, interpretation of the data requires that we measure certain properties of the sample. In this section we will discuss some of the most common measurements associated with a numerical sample.

### Measures of Central Tendency

One of the most useful measurements we can make when we are given a set of real numerical data is to find the "center" of the data. This center should be representative of the entire set of numbers. The two most commonly used measures of this center are the mean and the median of the data.

The *mean** of a set of numbers is the average of those numbers. We find the sample mean by adding all the data points in the set and dividing this sum by the number of data points in the set. It is customary to denote the sample mean by the symbol $\bar{x}$ (read $x$ bar). Thus,

$$\bar{x} = \frac{1}{n}(x_1 + x_2 + \cdots + x_n), \tag{7.1}$$

the average of the $n$ values $x_1, x_2, \ldots, x_n$. For example, the mean of the sample {65, 82, 66, 44, 80, 66, 85, 86, 42, 38} is

$$\begin{aligned}\bar{x} &= [65 + 82 + 66 + 44 + 80 + 66 + 85 + 86 + 42 + 38]/10\\ &= \frac{654}{10} = 65.4.\end{aligned}$$

The mean has the advantage of taking *all* the sample points into consideration. Sometimes, however, this can be a disadvantage because it can be greatly influenced by a few exceptionally large or small values. For example, the sample

$$2, 4, 5, 6, 6, 7, 8, 8, 1200$$

has mean $\bar{x} = 1246/9 = 138.444\ldots$, although 8 of the 9 values are smaller than $\bar{x}$. It is hard to justify a claim that $\bar{x} = 138.444\ldots$ is representative of the data in the sample. In this case, it may be more meaningful to use a different measure of central tendency.

If a random sample has been selected from a larger population, it is generally the case that the mean of the sample $\bar{x}$ differs from the mean $\mu$ of the entire population. However, if the sample has been carefully selected to be representative of the entire population, the value $\bar{x}$ should be a good approximation to $\mu$. For example, if we consider the first 30 pairs of random digits in Exercise 4 on p. 317 we obtain the sample mean $\bar{x} = 50.7$. This is in close agreement with the population mean $\mu = 49.5$, since each of the integers 0 to 99 is equally likely. (Why is $\mu = 49.5$?)

Suppose we have organized the data in the sample in increasing order. The *median* of the sample is:

1. The middle number in the list, if the sample contains an odd number of values.
2. The average of the two middle numbers of the list, if the sample has an even number of values.

For example, the sample

$$2, 4, 5, 6, 6, 7, 8, 8, 1200$$

---

* In this case we are referring to the *arithmetic mean* of the data points. There are other means (such as the geometrical mean) but these will not be considered.

has the number 6 as its median, which is more representative than the mean $\bar{x} = 138.444\ldots$, while the sample

$$5, 5, 11, 17, 24, 31, 48, 55, 55, 60$$

has median

$$\frac{24 + 31}{2} = \frac{55}{2} = 27.5.$$

The main disadvantage of the median is that it only involves one or two sample points, instead of the entire sample. For example, the median of the sample

$$5, 5, 5, 5, 90, 95, 100$$

is 5. In this case, the mean $\bar{x} = 43.571428\ldots$ is more representative. It is clear that we must use caution in picking the measure of central tendency, and try to select the measurement that seems most appropriate for the given sample.

In the following example we see one possible application of the sample mean.

***Example 1*** Suppose your automobile odometer registers only miles. You wish to discover how far it is along your usual route from home to your place of work. To get an idea you record the distance traveled (to the nearest mile) on each of twenty trips, obtaining the sample

$$10, 10, 11, 10, 11, 11, 10, 10, 10, 11,$$

$$11, 10, 10, 10, 10, 11, 11, 10, 10, 10,$$

This implies that the distance $\mu$ between home and your job lies in the interval $10 < \mu < 11$. (If $\mu \leq 10$, then no odometer reading would ever be 11, while $\mu \geq 11$ prevents the odometer reading from ever being 10.) Although the sample is not very large, the sample mean furnishes a good approximation of the actual distance $\mu$:

$$\mu = \bar{x} = \frac{207}{20} = 10.35.$$

If the sample were larger, say, 100 data points, we would anticipate that the sample mean would be an extremely good approximation of $\mu$. The proof of this law (the Law of Large Numbers) is beyond the scope of this text.

**Measures of Variation**

Consider the two samples

a) 10, 12, 15, 19, 19;

b) 2, 8, 15, 21, 29.

Each of these samples has a mean and a median equal to 15, yet the data in sample (b) are spread out much more than those in sample (a). Another kind of measure is needed to describe the spread of the data. The two most commonly used measures of this spread are the range and the variance of the data.

The *range* of a set of numbers is simply the difference between the largest and the smallest numbers in the sample. Thus, the range of sample (a) is 9, since the largest number is 19 and the smallest is 10, while the range of sample (b) is 27. The range is easily calculated but suffers from the disadvantage of taking only the extremes into account. Thus, the effect of the other data points is ignored. For example, the sample

c) 2, 14, 15, 15, 29

has a mean and a median equal to 15 and a range of 27, yet it has more points clustered near 15 than sample (b) does.

A more useful measure of variation of a sample is the *variance*, which involves the squares of the differences between each of the data points and the sample mean. It is customary to denote the sample variance by the symbol $s^2$, and to define* it by

$$s^2 = \frac{1}{n}[(x_1 - \bar{x})^2 + (x_2 - \bar{x})^2 + \cdots + (x_n - \bar{x})^2]. \tag{7.2}$$

Using Eq. (7.2) we have the variance

$$s^2 = \frac{1}{5}[(10-15)^2 + (12-15)^2 + (15-15)^2 + (19-15)^2 + (19-15)^2]$$

$$= \frac{1}{5}[(-5)^2 + (-3)^2 + 0^2 + (4)^2 + (4)^2]$$

$$= \frac{66}{5} = 13.2$$

* Most statisticians prefer to divide the sum by $(n - 1)$ instead of $n$ because the mean $\bar{x}$ already depends on all the data points. Thus, for example, if the sample consists of only two data points and we know that $x_1 = 2$ and $\bar{x} = 4$, the other sample point is already completely determined: $x_2 = 6$. Hence, only *one* of the two squared differences is important; the other is a consequence of the first. This situation carries over to samples with $n$ data points: Once the first $(n - 1)$ squared differences are known, the $n$th squared difference is completely determined.

for sample (a);

$$s^2 = \frac{1}{5}[(2 - 15)^2 + (8 - 15)^2 + (15 - 15)^2 + (21 - 15)^2 + (29 - 15)^2]$$

$$= \frac{1}{5}[(-13)^2 + (-7)^2 + 0^2 + 6^2 + 14^2]$$

$$= \frac{450}{5} = 90$$

for sample (b); and

$$s^2 = \frac{1}{5}[13^2 + 1^2 + 0^2 + 0^2 + 14^2]$$

$$= \frac{366}{5} = 73.2$$

for sample (c).

The biggest advantage of the variance is the fact that it considers all the sample points. Thus, it can distinguish between two samples whose extremes are equally far apart. In this case, samples (b) and (c) have the same range, but quite different variances. Furthermore, as is apparent from Eq. (7.2), the closer the data points cluster around the mean, the smaller the variance. Generally, larger values of $s^2$ indicate greater spreads of the data points from the mean. The main disadvantages include the problem of calculating the variance, and the fact that a few exceptionally large differences may substantially outweigh a large number of points tightly clustered on the mean [compare samples (a) and (c)]. It is again the task of the researcher to select the most representative measure of variation for the sample he or she is considering.

Another frequently used measure of variation is the nonnegative square root $s$ of the variance $s^2$, called the *standard deviation*:

$$s = \sqrt{\frac{1}{n}[(x_1 - \bar{x})^2 + (x_2 - \bar{x})^2 + \cdots + (x_n - \bar{x})^2]}. \tag{7.3}$$

For example, the standard deviations of samples (a), (b), and (c) are approximately 3.63, 9.49, and 8.56, respectively.

The variance $s^2$ may also be computed using the formula

$$s^2 = \frac{1}{n}[x_1^2 + x_2^2 + \cdots + x_n^2] - (\bar{x})^2. \tag{7.4}$$

We can see how Eq. (7.4) is obtained after using it with sample (a):

$$s^2 = \frac{1}{5}[10^2 + 12^2 + 15^2 + 19^2 + 19^2] - (15)^2$$

$$= \frac{1191}{5} - 225 = 13.2.$$

Now, to verify Eq. (7.4), square each of the terms in parentheses in Eq. (7.2), obtaining

$$s^2 = \frac{1}{n}[(x_1^2 - 2x_1\bar{x} + \bar{x}^2) + (x_2^2 - 2x_2\bar{x} + \bar{x}^2) + \cdots + (x_n^2 - 2x_n\bar{x} + \bar{x}^2)]. \tag{7.5}$$

Collecting terms that involve different powers of $\bar{x}$, we have

$$s^2 = \frac{1}{n}[(x_1^2 + x_2^2 + \cdots + x_n^2) - 2\bar{x}(x_1 + x_2 + \cdots + x_n) + n\bar{x}^2]. \tag{7.6}$$

But $x_1 + x_2 + \cdots + x_n = n\bar{x}$, so the middle term on the right side of Eq. (7.6) becomes $2n(\bar{x})^2$, yielding

$$s^2 = \frac{1}{n}[(x_1^2 + x_2^2 + \cdots + x_n^2) - 2n\bar{x}^2 + n\bar{x}^2]$$

$$= \frac{1}{n}[(x_1^2 + x_2^2 + \cdots + x_n^2) - n\bar{x}^2]$$

from which Eq. (7.4) follows.

As is the case with the sample mean $\bar{x}$ and the population mean $\mu$, the sample variance $s^2$ generally differs from the population variance $\sigma^2$. If the sample has been carefully selected as representative of the entire population, the value of the sample variance $s^2$ will closely approximate the population variance $\sigma^2$.

***Example 2*** The following weighing procedure is commonly used in the trucking industry: First you weigh the empty truck $T_1$; then you weigh the truck and cargo $T_2 + C_2$. The range, $(T_2 + C_2) - T_1$, is the weight of the cargo. While this procedure seems quite simple, it is subject to greater possible errors than simply weighing the cargo!

To understand why this occurs we must first realize that every weighing is only an approximation to the correct weight. Try the following experiment the next time you are in a chemistry laboratory and have an analytical balance handy. Select any object, weigh it, remove it from the balance, and weigh it again. You are very likely to get a different weight. If you weigh an object a large number of times, the mean weight will be a good approximation to the correct weight. This situation is similar to that in Example 1. (Why?)

If $C$ is the correct weight of the cargo, the error in the weight of the cargo is

$$(T_2 + C_2 - T_1) - C = (T_2 - T_1) + (C_2 - C).$$

Thus, not only do we have an error in weighing the cargo $C_2 - C$ but also an error from the two weighings of the truck $T_2 - T_1$. While it is possible that these errors may cancel each other (giving a very good approximation to the correct weight), it is equally likely that the errors will act in unison, producing a much larger error than would have resulted from a single weighing.

It can be argued that errors acting in unison should average with the errors canceling, leading to the same average error as weighing only the cargo. However, the *variance* of the double weighing is much greater than the variance of weighing only the cargo, since the *squares* of the errors will not average out.

**Mathematical Expectation** Both the mean and the variance of a sample space are examples of a very useful concept known as mathematical expectation. The *mathematical expectation*, or *expected value*, of a function g defined at each point $x_1, x_2, \ldots, x_n$ of a sample space is given by the expression

$$E[g(x)] = g(x_1)p_1 + g(x_2)p_2 + \cdots + g(x_n)p_n, \tag{7.7}$$

where $p_k$ is the probability that the event $x_k$ occurs, for $k = 1, \ldots, n$.

To see why the sample mean is an expected value, we assume that each point $x_1, x_2, \ldots, x_n$ has equal probability of occurring. This situation holds in uniform sample spaces. Hence, the probability that event $x_k$ occurs in a random sample of size $n$ is

$$p_k = \frac{1}{n}, \qquad k = 1, 2, \ldots, n,$$

since $x_k$ is only one of the $n$ points in the sample space. Letting $g(x) = x$, we have

$$\begin{aligned} E[x] &= x_1p_1 + x_2p_2 + \cdots + x_np_n \\ &= \frac{1}{n}(x_1 + x_2 + \cdots + x_n) = \bar{x}. \end{aligned}$$

Similarly, letting

$$g(x) = (x - \bar{x})^2$$

gives us

$$\begin{aligned} E[g(x)] &= (x_1 - \bar{x})^2p_1 + (x_2 - \bar{x})^2p_2 + \cdots + (x_n - \bar{x})^2p_n \\ &= \frac{1}{n}[(x_1 - \bar{x})^2 + (x_2 - \bar{x})^2 + \cdots + (x_n - \bar{x})^2], \end{aligned}$$

so that the sample variance

$$s^2 = E[(x - \bar{x})^2].$$

Of course, it is not necessary that all the probabilities $p_k$ be the same. We will see several examples in this and the following section, where the probabilities $p_k$ differ.

One of the most important uses of expected values is in the making of decisions. Expected values are particularly useful in situations where a numerical value can be assigned to each event, such as in business, economics, games of chance, etc. The following examples illustrate some typical applications.

***Example 3*** Gambling at roulette is not a fair game: The house has a definite advantage no matter how you play. Suppose, however, that you decide to play on black. The roulette wheel contains 38 slots: 18 red, 18 black, and 2 green. If you bet \$1 on black, you win \$1 if black comes up but lose your money otherwise. Thus, the sample space consists of only the two events {black} and {red or green}. Each event has an associated payoff $x_1 = +1$ for {black} and $x_2 = -1$ for {red or green}. This is the amount you win or lose depending on the outcome of the wheel. The probability $p_1$ that black occurs is

$$p_1 = \frac{18}{38},$$

since only 18 of the 38 slots are black. Similarly,

$$p_2 = \frac{20}{38}.$$

Then the *expected value* of betting $1 on black is

$$\begin{aligned} E[x] &= x_1p_1 + x_2p_2 \\ &= (+1)\frac{18}{38} + (-1)\frac{20}{38} \\ &= \frac{-2}{38} = \frac{-1}{19} \approx -0.05263. \end{aligned}$$

What this means is that in the long run gamblers can expect to lose $1 every 19 times that they play, so that it costs them approximately 5.263 cents per game to play roulette.

***Example 4*** Another common gambling game is Chuck-a-Luck. It's a favorite of hustlers, who lead their opponents to believe that they have the advantage. The usual explanation of the game goes something like this.

"Say Bob, I know another game we can play with dice that will give you a chance to get all your money back. It's played with three dice, all of which are tossed at the same time. You pay a buck to play and select one of the spots on a die. If it comes up on one die, you win your dollar back; if it comes up on two of the dice, you win two dollars; and if it comes up on all three, you get three dollars."

"I'm sorry, Joe, I don't want to lose any more money. I've lost enough already."

"But the odds are in your favor! You'll be able to get all your dough back in no time!"

"I don't see why you say the game gives me the advantage."

"Well look, each die has one chance in six of coming up on the spot you selected. There are three of them, so it's even money your spot will turn up. And some of those times it will come up twice or three times, so you'll win back what you lost!"

"Well, OK. Let's play!"

Where is the flaw in the argument? We can figure out this game if we use expected values. To do so we must examine each of the four possible events that occur on each toss. Suppose we select sixes. Then the outcomes of each toss are

{no sixes}, {one six}, {two sixes}, {three sixes}.

The corresponding payoffs are

$$x_1 = -1, \qquad x_2 = +1, \qquad x_3 = +2, \qquad x_4 = +3,$$

respectively. What is the probability of each event?

To find the probability of each event we observe that by the sequential counting principle there are $6^3 = 216$ possible outcomes. Of these we have

125 ways of getting {no sixes},
75 ways of getting {one six},
15 ways of getting {two sixes},
1 way of getting {three sixes}.

If you had trouble getting these values, imagine that there are a red die, a blue die, and a white die. To determine the number of ways that {two sixes} is possible, imagine first that the red and blue dice are sixes but the white one is not. There are five possible outcomes in this situation. Now imagine that the red and white dice are sixes but the blue one is not. Finally, imagine that the red die is the one without a six. This gives us 15 possible outcomes. A similar strategy gives us the {one six} outcome. The {no six} outcome is obtained by the sequential counting principle or by subtraction. Hence

$$p_1 = \frac{125}{216}, \qquad p_2 = \frac{75}{216}, \qquad p_3 = \frac{15}{216}, \qquad p_4 = \frac{1}{216}.$$

Now we can find the expected value

$$\begin{aligned} E[x] &= x_1p_1 + x_2p_2 + x_3p_3 + x_4p_4 \\ &= (-1)\frac{125}{216} + (1)\frac{75}{216} + (2)\frac{15}{216} + (3)\frac{1}{216} \\ &= \frac{-17}{216} \approx -0.0787. \end{aligned}$$

In other words, in the long run it costs the unwary player almost 8 cents each time he or she plays Chuck-a-Luck. Playing roulette is a better gamble!

***Example 5*** Truck farmers in western Montana must gamble against nature every year. The problem is particularly acute with certain crops, like tomatoes, that are extremely sensitive to frost. Generally only 40% of the crop of 300 flats per acre has ripened on the vine by September 1, but frosts become a definite possibility at that time. Each additional week that the tomatoes remain on the vine means that an additional 20% of the crop will ripen, but a killing frost is certain by September 23. Vine-ripened tomatoes sell at \$4 per flat and green tomatoes sell at \$2 per flat. The longer the farmer can keep the tomatoes on the vine the more he or she will earn, but if a frost occurs, the remaining crop will be lost. For simplicity, we will assume that the farmer must decide whether or not to harvest all the remaining crop on September 1, 8, 15, or 22.

There are several alternatives that the farmer can take. All the crop can be harvested September 1, yielding \$840 per acre. If the farmer takes a chance and waits, he or she may earn only \$480 per acre if a frost occurs before September 8, but \$960 per acre if no frost occurs and all the crop is picked at that time. Or the farmer may take a chance and wait until September 15 or even September 22 to harvest the crop. If it's the right choice, the profits will increase; if not, the remaining crop is lost. The alternatives are shown in Fig. 7.6, together with the probabilities of having a frost each of the given weeks.

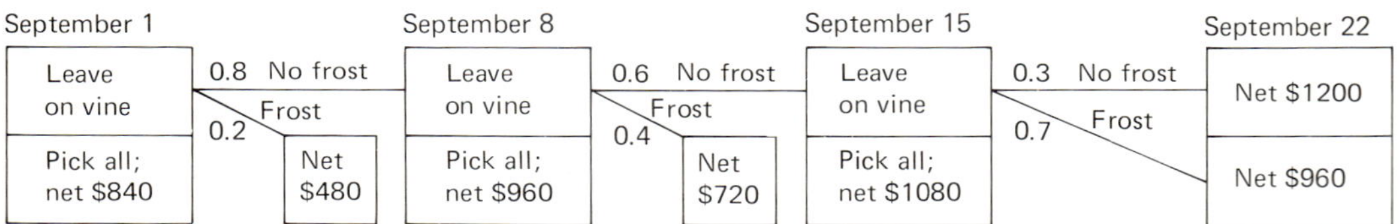


**Figure 7.6**

In this problem we make the final decision by considering each possible strategy. Each strategy is determined by the action that the farmer takes on the four dates in question. We use the definition to calculate the expected value of each strategy. Thus, since the probability is $p = 1$ that we get \$840 by picking all the crop on September 1, we have

$$E[\{\text{pick all Sept. 1}\}] = 840\,(1) = \$840.$$

To calculate the expected value of picking all the crop on September 8, we consider the two alternatives of whether or not it frosts. The probability of a frost is $p_1 = 0.2$ with a payoff of $x_1 = \$480$ per acre, while there is probability $p_2 = 0.8$ of no frost with payoff $x_2 = \$960$ per acre. Hence,

$$E[\{\text{pick all Sept. 8}\}] = 480(0.2) + 960(0.8) = \$864.$$

When we consider the strategy of picking the crop on September 15, we have three possible outcomes:

$$\{\text{frost by Sept. 8}\}, \qquad \{\text{frost by Sept. 15}\}, \qquad \{\text{no frost}\},$$

with payoffs

$$x_1 = \$480/\text{acre}, \qquad x_2 = \$720/\text{acre}, \qquad x_3 = \$1080/\text{acre},$$

and probabilities (using the sequential counting principle)

$$p_1 = 0.2, \qquad p_2 = (0.8)(0.4) = 0.32, \qquad p_3 = (0.8)(0.6) = 0.48,$$

respectively. Thus,

$$\begin{aligned} E[\{\text{pick all Sept. 15}\}] &= 480(0.2) + 720(0.32) + 1080(0.48) \\ &= \$844.80. \end{aligned}$$

Similarly,

$$\begin{aligned} E[\{\text{pick all Sept 22}\}] &= 480(0.2) + 720(0.32) + 960(0.48)(0.7) \\ &\quad + 1200(0.48)(0.3) \\ &= \$821.76. \end{aligned}$$

Therefore, the best strategy is to pick the entire crop on September 8.

---

## EXERCISES 7.2

*In Exercises 1–12, calculate the mean, median, range, and variance of each of the given sets of data.*

**1.** 52, 63, 55, 57, 58, 60, 54

**2.** 8, 5, 11, 12, 13, 7

**3.** 51, 39, 49, 48, 53, 39, 29, 51

**4.** 58, 54, 68, 78, 86, 82, 69, 68

**5.** 67, 58, 63, 69, 72, 65, 61, 60

**6.** 0, 6, 8, 10, 6, 7, 12, 8

**7.** The first two lines of numbers in Exercise 4 on p. 317.

**8.** The high temperature data in Table 7.1.

**9.** The high temperature data in Exercise 1 on p. 317.

**10.** The per capita incomes in Exercise 2 on p. 317.

**11.** The grant budgets in Exercise 3 on p. 317.

**12.** The random numbers in Exercise 4 on p. 317.

**13.** Suppose a given roulette wheel has only 37 slots (the double zero is missing): 18 red, 18 black, and 1 green. What is the expected value of betting $1 on black?

**14.** Bob and Joe are flipping a coin. Bob pays Joe $2 every time tails comes up. Joe pays Bob $1 for the first heads, $2 for the second *consecutive* heads, $3 for the third *consecutive* heads, and so forth. What is Bob's expected value on this game?

**15.** A die is cast and a gambler agrees to collect $9 if a 1 or a 6 come up. Otherwise he is to pay the number (in dollars) of spots that show on the die. What is the expected value of this game for the gambler?

**16.** Suppose a gambler collects $5 if a 1, 2, or 3 comes up in a toss of a die but loses the amount (in dollars) shown on the die otherwise. What is the expected value of this game?

**17.** As a gimmick a refrigerator dealer offers a free chance on a $5000 automobile to the first 200 customers who buy the top-of-the-line side-by-side model. Suppose the same refrigerator can be bought at another store for $40 less. What is the expected value of each free chance?

**18.** The senior citizens are having a raffle in which the top prize is a color TV retailing at $600, the second prize is a black-and-white TV ($200), and the third prize is a portable radio ($50). One thousand $1 tickets are sold. What is the expected value of each ticket?

**19.** Calculate the best strategy in Example 5 if vine-ripened tomatoes sell at $5 per flat and all other conditions remain the same.

**20.** Calculate the best strategy in Example 5 if green tomatoes sell at $1 per flat and all other conditions are unchanged.

**21.** An employee reaches retirement and must elect one of the following retirement pay options:

a) $3000 per year for the rest of her life;
b) $6000 per year for the next ten years or until she dies, whichever comes first;
c) A lump sum of $30,000;
d) $4500 per year for the next ten years, regardless of when she dies.

Suppose she has a 5% chance of dying in each of the following years. What option gives her the greatest expected value?

**22.** Suppose in Exercise 21 that she has an 8% chance of dying in each of the ensuing years. Which option should she select?

---

## 7.3 THE BINOMIAL DISTRIBUTION

In this section we will discuss a particularly useful theoretical model called the binomial distribution, and illustrate how this model can be applied to everyday situations.

In many $k$-stage experiments, we are only interested in determining whether or not a definite result occurs at each stage. Is the light bulb we are testing defective or not? Is tails the result of a certain toss of a coin? Did the patient suffer a side effect from taking a certain medicine or not? Experiments for which the result at each stage is (1) independent of the result of all other stages, and (2) one of two alternatives, are called *Bernoulli experiments*, and each stage of the experiment is usually called a *Bernoulli trial.** Thus, recording the outcome of tossing a coin $k$ times is a Bernoulli experiment and each toss is a Bernoulli trial. It is customary to call one of the alternatives of a Bernoulli trial a *success* and the other a *failure.* In the coin-tossing experiment we might label each tails a success and each heads a failure. The choice of label is ours and need not indicate any particular advantage of one alternative over the other.

Consider the following Bernoulli experiment. We toss a fair die four times. Each time the die comes up with a 6, we consider the toss a success; otherwise, we consider it a failure. Since each trial yields either a success or a failure, each outcome of the Bernoulli experiment can have from 0 to 4 successes.

Now let $x_k$ be the event that the Bernoulli experiment has exactly $k$ successes, $k = 0, 1, 2, 3$, or 4. What probabilities does $x_k$ have of occurring?

On each toss of the die the probability of a success is $p = 1/6$, since 6 is only one of the six faces. Suppose we wish to find the probability that $x_1$ occurs, that is, that one success occurs. Then the Bernoulli experiment consists of three failures and one success so that it must have occurred in one of the following four orders:

$$\{S\,F\,F\,F\}, \quad \{F\,S\,F\,F\}, \quad \{F\,F\,S\,F\}, \quad \{F\,F\,F\,S\}.$$

By the tree multiplication property we know that each of these outcomes occurs with probability

$$p(1-p)^3,$$

since it involves one success (with probability $p$ of occurring) and three failures [each with probability $(1 - p)$]. Hence,

$$p(x_1) = 4p(1-p)^3.$$

Similarly, for $x_2$ to occur, one of the following six orders is required:

$$\{S\,S\,F\,F\}, \quad \{S\,F\,S\,F\}, \quad \{S\,F\,F\,S\},$$
$$\{F\,S\,S\,F\}, \quad \{F\,S\,F\,S\}, \quad \{F\,F\,S\,S\},$$

---

* After Jacob Bernoulli (1654–1705), a noted Swiss mathematician and astronomer.

and each has probability $p^2(1 - p)^2$ of occurring. Thus,

$$p(x_2) = 6p^2(1 - p)^2.$$

Now, why are there six possible orders in which the Bernoulli experiment could have two successes? We can specify each order by stating which two of the four trials were successes. Hence, we are simply selecting two different objects from a set of four objects without regard to order. This situation should be familiar, since this is exactly what we did when we took combinations of two objects from a set of four different objects. Thus, the number of possible orders in which the Bernoulli experiment can have two successes is $C_{4,2} = 6$.

Now consider the situation when $k$ successes occur. This can happen in $C_{4,k}$ ways and each of these outcomes has probability

$$p^k(1 - p)^{4-k}$$

of occurring. Hence,

$$p(x_k) = C_{4,k}p^k(1 - p)^{4-k},$$

or

$$p(x_k) = C_{4,k}\left(\frac{1}{6}\right)^k\left(\frac{5}{6}\right)^{4-k}.$$

In general, suppose a Bernoulli experiment consists of $n$ Bernoulli trials and $x_k$ denotes the event that $k$ of the trials are successes. If the probability of success in each trial is $p$, then the probability that the event $x_k$ occurs is

$$p(x_k) = C_{n,k}p^k(1 - p)^{n-k}. \tag{7.8}$$

We will now do something unusual. We will construct a new sample space whose sample points are the $x_k$ defined above. Note that this new sample space is quite different from the sample space of the Bernoulli experiment, since each point in the new space is an event consisting of one or *more* points in the Bernoulli experiment. We call this new sample space

$$\{x_0, x_1, x_2, \ldots, x_n\},$$

together with the probabilities

$$p(x_k) = C_{n,k}p^k(1 - p)^{n-k},$$

the *binomial distribution* associated with the original Bernoulli experiment. We will find that binomial distributions have very interesting and useful properties.

Before proceeding, let's calculate the binomial distribution associated with the dice-tossing experiment with which this section started. We have

$$p(x_0) = C_{4,0}\left(\frac{1}{6}\right)^0\left(\frac{5}{6}\right)^4 = \frac{1(5)^4}{6^4} = \frac{625}{1296} \approx 0.4823;$$

$$p(x_1) = C_{4,1}\left(\frac{1}{6}\right)^1\left(\frac{5}{6}\right)^3 = \frac{4(5)^3}{6^4} = \frac{500}{1296} \approx 0.3858;$$

$$p(x_2) = C_{4,2}\left(\frac{1}{6}\right)^2\left(\frac{5}{6}\right)^2 = \frac{6(5)^2}{6^4} = \frac{150}{1296} \approx 0.1157;$$

$$p(x_3) = C_{4,3}\left(\frac{1}{6}\right)^3\left(\frac{5}{6}\right)^1 = \frac{4(5)}{6^4} = \frac{20}{1296} \approx 0.0154;$$

$$p(x_4) = C_{4,4}\left(\frac{1}{6}\right)^4\left(\frac{5}{6}\right)^0 = \frac{1(1)}{6^4} = \frac{1}{1296} \approx 0.0008.$$

We can plot the events and their associated probabilities in a line chart, as shown in Fig. 7.7. Note that the sum of these probabilities equals 1 since all possible outcomes of the Bernoulli experiment are considered:

$$\begin{aligned} p(x_0) + p(x_1) + p(x_2) + p(x_3) + p(x_4) &= \frac{625 + 500 + 150 + 20 + 1}{1296} \\ &= \frac{1296}{1296} = 1. \end{aligned}$$

This will also be the case with the general situation.

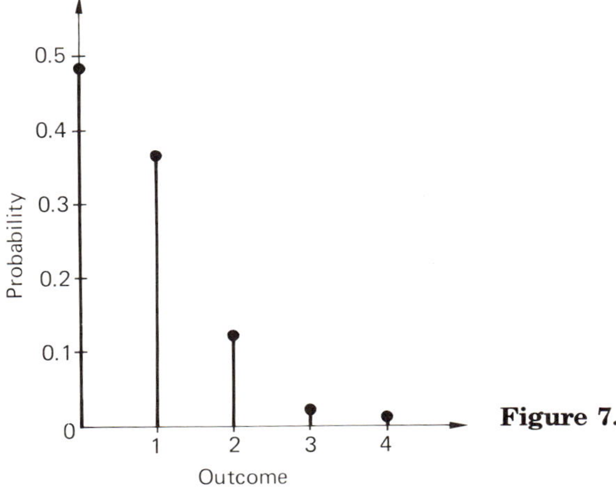


**Figure 7.7**

If we denote each sample point $x_k$ by its number of successes $k$, we can determine the mean $\mu$ and variance $\sigma^2$ of the binomial distribution. We can use the symbols $\mu$ and $\sigma^2$ instead of $\bar{x}$ and $s^2$, since we will consider the *entire population* and not just a representative sample as we did in the examples in Section 7.2. As was shown in Section 7.2, we have

$$\mu = E[x] = x_0p_0 + x_1p_1 + x_2p_2 + x_3p_3 + x_4p_4$$
$$= \frac{0(625) + 1(500) + 2(150) + 3(20) + 4(1)}{1296} = \frac{864}{1296} = \frac{2}{3},$$

where each $p_k = p(x_k)$ was calculated above, and

$$\sigma^2 = E[(x-\mu)^2] = E[(x-\tfrac{2}{3})^2]$$
$$= (x_0-\mu)^2p_0 + (x_1-\mu)^2p_1 + (x_2-\mu)^2p_2 + (x_3-\mu)^2p_3 + (x_4-\mu)^2p_4$$
$$= \frac{(0-\frac{2}{3})^2(625) + (1-\frac{2}{3})^2(500) + (2-\frac{2}{3})^2(150) + (3-\frac{2}{3})^2(20) + (4-\frac{2}{3})^2}{1296}$$
$$= \frac{720}{1296} = \frac{5}{9}.$$

Actually there are much simpler formulas for finding the mean and variance of a binomial distribution:

$$\mu = np \quad \text{and} \quad \sigma^2 = np(1-p). \tag{7.9}$$

In the example above, we have $n = 4$ and $p = \frac{1}{6}$, so that

$$\mu = 4\left(\frac{1}{6}\right) = \frac{2}{3} \quad \text{and} \quad \sigma^2 = 4\left(\frac{1}{6}\right)\left(\frac{5}{6}\right) = \frac{5}{9}.$$

The following proof verifies the first identity in (7.9). The second identity is left as an exercise (see Exercise 23 on p. 339). By the definition of the mean, we have

$$\mu = x_0p_0 + x_1p_1 + x_2p_2 + \cdots + x_np_n$$
$$= 0\cdot p_0 + 1\cdot p_1 + 2\cdot p_2 + \cdots + np_n.$$

Substituting the expression on the right side of Eq. (7.8) for $p_k = p(x_k)$, we get

$$\mu = C_{n,1}p(1-p)^{n-1} + 2C_{n,2}p^2(1-p)^{n-2} + \cdots + nC_{n,n}p^n(1-p)^0. \tag{7.10}$$

Using the factorial notation for combinations (see Eq. 6.32), we see that the $k$th term in this sum is

$$kC_{n,k}p^k(1-p)^{n-k} = \frac{(k)\,n!}{k!\,(n-k)!}p^k(1-p)^{n-k}$$
$$= npC_{n-1,k-1}p^{k-1}(1-p)^{(n-1)-(k-1)}.$$

Thus we may factor $np$ from each term in the sum (7.10), getting

$$\mu = np[C_{n-1,0}p^0(1-p)^{n-1} + C_{n-1,1}p(1-p)^{n-2} + \cdots + C_{n-1,n-1}p^{n-1}(1-p)^0]. \tag{7.11}$$

But the terms in the sum are simply the probabilities of each event of a binomial distribution involving $(n-1)$ trials each with probability $p$ of success. Since the sum consists of all possible events that can occur, at least one of them must occur, so the probability that one of them occurs is equal to 1. Thus, the term in brackets equals 1 so that $\mu = np$.

The following examples illustrate some of the applications of the binomial distribution.

***Example 1*** Airlines deliberately overbook flights because they know that many passengers will not notify them of a change in plans. Suppose past experience has determined that 15% of all passengers making reservations will not show up for the flight. What is the probability that if 100 people make reservations for a flight, at least 97 will show?

In this problem we must find the probability of at least 97 successes. Thus we must add the probabilities

$$p(x_{97}) + p(x_{98}) + p(x_{99}) + p(x_{100}),$$

where $x_k$ is the event of $k$ successes occurring. Since

$$p(x_k) = C_{n,k}p_k(1-p)^{n-k},$$

$n = 100$, and $p = 0.85$ (since 85% show and each show is a success), we have

$$p(x_{97}) = C_{100,97}(0.85)^{97}(0.15)^3 \approx 7.7735536 \times 10^{-5};$$
$$p(x_{98}) = C_{100,98}(0.85)^{98}(0.15)^2 \approx 1.3484736 \times 10^{-5};$$
$$p(x_{99}) = C_{100,99}(0.85)^{99}(0.15)^1 \approx 0.1543707 \times 10^{-5};$$
$$p(x_{100}) = C_{100,100}(0.85)^{100}(0.15)^0 \approx 0.0087477 \times 10^{-5}.$$

Adding these together we have

$$p(\text{at least 97 show}) \approx 9.2851456 \times 10^{-5}$$

or 0.000092851456%: Indeed, the probability that at least 93 will show is only about 1%. For this reason, overbooking is an economic necessity.

***Example 2*** Many companies worry about the quality of the product they sell. By sampling randomly chosen items from each batch, they can control the quality of the product.

Suppose a company wishes to ensure that no more than 2% of the items it produces are defective. A sample of 10 items is chosen from a large batch and two of the items are found to be defective. Does this indicate that more care should be taken in the production process?

Let's consider each item a failure if it is defective. Then we need to find the probability of the event $x_8$ (8 successes) in a sample of size $n = 10$, where the probability of each success is

$$p = 0.98 \text{ (or 98\%)}.$$

Hence,

$$\begin{aligned} p(x_8) &= C_{10,8}p^8(1-p)^2 \\ &= \frac{10!}{8!\,2!}(0.98)^8(0.02)^2 \approx 0.0153, \end{aligned}$$

which means that this occurrence could happen only about $1\frac{1}{2}\%$ of the time if no more than 2% of the items in any batch are defective. While this event is not highly unusual, it might call for a more careful scrutiny.

Suppose on the other hand that three items were defective. Now

$$\begin{aligned} p(x_7) &= C_{10,7}p^7(1-p)^3 \\ &= \frac{10!}{7!\,3!}(0.98)^7(0.02)^3 \approx 0.0008334, \end{aligned}$$

or about 0.08%, an event so unusual that trouble is definitely indicated.

***Example 3*** During World War II the Nazis committed a serious blunder by numbering their tanks sequentially as they came off the production line. By keeping track of the serial numbers of captured and destroyed German tanks, the Allies were able to make an accurate estimate of the total tank force and annual output.

To see how this is possible, we assume that the Nazi war machine has produced $N$ tanks, numbered by the serial numbers $1, 2, 3, \ldots, N$. Suppose that we have captured or destroyed 50 tanks and recorded their serial numbers. *What is the probability that the highest of these serial numbers is within 5% of $N$?* (This is an arbitrary percentage; any percentage could have been selected.)

To answer that question let's find the probability of the complementary event, that is, that every single one of the serial numbers is *not* within 5% of $N$. (If the highest serial number is not within 5% of $N$, neither are the others!) We will call each serial number a failure if it is not within 5% of $N$. Hence we wish to find the probability of the event $x_0$, that is, no successes. Now the probability of success is $p = 0.05$ since only 5% of the serial numbers are successes, so we have

$$p(x_0) = C_{50,0}p^0(1 - p)^{50} = (0.95)^{50} \approx 0.07694.$$

Thus, the probability of the complementary event, that the highest serial number is within 5% of $N$, is

$$p(x_0)' = 1 - p(x_0) \approx 1 - 0.07694 = 0.92306.$$

Thus there is at least a 92% probability that the highest serial number recorded is within 5% of $N$. Suppose the highest serial number we have found is $M$; then we say that we have a 92% confidence level that

$$0.95N \leq M \leq N$$

or, equivalently, we say that

$$M \leq N \leq \frac{M}{0.95}$$

is a 92% *confidence interval*. For example, if $M = 1792$, we could be 92% sure that $N$ lies in the interval

$$1792 \leq N \leq \frac{1792}{0.95} < 1887.$$

As the number of captured or destroyed tanks increased, we could make better estimates.

---

## EXERCISES 7.3

*In Exercises 1–10, find all the probabilities associated with the events $x_k$ ($k = 0, 1, 2, \ldots, n$) for the given $n$ and probability of success $p$. (Hint: Use Eq. 7.8.) Also find the mean and variance.*

**1.** $n = 3, p = 1/3$

**2.** $n = 4, p = 1/2$

**3.** $n = 4, p = 1/3$

**4.** $n = 5, p = 1/4$

**5.** $n = 5, p = 2/5$

**6.** $n = 6, p = 0.8$

**7.** $n = 7, p = 0.4$

**8.** $n = 8, p = 0.2$

**9.** $n = 9, p = 0.1$

**10.** $n = 10, p = 1/4$

11. Using the data given in Example 1, determine the probability that if 100 people make reservations for a flight at least 90 will show; that 85 will show.

12. Holiday Inns have determined by past experience that 25% of those making reservations do not keep them. Suppose an Inn manager has 20 rooms available. What is the probability that of 20 people making reservations at least 16 will keep them?

13. In Exercise 12, suppose the Inn manager deliberately overbooks the 20 available rooms and agrees to reserve rooms for 22 people. What is the probability that *more* than 20 people will claim their reservation?

14. Suppose that the airline in Example 1 (see p. 335) has 100 seats available on a flight, but accepts 110 reservations. What is the probability that they will have to pay a penalty for not being able to carry a passenger with a confirmed reservation? How often will they have to pay two passengers a penalty?

15. Find all the probabilities for the binomial distribution in Example 2 (see p. 336). What is the most likely outcome?

16. Suppose the company in Example 2 is under a new management that is more concerned with the quality of their product. The new campaign calls for at most 1 defective item in each batch of 500. How likely is a sample of 10 items to contain 2 defective items? Would you recommend a check of the production process if such a random sample were selected?

17. Suppose the manufacturer in Example 2 instructs the inspectors to sample 15 items and report any instances in which 3 or more defective items are found. How often will that happen if no more than 2% of the items are defective?

18. In Exercise 17 assume that 3% of all items are defective. How often does a sample of 15 contain 3 defective items?

19. Suppose 100 tank serial numbers have been obtained. How certain are you that the highest serial number is within 5% of the largest number produced?

20. Suppose 100 tank serial numbers have been obtained. How certain are you that the highest serial number is within 2% of the largest number $N$ of tanks produced? If the highest number found is 4769, determine the confidence interval $N$ for the largest number.

21. Suppose 200 tank serial numbers are known, and the largest found is 4769. What is the probability that $N$ lies in the interval $4769 \le N \le 4769/0.99 < 4818$?

22. Each county in Montana sells auto tags that have been numbered sequentially, except for the first one or two digits, which identify the county. How

many license tags should you look at to have a 90% probability of seeing one tag that is within 5% of $N$?

***23.** Prove that $\sigma^2 = np(1 - p)$. (*Hint:* $k = (k - 1) + 1$.)

## 7.4 THE NORMAL DISTRIBUTION

The binomial distribution is a *discrete* distribution, since each event is determined by the number of successes in a fixed sequence of trials. Thus, if the number of trials is $n$, the only possible outcomes are the integers $k = 0, 1, 2, \ldots, n$. In Section 7.3 we determined a formula (Eq. 7.8) for the probability of each sample point and noted that the sum of these probabilities is 1. This indicates that the sum of the areas of the rectangles in the histogram for this distribution is 1. We will now consider the result of letting the number of trials $n$ increase without bound.

To see what problems arise, notice that Eq. (7.8)

$$p(x_k) = C_{n,k}p^k(1 - p)^{n-k}$$

becomes progressively harder to calculate as $n$ increases. This is true even when we let $p = 1 - p = \frac{1}{2}$, simplifying the formula (in factorial form) to

$$p(x_k) = \frac{n!}{k!(n - k)!}\left(\frac{1}{2}\right)^n,$$

since the number of calculations will increase with $n$. There is one important property that persists: *The area of the histogram is still equal to* 1. Figure 7.8 illustrates the shape of two different binomial distributions. Note that in each case, the area of the rectangle in the histogram resembles the area under the bell-shaped curve in Fig. 7.9. This resemblance becomes more pronounced as $n$ increases. Distributions such

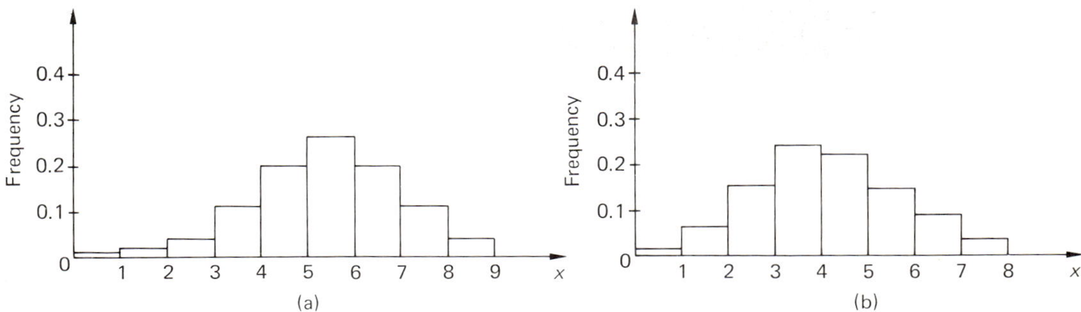


**Fig. 7.8**
(a) $p = 0.5$, $n = 10$; (b) $p = 0.3$, $n = 12$.

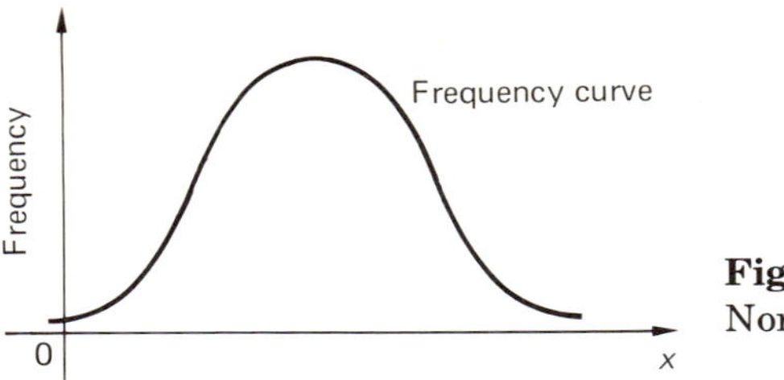


**Fig. 7.9**
Normal distribution.

as the curve in Fig. 7.9 that have an unbroken smooth frequency curve for a "histogram" are called *continuous distributions.*

The continuous distribution that is illustrated in Fig. 7.9 is called a *normal distribution,* primarily because many things in nature are distributed in this way. The principal characteristics of the normal distribution are the following.

1. *It has a symmetrical "bell" shape.*
2. *The mean and median are both located along the vertical "center" line of the distribution.*
3. *The area under the curve equals* 1 (*or* 100%), *since it represents the probability of occurrence of all possible outcomes.*

There are many different normal distributions, since their means and variances can differ. Since the variance measures the spread of a distribution, small variances have tall and thin "bell" shapes, while a large variance has a short and wide curve (see Fig. 7.10). When working with a given normal distribution, it is customary to compare it with the *standard normal distribution,* which has mean $\mu = 0$ and variance $\sigma^2 = 1$. Any other normal distribution in the variable $x$ can be converted to a standard normal distribution in the variable $z$ by setting

$$z = \frac{x - \mu}{\sigma}, \tag{7.12}$$

where $\mu$ and $\sigma^2$ are the mean and variance of the given normal distribution. This procedure is called *standardizing* a normal distribution. We will now see why standardizing is useful.

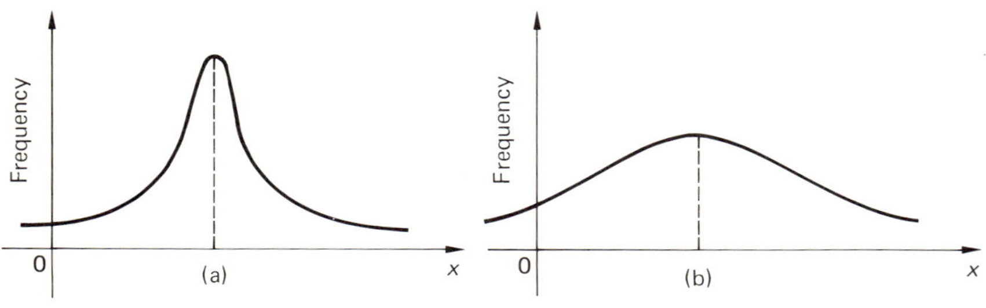


**Fig. 7.10**
Two normal distributions: (a) small; (b) large.

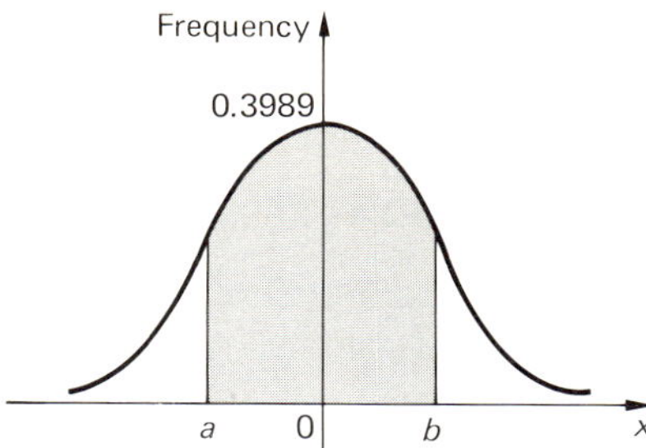


**Fig. 7.11**
A standard normal distribution.

Figure 7.11 illustrates the standard normal distribution. Suppose we wish to find the shaded area between $z = a$ and $z = b$ in the figure. *This area represents the probability that the event z will be between the values a and b.* Table 7.4 contains values of certain areas of the standard normal distribution. The following example illustrates the use of this table.

**Table 7.4**
**Area *A* under the curve between 0 and *z* for the standard normal distribution**

| $z$ | $A$ | $z$ | $A$ | $z$ | $A$ | $z$ | $A$ | $z$ | $A$ |
|---|---|---|---|---|---|---|---|---|---|
| 0.05 | 0.0199 | 0.55 | 0.2088 | 1.05 | 0.3531 | 1.55 | 0.4394 | 2.05 | 0.4798 |
| 0.10 | 0.0398 | 0.60 | 0.2257 | 1.10 | 0.3643 | 1.60 | 0.4452 | 2.10 | 0.4821 |
| 0.15 | 0.0596 | 0.65 | 0.2422 | 1.15 | 0.3749 | 1.65 | 0.4505 | 2.15 | 0.4842 |
| 0.20 | 0.0793 | 0.70 | 0.2580 | 1.20 | 0.3849 | 1.70 | 0.4554 | 2.20 | 0.4861 |
| 0.25 | 0.0987 | 0.75 | 0.2734 | 1.25 | 0.3944 | 1.75 | 0.4599 | 2.25 | 0.4878 |
| 0.30 | 0.1179 | 0.80 | 0.2881 | 1.30 | 0.4032 | 1.80 | 0.4641 | 2.30 | 0.4893 |
| 0.35 | 0.1368 | 0.85 | 0.3023 | 1.35 | 0.4115 | 1.85 | 0.4678 | 2.35 | 0.4906 |
| 0.40 | 0.1554 | 0.90 | 0.3159 | 1.40 | 0.4192 | 1.90 | 0.4713 | 2.40 | 0.4918 |
| 0.45 | 0.1736 | 0.95 | 0.3289 | 1.45 | 0.4265 | 1.95 | 0.4744 | 2.45 | 0.4929 |
| 0.50 | 0.1915 | 1.00 | 0.3413 | 1.50 | 0.4332 | 2.00 | 0.4772 | 2.50 | 0.4938 |

***Example 1*** Find the areas under the standard normal distribution for the intervals:

a) $0.15 \leq z \leq 1.95$;
b) $-1.25 \leq z \leq 2.15$;
c) $z \leq 0.55$;
d) $z \leq -2.35$.

To solve (a) we use Table 7.4 to calculate the area from 0 to 1.95 and then subtract the area from 0 to 0.15. Thus we have

$$A(1.95) - A(0.15) = 0.4744 - 0.0596 = 0.4148.$$

(b) Here we use the symmetry of the curve. The area from 0 to 1.25 is exactly equal to the area from $-1.25$ to 0, so that the total area is the *sum* of the area from 0 to 2.15 and the area from $-1.25$ to 0. Thus,

$$A(2.15) + A(1.25) = 0.4842 + 0.3944 = 0.8786.$$

(c) In this situation we have to add the area from 0 to 0.55 to the area from $-\infty$ to 0. Since the median occurs at $z = 0$, half the possible events lie in the interval from $-\infty$ to 0. Hence the area for that half of the curve is 0.5. Thus the total area is

$$0.5 + A(0.55) = 0.5 + 0.2088 = 0.7088.$$

(d) Again we use the fact that the area from $-\infty$ to 0 is 0.5, but this time we subtract the area from 0 to 2.35, since by symmetry it equals the area from $-2.35$ to 0. Hence,

$$0.5 - A(2.35) = 0.5 - 0.4906 = 0.0094.$$

***Example 2*** Suppose the high temperature in Denver during the month of June is normally distributed with mean 87.25°F and variance 30.75. What is the probability that on a given day the high temperature will not exceed 90°F?

In this situation we must standardize our information. Since $\mu = 87.25$ and $\sigma^2 = 30.25$, we use the substitution

$$z = \frac{x - \mu}{\sigma} = \frac{x - 87.25}{\sqrt{30.25}} = \frac{x - 87.25}{5.5}.$$

Hence, if $x = 90°$ we get

$$z = \frac{90 - 87.25}{5.5} = \frac{2.75}{5.5} = 0.5 .$$

Thus, we need to discover the probability that $z$ lies in the interval $z \leq 0.5$. Since the probability that $z$ lies in the interval from $-\infty$ to 0 is 0.5, we have, using Table 7.4,

$$p(x \leq 90°\text{F}) = p(z \leq 0.5) = 0.5 + 0.1915 = 0.6915.$$

Hence, on approximately 69.15% of the days in June the temperature does not exceed 90°F.

Similarly, the probability that the temperature will exceed 79°F is

$$p(x \geq 79°F) = p\left(z \geq \frac{79 - 87.25}{5.5}\right) = p(z \geq -1.5)$$
$$= 0.5 + 0.4332 = 0.9332.$$

## EXERCISES 7.4

*In Exercises 1–12, find the probability that the event $x$ occurs in the given interval, assuming $x$ is normally distributed with mean $\mu$ and variance $\sigma^2$.*

**1.** $-1 \le x \le 1,\ \mu = 0,\ \sigma^2 = 1$

**2.** $-2 \le x \le 2,\ \mu = 0,\ \sigma^2 = 1$

**3.** $|x| \le 0.5,\ \mu = 0, \sigma^2 = 1$

**4.** $|x| \le 1, \mu = 1, \sigma^2 = 1$

**5.** $28 \le x \le 32, \mu = 30, \sigma^2 = 4$

**6.** $0 \le x \le 100, \mu = 30, \sigma^2 = 4900$

**7.** $x \le 20, \mu = 30, \sigma^2 = 25$

**8.** $x \ge 20, \mu = 30, \sigma^2 = 36$

**9.** $1 \le x \le 7, \mu = -2, \sigma^2 = 16$

**10.** $-1 \le x \le 7, \mu = 2, \sigma^2 = 16$

**11.** $|x - 1| \le 2, \mu = 3, \sigma^2 = 9$

**12.** $x \ge 2, \mu = 3, \sigma^2 = 9$

**13.** Assume that the sample of high temperatures for Stapleton International Airport in Table 7.1 is representative of the actual June temperatures. Find the sample mean and variance and assume that the high temperature for June is normally distributed. Use linear interpolation and Table 7.4 to predict the probability that the temperature will not exceed 90°F. Does it agree with the observed data?

**14.** Repeat Exercise 13 using the data given for May in Exercise 1 on p. 317.

a) How likely is it that the temperature will not exceed 80°F?
b) How likely is it that the temperature will exceed 65°F?
c) How likely is it that the temperature will be in the 70s?

**15.** Suppose that the 1973 U.S. personal income per capita is normally distributed. Use the sample in Exercise 2 on p. 317 to estimate the mean and variance. What percentage of the population had a per-capita income in excess of \$5000? In excess of \$4000?

**16.** Suppose the weights of 50-lb bags of Friskies dog food are normally distributed with mean $\mu = 50.3$ and variance $\sigma^2 = 0.09$. How likely is it that a bag weighs less than 50 lb?

**17.** A beer bottler is set to fill each bottle with an average of 12.2 ounces, with a variance of 0.01 ounce. What is the probability that a bottle will contain less than 12 ounces of beer?

**18.** The Red Midget Packing Company cans green peas. Each can is supposed to contain 15.5 oz of peas in brine. Red Midget has set its equipment to deliver an average of 15.6 oz per can, but the variance is 1 oz per can. What percentage of the cans contain as little as 14 oz of peas and brine? What percentage contain over 16 oz?

**19.** Suppose the scores on a multiple-choice biology exam are normally distributed with mean 75 and variance 36. If an A is a 90, what percentage of the class got an A?

**20.** A teacher decides to assign grades based on a 100-point test that is normally distributed with mean 75 and variance 25. If the teacher decides to assign 10% A, 15% B, 40% C, 20% D, and 15% F, find the test scores for each grade.

## 7.5 MARKOV CHAINS

In Section 7.3 we discussed the outcomes of a sequence of independent trials, where the result of each trial was either a success or a failure. We will now study other sequences of trials in which the result of each trial *need not* be independent of the previous trial. That is, we will discuss situations in which the result of a given trial can depend on the result of the previous trial.

For example, suppose Mary shops at the grocery store once a week. If she buys a 10-lb sack of potatoes, there is an 80% probability that she will not buy another 10-lb sack the following week. However, if she doesn't buy a 10-lb sack of potatoes, the probability that she will buy one the following week is 60%. We can represent the two situations in matrix notation:

| | | Next week's purchases | |
|---|---|---|---|
| | | Potatoes | No potatoes |
| This week's purchases | Potatoes | 0.2 | 0.8 |
| | No potatoes | 0.6 | 0.4 |

$$\begin{bmatrix} 0.2 & 0.8 \\ 0.6 & 0.4 \end{bmatrix}. \tag{7.13}$$

The matrix of probabilities in (7.13) is called a *transition matrix.* Transition matrices have the property that *the sum of the entries of any row is* 1, since each row consists of all possible outcomes of the next trial. The transition matrix (7.13) clearly reveals that the result of next week's shopping depends on what was purchased this week, so the two trials are not independent.

Assuming that Mary initially purchased a 10-lb sack of potatoes, what is the probability that she will purchase another sack one week later, two weeks later, or three weeks later? We can solve this problem by constructing a tree diagram as we did in Chapter 6, but it is more convenient to use the matrix (7.13). The repeated nature of the experiment is similar to the transportation problems we solved in Chapter 4 by matrix multiplication. This analogy suggests that we should compare matrix multiplications with the results from a tree diagram.

Since we are assuming that Mary initially purchased a 10-lb sack of potatoes, during the initial (or 0th) week the probability that potatoes were purchased is 1. Our problem now is to find a way of multiplying this initial situation, or *state*, by the transition matrix (7.13). Since (7.13) is a $2 \times 2$ matrix, only vectors with two entries can be multiplied with it, so how can we use our initial state to multiply the matrix? The answer is to construct a row vector consisting of both the initial probability that potatoes were purchased and the initial probability that potatoes were not purchased, that is,

| Purchased | Not purchased |
|---|---|
| [1 | 0]. |

$$[1 \quad 0]. \tag{7.14}$$

Of course, one entry in (7.14) is redundant, but since this allows us to use matrix multiplication we do not mind carrying along some unnecessary information. Now, we multiply (7.14) by (7.13), obtaining

$$[1 \quad 0]\begin{bmatrix} 0.2 & 0.8 \\ 0.6 & 0.4 \end{bmatrix} = [0.2 \quad 0.8]. \tag{7.15}$$

Note that the resulting vector consists of two entries: The first is the probability that potatoes were purchased and the second that they were not purchased at the end of the first week. This corresponds directly to the actual situation since we know that initially potatoes were purchased.

Now let's multiply this new vector by the transition matrix (7.13), getting

$$[0.2 \quad 0.8]\begin{bmatrix} 0.2 & 0.8 \\ 0.6 & 0.4 \end{bmatrix} = [0.52 \quad 0.48]. \tag{7.16}$$

At the same time consider the tree shown in Fig. 7.12. Adding the two probabilities that a purchase is made during the second week yields

$$0.04 + 0.48 = 0.52,$$

which agrees with the first value on the right side of Eq. (7.16). Similarly, the probability that no purchase is made is 0.16 + 0.32 = 0.48.

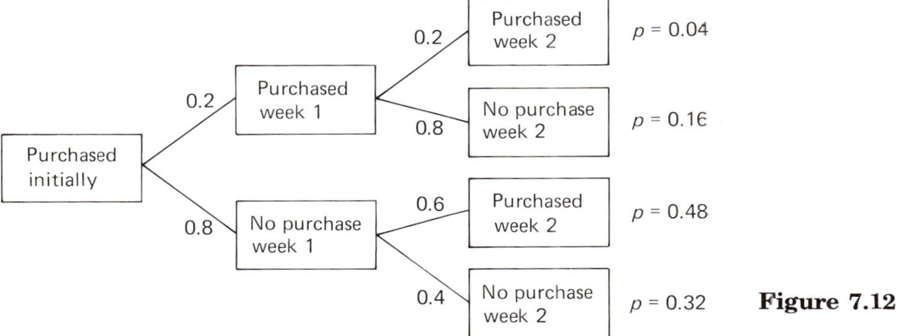


**Figure 7.12**

Repeating the multiplication process once more gives us

$$[0.52 \quad 0.48]\begin{bmatrix} 0.2 & 0.8 \\ 0.6 & 0.4 \end{bmatrix} = [0.392 \quad 0.608], \tag{7.17}$$

which are the probabilities of making and not making a purchase in the third week, respectively.

The row vector of entries is called a *state vector* for the problem, and indicates the probabilities that the particular state occurs.

Let's denote the transition matrix by the letter $A$, and the initial state by the symbol $S_0$. Then the $n$th state vector $S_n$ is obtained by multiplying

$$S_n = S_0 A^n. \tag{7.18}$$

***Example 1*** Let Mary's potato purchasing be given by the transition matrix $A$ in Eq. (7.13) and suppose that Mary initially purchases potatoes. Find the 5th state vector for this problem.

We can use Eq. (7.18) and write $S_5 = S_0 A^5$, but in this case it is easier to use $S_3$, which is the right side of Eq. (7.17), and multiply it twice by $A$. Hence,

$$S_4 = S_3 A = [0.392 \quad 0.608]\begin{bmatrix} 0.2 & 0.8 \\ 0.6 & 0.4 \end{bmatrix} = [0.4432 \quad 0.5568],$$

and (to four decimal places)

$$S_5 = S_4 A = [0.4432 \quad 0.5568]\begin{bmatrix} 0.2 & 0.8 \\ 0.6 & 0.4 \end{bmatrix} = [0.4227 \quad 0.5773].$$

If we continue in this fashion, we obtain

$$\begin{aligned} S_6 &= [0.4309 \quad 0.5691], \\ S_7 &= [0.4276 \quad 0.5724], \\ S_8 &= [0.4290 \quad 0.5710], \\ S_9 &= [0.4284 \quad 0.5716], \\ S_{10} &= [0.4286 \quad 0.5714], \\ S_{11} &= [0.4286 \quad 0.5714], \end{aligned}$$

and we see that the state vectors *stabilize* with probability 0.4286 that potatoes will be purchased and probability 0.5714 that they will not be purchased in any given week. This stable state vector is called a *steady-state vector* or a *fixed-point vector* for the *Markov* chain*, the name that is given to problems of this type. A Markov chain is any process in which the probabilities for the next state are completely determined by the present state. Symbolically, we may write

$$S_{n+1} = S_n A.$$

---

* Andrei A. Markov (1856–1922) was one of the most important Russian mathematicians. His work forms the foundation of the theory of stochastic processes.

Do all Markov chains have a steady-state vector? Surprisingly, the answer is yes. *Every Markov chain has at least one steady-state vector.* Unfortunately *some Markov chains have more than one steady-state vector,* as the following example illustrates.

***Example 2*** Let the transition matrix be

$$A = \begin{bmatrix} 1 & 0 & 0 \\ 0 & 1 & 0 \\ 1/4 & 1/4 & 1/2 \end{bmatrix}.$$

Then

$$[1 \quad 0 \quad 0]\begin{bmatrix} 1 & 0 & 0 \\ 0 & 1 & 0 \\ 1/4 & 1/4 & 1/2 \end{bmatrix} = [1 \quad 0 \quad 0].$$

and

$$[0 \quad 1 \quad 0]\begin{bmatrix} 1 & 0 & 0 \\ 0 & 1 & 0 \\ 1/4 & 1/4 & 1/2 \end{bmatrix} = [0 \quad 1 \quad 0].$$

Hence $[1 \quad 0 \quad 0]$ and $[0 \quad 1 \quad 0]$ are both steady-state vectors. Indeed $(p, 1 - p, 0)$ is a steady-state vector for any $0 \leq p \leq 1$.

---

As this example indicates, it is possible to have infinitely many steady-state vectors. However, there is one class of Markov chains that always has a unique steady-state vector. These are the regular Markov chains.

**DEFINITION 7.1** **A Markov chain is *regular* if some power of its transition matrix has all entries positive.**

In other words, if we multiply $A$ by itself repeatedly and find that $A^n$ has all entries positive (*not zero*) for some integer $n = 1, 2, \ldots,$ then *that Markov chain is regular and the transition matrix A has only one steady-state vector.** For example, the matrix showing Mary's potato purchasing is regular since

$$A = \begin{bmatrix} 0.2 & 0.8 \\ 0.6 & 0.4 \end{bmatrix}$$

---

* Note that we have *not* said that every Markov chain that has a unique steady-state vector is regular (see Exercise 19 on p. 352).

already has all its entries positive, while the matrix in Example 2 satisfies

$$\begin{bmatrix} 1 & 0 & 0 \\ 0 & 1 & 0 \\ \frac{1}{4} & \frac{1}{4} & \frac{1}{2} \end{bmatrix}^n = \begin{bmatrix} 1 & 0 & 0 \\ 0 & 1 & 0 \\ \frac{2^n - 1}{2^{n+1}} & \frac{2^n - 1}{2^{n+1}} & \frac{1}{2^n} \end{bmatrix}$$

so that zeros are *always* present. (Convince yourself of this identity by doing several multiplications.) Thus, Example 2 is not a regular Markov chain.

***Example 3*** Stingy Rent-A-Car has two offices: one at Dulles Airport and one at National Airport in Washington, D.C. Stingy has noticed that every car rented at Dulles is returned to the office at National, while 2/3 of the cars rented at National are returned to that office. How should Stingy Rent-A-Car distribute its fleet of 60 cars between the two airports?

The transition matrix $A$ is given by

| | | Returned to | |
|---|---|---|---|
| | | National | Dulles |
| Rented at | National | 2/3 | 1/3 |
| | Dulles | 1 | 0 |

Now

$$A^2 = \begin{bmatrix} 2/3 & 1/3 \\ 1 & 0 \end{bmatrix} \cdot \begin{bmatrix} 2/3 & 1/3 \\ 1 & 0 \end{bmatrix} = \begin{bmatrix} 7/9 & 2/9 \\ 2/3 & 1/3 \end{bmatrix},$$

which has all entries positive. Therefore, this is a regular Markov chain and has a unique steady-state vector.

While we could try to find the steady-state vector by multiplying some initial state repeatedly by $A$, it is easier to solve the problem algebraically. Let the steady-state vector be $S = [p \quad 1 - p]$. Then, since $SA = S$, we have

$$S(A - I) = SA - S = 0,$$

where $I$ is the $2 \times 2$ identity matrix. Calculating $A - I$, we have

$$[p \quad 1 - p] \begin{bmatrix} -1/3 & 1/3 \\ 1 & -1 \end{bmatrix} = [0 \quad 0]$$

or

$$\left[ -\frac{p}{3} + (1 - p) \quad \frac{p}{3} - (1 - p) \right] = [0 \quad 0].$$

Hence, using either entry of the vector, we have

$$\tfrac{4}{3}p - 1 = 0,$$

so that $p = 3/4$. Thus, the steady-state vector is

$$S = [3/4 \quad 1/4].$$

This vector indicates that the probability that a particular car is located at National Airport is 3/4. Thus, if Stingy Rent-A-Car stations 3/4 of its fleet at National Airport, the ratio of cars at National and Dulles will remain (approximately) constant in the long run. There may, however, be other compelling reasons for not doing so; for example, it might be the case that more rentals occur at Dulles.

---

One of the startling implications of regular Markov chains is that *regardless of the initial state*, repeated multiplication by $A$ will eventually lead to the steady-state vector. (We omit the proof of this theorem.)

Let's see how closely we approximated the steady-state vector in Example 1. To obtain the steady-state vector exactly, we solve the algebraic equation

$$S(A - I) = 0,$$

where $S = [p \quad 1 - p]$ and $A$ is given in Eq. (7.13). Hence,

$$[p \quad 1 - p]\begin{bmatrix} -0.8 & 0.8 \\ 0.6 & -0.6 \end{bmatrix} = [0 \quad 0]$$

or

$$[-0.8p + 0.6(1 - p) \quad 0.8p - 0.6(1 - p)] = [0 \quad 0].$$

Thus, using either entry of the vectors, we have

$$1.4p - 0.6 = 0$$

or

$$p = \frac{0.6}{1.4} = \frac{3}{7} \approx 0.428571428571\ldots.$$

Therefore, we see that repeated multiplication by the transition matrix $A$ led to a good approximation of the steady-state vector.

Example 2 illustrates that a nonregular Markov chain need not have this property. In that example the initial state determines the steady-state vector that is obtained.

***Example 4*** Van owners of the three most popular types of vans were interviewed. Of the Volkswagen owners, 70% indicated that they would buy another VW, 20% said they would buy a Ford van, and 10% said they would buy a Chevy van. Of the Ford van owners, 20% indicated that they would switch to a VW and another 20% said they would switch to a Chevy. Of the Chevy van owners, 40% would try a Ford next time and 10% would switch to a VW. What is the long-term trend for the van market?

The transition matrix $A$ is given by

$$\begin{array}{ll} & \quad\text{Future van} \\ & \begin{array}{ccc}\text{VW} & \text{Ford} & \text{Chevy}\end{array} \\ \text{Present van} \begin{array}{l}\text{VW}\\ \text{Ford}\\ \text{Chevy}\end{array} & \begin{bmatrix} 0.7 & 0.2 & 0.1 \\ 0.2 & 0.6 & 0.2 \\ 0.1 & 0.4 & 0.5 \end{bmatrix}, \end{array}$$

so the Markov chain is regular since all entries are positive. To find the steady-state vector $S$, assume that the long-run proportions of VW, Ford, and Chevy owners are $p$, $q$, and $1 - p - q$, respectively. Since $SA = S$, we have $S(A - I) = 0$, so that

$$[p \quad q \quad 1-p-q]\begin{bmatrix} -0.3 & 0.2 & 0.1 \\ 0.2 & -0.4 & 0.2 \\ 0.1 & 0.4 & -0.5 \end{bmatrix} = [0 \quad 0 \quad 0].$$

Hence, the first two entries of the product equal

$$-0.3p + 0.2q + 0.1(1 - p - q) = 0$$
$$0.2p - 0.4q + 0.4(1 - p - q) = 0,$$

so that

$$-0.4p + 0.1q = -0.1$$
$$-0.2p - 0.8q = -0.4.$$

Multiplying all entries by $-10$ yields the system

$$\begin{aligned} 4p - \phantom{8}q &= 1 \\ 2p + 8q &= 4. \end{aligned} \qquad (7.19)$$

Hence

$$\begin{bmatrix} 4 & -1 \\ 2 & 8 \end{bmatrix} \cdot \begin{bmatrix} p \\ q \end{bmatrix} = \begin{bmatrix} 1 \\ 4 \end{bmatrix}$$

and inverting the matrix by the methods shown in Section 4.4 yields

$$\frac{1}{34}\begin{bmatrix} 8 & 1 \\ -2 & 4 \end{bmatrix} \cdot \begin{bmatrix} 4 & -1 \\ 2 & 8 \end{bmatrix} \cdot \begin{bmatrix} p \\ q \end{bmatrix} = \frac{1}{34}\begin{bmatrix} 8 & 1 \\ -2 & 4 \end{bmatrix}\begin{bmatrix} 1 \\ 4 \end{bmatrix},$$

or

$$\begin{bmatrix} p \\ q \end{bmatrix} = \begin{bmatrix} 12/34 \\ 14/34 \end{bmatrix}.$$

Hence the eventual proportions of VW, Ford, and Chevy vans are 12/34, 14/34, and 8/34, respectively.

---

## EXERCISES 7.5

*In Exercises* 1–10, *determine if each given matrix is a regular transition matrix.*

**1.** $\begin{bmatrix} 1/2 & -1/2 \\ 1/4 & 3/4 \end{bmatrix}$

**2.** $\begin{bmatrix} 1/2 & 1/4 \\ 1/3 & 2/3 \end{bmatrix}$

**3.** $\begin{bmatrix} 1/3 & 2/3 \\ 0 & 1 \end{bmatrix}$

**4.** $\begin{bmatrix} 1/3 & 2/3 \\ 1 & 0 \end{bmatrix}$

**5.** $\begin{bmatrix} 1 & 0 & 0 \\ 1/2 & 1/4 & 1/2 \\ 0 & 0 & 1 \end{bmatrix}$

**6.** $\begin{bmatrix} 1 & 0 & 1 \\ 0 & 1 & 0 \\ 1/3 & 1/3 & 1/3 \end{bmatrix}$

**7.** $\begin{bmatrix} 0 & 1 & 0 \\ 1/4 & 1/2 & 1/4 \\ 0 & 1 & 0 \end{bmatrix}$

**8.** $\begin{bmatrix} 1/2 & 1/2 & 0 \\ 1/3 & 0 & 2/3 \\ 0 & 1/4 & 3/4 \end{bmatrix}$

**9.** $\begin{bmatrix} 1 & 0 & 0 & 0 \\ 0 & 0 & 1 & 0 \\ 1/4 & 1/4 & 1/4 & 1/4 \\ 1/2 & 1/2 & 0 & 0 \end{bmatrix}$

**10.** $\begin{bmatrix} 1/3 & 1/3 & 0 & 1/3 \\ 1/4 & 0 & 1/2 & 1/4 \\ 0 & 2/3 & 1/6 & 1/6 \\ 1/2 & 1/4 & 1/4 & 0 \end{bmatrix}$

**11.** The weather department in a certain city has noticed that if it is raining there is a 60% probability that it will rain the next day, and if it is sunny there is a 70% probability that it will be sunny the next day. Find the probability that five days from now it will be sunny if it is raining today.

**12.** In Exercise 11, find the probability that it will be sunny a year from today. (*Hint*: 365 days should be a sufficiently long time to approximate the steady-state vector extremely well.)

**13.** Yellowstone National Park officials relocate bears that make a habit of scavenging in campgrounds. The probability that a bear that has scavenged before will return to a campground is 80%. Wild bears are generally very shy around people, so only 10% of them will ever approach a campground. In the long run, what percentage of the bears are campground scavengers?

**14.** In Example 4 (see p. 350) suppose it is discovered that the percentage of VW owners who would buy a Ford is 30% and the percentage who would buy another VW is 60%, all other percentages remaining the same. What is the eventual proportion of owners of the three types of vans?

**15.** Bob makes all his purchases once a week. He never buys more than two six-packs of beer and never buys the same amount two weeks in a row. If he skips the beer one week, there is a 60% probability that he will buy two six-packs the next time. If he buys one six-pack there is a 30% probability that he will skip the beer the next time, but if he buys two six-packs the probability that he will skip the next time is 70%. Suppose initially he buys two six-packs of beer. What is the probability that four weeks later he will again buy two six-packs? In the long run, what is his most frequent purchase?

**16.** The weather department classifies days as being hot or cold and wet or dry. The following transition matrix has been found.

| | | Tomorrow | | | |
|---|---|---|---|---|---|
| | | Hot and dry | Hot and wet | Cold and dry | Cold and wet |
| Today | Hot and dry | 0.5 | 0.2 | 0.2 | 0.1 |
| | Hot and wet | 0.3 | 0.4 | 0.1 | 0.2 |
| | Cold and dry | 0.4 | 0 | 0.4 | 0.2 |
| | Cold and wet | 0 | 0.1 | 0.3 | 0.6 |

a) Show that this Markov chain is regular.
b) If today is cold and wet, what is the probability that three days from now it will be hot and dry?
c) Sunday dawns hot and dry. What is the probability that the next Sunday will also be hot and dry?
d) Find the long-run probabilities of the four events. Which is the most common?

**17.** A TV preference questionnaire is tabulated, and owners are asked to name the type of TV they own and indicate what kind they plan to purchase next time. The following data are compiled.

| | | Plan to purchase | | |
|---|---|---|---|---|
| | | RCA | Zenith | Other |
| Now own | RCA | 0.6 | 0.3 | 0.1 |
| | Zenith | 0.3 | 0.7 | 0 |
| | Other | 0.4 | 0.3 | 0.3 |

If the present percentages of RCA and Zenith owners are 30% and 20%, respectively, what is the eventual proportion of TV ownership?

**18.** Of voters sampled, 60% of those who voted Democratic in the last election intend to vote Democratic next time, 30% intend to vote Independent, and 10% will vote Republican. Of the Republicans, 80% will stay faithful and the other 20% will vote Independent. The Independents will lose 50% to the Democrats and 20% to the Republicans. What are the long-run percentages of voters in these three categories?

**19.** Show that the matrix

$$A = \begin{bmatrix} 0 & 1 & 0 \\ \frac{1}{3} & 0 & \frac{2}{3} \\ 0 & 1 & 0 \end{bmatrix}$$

is not regular but has a unique steady-state vector.

**20.** Compute $A^2$ using the matrix in Exercise 19. Does the resulting matrix have a unique steady-state vector?

---

## 7.6 GAME THEORY

Rutgers and Cornell are playing football against each other. Rutgers has the ball on first down and the quarterback must decide which of three plays to use: end run, quarterback sneak, or pass. Similarly, the captain of Cornell's defensive squad must decide which of three defenses to use: blitz, prevent, or normal defense. The blitz is particularly effective against quarterback sneaks and passes, the prevent defense is most useful against passes, and the normal defense is most effective against quarterback sneaks. On an average, the Rutgers team can expect to gain on each play the number of yards shown in the matrix below.

| | | Cornell | | |
|---|---|---|---|---|
| | | Blitz | Normal | Prevent |
| Rutgers | End run | 4 | 4 | 5 |
| | QB sneak | −2 | 3 | 5 |
| | Pass | −5 | 10 | 2 |

$$\begin{bmatrix} 4 & 4 & 5 \\ -2 & 3 & 5 \\ -5 & 10 & 2 \end{bmatrix}. \tag{7.20}$$

What plays should the Rutgers quarterback and Cornell defensive captain call?

We can use the matrix (7.20) and some of the methods of *game theory* to supply the proper play for each team. We will continue using this example in illustrating the principles of game theory.

We call (7.20) the *payoff matrix* for this *game*. Each of the three plays that the Rutgers quarterback can select is called a *pure strategy* for Rutgers. Similarly, the three defensive alignments Cornell can choose are pure strategies for Cornell. If we examine the rows of the payoff matrix, we may notice that every entry in the first row is at least as large as every entry in the second row. This means Rutgers can always make more yards (on the average) by trying an end run rather than a quarterback sneak. The quarterback should only call a quarterback sneak in the most unusual of circumstances. Thus, for all practical purposes Rutgers can ignore this pure strategy.

Similarly, on examining the columns of the payoff matrix we see that each entry in the first column is no more than the corresponding entry in the second column. Since the entries are yards *lost*, Cornell is better off blitzing than using the normal defense. Thus, Cornell can ignore that pure strategy. Deleting the second row and column in (7.20) yields a new payoff matrix:

| | | Cornell | |
|---|---|---|---|
| | | Blitz | Prevent |
| Rutgers | End run | 4 | 5 |
| | Pass | −5 | 2 |

$$\begin{bmatrix} 4 & 5 \\ -5 & 2 \end{bmatrix}, \tag{7.21}$$

The process of reducing the original payoff matrix by eliminating certain pure strategies to obtain (7.21) is called *dominance*. Row 1 is said to *dominate* row 2, while column 1 dominates column 2. Note the following.

1. Row $A$ dominates row $B$ if all entries in row $A$ are *greater than or equal* to the entries in row $B$.
2. Column $A$ dominates column $B$ if all entries in column $A$ are *less than or equal* to the entries in column $B$.

Now applying dominance to the payoff matrix (7.21) we see that row 1 dominates row 2, while column 1 dominates column 2. Hence (7.21) reduces to the pure strategies

| | Cornell<br>Blitz |
|---|---|
| Rutgers End run | (4) |

Therefore Rutgers should try an end run and Cornell should blitz. The payoff that Rutgers obtains from Cornell as a result of this optimal pair of pure strategies is called the *value* of the game and is denoted by the letter $v$. In this case, $v = 4$ (yards).

Not all games can be reduced to a single pair of pure strategies by the concept of dominance. For those that can be reduced, however, there is a simple and effective way of finding this solution.

**Saddle Point Rule**

*Step* 1. To the right of each row write the smallest value in that row (called the row minimum). Then take the maximum $R$ of these numbers.

*Step* 2. Below each column write the largest value in that column (called the column maximum). Then take the minimum $C$ of these numbers.

*Step* 3. If $R = C$, then each of the pure strategies corresponding to these values is optimal for each player. The entry (or entries) in the payoff matrix corresponding to these strategies is called a *saddle point*. Furthermore, the value of the game $v = R = C$.

If the saddle point rule yields a single saddle point, this entry is the value of the game, and the payoff matrix can be reduced to this single entry.

***Example 1*** Let's use the saddle point rule to find the solution of the payoff matrix (7.20):

$$\begin{bmatrix} \textcircled{4} & 4 & 5 \\ -2 & 3 & 5 \\ -5 & 10 & 2 \end{bmatrix} \quad \begin{matrix} \text{Row minimum} \\ 4 \\ -2 \\ -5 \end{matrix} \quad \text{Maximum } R = 4$$

$$\text{Column maximum} \quad 4 \quad 10 \quad 5$$

$$\text{Minimum } C = 4$$

Hence $v = R = C = 4$ and the circled entry is the saddle point for this game.

---

Some payoff matrices have no saddle point.

***Example 2*** Consider the following payoff matrix.

$$\begin{bmatrix} 2 & 3 \\ 4 & -1 \end{bmatrix} \quad \begin{matrix} \text{Row minimum} \\ 2 \\ -1 \end{matrix} \quad \text{Maximum } R = 2$$

$$\text{Column maximum} \quad 4 \quad 3$$

$$\text{Minimum } C = 3$$

Hence $R \neq C$ and there is no saddle point. We will discuss later what can be done in situations of this type.

---

Some payoff matrices have more than one saddle point.

***Example 3*** Consider the following matrix.

$$\begin{bmatrix} \textcircled{4} & 5 & \textcircled{4} \\ 1 & 2 & -2 \\ 4 & 6 & 3 \end{bmatrix} \quad \begin{matrix} \text{Row minimum} \\ 4 \\ -2 \\ 3 \end{matrix} \quad \text{Maximum } R = 4$$

$$\text{Column maximum} \quad 4 \quad 6 \quad 4$$

$$\text{Minimum } C = 4$$

Here $v = R = C = 4$, but the value 4 corresponds to the first row and *both* column 1 and column 3. Thus the two circled entries are saddle points. In this case, the row player should always select the pure strategy of row 1, while the column player can select either of the pure strategies of column 1 and column 3.

To understand why the saddle point rule yields the optimal solution(s) to games with saddle points, note that each row minimum is the worst that the row player can do by playing that row. Thus, by taking the maximum of these row minimums the row player obtains the largest possible payoff against the column player's best countermoves. Similarly, the column player makes the smallest possible payoff against the row player's best countermoves.

---

We now turn to games for which there are no saddle points. In these games a *mixed strategy* is required as the following example illustrates.

***Example 4*** On 4 January 1912 the antarctic explorer Robert Falcon Scott had to make an extremely difficult decision. Because the motor sledges he had planned to use in his trip to the south pole had lasted only one week, and only ten of the nineteen Manchurian ponies had survived the trip, he and his party found themselves 150 miles from the pole. Their supplies were just enough to take a five-man team to the pole and back to safety if the weather remained good. A long storm after they reached the pole, however, would prevent them from ever reaching their base at One Ton Depot. The decision then was whether to continue to the pole chancing death, or to turn back in bitter disappointment:

| | | Future weather | |
|---|---|---|---|
| | | Good | Stormy |
| Scott's decision | Turn back | Bitter disappointment in rest of life | Satisfaction over saving lives |
| | Keep going | Personal glory | Death for entire group |

To use game theory, we must assign a numerical value to each entry in the payoff matrix. Clearly, personal glory and satisfaction over saving lives are favorable outcomes, hence must be assigned positive values. The other two events are unfavorable for Scott and should be assigned negative values. But the actual amounts to be assigned depend on the individual making the decision and his or her assessment of their worth. Each individual may assign different values to each entry; this is often the case when attempting to solve social and political problems. Once the entries in the payoff matrix have been agreed on, it is possible to use game theory to analyze the problem. Suppose in this case that Scott had chosen the following values:

$$\begin{array}{cc} & \text{Future weather} \\ & \begin{array}{cc}\text{Good} & \text{Stormy}\end{array} \\ \text{Scott's decision}\quad \begin{array}{l}\text{Turn back}\\ \text{Keep going}\end{array} & \begin{bmatrix} -10 & 5 \\ 40 & -100 \end{bmatrix}. \end{array} \tag{7.22}$$

Let's examine the payoff matrix (7.22) to see if it has a saddle point:

$$\begin{array}{lcc|c} & & & \text{Row minimum} \\ & \begin{bmatrix} -10 & 5 \\ 40 & -100 \end{bmatrix} & & \begin{array}{c} -10 \\ -100 \end{array} \quad \text{Maximum } R = -10 \\ \text{Column maximum} & 40 \quad\quad 5 & & \end{array}$$

Minimum $C = 5$

Since $C \neq R$ there is no saddle point, so we will have to design a *mixed strategy* for this game. By this we mean that each pure strategy is assigned a certain probability of being selected in such a way that the largest possible expected payoff is obtained regardless of the mixed strategy the opponent may select. In this example Scott had no idea what nature was planning to do. Instead, probabilities should be assigned to each option ("turn back" or "keep going") that maximize the *expected value* of the game regardless of nature's choice. The problem, now, is to decide how to assign the probabilities. Let's do several calculations to get the general idea and see how we might optimize the expected value. Suppose we assign each option a 50% probability, and suppose that nature will select good weather with probability $q = 0.3$. (In general, $q$ is unknown.) Then the probability of occurrence of each of the four possible events ("turn back and good weather," "turn back and stormy weather," "keep going and good weather," "keep going and stormy weather") is shown in the matrix

$$\begin{array}{l} \\ \text{Turn back} \\ \text{Keep going} \end{array} \begin{array}{c} \text{Good weather} \quad \text{Stormy weather} \\ \begin{bmatrix} 0.5(0.3) & \quad 0.5(0.7) \\ 0.5(0.3) & \quad 0.5(0.7) \end{bmatrix}. \end{array}$$

The *expected value* of the matrix game is obtained by multiplying each of these probabilities by the corresponding entry in the payoff matrix (7.22):

$$\begin{aligned} E[x] &= (-10)(0.5)(0.3) + (5)(0.5)(0.7) \\ &\quad + (40)(0.5)(0.3) + (-100)(0.5)(0.7) \\ &= -1.5 + 1.75 + 6 - 35 = -28.75. \end{aligned}$$

On the other hand, suppose we assign "turning back" an 80% probability. If nature again selects good weather with probability $q = 0.3$, we have the matrix of probabilities

$$\begin{array}{l} \\ \text{Turn back} \\ \text{Keep going} \end{array} \begin{array}{c} \text{Good weather} \quad \text{Stormy weather} \\ \begin{bmatrix} 0.8(0.3) & \quad 0.8(0.7) \\ 0.2(0.3) & \quad 0.2(0.7) \end{bmatrix}. \end{array}$$

Hence the expected value for this mixed strategy is

$$E[x] = (-10)(0.24) + 5(0.56) + 40(0.06) + (-100)(0.14) = -11.2.$$

Thus the second strategy is better than the first. Of course if nature's probability of having good weather is other than $q = 0.3$, the first strategy might be better. Since we cannot try all possible mixed strategies, we need a method for finding the optimal strategy. For $2 \times 2$ matrices there is a simple formula.

**MIXED STRATEGY RULE**

> **Suppose the payoff matrix**
>
> $$\begin{bmatrix} a & b \\ c & d \end{bmatrix}, \qquad a + d - b - c \neq 0$$
>
> **has no saddle points. Then the optimal strategy for the row player is to play the first row with probability**
>
> $$p = \frac{d - c}{a + d - b - c}.$$
>
> **Similarly, the optimal strategy for the column player is to play the first column with probability**
>
> $$q = \frac{d - b}{a + d - b - c}.$$

If we apply the mixed strategy rule to Scott's decision, we find that the optimal strategy is to turn back with probability

$$p = \frac{-100 - 40}{-10 - 100 - 40 - 5} = \frac{-140}{-155} = \frac{28}{31} \approx 0.9032$$

and keep going with probability

$$1 - p = \frac{3}{31} \approx 0.0968.$$

Thus, the best strategy suggests turning back over 90% of the time!

If nature could optimize its strategy, that is, if nature wished to be perverse, then good weather would occur with probability

$$q = \frac{-100 - 5}{-10 - 100 - 40 - 5} = \frac{-105}{-155} = \frac{21}{31} \approx 0.6774,$$

while bad weather would occur with probability $1 - q \approx 0.3226$. Of course, nature is unlikely to be playing games, but the probabilities $q$ and $1 - q$ yield the least expected value for this situation. The expected value is

$$E[x] = (-10)\frac{28}{31}\left(\frac{21}{31}\right) + 5\left(\frac{28}{31}\right)\frac{10}{31} + 40\left(\frac{3}{31}\right)\frac{21}{31} + (-100)\frac{3}{31}\left(\frac{10}{31}\right)$$

$$= \frac{-5880 + 1400 + 2520 - 3000}{961} = \frac{-4960}{961} \approx -5.1613,$$

and represents the most that Scott can expect from this situation if the weather acts perversely.

As we have seen, the odds favor turning back almost 10 to 1 over going on. Scott chose to go on, trusting in luck and the courage and endurance of his men. On 21 March 1912 a continuous gale trapped them eleven miles away from One Ton Depot. The gale continued day after day. Eventually cold and lack of food ended their ordeal. Scott's last entry in his diary on 29 March 1912 ends with the words "For God's sake look after our people."

***Example 5*** Joe, our enterprising gambler, has a new game for his favorite "pigeon," Bob:

"Bob, let's play 'One-or-two-fingers.' In this game each of us holds out one or two fingers simultaneously. If the number of fingers is even, I pay you that amount in dollars. If it is odd you pay me that amount in dollars!"

"Oh, Joe, not another one of these games!"

"But Bob, this one is fair. The most I can win is $3 and the average of what you can win is also $3."

Is Bob being taken?

Joe's payoff matrix for this game is

$$\begin{array}{ccc} & & \text{Bob's play} \\ & & \begin{array}{cc}\text{One} & \text{Two}\end{array} \\ \text{Joe's play} & \begin{array}{c}\text{One}\\ \text{Two}\end{array} & \begin{bmatrix} -2 & 3 \\ 3 & -4 \end{bmatrix}. \end{array}$$

The game has no saddle points since $R = -2$ and $C = 3$, so a mixed strategy is called for. Joe should play one finger with probability

$$p = \frac{d - c}{a + d - b - c} = \frac{-4 - 3}{-2 - 4 - 3 - 3} = \frac{7}{12},$$

and two fingers with probability $1 - p = 5/12$. Bob almost surely does not know his best strategy, but in this game it doesn't matter. Regardless of how he plays Joe is always the winner! Suppose Bob plays one finger with probability $p$. Then, Joe's expected value is

$$\begin{aligned} E[x] &= (-2)\frac{7}{12}(p) + 3\left(\frac{7}{12}\right)(1 - p) + 3\left(\frac{5}{12}\right)p + (-4)\frac{5}{12}(1 - p) \\ &= \frac{-14p + 21 - 21p + 15p - 20 + 20p}{12} = \frac{1}{12}. \end{aligned}$$

Hence Joe will be $1 ahead on the average every 12 games, regardless of what Bob might do. This is a better return for him than playing roulette or Chuck-a-Luck.

***Example 6*** Consider the following game: Richard and Charles each select a face of a die. If the sum of the spots is 2, 3, 4, 6, or 7, Charles pays Richard \$1. Otherwise Richard pays Charles \$1. Is this game *fair*, that is, is the expected value of this game equal to zero?

This time the payoff matrix is $6 \times 6$:

$$\text{Richard plays}\quad \begin{array}{c} \\ 1\\2\\3\\4\\5\\6 \end{array}\overset{\text{Charles plays}}{\begin{array}{cccccc} 1 & 2 & 3 & 4 & 5 & 6 \\ \end{array}} \begin{bmatrix} 1 & 1 & 1 & -1 & 1 & 1 \\ 1 & 1 & -1 & 1 & 1 & -1 \\ 1 & -1 & 1 & 1 & -1 & -1 \\ -1 & 1 & 1 & -1 & -1 & -1 \\ 1 & 1 & -1 & -1 & -1 & -1 \\ 1 & -1 & -1 & -1 & -1 & -1 \end{bmatrix}.$$

By column dominance we see that column 6 dominates columns 1, 2, 3, and 5, which reduces the payoff matrix to

$$\text{Richard plays}\quad \begin{array}{c} 1\\2\\3\\4\\5\\6 \end{array} \overset{\begin{array}{c}\text{Charles plays}\\ 4 \quad 6\end{array}}{\begin{bmatrix} -1 & 1 \\ 1 & -1 \\ 1 & -1 \\ -1 & -1 \\ -1 & -1 \\ -1 & -1 \end{bmatrix}}.$$

But now 2 dominates rows 3, 4, 5, and 6. Actually rows 2 and 3 are identical so either can be canceled. We have reduced the payoff matrix to the $2 \times 2$ matrix:

| | Charles plays 4 | 6 | Row minimum | |
|---|---|---|---|---|
| Richard plays 1 | $-1$ | 1 | $-1$ | Maximum $R = -1$ |
| 2 | 1 | $-1$ | $-1$ | |
| Column maximum | 1 | 1 | | |
| | Minimum $C = 1$ | | | |

Since there is no saddle point we use a mixed strategy. Richard should play a 1 with probability

$$p = \frac{-1 - 1}{-1 - 1 - 1 - 1} = \frac{2}{4} = 0.5.$$

Similarly, the best strategy for Charles is to play a 4 with probability

$$q = \frac{-1 - 1}{-1 - 1 - 1 - 1} = \frac{2}{4} = 0.5.$$

The expected value of the game is

$$E[x] = -1(0.5)(0.5) + 1(0.5)(0.5) + 1(0.5)(0.5) + (-1)(0.5)(0.5) = 0.$$

So the game is fair.

---

### EXERCISES 7.6

*In Exercises 1–12, use dominance to reduce the matrix game as much as possible. Find the saddle point (if any) or the mixed strategy if no saddle points exist.*

**1.** $\begin{bmatrix} 5 & 4 & 6 \\ 1 & 2 & 2 \\ -1 & 3 & 0 \end{bmatrix}$ **2.** $\begin{bmatrix} 4 & 2 & 3 \\ 6 & 4 & 3 \\ -2 & 5 & 1 \end{bmatrix}$ **3.** $\begin{bmatrix} 3 & 4 & 2 \\ 5 & 3 & 2 \\ 3 & 2 & 1 \end{bmatrix}$

**4.** $\begin{bmatrix} 1 & 2 & 9 \\ 4 & 3 & 8 \\ 5 & 6 & 7 \end{bmatrix}$ **5.** $\begin{bmatrix} -2 & 1 & -2 \\ 5 & -3 & 4 \\ 4 & -4 & 3 \end{bmatrix}$ **6.** $\begin{bmatrix} -2 & 3 & 4 \\ 2 & -7 & 0 \\ 3 & -4 & 1 \end{bmatrix}$

**7.** $\begin{bmatrix} 4 & 2 & 2 & 4 \\ 2 & 1 & 6 & 0 \\ 1 & 2 & 7 & -1 \\ 5 & 0 & 1 & 5 \end{bmatrix}$ **8.** $\begin{bmatrix} 4 & -8 & -2 & -1 \\ -1 & 2 & 2 & 1 \\ 6 & 4 & 5 & 3 \\ 5 & 2 & 3 & 2 \end{bmatrix}$ **9.** $\begin{bmatrix} 4 & 1 & 5 & 1 \\ -1 & -2 & 0 & -1 \\ 1 & 0 & 1 & -3 \\ 3 & 1 & 2 & 1 \end{bmatrix}$

**10.** $\begin{bmatrix} 4 & -1 & 2 & 7 \\ 0 & -1 & -2 & 5 \\ 2 & -1 & 3 & -3 \\ 6 & -1 & 9 & 8 \end{bmatrix}$ **11.** $\begin{bmatrix} -1 & 2 & -6 & 0 \\ 5 & -2 & 4 & 5 \\ 1 & 3 & -5 & 2 \\ 4 & -3 & 3 & 4 \end{bmatrix}$ **12.** $\begin{bmatrix} 4 & 3 & 3 & -8 \\ -2 & 0 & -3 & 5 \\ 0 & 2 & -4 & 4 \\ 4 & 5 & 4 & -6 \end{bmatrix}$

**13.** Suppose Rutgers plays Colgate with the payoff matrix

| | | Colgate | | |
|---|---|---|---|---|
| | | Blitz | Normal | Prevent |
| | End run | 7 | 4 | 4 |
| Rutgers | QB sneak | −5 | 2 | 4 |
| | Pass | −2 | 8 | 5 |

on offense. What offensive and defensive strategies would you select?

**14.** Suppose the payoff matrix in Exercise 13 is

$$\begin{bmatrix} 6 & 3 & 8 \\ -3 & 2 & 5 \\ -5 & 10 & 0 \end{bmatrix}.$$

What strategies would you recommend in this case?

**15.** Let Rutgers' offensive options be given against Cornell in the payoff matrix below. What strategy would you recommend?

| | | Cornell | | |
|---|---|---|---|---|
| | | Blitz | Normal | Prevent |
| Rutgers | End run | 5 | 3 | 7 |
| | QB sneak | −2 | 2 | 4 |
| | Short pass | −5 | 12 | 8 |
| | Long pass | −8 | 10 | 5 |

**16.** Solve Exercise 15 for the payoff matrix

$$\begin{bmatrix} 6 & 7 & 6 \\ -2 & 5 & 4 \\ 6 & 12 & 6 \\ -5 & 10 & 6 \end{bmatrix}.$$

**17.** Joe manages to sucker Bob into playing Two-or-Three Fingers. They agree that Bob pays Joe whenever the number of fingers is odd and Joe pays Bob whenever it is even. The amount paid in dollars is the total number of fingers shown. What is the expected value of this game?

**18.** Joe and Bob are playing One-or-Two Fingers but the payoff is one dollar more than the number of fingers shown. What is the expected value of this game?

**19.** Richard and Charles are playing the same dice game given in Example 6 except that Richard wins one dollar when the total is 3, 4, 5, or 6, and Charles wins one dollar when the total is 8, 9, 10, or 11. The numbers 2, 7, and 12 are ties and nobody wins. What is the best strategy that Richard can use? That Charles can use?

**20.** In Example 5 in Section 7.2 we discussed a strategy optimizing the yield on a tomato crop in western Montana. Suppose no information is known about the weather and a farmer must decide whether to pick the crop on September 1 or September 8. If the crop is picked on September 1 the farmer receives \$840 per acre. If it is picked on September 8, the earnings are \$960 per acre if it doesn't frost and \$480 per acre if it does. What strategy should the farmer use?

**21.*** Christopher Columbus faced a hard decision on his first voyage. His crew, threatening mutiny, demanded that he turn back to Spain before he dropped off the end of what they thought was a flat earth. Columbus knew the world was round, but feared that mutiny would occur if land was not

---

* This problem is based on an article by Leonid Hurwicz in *Mathematical Thinking in Behavioral Sciences* (San Francisco: Freeman, 1968).

sighted very soon. The payoff matrix has the following form:

| | | Location of land | |
|---|---|---|---|
| | | Near | Far |
| Columbus's decision | Turn back | Bitter disappointment | Satisfaction of preventing mutiny |
| | Keep going | Fame and fortune | Mutiny and death for Columbus |

Assign to the entries those values that *you* feel they deserve and find the optimal strategy. Would you have discovered America?

**22.** At 6 P.M. on 18 June 1815 the battle of Waterloo was at its height, and the Duke of Wellington had committed the last reserves of his "infamous army" of British, Dutch, Belgian, and German recruits. Marshal Ney of France, Napoleon's "bravest of the brave," had seized La Haye Sainte and had backed Wellington to his last line of defense. The French were massing to crush the English before the Prussians under General Bülow could arrive to help Wellington. The choices were to retreat or hold. If the Prussians got there before the French attack, Wellington believed he would be victorious. If they did not arrive in time, he would be defeated. His payoff matrix was

| | | Prussians | |
|---|---|---|---|
| | | In time | Not in time |
| Wellington's decision | Retreat | Napoleon traps Prussians | Battle a draw |
| | Hold | Napoleon defeated | Wellington defeated |

Assign entries to the matrix as if you were Wellington and find the optimal strategy. Would you have won the battle of Waterloo?

---

## CHAPTER VOCABULARY

**Bernoulli experiment**
**Bernoulli trial**
**binomial distribution** $C_{n,k}p^k(1-p)^{n-k}$
**confidence interval**
**continuous distribution**
**cumulative frequency**
**cumulative frequency polygon**
**discrete distribution**
**dominance (row or column)**
**expected value** $E(x)$
**expected value of a game**
**fair game**
**frequency**
**frequency polygon**
**game theory**
**grouped cumulative frequency**
**grouped data**
**grouped frequency**
**histogram**
**line chart**

**Markov chain**
**Markov process**
**median (of a sample)**
**mixed strategy**
**mixed strategy rule**
**normal distribution**
**payoff matrix**
**population mean** $\mu$
**population variance** $\sigma^2$
**pure strategy**
**random sample**
**range (of a sample)**
**regular Markov chain**
**saddle point**
**saddle point rule**
**sample**
**sample mean** $\bar{x}$
**sample space**
**sample variance** $s^2$
**standard deviation** $s$
**standard normal distribution**
**state (of the process)**
**state vector**
**steady-state vector**
**transition matrix**

# PART III CALCULUS

Two questions often arise in connection with an arithmetic progression:

1. What is the $n$th term of the progression?
2. What is the sum of the first $n$ terms of the progression?

If $n$ is small we can easily obtain the $n$th term, and add the terms together. But if $n$ is large this method is too time-consuming. We shall determine two formulas that give the answer to these two questions.

Suppose the sequence

$$x_1, x_2, x_3, \ldots, x_n \tag{8.4}$$

is an arithmetic progression with difference $d$. Since $x_2$ is obtained by adding $d$ to $x_1$, and $x_3$ is obtained by adding $d$ to $x_2$ we have that

$$x_3 = x_2 + d = (x_1 + d) + d$$

or

$$x_3 = x_1 + 2d.$$

Similarly, $x_4$ is obtained by adding $d$ to $x_3$, or

$$x_4 = x_3 + d = (x_1 + 2d) + d = x_1 + 3d.$$

In general, we will have to add $d$ to $x_1$ $(n-1)$ times to get $x_n$:

$$x_n = x_1 + (n - 1)d. \tag{8.5}$$

***Example 1*** What is the 999th term of the arithmetic progression (8.1)?

The first term of (8.1) is $x_1 = 1$ and the difference $d = 2$. Hence, the 999th term is

$$x_{999} = x_1 + (999 - 1)d = 1 + (998)2 = 1997.$$

---

A practical application of arithmetic progressions is in the use of *simple-interest accounts,* in which the amount in the account is increased each year by a fixed percentage of the original amount deposited. The original deposit is usually called the *principal,* while the fixed percentage used in determining the increase is referred to as the (simple) *interest.*

***Example 2*** Suppose \$1000 is deposited in an account that pays 6% simple interest. How much money will be in the account after 8 years?

Let $x_n$ be the amount in the account at the end of $n$ years. Since the account pays *simple interest*, the amount at the end of any year is obtained by adding 6% of the principal $P = 1000$ to the amount at the end of the previous year. Thus, the difference in the amounts in two consecutive years is always $d = 0.06(1000) = \$60$, so that the sequence

$$x_0, x_1, x_2, \ldots, x_n$$

is an arithmetic progression, where $P = x_0 = \$1000$ is the initial amount in the account. Now $x_1$ differs from the principal $x_0$ by \$60, so that

$$x_1 = x_0 + 60 = \$1060.$$

Similarly,

$$x_2 = x_1 + 60 = x_0 + 2(60) = \$1120;$$

adding \$60 eight times to $x_0$ we have

$$x_8 = x_0 + 8(60) = \$1480.$$

---

In general, if a principal $P = x_0$ has been deposited into an account paying $i\%$ simple interest annually, then the amount $A$ in the account at the end of $n$ years is

$$A = x_n = x_0 + n(x_0 i) = P + n(Pi),$$

or

$$A = P(1 + ni). \tag{8.6}$$

Before finding the sum of the first $n$ terms of *any* arithmetic progression, it is useful to find this sum for the arithmetic progression

$$1, 2, 3, \ldots, n - 2, n - 1, n. \tag{8.7}$$

This is easily done by adding the terms from two copies of this progression, one written in reverse order, as shown below:

$$\begin{aligned} 1 + n &= n + 1 \\ 2 + (n - 1) &= n + 1 \\ 3 + (n - 2) &= n + 1 \\ &\vdots \\ n + 1 &= n + 1. \end{aligned}$$

Thus, we have $n$ sums, each equal to $(n + 1)$ so that *twice* the sum of this sequence equals

$$2S_n = n(n + 1).$$

Hence, the sum of the $n$ terms of the sequence (8.7) is

$$S_n = 1 + 2 + 3 + 4 + \cdots + (n - 1) + n = \frac{n(n + 1)}{2}. \tag{8.8}$$

***Example 3*** What is the sum of all the integers from 27 to 98?

We subtract the sum of the integers from 1 to 26 from the sum from 1 to 98. The sum from 1 to 98 equals, by Eq. (8.8),

$$S_{98} = \frac{98(98 + 1)}{2} = 4851,$$

while the sum from 1 to 26 equals

$$S_{26} = \frac{26(26 + 1)}{2} = 351.$$

Therefore, the sum of the integers from 27 to 98 is equal to 4851 − 351 = 4500.

---

We now add the first $n$ terms of an arbitrary arithmetic progression with difference $d$:

$$S_n = x_1 + x_2 + x_3 + \cdots + x_n.$$

Again adding terms from two copies of this sum, one written in reverse order, we have, by Eq. (8.5),

$$x_1 + x_n = x_1 + x_1 + (n - 1)d = 2x_1 + (n - 1)d,$$

$$x_2 + x_{n-1} = x_1 + d + x_1 + (n - 2)d = 2x_1 + (n - 1)d,$$

$$\vdots$$

so that

$$2S_n = n[2x_1 + (n - 1)d], \tag{8.9}$$

or

$$S_n = n\left[x_1 + \frac{(n-1)}{2}d\right]. \tag{8.10}$$

Note also that since the term in brackets in Eq. (8.9) equals $x_1 + x_n$, we also have

$$2S_n = n[x_1 + x_n],$$

so that

$$S_n = n\left[\frac{x_1 + x_n}{2}\right]. \qquad \textbf{(8.11)}$$

In other words, the sum is $n$ times the average of the first and last terms.

***Example 4*** Find the sum of the arithmetic progression

$$4 + 7 + 10 + 13 + \cdots + 94{,}123.$$

Regardless of which formula we use, we must first find the number of terms in the progression. Here $x_1 = 4$, $d = 3$, and $x_n = 94{,}123$, so by Eq. (8.5)

$$94{,}123 = 4 + (n - 1)3.$$

Subtracting 4 from both sides and dividing by 3, we have

$$n - 1 = (1/3)[94123 - 4] = 31{,}373,$$

so that $n = 31{,}374$. Now, replacing $x_1$, $x_n$, and $n$ in Eq. (8.11) we obtain

$$S_n = 31{,}374\left[\frac{4 + 94123}{2}\right] = 1{,}476{,}570{,}249.$$

**Geometric Progressions** A geometric progression is a sequence of numbers in which each succeeding number is obtained by *multiplying* the preceding number by a fixed quantity $r$, called the *ratio*.

This rule is represented by the formula

$$x_{n+1} = x_n r, \qquad \textbf{(8.12)}$$

where $x_n$ is the $n$th term of the sequence. We notice that the sequence (8.2) is a geometric progression with ratio $r = 2$.

The same questions we asked for arithmetic progression may again be posed concerning geometric progressions. We can obtain the $n$th term in a geometric progression if we know the first term $x_1$ and the ratio $r$, by using the formula

$$x_n = x_1 r^{n-1}. \tag{8.13}$$

This formula is easily obtained by applying Eq. (8.12) repeatedly:

$$x_n = x_{n-1}r = (x_{n-2}r)r = x_{n-2}r^2 = (x_{n-3}r)r^2 = x_{n-3}r^3 = \cdots = x_1 r^{n-1}.$$

***Example 5*** Find the 32nd term in a geometric progression having the first term $x_1 = 0.01$ and ratio $r = 2$.

$$x_{32} = x_1 r^{31} = 0.01(2)^{31} = 21{,}474{,}836.48$$

---

Compound interest is a useful application of geometric progressions. In a compound-interest account the amount is increased each year by a fixed percentage (called the compound *interest*) of the total amount then in the account.

***Example 6*** Suppose \$1000 is deposited in a bank that pays 6% compound interest annually. How much is in the account after 5 years?

Let $x_n$ be the amount in the account after $n$ years. We can let the *principal* $P = x_0 = \$1000$, since this is the initial amount in the account. The amount $x_n$ is determined by adding 6% of $x_{n-1}$ to $x_{n-1}$, that is,

$$x_n = x_{n-1} + 0.06x_{n-1} = x_{n-1}(1.06).$$

Thus, each succeeding year's amount is determined by the ratio $r = 1.06$, yielding a geometric progression

$$x_0, x_1, x_2, \ldots, x_n.$$

Since $x_1 = x_0 r$, we may replace $x_1$ in Eq. (8.13), getting

$$x_n = x_0 r^n. \tag{8.14}$$

Then

$$x_5 = x_0 r^5 = 1000(1.06)^5 = \$1338.23.$$

---

We can restate Eq. (8.14) in financial terms by noting that the ratio $r$ is simply 1 plus the interest (expressed as a decimal).

If the principal $P$ is deposited in an account that pays $i$% interest compounded annually, then the amount $A$ in the account at the end of $n$ years is

$$A = P(1 + i)^n. \tag{8.15}$$

We now turn to the problem of finding the sum of the first $n$ terms of a geometric progression. Let the sum be given by

$$S_n = x_1 + x_2 + x_3 + \cdots + x_n.$$

Multiplying each term by $r$ and subtracting the result from $S_n$, we have

$$S_n - rS_n = x_1 + (x_2 - x_1 r) + (x_3 - x_2 r) + \cdots + (x_n - x_{n-1} r) - x_n r.$$

By Eq. (8.12) the terms in parentheses all cancel, leaving

$$S_n(1 - r) = x_1 - x_n r,$$

or

$$S_n = \frac{x_1 - x_n r}{1 - r}. \qquad \textbf{(8.16)}$$

If we replace $x_n$ in Eq. (8.16) by the value given in Eq. (8.13), we also have

$$S_n = x_1\left(\frac{1 - r^n}{1 - r}\right) = x_1\left(\frac{r^n - 1}{r - 1}\right). \qquad \textbf{(8.17)}$$

***Example 7*** Find the sum of the first 20 terms of the geometric progression

$$3, 6, 12, 24, \ldots.$$

In this case $x_1 = 3$, $r = 2$, and $n = 20$. By Eq. (8.17) the sum of the first 20 terms equals

$$S_{20} = 3\left(\frac{2^{20} - 1}{2 - 1}\right) = 3(2^{20} - 1) = 3{,}145{,}725.$$

---

A practical application of Eq. (8.17) is given in the following example.

***Example 8*** Suppose \$1000 is deposited each year in an annuity that pays 6% interest annually on all previous deposits and interest. What will be the amount in the annuity at the end of 20 years?

To solve this problem we must understand that the \$1000 deposited the first year has compounded at 6% for 19 years, while the \$1000 deposited the second year has compounded at 6% for 18 years, and so on.

The amount deposited during the 20th year earns no interest. Thus we must add each of these amounts to get the value of the annuity, that is,

$$S_{20} = 1000 + 1000(1.06) + \cdots + 1000(1.06)^{18} + 1000(1.06)^{19}.$$

Thus $x_1 = 1000$, $r = 1.06$, and $n = 20$, so using Eq. (8.17) we have

$$S_{20} = 1000\left(\frac{(1.06)^{20} - 1}{1.06 - 1}\right) = 1000\left(\frac{2.20714}{0.06}\right) = \$36{,}785.59.$$

***Example 9*** Certain diseases, such as *retinoblastoma* (a kind of cancer of the eye),* depend on the mutation of a normal allele $a$ into a dominant gene $A$. Suppose such a mutation occurs in one of every 100,000 births in each generation, and that back mutations ($A$ into $a$) never occur. Since the gene $A$ is very rare, we can neglect those individuals with genotype $AA$, and assume that all affected individuals have genotype $Aa$. Suppose, further, that the disease does not affect the victim until after his or her reproductive years. Then only half the children of an affected individual will inherit the gene $A$ since the mate will generally be unaffected due to the rarity of the disease. We wish to find the rate of occurrence of the disease in the $n$th generation starting with zero inherited cases in an original generation, which we number the 0th generation.

The rate of occurrence of the disease in the first generation is $m = 1/100{,}000$, since the disease is entirely due to mutation. However, for the second generation we must add the inherited rate to the mutation rate $m$. The *inherited rate* is obtained by multiplying the rate of occurrence of the previous generation by $r = 1/2$, since only half of the children of an affected individual will inherit the disease. Thus, the rate of occurrence for the second generation is

$$m + mr = \frac{3}{200{,}000}.$$

Similarly, the rate of occurrence in the third generation is obtained by adding half the rate of occurrence of the second generation (inherited rate) to the mutation rate:

$$m + (m + mr)r = m + mr + mr^2 = \frac{7}{400{,}000}.$$

Hence, the rate of occurrence for the $n$th generation will be

$$m + mr + mr^2 + \cdots + mr^{n-1} = m\left(\frac{1 - r^n}{1 - r}\right).$$

---

* See J. V. Neel and W. J. Schull, *Human Heredity*, 3rd ed. (Chicago: University of Chicago Press, 1958, p. 333).

according to Eq. (8.17). For example, after ten generations the rate of occurrence is

$$\frac{1}{100{,}000}\left(\frac{1-(\frac{1}{2})^{10}}{1-\frac{1}{2}}\right)=1.99805\times 10^{-5},$$

almost double the mutation rate.

---

## EXERCISES 8.1

*In Exercises 1–8, find the indicated term of the arithmetic progression with first term $x_1$ and difference d.*

**1.** $x_{37}; x_1 = 1, d = 3$

**2.** $x_{29}; x_1 = 1, d = -1/5$

**3.** $x_{41}; x_1 = 3, d = 5$

**4.** $x_{62}; x_1 = 4, d = 3$

**5.** $x_{801}; x_1 = 18, d = -2/3$

**6.** $x_{123}; x_1 = -85, d = 1.1$

**7.** $x_{89}; x_1 = 5/7, d = 1/3$

**8.** $x_{101}; x_1 = 1.8, d = -0.34$

*In Exercises 9–14, find the sum of the given arithmetic progression.* (*Hint: Determine the number n of terms first.*)

**9.** $3 + 4 + 5 + \cdots + 187$

**10.** $1 + 3 + 5 + 7 + \cdots + 999$

**11.** $1 + 2 + 3 + \cdots + 1000$

**12.** $2 + 4 + 6 + \cdots + 10{,}000$

**13.** $3 + 8 + 13 + \cdots + 943$

**14.** $\frac{2}{7} + \frac{5}{7} + \frac{8}{7} + \cdots + \frac{842}{7}$

*In Exercises 15–20, find the indicated term of the geometric progression with first term $x_1$ and ratio r.*

**15.** $x_{15}; x_1 = 1, r = 2$

**16.** $x_9; x_1 = 4, r = -1$

**17.** $x_{12}; x_1 = 800, r = \frac{1}{2}$

**18.** $x_8; x_1 = 9, r = 0.1$

**19.** $x_{20}; x_1 = 4, r = 1.01$

**20.** $x_{10}; x_1 = 5, r = -3$

*In Exercises 21–24, find the sum of the given geometrical progression using Eq.* (8.16).

**21.** $2 + 4 + 8 + \cdots + 32768$

**22.** $3 - \frac{3}{2} + \frac{3}{4} - \frac{3}{8} + \cdots + \frac{3}{1024}$

**23.** The first 20 terms of $1 + 1.5 + 2.25 + 3.75 + \cdots$.

**24.** The first 20 terms of $1 + 3 + 9 + 27 + \cdots$.

*In Exercises 25–30, find the amount A obtained by depositing the principal P at i% annual interest for n years if:*

a) The interest is simple.
b) The interest is compounded annually.

**25.** $P = \$1000, i = 6\%, n = 12$ **26.** $P = \$1200, i = 6\%, n = 15$

**27.** $P = \$800, i = 12\%, n = 10$ **28.** $P = \$5000, i = 12\%, n = 20$

**29.** $P = \$4800, i = 9\%, n = 9$ **30.** $P = \$1250, i = 8\%, n = 12$

**31.** In an effort to eliminate the dragons in his kingdom, King Mark offers a reward of one gold coin for the first dragon, two for the second, four for the third, and so on, doubling the amount for each additional dragon. St. George sets out and slays 30 dragons. How many gold coins must King Mark give to St. George as his reward?

**32.** Let $P$ be the yearly deposit in an annuity that pays $i\%$ annual interest on all previous deposits and interest. Show that the amount $A$ in the annuity after $n$ years is given by the equation

$$A = P\frac{(1 + i)^n - 1}{i}. \tag{8.18}$$

**33.** Use Eq. (8.18) to find the amount that would result from depositing \$100 yearly at 8% annual interest for 10 years.

**34.** Use Eq. (8.18) to find the amount that would result from depositing \$10 monthly at 6% annual interest compounded quarterly for 10 years. (*Hint:* Use quarterly time periods.)

**35.** Suppose in Example 9 that the disease reduced the reproductive ability of the affected individuals to 80% that of normal individuals. This might be due to early sterility or disfiguration of diseased individuals. The inherited rate will now be 80% what it was in Example 9. What will be the rate of occurrence in the 10th generation, if all the other conditions remain the same?

---

## 8.2 DISCRETE MODELS IN BUSINESS AND THE BIOLOGICAL SCIENCES

We have seen many situations in which real problems were reduced to mathematical formulas. Such simplifications are called *models* of the real world.

A model is a mathematical approximation of a real situation. To be useful, it must involve only the most important factors, eliminating all others to reduce the complexity present in the actual problem. In addition, it must be possible to solve the model mathematically and reinterpret the solution in terms of the real situation. Designing a good model can be very difficult, but the rewards that such a model provides may be substantial.

In many practical situations we are limited to collecting data at certain fixed values of one of the factors involved. This may be due not only to limitations of the equipment that gathers the data, but also to the nature of the data that are being collected. For example, when measuring the tread wear of a particular type of automobile tire, we are limited to the finite set of all such tires. Moreover, even if it were possible to constantly determine each tire's wear, such precision is impractical and unnecessary. Instead, we would probably be willing to accept the results obtained by testing the wear of a very limited number of these tires (say, 20 to 50) at fixed intervals of usage (say, every 1000 miles). Thus, in this model we would replace the continual wear in the real situation by a sequence of numbers indicating the wear after each 1000 miles. Any mathematical model in which one of its factors is determined only at certain fixed values is called a *discrete model.*

In this section we illustrate some typical discrete models applied to a variety of situations. In all these examples one of the variables is determined only at equally spaced intervals.

***Example 1*** A patient in a hospital is suddenly placed on oxygen ($O_2$). We wish to study how fast the nitrogen ($N_2$) in his system washes out. Suppose his total lung capacity at rest is 6.2 liters and the volume of gas inspired and expired during each respiratory cycle (tidal volume) is 0.6 liters (see Fig. 8.1). Nitrogen, which constitutes 78% of the composition (by volume) of air, has a very low solubility in blood, so we can neglect any exchange of nitrogen between the circulatory and respiratory systems.

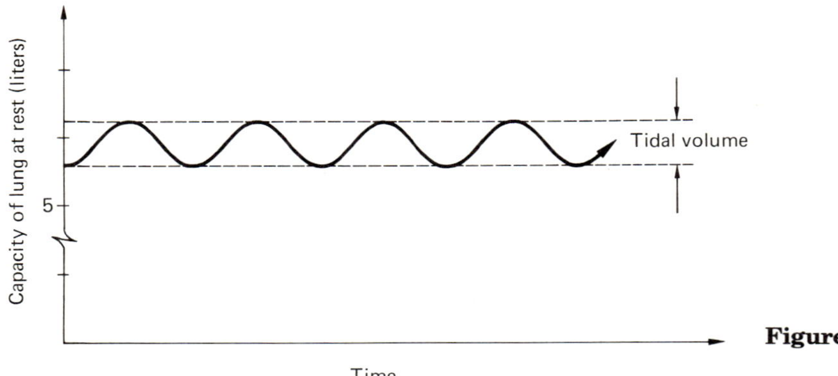


**Figure 8.1**

Let $x_n$ be the concentration of nitrogen in the patient's lungs at the end of the $n$th inspiration. At this time the amount of nitrogen in the lungs is $(6.2x_n)$. After expiration, the volume of gas remaining in the lungs is $6.2 - 0.6 = 5.6$ liters, of which $(5.6x_n)$ is nitrogen. During the next inspiration the amount of nitrogen at the end of inspiration is $(6.2x_{n+1})$ and must equal that at the beginning of inspiration, so that

$$6.2x_{n+1} = 5.6x_n. \tag{8.19}$$

Equation (8.19) can be used to determine the nitrogen washout by letting $x_0 = 0.78$, the percentage of nitrogen in the air. Dividing both

sides of Eq. (8.19) by 6.2, we obtain

$$x_{n+1} = \frac{5.6}{6.2} x_n, \tag{8.20}$$

which indicates that we are dealing with a geometric progression with ratio $r = 5.6/6.2 \approx 0.9032$. By Eq. (8.14) we have

$$x_n = x_0 r^n = 0.78(0.9032)^n, \tag{8.21}$$

so that after 27 breaths the concentration of nitrogen in the lungs is

$$x_{27} = 0.78(0.9032)^{27} \approx 0.05,$$

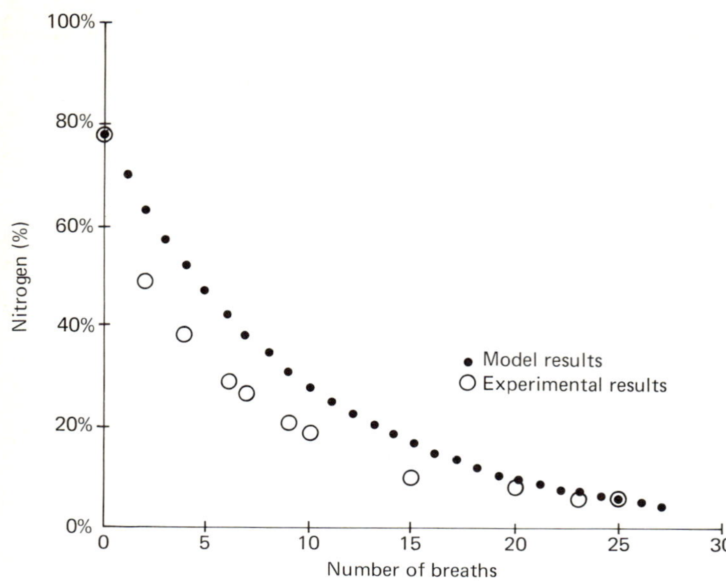


**Fig. 8.2**
Pulmonary nitrogen washout in humans.

or 5%. In Fig. 8.2 we compare the results of this model with experimentally obtained data.* The differences may be due to abnormally large initial inspirations.

***Example 2*** Suppose we borrow \$40,000 at 9% annual interest (compounded monthly), and wish to repay the principal and interest by making equal monthly payments for 30 years. This is the most common way of repaying a mortgage, and is called *amortizing* the loan. The problem is to determine the amount $A$ of the monthly payment.

A formula for determining the quantity $A$ can easily be developed if we notice that amortization can be reduced to an annuity and compound interest problem, with which we are already familiar. The whole process can be simplified if we realize that our monthly payment is in reality an annuity, at 9% annual interest (compounded monthly), which at the end of 30 years must equal the principal plus interest obtained by compounding \$40,000 monthly at 9% annual interest.

Let $i = 0.09$, the annual interest; then the monthly interest is $i/12 = 0.0075$, and the ratio $r$ that we must use in computing our monthly interest is

$$r = 1 + \frac{i}{12} = 1.0075. \tag{8.22}$$

---

*P. L. Altman and D. S. Dittmer, eds., *Respiration and Circulation*. Bethesda, Md.: Federation of American Societies for Experimental Biology, 1971, p. 132.

The number of payments (or times that the principal $P = x_0 = \$40{,}000$ is compounded) is

$$n = 30(12) = 360. \tag{8.23}$$

Using Eq. (8.14) with $P = x_0$, we find that the amount obtained by compounding is

$$\begin{aligned} x_n &= Pr^n \\ &= 40{,}000(1.0075)^{360} \\ &= \$589{,}223.04, \end{aligned} \tag{8.24}$$

which is almost 15 times the principal borrowed!

To find the value of the annuity, we proceed as in Example 8 in Section 8.1 and use Eq. (8.17) with the monthly payment $A$ replacing the quantity $x_1$, that is,

$$\begin{aligned} S_n &= A\left(\frac{r^n - 1}{r - 1}\right) \\ &= A\left(\frac{(1.0075)^{360} - 1}{1.0075 - 1}\right) \\ &= A(1830.74). \end{aligned} \tag{8.25}$$

Equating the results in Eqs. (8.24) and (8.25), we have

$$589{,}223.04 = A(1830.74),$$

so that

$$A = \$321.85. \tag{8.26}$$

To obtain a formula for finding $A$ in all situations, we equate the second terms of Eqs. (8.24) and (8.25):

$$Pr^n = A\left(\frac{r^n - 1}{r - 1}\right). \tag{8.27}$$

Multiplying both sides of (8.27) by $(r - 1)/(r^n - 1)$, we obtain

$$A = \frac{Pr^n(r - 1)}{r^n - 1}. \tag{8.28}$$

Dividing numerator and denominator of the right-hand side of Eq. (8.28) by $r^n$, we finally get

$$A = \frac{P(r - 1)}{1 - r^{-n}}. \tag{8.29}$$

Equation (8.29) can also be rewritten in terms of the monthly interest rate $i/12$ by replacing $r$ by $r = 1 + i/12$, so that

$$A = \frac{P[(1 + i/12) - 1)]}{1 - (1 + i/12)^{-n}}$$

or

$$A = \frac{P(i/12)}{1 - (1 + i/12)^{-n}}. \tag{8.30}$$

***Example 3*** How much should be paid monthly for four years in order to repay a \$5000 auto loan at 12% annual interest compounded monthly?

Here $P = 5000$, $i = 0.12$, and $n = 4(12) = 48$, so that Eq. (8.30) yields

$$A = \frac{5000(0.12/12)}{1 - (1 + 0.12/12)^{-48}}$$

$$= \frac{5000(0.01)}{1 - (1.01)^{-48}} = \$131.67.$$

***Example 4*** Suppose that after making 40 monthly payments for the loan in Example 3, we wish to pay off the remaining debt with a single payment. How much should we pay?

To solve this problem we must recall that the principal is compounded monthly, while the monthly payments act as an annuity. Thus, the outstanding debt is the difference between the values of the principal plus interest and the annuity. Letting $m$ be the number of monthly payments that were made, we find that debt is therefore

$$R = Pr^m - A\left(\frac{r^m - 1}{r - 1}\right). \tag{8.31}$$

The difference $P - R$ is called the *equity*, and represents that part of the principal that the borrower has repaid.

Setting $P = 5000$, $r = 1 + i/12 = 1.01$, and $m = 40$, we obtain from Eq. (8.31) and the solution of Example 3:

$$R = 5000(1.01)^{40} - 131.67\left(\frac{(1.01)^{40} - 1}{(1.01) - 1}\right)$$

$$= 5000(1.4889) - 131.67(48.8864)$$

$$= 7444.32 - 6436.87 = \$1007.45.$$

The growth of populations affords another example of a discrete model. The next example is a highly simplified illustration of this process.

***Example 5*** Suppose we divide the population into four groups: youths (0–19 years of age), adults (20–39 years of age), middle-agers (40–59 years of age), and retirees (60–79 years of age). *We assume that only the adult group reproduces, and that all deaths occur in the retiree group.* Suppose $r$ is the average reproductive rate of each individual during a lifetime, and that each generation is 20 years apart in age. Let $Y_n$, $A_n$, $M_n$, and $R_n$ represent the population size of each of the four groups, respectively, $n$ generations after an original population whose four groups were $Y_0$, $A_0$, $M_0$, and $R_0$. To study the growth of the population, we must determine the effects of each group in one generation on the next generation. Clearly each youth in the $(n - 1)$st generation becomes an adult in the $n$th generation, that is,

$$A_n = Y_{n-1}. \tag{8.32}$$

The youths of the $n$th generation are completely determined by the adults in the $(n - 1)$st generation, since these are the only ones who reproduce. Thus,

$$Y_n = rA_{n-1}. \tag{8.33}$$

Finally each adult becomes a middle-ager, and each middle-ager becomes a retiree, so that

$$M_n = A_{n-1} \tag{8.34}$$

and

$$R_n = M_{n-1}. \tag{8.35}$$

To determine the population size during the $n$th generation we must add together the sizes of each group, getting

$$P_n = Y_n + A_n + M_n + R_n. \tag{8.36}$$

Replacing $Y_n$, $M_n$, and $R_n$ by their values in Eqs. (8.33)–(8.35), we have

$$P_n = rA_{n-1} + A_n + A_{n-1} + M_{n-1}. \tag{8.37}$$

Since $M_{n-1} = A_{n-2}$, Eq. (8.37) becomes

$$P_n = A_n + (1 + r)A_{n-1} + A_{n-2}, \tag{8.38}$$

so the population can be completely determined simply by keeping track of the values in the sequence $A_0, A_1, A_2, \ldots$.

To see how this model works, we shall use the data given in Table 8.1.

**Table 8.1**

U.S. population, July 1967 (U.S. Bureau of the Census)

| Age | Male | Female | Total |
|---|---|---|---|
| 0–19 | 39,620,000 | 38,235,000 | 77,855,000 |
| 20–39 | 24,850,000 | 25,049,000 | 49,899,000 |
| 40–59 | 21,577,000 | 22,946,000 | 44,523,000 |
| 60 and over | 11,899,000 | 14,945,000 | 26,844,000 |
| Totals | 97,946,000 | 101,175,000 | 199,121,000 |

*Source: Information Please Almanac*, 1969, p. 614.

Assuming the year 1967 was the $n$th generation we have

$$Y_n = 77{,}855{,}000, \qquad A_n = 49{,}899{,}000,$$
$$M_n = 44{,}523{,}000, \qquad R_n = 26{,}844{,}000.$$

To make matters simple we shall assume $r = 1$, that is, on the average each individual produces another. (This is the aim of the Zero Population Growth movement.) Using Eqs. (8.32)–(8.35), we have

$$Y_{n+1} = rA_n = 49{,}899{,}000, \qquad A_{n+1} = Y_n = 77{,}855{,}000,$$
$$M_{n+1} = A_n = 49{,}899{,}000, \qquad R_{n+1} = M_n = 44{,}523{,}000,$$

so we may predict that the population in the next generation (in the year 1987) will be

$$P_{n+1} = Y_{n+1} + A_{n+1} + M_{n+1} + R_{n+1} = 222{,}176{,}000.$$

Equation (8.38) can be used to calculate the population in the $(n + 2)$nd generation (in the year 2007) since we know that $A_n = 49{,}899{,}000$, $A_{n+1} = 77{,}855{,}000$, and $A_{n+2} = Y_{n+1} = 49{,}899{,}000$. Substituting $n + 2$ for each $n$ in Eq. (8.38), we have

$$\begin{aligned} P_{n+2} &= A_{n+2} + (1 + r)A_{n+1} + A_n \\ &= 49{,}899{,}000 + 2(77{,}855{,}000) + 49{,}899{,}000 \\ &= 255{,}508{,}000. \end{aligned}$$

Since $A_{n+3} = Y_{n+2} = rA_{n+1} = 77{,}855{,}000$, we also find that $P_{n+3} = 255{,}508{,}000$, showing that the size of the U.S. population would stabilize at 255,508,000 under Zero Population Growth in 40 years.

***Example 6*** In general, human populations have a higher reproductive rate $r$ than that proposed under Zero Population Growth. For example, the birth rate per 1000 individuals for each year between 1948 and 1967 averaged 23.3. To find $r$, we multiply (23.3/1000) by the average expected lifespan in the United States (in 1967), which was 70.5 years, obtaining

$$r = \left(\frac{23.3}{1000}\right)70.5 = 1.643.$$

Using this value we obtain

$$Y_{n+1} = rA_n = 81{,}984{,}000, \qquad A_{n+1} = Y_n = 77{,}855{,}000,$$

$$M_{n+1} = A_n = 49{,}899{,}000, \qquad R_{n+1} = M_n = 44{,}523{,}000,$$

and a total population of $P_{n+1} = 254{,}261{,}000$ in 1987 and $P_{n+2} = 337{,}654{,}000$ in 2007.

***Example 7*** Suppose a gambler bets one dollar on black each time a roulette game is played. A typical roulette wheel contains 18 blacks, 18 reds, a zero, and a double zero. If the wheel is honest, the probability that the casino will win is $p = 20/38 = 0.5263$, while the probability that the gambler will win is $q = 18/38 = 0.4737$ each time the game is played. Let $P_n$ be the probability that the casino will bankrupt the gambler when the casino has $n$ dollars, and suppose that initially the gambler had $M$ dollars and the casino $N$ dollars.

To find $P_n$ for all values $n$ between 0 and $M + N$, we first obtain a *recursion equation*, that is, a formula that defines the $n$th term of the sequence $P_0, P_1, P_2, \ldots, P_{M+N}$ by using other terms in the same sequence. If the casino has $n$ dollars at the beginning of any given game, the probability of its having $(n - 1)$ dollars at the end of the game is $q$, while the probability of its having $(n + 1)$ dollars is $p$. Thus,

$$P_n = qP_{n-1} + pP_{n+1}. \tag{8.39}$$

It is not difficult to verify that Eq. (8.39) has the solution (whenever $p \neq q$)

$$P_n = \frac{1 - (q/p)^n}{1 - (q/p)^{M+N}}. \tag{8.40}$$

To do so, we substitute Eq. (8.40) in the right-hand side of Eq. (8.39) with $n$ replaced first by $n - 1$ and then by $n + 1$:

$$\begin{aligned} qP_{n-1} + pP_{n+1} &= q\left(\frac{1 - (q/p)^{n-1}}{1 - (q/p)^{M+N}}\right) + p\left(\frac{1 - (q/p)^{n+1}}{1 - (q/p)^{M+N}}\right) \\ &= \frac{q - q(q/p)^{n-1} + p - p(q/p)^{n+1}}{1 - (q/p)^{M+N}}. \end{aligned} \tag{8.41}$$

Since $p + q = \frac{20}{38} + \frac{18}{38} = \frac{38}{38} = 1$, we can rewrite Eq. (8.41) as

$$\begin{aligned} qP_{n-1} + pP_{n+1} &= \frac{1 - (q^n/p^{n-1}) - (q^{n+1}/p^n)}{1 - (q/p)^{M+N}} \\ &= \frac{1 - p(q/p)^n - q(q/p)^n}{1 - (q/p)^{M+N}} \\ &= \frac{1 - (p + q)(q/p)^n}{1 - (q/p)^{M+N}} \\ &= \frac{1 - (q/p)^n}{1 - (q/p)^{M+N}} = P_n. \end{aligned}$$

Thus, Eq. (8.40) is the solution of the recursion equation (8.39). To consider a typical situation, suppose the gambler has \$10 and the casino \$100. (Remember, the bet is only one dollar on each roll. If the table limits are larger, say \$1000, we use the same calculations for units of \$1000.) Since

$$\frac{q}{p} = \frac{18/38}{20/38} = \frac{18}{20} = 0.9,$$

we see that the probability that the gambler will be wiped out is

$$\begin{aligned} P_{100} &= \frac{1 - (0.9)^{100}}{1 - (0.9)^{110}} = \frac{1 - 0.00002656}{1 - 0.00000926} \\ &= \frac{0.99997344}{0.99999074} \\ &= 0.9999827. \end{aligned}$$

If the gambler has \$100, the odds improve slightly:

$$\begin{aligned} P_{100} &= \frac{1 - (0.9)^{100}}{1 - (0.9)^{200}} = \frac{0.99997344}{0.99999999} \\ &= 0.99997344, \end{aligned}$$

but the slight edge the casino has on each roll will, in almost every case, bankrupt the gambler if he or she continues to play.

---

**EXERCISES 8.2**

*In Exercises 1–8, amortize the loan P at i% annual interest (compounded monthly) over a period of n years.*

**1.** $P = 1250,\ i = 18\%,\ n = 2$

**2.** $P = 2500,\ i = 12\%,\ n = 3$

**3.** $P = 5000,\ i = 12\%,\ n = 4$

**4.** $P = 10{,}000,\ i = 10\%,\ n = 5$

**5.** $P = 30{,}000$, $i = 9\%$, $n = 30$

**6.** $P = 30{,}000$, $i = 9\%$, $n = 20$

**7.** $P = 30{,}000$, $i = 8\frac{3}{4}\%$, $n = 30$

**8.** $P = 50{,}000$, $i = 8\frac{1}{2}\%$, $n = 30$

**9.** To find the effect of different interest rates on the payment, suppose we amortize \$30,000 with monthly payments over 30 years at five different interest rates: $i = 7\frac{1}{2}\%, 8\%, 8\frac{1}{2}\%, 9\%$, and $9\frac{1}{2}\%$. What is the *average* monthly increase in the payment caused by each $\frac{1}{2}\%$ increase in the interest rate?

**10.** To discover the effect of different loan periods on the monthly payment, amortize \$30,000 at 9% annual interest (compounded monthly) over $n =$ 20, 25, and 30 years. Did the small size of the increase surprise you? Calculate the amount that is paid over 20, 25, and 30 years.

*The monthly payment needed to amortize a* $P = \$40{,}000$ *loan at* $i = 9\%$ *annual interest (compounded monthly) over* 30 *years is* $A = \$321.85$. *In Exercises* 11–16, *find the equity that is attained by paying the loan for* $m$ *years.*

**11.** $m = 1$

**12.** $m = 5$

**13.** $m = 10$

**14.** $m = 15$

**15.** $m = 20$

**16.** $m = 25$

**17.** In Example 5 (see p. 383), suppose that $r = 0.8$. Predict the population of the United States in the years 1987, 2007, and 2027.

**18.** Suppose the birth rate per 1000 individuals averages 17.5 each year after 1967. Predict the population of the United States in the years 1987, 2007, and 2027.

**19.** If a casino and a gambler each have \$10, what is the probability that the gambler is wiped out if he or she plays roulette and bets one dollar on red each time?

**20.** Suppose the table limit is \$1000, the gambler has \$5000, and the casino has \$25,000. What is the probability that the gambler loses all his or her money playing table limits at roulette?

**★21.** Observe that Eq. (8.40) is undefined if $p = q = \frac{1}{2}$. Verify that the solution of the recurrence equation

$$P_n = \tfrac{1}{2}P_{n-1} + \tfrac{1}{2}P_{n+1}$$

is

$$P_n = \frac{n}{M + N}.$$

**22.** Using the result of Exercise 21, find the probability that the casino will bankrupt the gambler in a fair game when the gambler has \$1000 and the casino has \$1,000,000.

**★23.** Penguins and albatross compete for the same nesting areas on South Georgia Island. Each nesting bird will peck at any other bird within pecking range. Assume that $N$ nesting spots are available and the penguins control $n$ of them, while the albatross control the remainder. Suppose that each hour there is competition for one nesting site between the two species, and that the penguins have probability $p$ of being successful in controlling the contested nesting site. Find an expression for the probability $P_n$ that the albatross will eventually lose all their nesting sites when the penguins control $n$ nests.

---

## 8.3 LIMIT OF A SEQUENCE

Consider the geometric progression

$$\frac{1}{2}, \frac{1}{4}, \frac{1}{8}, \frac{1}{16}, \ldots, \tag{8.42}$$

with common ratio $r = 1/2$. Since $x_1 = 1/2$, we find using Eq. (8.13) that

$$x_n = x_1 r^{n-1} = \frac{1}{2}\left(\frac{1}{2}\right)^{n-1} = \left(\frac{1}{2}\right)^n$$

for each positive integer $n$. Note that as $n$ increases, the values $x_n$ get closer and closer to zero, since each succeeding term is half the previous one. For example, if $n = 10$,

$$x_{10} = \left(\frac{1}{2}\right)^{10} \approx 0.00097656,$$

while if $n = 20$,

$$x_{20} = \left(\frac{1}{2}\right)^{20} \approx 0.00000095.$$

In fact, by choosing $n$ large enough, we can get $x_n$ as close to zero as we wish, even though no value $x_n$ ever equals zero. We indicate this

property of the sequence (8.42) by writing

$$\lim_{n\to\infty} x_n = \lim_{n\to\infty} \left(\frac{1}{2}\right)^n = 0,$$

which is read "*the limit of the sequence $x_n$ is zero as n increases without bound.*" Alternatively, we may say that the sequence $x_n$ *converges* to zero as *n tends to infinity.*

More precisely, the notation

$$\lim_{n\to\infty} x_n = A \tag{8.43}$$

means that no matter how small a positive number $\varepsilon > 0$ we might select, we can find an integer $N$ such that the difference between $x_n$ and $A$ is less than $\varepsilon$,

$$|x_n - A| < \varepsilon,$$

for all subscripts $n > N$.

Any sequence of numbers may have a limit. Consider the sequence

$$\frac{1}{2}, \frac{2}{3}, \frac{3}{4}, \frac{4}{5}, \ldots,$$

whose $n$th term is given by the equation $x_n = (n - 1)/n$. Since

$$x_{n+1} - x_n = \frac{n}{n+1} - \frac{n-1}{n} = \frac{1}{n(n+1)}$$

and

$$\frac{x_{n+1}}{x_n} = \frac{n^2}{n^2 - 1},$$

the sequence is neither an arithmetic nor a geometric progression. However, the difference between 1 and $x_n$ is $1/n$, which approaches zero as $n$ increases without bound. Thus,

$$\lim_{n\to\infty} x_n = \lim_{n\to\infty} \frac{n-1}{n} = 1.$$

***Example 1*** Does the sequence

$$x_n = \frac{n-1}{n+2}$$

have a limit as $n$ increases without bound?

A useful trick in solving this kind of problem is to divide both the numerator and denominator by the highest power of $n$ in the denominator. In this case, we divide both numerator and denominator by $n$ to obtain

$$x_n = \frac{n-1}{n+2} = \frac{(n-1)/n}{(n+2)/n} = \frac{1-1/n}{1+2/n}. \tag{8.44}$$

Note that $x_n$ is not changed by these divisions since in effect we are dividing $x_n$ by $1\,(= n/n)$. But $1/n$ and $2/n$ both approach 0 as $n$ increases without bound, so that

$$\lim_{n\to\infty} x_n = \lim_{n\to\infty} \frac{1-1/n}{1+2/n} = 1.$$

***Example 2*** Find

$$\lim_{n\to\infty} \frac{2n^2+4n-3}{5n^2+3n+7}.$$

Dividing both numerator and denominator by $n^2$, we obtain

$$\lim_{n\to\infty} \frac{2n^2+4n-3}{5n^2+3n+7} = \lim_{n\to\infty} \frac{2+(4/n)-(3/n^2)}{5+(3/n)+(7/n^2)}.$$

As $n$ increases without bound, the quantities $4/n$, $3/n^2$, $3/n$, and $7/n^2$ all approach zero, so that the fraction approaches 2/5. Thus,

$$\lim_{n\to\infty} \frac{2n^2+4n-3}{5n^2+3n+7} = \frac{2}{5}.$$

***Example 3*** What is the limit of the sequence

$$x_n = \frac{n^2+1}{n}$$

as $n$ increases without bound?

Dividing numerator and denominator both by $n$ yields

$$x_n = \frac{(n^2+1)/n}{n/n} = \frac{n+1/n}{1} = n + \frac{1}{n}.$$

However, in this case $x_n$ does not approach any fixed number as $n$ increases without bound. In fact, $x_n$ is always larger than $n$ and must also increase without bound as $n$ does. When this occurs we say that the sequence *diverges* and that

$$\lim_{n\to\infty} x_n = \lim_{n\to\infty} \frac{n^2+1}{n} \quad \text{does not exist.}$$

**Example 4** Does the sequence $x_n = (-1)^n$ have a limit?

When $n$ is even, $x_n = 1$, and when $n$ is odd, $x_n = -1$. Thus no *one* number is approached by $x_n$ as $n$ increases without bound. Hence,

$$\lim_{n\to\infty} x_n = \lim_{n\to\infty} (-1)^n \quad \text{does not exist.}$$

Such a sequence, in which the terms alternate signs, is called an *alternating sequence.*

**Example 5** Does the sequence

$$x_n = \frac{3n + (-1)^n}{n} \tag{8.45}$$

have a limit as $n$ increases without bound?

Dividing numerator and denominator both by $n$ yields

$$x_n = \frac{[3n + (-1)^n]/n}{n/n} = 3 + \frac{(-1)^n}{n}.$$

Although $(-1)^n/n$ alternates in sign, the values all approach zero as $n$ tends to infinity. Thus,

$$\lim_{n\to\infty} x_n = \lim_{n\to\infty} 3 + \frac{(-1)^n}{n} = 3.$$

The first ten terms in the sequence (8.45)

$$2, \frac{7}{2}, \frac{8}{3}, \frac{13}{4}, \frac{14}{5}, \frac{19}{6}, \frac{20}{7}, \frac{25}{8}, \frac{26}{9}, \frac{31}{10}$$

are shown in Fig. 8.3.

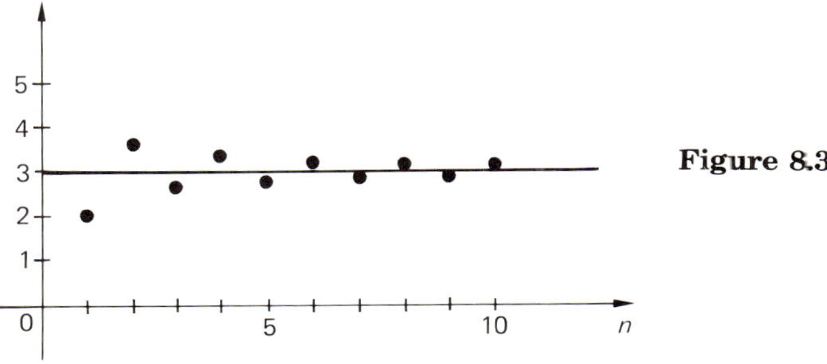


**Figure 8.3**

---

We can also use limits to calculate the sum of the terms of an infinite geometric progression.

***Example 6*** Find the sum of the terms of the geometric progression

$$\frac{1}{2}, \frac{1}{4}, \frac{1}{8}, \frac{1}{16}, \ldots. \tag{8.46}$$

Here we have $x_1 = r = 1/2$, so that again we have

$$x_n = x_1 r^{n-1} = \left(\frac{1}{2}\right)^n.$$

By Eq. (8.17) we know that the sum of the first $n$ terms of the progression (8.46) is

$$\begin{aligned} S_n &= x_1\left(\frac{1 - r^n}{1 - r}\right) = \frac{1}{2}\left(\frac{1 - (\frac{1}{2})^n}{1 - \frac{1}{2}}\right) \\ &= 1 - \left(\frac{1}{2}\right)^n. \end{aligned}$$

Since we wish to find the sum of the terms as $n$ increases without bound, we take the limit of $S_n$ as $n$ tends to infinity, that is,

$$\lim_{n\to\infty} S_n = \lim_{n\to\infty} 1 - \left(\frac{1}{2}\right)^n = 1.$$

Thus, the sum of the terms in the infinite geometric progression (8.46) equals 1. We can write

$$\frac{1}{2} + \frac{1}{4} + \frac{1}{8} + \frac{1}{16} + \cdots = 1. \tag{8.47}$$

We can justify Eq. (8.47) by noticing in Fig. 8.4 that each succeeding interval is half the length of the remaining portion of the interval $0 \le x \le 1$.

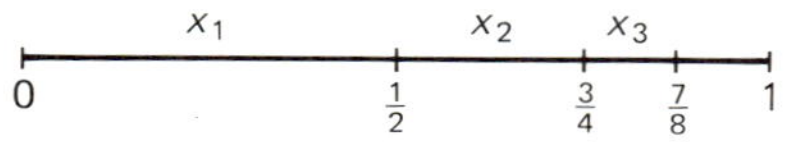

**Figure 8.4**

---

This procedure may be used to find the sum of any geometric progression whose ratio $r$ satisfies $|r| < 1$. If $|r| < 1$, we note that

$$|r|^{n+1} = |r| \cdot |r|^n < 1 \cdot |r|^n = |r|^n,$$

so that the values of the sequence $|r|^n$ always decrease. Repeatedly multiplying a number $|r| < 1$ by itself yields a sequence that converges to zero. Indeed

$$\lim_{n\to\infty} r^n = 0, \qquad \text{if } |r| < 1, \tag{8.48}$$

regardless what the sign of $r$ might be. Using Eq. (8.17), we then obtain

$$\lim_{n\to\infty} S_n = \lim_{n\to\infty} x_1\left(\frac{1 - r^n}{1 - r}\right) = \frac{x_1}{1 - r}, \qquad \text{if } |r| < 1, \tag{8.49}$$

since the $r^n$ term tends to zero. Therefore,

$$x_1 + x_1 r + x_1 r^2 + \cdots = \frac{x_1}{1 - r}, \qquad \text{whenever } |r| < 1. \tag{8.50}$$

If $|r| \geq 1$, we no longer obtain this result since the limit (8.48) no longer holds.

***Example 7*** Find the infinite sum

$$4 - \frac{8}{3} + \frac{16}{9} - \frac{32}{27} + \cdots.$$

In this example $x_1 = 4$ and $r = -2/3$. Using Eq. (8.50) we have

$$4 - \frac{8}{3} + \frac{16}{9} - \frac{32}{27} + \cdots = \frac{4}{1 - (-2/3)} = \frac{4}{(5/3)} = \frac{12}{5}.$$

***Example 8*** In Example 9 in Section 8.1 we studied the effect of the mutation of a normal allele $a$ into a dominant gene $A$. The rate of occurrence in the $n$th generation was found to be (see Eq. 8.17)

$$m + mr + mr^2 + \cdots + mr^{n-1} = m\left(\frac{1 - r^n}{1 - r}\right),$$

with $r = 1/2$. Letting $n$ tend to infinity, we find that the fraction of the population having the disease will approach

$$\lim_{n\to\infty} m\left(\frac{1 - (\frac{1}{2})^n}{1 - \frac{1}{2}}\right) = \frac{m}{1 - \frac{1}{2}} = 2m,$$

that is, double the mutation rate.

***Example 9*** Find the infinite sum

$$13 + \frac{13}{100} + \frac{13}{(100)^2} + \frac{13}{(100)^3} + \cdots.$$

In this example $x_1 = 13$ and $r = 1/100$, so by Eq. (8.50)

$$13 + \frac{13}{100} + \frac{13}{(100)^2} + \cdots = \frac{13}{1 - \dfrac{1}{100}} = \frac{1300}{99}.$$

Note that the infinite sum can be written as a repeating decimal:

$$13 + \frac{13}{100} + \frac{13}{(100)^2} + \cdots = 13 + 0.13 + 0.0013 + \cdots = 13.131313\cdots.$$

Hence, this shows that

$$13.131313\cdots = \frac{1300}{99}.$$

---

## EXERCISES 8.3

*In Exercises 1–20, find the limit, if it exists.*

**1.** $\lim_{n\to\infty} \frac{n+1}{n}$

**2.** $\lim_{n\to\infty} \frac{n-2}{2n+4}$

**3.** $\lim_{n\to\infty} \frac{n-(-1)^n}{n+(-1)^n}$

**4.** $\lim_{n\to\infty} \frac{(-1)^n n}{n+1}$

**5.** $\lim_{n\to\infty} \frac{3n+2}{5n+7}$

**6.** $\lim_{n\to\infty} \frac{n+8}{n^2+7n-2}$

**7.** $\lim_{n\to\infty} \frac{n^2-3n+4}{2n^2+3n+5}$

**8.** $\lim_{n\to\infty} \frac{n^2}{n+1}$

**9.** $\lim_{n\to\infty} \frac{2n^2+7n+8}{n+3}$

**10.** $\lim_{n\to\infty} \frac{n^2-(-1)^n}{n^2}$

**11.** $\lim_{n\to\infty} \frac{n^4+2n^2+1}{(n+1)^4}$

**12.** $\lim_{n\to\infty} \frac{n^4-1}{n^3-1}$

**13.** $\lim_{n\to\infty} \frac{3n^2-5n+4}{4n^2+7n-8}$

**14.** $\lim_{n\to\infty} \frac{(n+2)(n+3)}{2(n+4)(n+1)}$

**15.** $\lim_{n\to\infty} \frac{(2n+3)(5n+2)}{(4n+1)(3n+2)}$

**16.** $\lim_{n\to\infty} \frac{(-1)^n n}{n^2+1}$

**17.** $\lim_{n\to\infty} \frac{(-1)^n n^2}{n^3+1}$

**18.** $\lim_{n\to\infty} \left[1 - \frac{(-1)^n}{n}\right]$

**19.** $\lim_{n\to\infty} [1-(-1)^n]$

**20.** $\lim_{n\to\infty} (-1)^n n$

*In Exercises 21–26, find each infinite sum.*

**21.** $2 + \frac{2}{3} + \frac{2}{9} + \frac{2}{27} + \cdots$

**22.** $3 - \frac{3}{7} + \frac{3}{49} - \frac{3}{343} + \cdots$

**23.** $4 + \frac{8}{5} + \frac{16}{25} + \frac{32}{125} + \cdots$

**24.** $2 - \frac{6}{5} + \frac{18}{25} - \frac{54}{125} + \cdots$

**25.** $7 + 2 + \frac{4}{7} + \frac{8}{49} + \cdots$

**26.** $3 + 0.3 + 0.03 + 0.003 + \cdots$

**27.** Suppose that 60 cents of every dollar earned in a certain city is spent in that city. If one of the city companies has a \$1,200,000 yearly payroll, how much revenue for the city will this amount generate? (*Hint:* 60% of the

0.60(1,200,000) = \$720,000 spent by company employees will again be spent in the city, and so on. The fraction total revenue per payroll is called the *multiplier* by economists.)

**28.** On an average, 85% of every dollar earned in the U.S. is spent on American products and services. In an effort to curb a recession, the government creates 10,000 jobs at an average salary of \$9000. What is the multiplier effect of this program?

## 8.4 LIMITS OF A FUNCTION

We will now expand the concept of a limit to include limits of functions. As we shall discover, it is possible to find the limit of an expression as the independent real variable $x$ approaches any real number or $\pm\infty$.

As an example consider the function

$$f(x) = \frac{x^2 + x - 2}{x^2 - 1}. \tag{8.51}$$

This function exists at all real numbers $x \neq \pm 1$, since these are the only real values that make the denominator vanish. We may now ask if the values of $f(x)$ approach any particular number as the values of $x$ approach $a$, where $a$ is some real number or $\pm\infty$.

If $a = 0$, we select several values of $x$ *close to and on either side of* $a = 0$ and evaluate $f(x)$ at these choices. The following table shows the results of a few such choices:

| $x$ | −0.1 | −0.01 | −0.001 | 0.001 | 0.01 | 0.1 |
|---|---|---|---|---|---|---|
| $f(x)$ | 2.1111 | 2.0101 | 2.0010 | 1.9990 | 1.9901 | 1.9091 |

The table indicates that $f(x)$ tends to 2 as $x$ approaches 0. In this case we can also substitute the value $x = 2$ into Eq. (8.51) and obtain $f(0) = -2/-1 = 2$. Thus, we can write

$$\lim_{x\to 0} f(x) = \lim_{x\to 0} \frac{x^2 + x - 2}{x^2 - 1} = 2.$$

The problem is a little more difficult if $a = 1$. We can no longer substitute $x = 1$ into Eq. (8.51) to obtain the answer, since the denominator of $f(x)$ vanishes at $x = 1$, indicating that $f(x)$ is not defined at $x = 1$. However, it may still have a limit as $x$ approaches 1. Again selecting several values of $x$ close to and on either side of $a = 1$, we obtain:

| $x$ | 0.9 | 0.99 | 0.999 | 1.001 | 1.01 | 1.1 |
|---|---|---|---|---|---|---|
| $f(x)$ | 1.5263 | 1.5025 | 1.5003 | 1.4998 | 1.4975 | 1.4762 |

As we can see from this table, the values of $f(x)$ approach 1.5 as $x$ tends to 1. Therefore, we can write

$$\lim_{x \to 1} f(x) = \lim_{x \to 1} \frac{x^2 + x - 2}{x^2 - 1} = 1.5.$$

If $a = -1$, we again encounter a value at which $f(x)$ is not defined. To see whether $f(x)$ has a limit as $x$ approaches $-1$ we again construct a table of values of $x$ close to and on either side of $a = -1$:

| $x$ | $-1.1$ | $-1.01$ | $-1.001$ | $-0.999$ | $-0.99$ | $-0.9$ |
|---|---|---|---|---|---|---|
| $f(x)$ | $-9$ | $-99$ | $-999$ | 1001 | 101 | 11 |

In this case, we see that as we approach $a = -1$ with $x$-values less than $-1$, the corresponding $f(x)$-values tend to $-\infty$. On the other hand, as we approach $a = -1$ with values of $x$ greater than $-1$, $f(x)$ approaches $+\infty$. Since no one number is approached, we write

$$\lim_{x \to -1} f(x) = \lim_{x \to -1} \frac{x^2 + x - 2}{x^2 - 1} \quad \text{does not exist.}$$

The reason for the behavior we have seen exhibited by the function $f(x)$ is easier to understand if we factor the numerator and denominator, obtaining

$$f(x) = \frac{(x + 2)(x - 1)}{(x + 1)(x - 1)}. \tag{8.52}$$

Clearly, $f(x)$ is not defined at $x = \pm 1$. At all other values of $x$, the terms $(x - 1)$ in the numerator and denominator will cancel. Thus $f(x)$ will behave exactly as the function

$$g(x) = \frac{x + 2}{x + 1}, \tag{8.53}$$

except at $x = +1$, where $g(x)$ is defined and $f(x)$ is not defined. To obtain

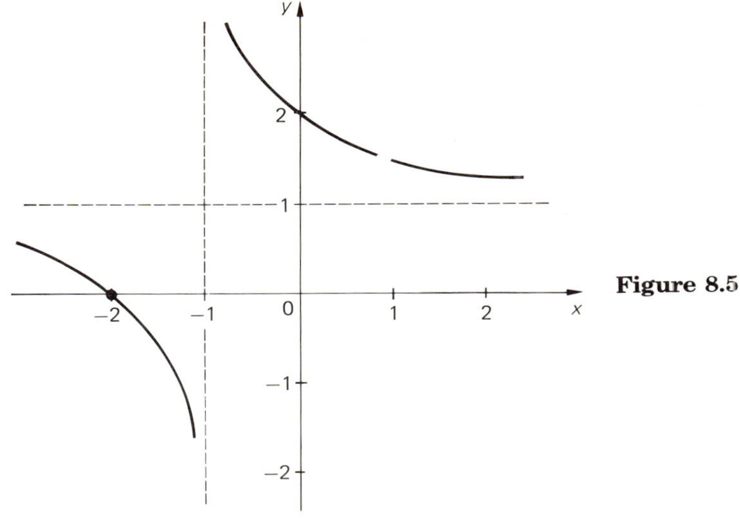


**Figure 8.5**

the graph of $f(x)$ we need only graph $g(x)$ and *remove the point where the graph crosses the vertical line* $x = 1$ (see Fig. 8.5).

In summary, if $a$ is a real number, the equation

$$\lim_{x \to a} f(x) = A \tag{8.54}$$

states that the function values $f(x)$ approach the single real number $A$ as the values $x$ tend to $a$. In other words, the difference $|f(x) - A|$ can be made smaller than any preassigned positive number by choosing $x$ so that $|x - a|$ is sufficiently small. If no such number $A$ can be found, we simply write

$$\lim_{x \to a} f(x) \quad \text{does not exist.}$$

Finally, let's consider what happens to the function in Eq. (8.51) as $x$ increases without bound. If we use the trick of dividing both the numerator and the denominator by the highest power of $x$ in the denominator (see Example 1 in Section 8.3), we obtain

$$f(x) = \frac{(x^2 + x - 2)/x^2}{(x^2 - 1)/x^2} = \frac{1 + (1/x) - (2/x^2)}{1 - (1/x^2)}. \tag{8.55}$$

But $1/x$ and $1/x^2$ tend to zero as $x$ increases without bound, so that $f(x)$ approaches 1. We can write this discussion mathematically as follows:

$$\begin{aligned}\lim_{x \to \infty} f(x) &= \lim_{x \to \infty} \frac{x^2 + x - 2}{x^2 - 1} \\ &= \lim_{x \to \infty} \frac{1 + (1/x) - (2/x^2)}{1 - (1/x^2)} = 1.\end{aligned}$$

Similarly,

$$\lim_{x\to-\infty} f(x) = \lim_{x\to-\infty} \frac{1 + (1/x) - (2/x^2)}{1 - (1/x^2)} = 1.$$

Hence,

$$\lim_{x\to\infty} f(x) = A$$

means that $|f(x) - A|$ gets arbitrarily close to zero whenever $x$ is sufficiently large (since $x$ tends to infinity). Similarly,

$$\lim_{x\to-\infty} f(x) = A$$

means that $|f(x) - A|$ is arbitrarily close to zero whenever $(-x)$ is sufficiently large. If no such number $A$ can be found, we write

$$\lim_{x\to\infty} f(x) \text{ (or } \lim_{x\to-\infty} f(x)) \quad \text{does not exist.}$$

***Example 1*** Evaluate

$$\lim_{x\to-2} \frac{x^2 + 3x + 2}{x + 2}$$

if it exists.

If we substitute $x = -2$ in the equation

$$f(x) = \frac{x^2 + 3x + 2}{x + 2}, \tag{8.56}$$

we discover that *both* the numerator and the denominator vanish. This indicates that $x = -2$ is a root of the numerator *and* the denominator, and suggests that we can eliminate our difficulties by factoring $(x + 2)$ from the numerator and canceling the $(x + 2)$-terms in the numerator and denominator. When we factor, we find that

$$\lim_{x\to-2} \frac{x^2 + 3x + 2}{x + 2} = \lim_{x\to-2} \frac{(x + 2)(x + 1)}{(x + 2)} = \lim_{x\to-2} (x + 1) = -1,$$

since the function

$$g(x) = x + 1$$

tends to $(-1)$ as $x \to -2$.

***Example 2*** Evaluate

$$\lim_{x\to 3} \frac{2x^2 + 5x + 3}{x - 3}$$

if it exists.

Substituting $x = 3$ into the equation

$$f(x) = \frac{2x^2 + 5x + 3}{x - 3}, \tag{8.57}$$

we find that the denominator vanishes, while the numerator equals 36. Thus, as the $x$-values get closer to 3, the absolute value of $f(x)$ increases without bound. Hence the limit does not exist.

***Example 3*** Evaluate

$$\lim_{x\to 2} \frac{3x^3 + 4}{2x^2 - 5}$$

if it exists.

Setting $x = 2$ in the equation

$$f(x) = \frac{3x^3 + 4}{2x^2 - 5},$$

we obtain

$$f(2) = \frac{3(2)^3 + 4}{2(2)^2 - 5} = \frac{24 + 4}{8 - 5} = \frac{28}{3}.$$

This value is the limit, so that we have

$$\lim_{x\to 2} \frac{3x^3 + 4}{2x^2 - 5} = \frac{28}{3}.$$

---

These examples illustrated the following theorem.

**THEOREM 8.1** **Let $P(x)$ and $Q(x)$ be any two functions. Then, provided all limits exist:**

**a)** $\lim_{x\to a} [P(x) + Q(x)] = \lim_{x\to a} P(x) + \lim_{x\to a} Q(x).$

**(The limit of a sum is the sum of the limits.)**

b) $\lim_{x\to a} [P(x) \cdot Q(x)] = \lim_{x\to a} P(x) \cdot \lim_{x\to a} Q(x).$

**(The limit of a product is the product of the limits.)**

c) $\lim_{x\to a} \dfrac{P(x)}{Q(x)} = \dfrac{\lim_{x\to a} P(x)}{\lim_{x\to a} Q(x)}$, whenever $\lim_{x\to a} Q(x) \neq 0$.

**(The limit of a quotient is the quotient of the limits.)**

d) $\lim_{x\to a} (kP(x)) = k \lim_{x\to a} P(x).$

**(constants are not affected by the limiting process.)**

***Example 4*** Evaluate

$$\lim_{x\to 2} \frac{(x-3)(x^2+4)}{3(x^3+1)}.$$

By Theorem 8.1(c) we have

$$\lim_{x\to 2} \frac{(x-3)(x^2+4)}{3(x^3+1)} = \frac{\lim_{x\to 2} (x-3)(x^2+4)}{\lim_{x\to 2} 3(x^3+1)}.$$

Using parts (b) and (d), we obtain

$$\frac{\lim_{x\to 2} (x-3)(x^2+4)}{\lim_{x\to 2} 3(x^3+1)} = \frac{\lim_{x\to 2} (x-3) \cdot \lim_{x\to 2} (x^2+4)}{3 \lim_{x\to 2} (x^3+1)}. \tag{8.58}$$

Finally, using (a) and (b), we may rewrite the right-hand side of Eq. (8.58) as

$$\frac{\left[\left(\lim_{x\to 2} x\right) - 3\right] \cdot \left[\left(\lim_{x\to 2} x\right)^2 + 4\right]}{3\left[\left(\lim_{x\to 2} x\right)^3 + 1\right]} = \frac{[2-3] \cdot [(2)^2+4]}{3[(2)^3+1]} = \frac{-8}{27}.$$

---

Example 4 and the previous examples in this section show that the following procedure should be used in dealing with limits of quotients of polynomials when $a$ is a real number.

***Example 2*** Evaluate

$$\lim_{x \to 3} \frac{2x^2 + 5x + 3}{x - 3}$$

if it exists.

Substituting $x = 3$ into the equation

$$f(x) = \frac{2x^2 + 5x + 3}{x - 3}, \tag{8.57}$$

we find that the denominator vanishes, while the numerator equals 36. Thus, as the $x$-values get closer to 3, the absolute value of $f(x)$ increases without bound. Hence the limit does not exist.

***Example 3*** Evaluate

$$\lim_{x \to 2} \frac{3x^3 + 4}{2x^2 - 5}$$

if it exists.

Setting $x = 2$ in the equation

$$f(x) = \frac{3x^3 + 4}{2x^2 - 5},$$

we obtain

$$f(2) = \frac{3(2)^3 + 4}{2(2)^2 - 5} = \frac{24 + 4}{8 - 5} = \frac{28}{3}.$$

This value is the limit, so that we have

$$\lim_{x \to 2} \frac{3x^3 + 4}{2x^2 - 5} = \frac{28}{3}.$$

---

These examples illustrated the following theorem.

**THEOREM 8.1** **Let $P(x)$ and $Q(x)$ be any two functions. Then, provided all limits exist:**

**a)** $\lim_{x \to a} [P(x) + Q(x)] = \lim_{x \to a} P(x) + \lim_{x \to a} Q(x)$.

**(The limit of a sum is the sum of the limits.)**

b) $\lim_{x \to a} [P(x) \cdot Q(x)] = \lim_{x \to a} P(x) \cdot \lim_{x \to a} Q(x)$.

**(The limit of a product is the product of the limits.)**

c) $\lim_{x \to a} \dfrac{P(x)}{Q(x)} = \dfrac{\lim_{x \to a} P(x)}{\lim_{x \to a} Q(x)}$, whenever $\lim_{x \to a} Q(x) \neq 0$.

**(The limit of a quotient is the quotient of the limits.)**

d) $\lim_{x \to a} (kP(x)) = k \lim_{x \to a} P(x)$.

**(constants are not affected by the limiting process.)**

***Example 4*** Evaluate

$$\lim_{x \to 2} \frac{(x - 3)(x^2 + 4)}{3(x^3 + 1)}.$$

By Theorem 8.1(c) we have

$$\lim_{x \to 2} \frac{(x - 3)(x^2 + 4)}{3(x^3 + 1)} = \frac{\lim_{x \to 2} (x - 3)(x^2 + 4)}{\lim_{x \to 2} 3(x^3 + 1)}.$$

Using parts (b) and (d), we obtain

$$\frac{\lim_{x \to 2} (x - 3)(x^2 + 4)}{\lim_{x \to 2} 3(x^3 + 1)} = \frac{\lim_{x \to 2} (x - 3) \cdot \lim_{x \to 2} (x^2 + 4)}{3 \lim_{x \to 2} (x^3 + 1)}. \tag{8.58}$$

Finally, using (a) and (b), we may rewrite the right-hand side of Eq. (8.58) as

$$\frac{\left[\left(\lim_{x \to 2} x\right) - 3\right] \cdot \left[\left(\lim_{x \to 2} x\right)^2 + 4\right]}{3\left[\left(\lim_{x \to 2} x\right)^3 + 1\right]} = \frac{[2 - 3] \cdot [(2)^2 + 4]}{3[(2)^3 + 1]} = \frac{-8}{27}.$$

---

Example 4 and the previous examples in this section show that the following procedure should be used in dealing with limits of quotients of polynomials when $a$ is a real number.

***PROCEDURE*** Let $P(x)$ and $Q(x)$ be two polynomials, and suppose we wish to evaluate

$$\lim_{x \to a} \frac{P(x)}{Q(x)} \tag{8.59}$$

if it exists.

1. First we substitute $a$ for $x$ in each polynomial. If $Q(a) \neq 0$, we immediately obtain

$$\lim_{x \to a} \frac{P(x)}{Q(x)} = \frac{P(a)}{Q(a)}, \quad [Q(a) \neq 0].$$

2. If $Q(a) = 0$ but $P(a) \neq 0$, the limit (8.59) does not exist.
3. If both $P(a) = Q(a) = 0$, the value $x = a$ is a root of both polynomials, so $(x - a)$ should be factored from each polynomial and canceled. If $P(x) = (x - a)P_1(x)$ and $Q(x) = (x - a)Q_1(x)$, then

$$\lim_{x \to a} \frac{P(x)}{Q(x)} = \lim_{x \to a} \frac{(x - a)P_1(x)}{(x - a)Q_1(x)} = \lim_{x \to a} \frac{P_1(x)}{Q_1(x)}.$$

We will see in Chapter 9 that this last procedure is encountered in determining the derivative of a function. Because of the frequency with which we will use this procedure in Chapter 9, we give four more examples in which this situation occurs.

***Example 5*** Evaluate

$$\lim_{x \to 2} \frac{x^3 - 8}{x - 2}.$$

Since both the numerator and the denominator vanish at $x = 2$, we factor $(x - 2)$ from each, obtaining

$$\lim_{x \to 2} \frac{x^3 - 8}{x - 2} = \lim_{x \to 2} \frac{(x - 2)(x^2 + 2x + 4)}{(x - 2)}.$$

Canceling the $(x - 2)$ terms in the numerator and denominator, we have

$$\lim_{x \to 2} \frac{x^3 - 8}{x - 2} = \lim_{x \to 2} (x^2 + 2x + 4) = 4 + 4 + 4 = 12.$$

***Example 6*** Find

$$\lim_{x \to 1} \frac{\sqrt{x + 3} - 2}{x^2 - 1}.$$

Again both the numerator and denominator vanish when we set $x = 1$. This time we encounter some difficulty in factoring $(x - 1)$ from the numerator. In this case it helps to rationalize the numerator by multiplying numerator and denominator by $(\sqrt{x+3} + 2)$:

$$\begin{aligned}\frac{\sqrt{x+3}-2}{x^2-1}\cdot\frac{\sqrt{x+3}+2}{\sqrt{x+3}+2} &= \frac{(\sqrt{x+3})^2-2^2}{(x^2-1)(\sqrt{x+3}+2)}\\ &= \frac{x+3-4}{(x+1)(x-1)(\sqrt{x+3}+2)}\\ &= \frac{(x-1)}{(x+1)(x-1)(\sqrt{x+3}+2)}\\ &= \frac{1}{(x+1)(\sqrt{x+3}+2)}.\end{aligned}$$

Hence,

$$\begin{aligned}\lim_{x\to 1}\frac{\sqrt{x+3}-2}{x^2-1} &= \lim_{x\to 1}\frac{1}{(x+1)(\sqrt{x+3}+2)}\\ &= \frac{1}{(1+1)(\sqrt{1+3}+2)} = \frac{1}{2(4)} = \frac{1}{8}.\end{aligned}$$

***Example 7*** Evaluate

$$\lim_{x\to a}\frac{\sqrt{x}-\sqrt{a}}{x-a}.$$

We rationalize the numerator by multiplying numerator and denominator by $\sqrt{x} + \sqrt{a}$:

$$\begin{aligned}\lim_{x\to a}\frac{\sqrt{x}-\sqrt{a}}{x-a} &= \lim_{x\to a}\left(\frac{\sqrt{x}-\sqrt{a}}{x-a}\cdot\frac{\sqrt{x}+\sqrt{a}}{\sqrt{x}+\sqrt{a}}\right)\\ &= \lim_{x\to a}\frac{(\sqrt{x})^2-(\sqrt{a})^2}{(x-a)(\sqrt{x}+\sqrt{a})}\\ &= \lim_{x\to a}\frac{(x-a)}{(x-a)(\sqrt{x}+\sqrt{a})}\\ &= \lim_{x\to a}\frac{1}{\sqrt{x}+\sqrt{a}} = \frac{1}{2\sqrt{a}}.\end{aligned}$$

We may use other letters for the independent variable. In the following example, we assume that $x$ is fixed and $h$ is varying.

***Example 8*** Find

$$\lim_{h\to 0} \frac{(x+h)^3 - x^3}{h}.$$

Setting $h = 0$ forces both numerator and denominator to vanish. However, if we expand the $(x + h)^3$ term by using the binomial theorem, we obtain

$$\lim_{h\to 0} \frac{(x+h)^3 - x^3}{h} = \lim_{h\to 0} \frac{(x^3 + 3x^2h + 3xh^2 + h^3) - x^3}{h}.$$

Canceling the two $x^3$-terms in the numerator yields

$$\lim_{h\to 0} \frac{(x+h)^3 - x^3}{h} = \lim_{h\to 0} \frac{3x^2h + 3xh^2 + h^3}{h}.$$

We can factor $h$ from each term in the numerator and cancel $h$ in the numerator and the denominator to obtain

$$\lim_{h\to 0} \frac{(x+h)^3 - x^3}{h} = \lim_{h\to 0} \frac{h(3x^3 + 3xh + h^2)}{h}$$

$$= \lim_{h\to 0} (3x^2 + 3xh + h^2) = 3x^2.$$

---

**EXERCISES 8.4**

*Find all the limits that exist.*

**1.** $\lim_{x\to 0} (7x + 2)$

**2.** $\lim_{x\to 1} (4x - 3)$

**3.** $\lim_{x\to -3} (3x^2 + 9x)$

**4.** $\lim_{x\to 0} (2x + 1)(3x^2 + 4)$

**5.** $\lim_{x\to 1} \frac{x - 2}{x + 3}$

**6.** $\lim_{x\to 4} \frac{x^2 - 16}{x + 4}$

**7.** $\lim_{x\to 0} \frac{\sqrt{x + 4} - 2}{x + 3}$

**8.** $\lim_{x\to -2} \frac{\sqrt{2 - x} + 3}{x + 2}$

**9.** $\lim_{x\to 1} \frac{x^2 - 1}{x - 1}$

**10.** $\lim_{x\to 0} \frac{\sqrt{4 + x} + 2}{\sqrt{4 + x} - 2}$

**11.** $\lim_{x\to 2} \frac{\sqrt{2 + x} - 2}{x - 2}$

**12.** $\lim_{x\to 7} \frac{x^3 - (7)^3}{x - 7}$

**13.** $\lim_{x\to \infty} \frac{\sqrt{x + 3} - 2}{x - 5}$

**14.** $\lim_{x\to \infty} \frac{2x^3 + 5x^2 + 7x - 3}{3x^3 - 5x + 9}$

**15.** $\lim_{x\to \infty} \frac{\sqrt{x^2 + 1} + 5}{x + 6}$

**16.** $\lim_{x\to-\infty} \dfrac{3x^4 - 8}{x^3 + 7}$ **17.** $\lim_{x\to-\infty} \dfrac{\sqrt{x^4 + 1}}{2x^2 - 1}$ **18.** $\lim_{h\to 0} \dfrac{(x + h)^2 - x^2}{h}$

**19.** $\lim_{h\to 0} \dfrac{(x + h)^4 - x^4}{h}$ **20.** $\lim_{h\to 0} \dfrac{\sqrt{x + h} - \sqrt{x}}{h}$ **21.** $\lim_{h\to 0} \dfrac{\sqrt[3]{x + h} - \sqrt[3]{x}}{h}$

**22.** $\lim_{h\to 0} \dfrac{\dfrac{1}{x + h} - \dfrac{1}{x}}{h}$ **23.** $\lim_{x\to a} \dfrac{x^{3/2} - a^{3/2}}{x - a}$ **24.** $\lim_{x\to a} \dfrac{\left(\dfrac{1}{x^2}\right) - \left(\dfrac{1}{a^2}\right)}{x - a}$

---

## 8.5 CONTINUITY OF FUNCTIONS

A function $y = f(x)$ is said to be *continuous* if there are no gaps or jumps in its graph. For example, the function $y = x^2$, whose graph is shown in Fig. 8.6, is continuous since it is a smooth line. However, the function

$$f(x) = \begin{cases} -1, & \text{for } x < 0, \\ 1, & \text{for } x \geq 0, \end{cases}$$

has a jump, or *discontinuity*, at $x = 0$ (see Fig. 8.7).

It is easy to make a precise definition of continuity by using limits.

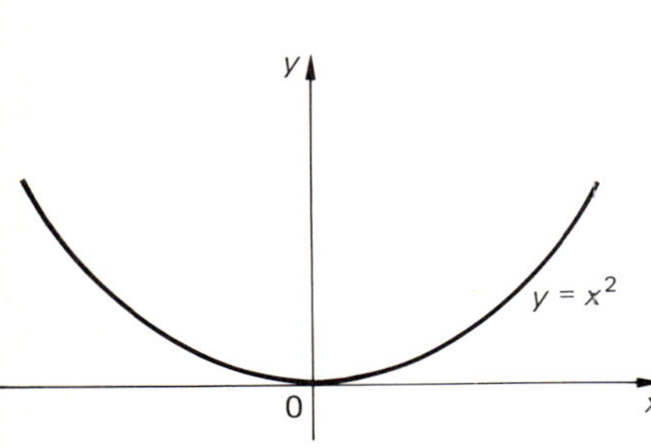


**Figure 8.6**

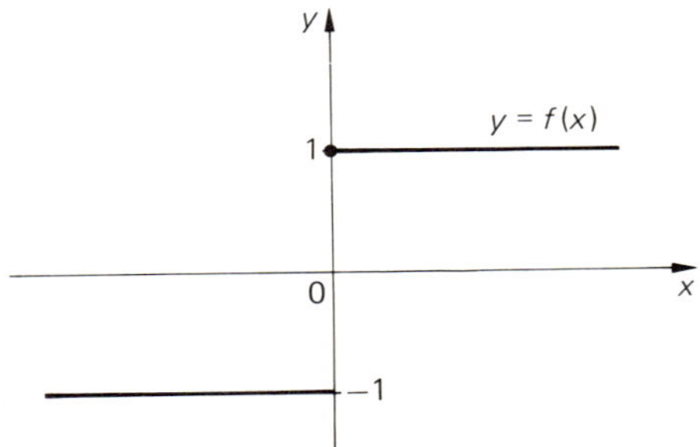


**Figure 8.7**

**DEFINITION 8.1** **Suppose the function $y = f(x)$ is defined at $x = a$. Then, $y = f(x)$ is *continuous at the point* $x = a$ if**

$$\lim_{x\to a} f(x) = f(a). \tag{8.60}$$

**A *continuous* function is one that is continuous at each point where it is defined.**

Graphically, this definition states that the points $(x, f(x))$ in the $xy$-plane approach the point $(a, f(a))$ as $x$ tends to $a$ (see Fig. 8.8). This prevents any possible jump at $x = a$. Note that three conditions are

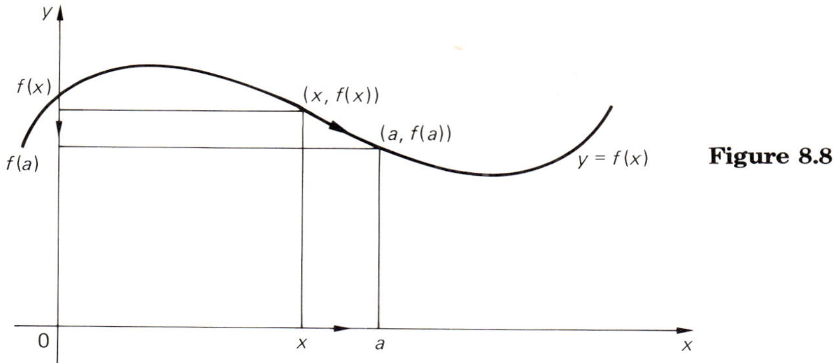


**Figure 8.8**

required of $y = f(x)$ for continuity at the point $x = a$:

1. The $\lim_{x \to a} f(x)$ must exist,
2. The function $y = f(x)$ must be defined at $x = a$, and
3. These two values must coincide, as indicated by Eq. (8.60).

If one or more of these conditions is not satisfied, the function is said to be *discontinuous at the point* $x = a$; that is, $x = a$ is called a point of *discontinuity* of the function $y = f(x)$. For example, the function

$$f(x) = \begin{cases} -1, & \text{for } x < 0, \\ 1, & \text{for } x \geq 0, \end{cases}$$

shown in Fig. 8.7, has a point of discontinuity at $x = 0$ because

$$\lim_{x \to 0} f(x) \quad \text{does not exist}$$

since the function values tend to $-1$ as we approach 0 through negative values of $x$, and $f(x)$ approaches 1 as $x > 0$ tends to 0. The following two examples show situations in which the other properties are violated.

***Example 1*** Let $f(x) = (x^2 - 1)/(x - 1)$, and suppose we wish to see whether this function is continuous at $x = 1$. No problem is experienced with the limit since

$$\lim_{x \to 1} \frac{x^2 - 1}{x - 1} = \lim_{x \to 1} \frac{(x + 1)(x - 1)}{x - 1}$$

$$= \lim_{x \to 1} (x + 1) = 2.$$

However, $f(1)$ is not defined, since division by zero is prohibited. Thus, $f(x)$ is not continuous at $x = 1$.

***Example 2*** Find all points of discontinuity of the function

$$f(x) = \begin{cases} x^3, & \text{for } x \neq 1, \\ 2, & \text{for } x = 1. \end{cases} \tag{8.61}$$

Since

$$\lim_{x \to 1} f(x) = \lim_{x \to 1} x^3 = 1,$$

and $f(1) = 2$, we see that $x = 1$ is a point of discontinuity. The graph of $y = f(x)$ shown in Fig. 8.9 indicates that this is the only point of discontinuity.

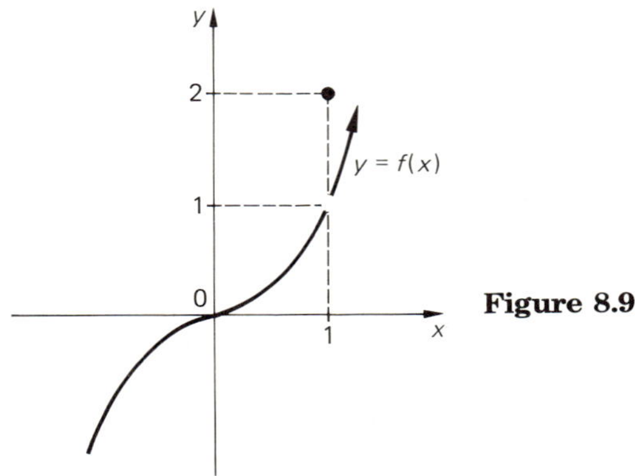


Figure 8.9

**THEOREM 8.2** **Let $P(x)$ and $Q(x)$ be polynomials. Then:**

**a) Both $P(x)$ and $Q(x)$ are continuous functions for all values $x$.**

**b) The quotient $P(x)/Q(x)$ is continuous except at the zeros of $Q(x)$, where the quotient is not defined.**

***Proof*** We will use the procedure that we developed in Section 8.4 (see Eq. 8.59). Take any polynomial $P(x)$ and consider the quotient $P(x) = P(x)/1$. Since the denominator is never zero, we see that

$$\lim_{x \to a} P(x) = P(a)$$

by simply substituting $a$ for each $x$ in $P(x)$. This proves part (a).

The proof of part (b) is also immediate, since the procedure shows

$$\lim_{x \to a} \frac{P(x)}{Q(x)} = \frac{P(a)}{Q(a)}$$

at all points $x = a$ that are not zeros of $Q(x)$. Since division by zero is not allowed, the quotient is not defined at the zeros of $Q(x)$. ■

***Example 3*** Is the function

$$f(x) = \begin{cases} \dfrac{x^2 - x - 2}{x^2 + x - 6}, & \text{for } x \neq 2, -3 \\ \dfrac{3}{5}, & \text{for } x = 2, -3 \end{cases}$$

continuous?

At $x = -3$ we have

$$\lim_{x \to -3} f(x) = \lim_{x \to -3} \frac{(x-2)(x+1)}{(x-2)(x+3)} = \lim_{x \to -3} \frac{x+1}{x+3},$$

which does not exist; thus $f(x)$ is discontinuous at $x = -3$. But

$$\lim_{x \to 2} f(x) = \lim_{x \to 2} \frac{x+1}{x+3} = \frac{3}{5} = f(2),$$

so $y = f(x)$ is continuous at $x = 2$.

If we study Example 3, we see that some points of discontinuity can be removed by properly defining the function at certain values. In this example, the discontinuity at $x = 2$ is removed by defining $f(2) = 3/5$. However, no matter how we define $f(-3)$ we cannot remove the discontinuity, since

$$\lim_{x \to -3} f(x) \quad \text{does not exist.}$$

Similar continuous and discontinuous functions often arise in real situations.

***Example 4*** A portion of the 1973 Tax Rate Schedule for married taxpayers is shown in Table 8.2. Letting $x$ be the taxable income and $f(x)$ the tax, we may reinterpret Table 8.2 in the following form:

**Table 8.2**

| If the taxable income is | the tax is |
|---|---|
| not over $1000 | 14% of the amount |
| over $1,000 but not over $2,000 | $140 + 15% of the excess over $1,000 |
| " $2,000 " " " $3,000 | $290 + 16% " " " " $2,000 |
| " $3,000 " " " $4,000 | $450 + 17% " " " " $3,000 |
| " $4,000 " " " $8,000 | $620 + 19% " " " " $4,000 |
| " $8,000 " " " $12,000 | $1,380 + 22% " " " " $8,000 |
| " $12,000 " " " $16,000 | $2,260 + 25% " " " " $12,000 |

$$f(x) = \begin{cases} (0.14)x, & \text{for } 0 \leq x \leq 1000, \\ (0.15)(x - 1000) + 140, & \text{for } 1000 < x \leq 2000, \\ (0.16)(x - 2000) + 290, & \text{for } 2000 < x \leq 3000, \\ (0.17)(x - 3000) + 450, & \text{for } 3000 < x \leq 4000, \\ (0.19)(x - 4000) + 620, & \text{for } 4000 < x \leq 8000, \\ (0.22)(x - 8000) + 1{,}380, & \text{for } 8000 < x \leq 12{,}000, \\ (0.25)(x - 12{,}000) + 2{,}260, & \text{for } 12{,}000 < x \leq 16{,}000. \end{cases} \quad \textbf{(8.62)}$$

To verify the continuity of the function (8.62) we must verify that Eq. (8.60) holds at the points $x = 1000, 2000, 3000, 4000, 8000$, and $12{,}000$. Now

$$\lim_{x \to 1000} f(x) = 140 = f(1000),$$

$$\lim_{x \to 2000} f(x) = 290 = f(2000),$$

$$\lim_{x \to 3000} f(x) = 450 = f(3000),$$

$$\lim_{x \to 4000} f(x) = 620 = f(4000),$$

$$\lim_{x \to 8000} f(x) = 1{,}380 = f(8000),$$

$$\lim_{x \to 12{,}000} f(x) = 2{,}260 = f(12{,}000).$$

Thus, $y = f(x)$ is continuous.

***Example 5*** The United States first-class postage rate of 15 November 1970 was 6¢ for each ounce or fraction of an ounce. This rate may be expressed as a function in terms of the weight $x$ in ounces. For weights of up to four ounces we would have

$$f(x) = \begin{cases} 6, & \text{for } 0 \leq x \leq 1, \\ 12, & \text{for } 1 < x \leq 2, \\ 18, & \text{for } 2 < x \leq 3, \\ 24, & \text{for } 3 < x \leq 4. \end{cases} \quad \textbf{(8.63)}$$

A graph of the function (8.63) (see Fig. 8.10) clearly shows discontinuities at $x = 1, 2, 3, 4, \ldots$, resulting from the nonexistence of the limit of $f(x)$ at these values. For example, as $x$ approaches 1 through values less than 1, the function $f(x)$ is constantly equal to 6. But if $x$ approaches 1 through larger values than 1, $f(x)$ is constantly equal to 12. Thus no limit exists at $x = 1$. Hence, the function (8.63) is discontinuous at the points $x = 1, 2, 3, 4, \ldots$.

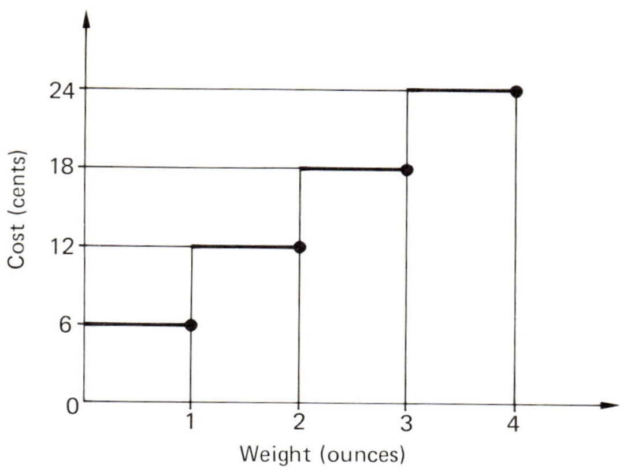


**Fig. 8.10**
The first-class letter rate on November 15, 1970.

***Example 6*** During the week of 25 October 1948 to 2 November 1948, the village of Donora, Pennsylvania, suffered a severe air pollution episode. Donora and the neighboring steel and zinc plants are located in a river valley. They experienced fog and a prolonged temperature inversion during this week. The sulfur oxides trapped by the inversion led to a five-fold increase in the death rate. Table 8.3 gives the percentage of persons severely affected by the air pollution during this period.

**Table 8.3**

| Age in years | 0–5 | 6–12 | 13–19 | 20–24 | 25–29 | 30–34 |
|---|---|---|---|---|---|---|
| Percent severely affected | 1.8 | 2.9 | 2.5 | 4.2 | 6.6 | 8.1 |

| Age in years | 35–39 | 40–44 | 45–49 | 50–54 | 55–59 | 60–64 | 65 and over |
|---|---|---|---|---|---|---|---|
| Percent severely affected | 9.1 | 10.9 | 14.6 | 18.5 | 20.4 | 29.4 | 28.9 |

*Source:* W. F. Ashe, *U.S. Technical Conference on Air Pollution*, L. C. McCabe, ed. (New York: McGraw-Hill, 1952).

Letting $x$ represent the age in years, and $f(x)$ the percentage of persons $x$ years old severely affected by the air pollution, we obtain the graph shown in Fig. 8.11. Note that a person not quite 13 years of age is listed as a 12-year-old, so that the graph exhibits discontinuities at $x = 6, 13, 20, 30, 35, 40, 45, 50, 55, 60$, and 65. Biological data of this type usually lead to discontinuities, but can still be quite useful in detecting trends. For example, in this case age is certainly a critical factor, and we can conclude that such air pollution episodes are extremely hazardous to older people.

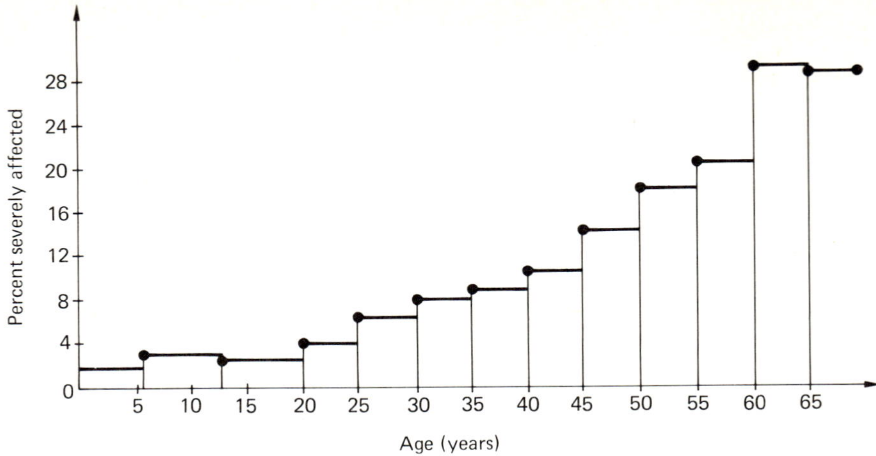


**Fig. 8.11**
Effects of air pollution in the Donora area from October 25 to November 2, 1948.

---

## EXERCISES 8.5

*Find all the points of discontinuity for each of the functions in Exercises* 1–16.

**1.** $f(x) = 5/x$

**2.** $f(x) = -3/(x^2 - 1)$

**3.** $f(x) = 2/(x - 3)$

**4.** $f(x) = \dfrac{x}{(x-1)(x+2)}$

**5.** $f(x) = \dfrac{2}{3x^2 - 4x + 1}$

**6.** $f(x) = \dfrac{4x - 2}{2x^2 - 7x + 3}$

**7.** $f(x) = \begin{cases} x + 1, & \text{for } x < 0 \\ 1 - x, & \text{for } x \geq 0 \end{cases}$

**8.** $f(x) = \begin{cases} 2x + 1, & \text{for } x < 1 \\ -x + 2, & \text{for } x \geq 1 \end{cases}$

**9.** $f(x) = \begin{cases} x^2, & \text{for } |x| < 2 \\ 4, & \text{for } |x| \geq 2 \end{cases}$

**10.** $f(x) = \begin{cases} \dfrac{3x - 1}{3x^2 - 4x + 1} & \text{for } x \neq 1/3 \\ -3/2, & \text{for } x = 1/3 \end{cases}$

**11.** $f(x) = \begin{cases} x^2, & \text{for } |x| < 1 \\ x + 1, & \text{for } x \geq 1 \\ x + 2, & \text{for } x \leq -1 \end{cases}$

**12.** $f(x) = \begin{cases} 2x + 1, & \text{for } x < 0 \\ x + 2, & \text{for } 0 \leq x \leq 1 \\ 2x + 1, & \text{for } 1 < x \leq 5 \\ x + 5, & \text{for } x > 5 \end{cases}$

**13.** $f(x) = \begin{cases} \dfrac{x}{|x|}, & \text{for } x \neq 0 \\ 1, & \text{for } x = 0 \end{cases}$

**14.** $f(x) = \begin{cases} \dfrac{x^2}{|x|}, & \text{for } x \neq 0 \\ 0, & \text{for } x = 0 \end{cases}$

**15.** $f(x) = \begin{cases} 1, & \text{for } x < 0 \\ x, & \text{for } 0 \leq x < 1 \\ x^2, & \text{for } 1 \leq x < 2 \\ x^3, & \text{for } x \geq 2 \end{cases}$

**16.** $f(x) = \begin{cases} \dfrac{x-4}{(x+2)(x-5)}, & \text{for } x < 1 \\ \dfrac{1}{4}, & \text{for } x = 1 \\ \dfrac{x}{5-x^2}, & \text{for } x > 1 \end{cases}$

**17.** Can $f(0)$ be redefined in Exercise 13 in such a way that the discontinuity is removed?

**18.** Find a constant $k$ so that when we redefine

$f(x) = k, \quad \text{for } x < 0$

in Exercise 12, the resulting function is continuous at $x = 0$.

**19.** The mean thickness of the tissue layer separating the alveoli and the capillaries in a rat depends on age. E. R. Weibel obtained the following data:*

| Age (days) | Thickness (microns) |
|---|---|
| 0 (birth) | 0.7 |
| 10 | 0.5 |
| 65 | 0.48 |

Find a continuous function that fits these data.

**20.** An adult male was asked to run at full speed until exhaustion, while the blood lactate was measured; this gave the following results:†

| Running time (minutes) | 0 | 1 | 2 | 2.58 |
|---|---|---|---|---|
| Blood lactate (in g/100 ml) | 10 | 68 | 112 | 160 |

Find a continuous function that fits these data.

* E. R. Weibel, "Postnatal growth of the lung and pulmonary gas-exchange capacity." In A. V. S. de Reuck and R. Porter (eds.), *Development of the Lung*, pp. 131–148. London: CIBA Foundation Symposium, 1965.

† S. Robinson et al., "Influence of fatigue on the efficiency of men during exhausting runs." *Journal of Applied Physiology* **12** (1958): 197–201.

## CHAPTER VOCABULARY

**amortizing** $A = \dfrac{P(i/12)}{1 - [1 + (i/12)]^{-n}}$

**annual interest rate** $i$

**annuity** $A = (P/i)[(1 + i)^n - 1]$

**arithmetic progression** $x_{n+1} = x_n + d$

**compound interest** $A = P(1 + i)^n$

**continuity at a point** $\lim_{x \to a} f(x) = f(a)$

**continuity on a set**

**converges**

**difference** $d$

**discontinuity at a point**

**discrete model**

**diverges**

**equity** $P - R$

**geometric progression** $x_{n+1} = x_n r$

**infinite sum**

**inherited rate**

**limit of a function** $\lim_{x \to a} f(x)$

**limit of a sequence** $\lim_{n \to \infty} x_n$

**model**

**multiplier (economic)** total revenue per payroll

**mutation rate**

**principal** $P$

**ratio** $r$

**recursion equation**

**sequence** (= ordered succession)

**simple interest** $A = P(1 + in)$

**tidal volume**

# CHAPTER 9 DIFFERENTIAL CALCULUS

## 9.1 DIFFERENCE QUOTIENTS, GROWTH RATES, AND MARGINAL CHANGE

You may have wondered why we placed so much emphasis in Section 8.4 on finding the limits of quotients in which both the numerator and denominator tend to zero as $x$ approaches $a$. The many examples and exercises exhibiting this behavior were designed to prepare you for differential calculus. Before making this statement precise, we present three real situations in which this behavior occurs and in which the limit has a useful physical meaning.

**1.** One hour out of Los Angeles on US Route 66, the traffic congestion disappears, and you settle down to steady driving. Over the next hour and a half you cover an additional 88 miles. What was your average velocity once you left the congestion behind?

To answer this question we must recall that

$$\text{Average velocity} = \frac{\text{distance traveled}}{\text{time elapsed}}.$$

Since we are only interested in the average velocity during the last hour and a half, we need only divide 88 by 1.5, obtaining

$$\text{Average velocity} = \frac{88}{1.5} = 58\tfrac{2}{3}\text{ mph.}$$

This simple calculation involves two obvious assumptions. The first concerns determining the elapsed time, which is *not* the time on the road (2.5 hours), but is the *difference* between the total time on the road and the time before the traffic congestion disappeared (1 hour):

$$\{\text{time elapsed}\} = \{\text{time on the road}\} - \{\text{time before congestion vanished}\}.$$

If we denote the time until congestion vanished by $t_0$ and the total time on the road by $t_0 + h$, we see that the time elapsed is

$$h = (t_0 + h) - t_0. \tag{9.1}$$

Similarly, the distance traveled is not the total distance since the trip began, but the distance since the congestion vanished. We can express this difference symbolically using functional notation if we let $s(t)$ denote the distance that has been covered when the time is $t$. Then the distance traveled since the congestion vanished is the *difference* between the total distance traveled, $s(t_0 + h)$, and the distance traveled in traffic congestion, $s(t_0)$, or

$$s(t_0 + h) - s(t_0). \tag{9.2}$$

Thus, the average velocity is given by the *difference quotient*:

$$\text{average velocity} = \frac{s(t_0 + h) - s(t_0)}{(t_0 + h) - t_0},$$

or

$$\text{average velocity} = \frac{s(t_0 + h) - s(t_0)}{h}. \tag{9.3}$$

Since $s(t_0 + h) = s(t_0) + 88$, $t_0 = 1$, and $h = 1.5$, we have

$$\begin{aligned}\text{average velocity} &= \frac{s(2.5) - s(1)}{(2.5) - (1)}\\ &= \frac{s(1) + 88 - s(1)}{1.5}\\ &= \frac{88}{1.5} = 58\tfrac{2}{3}\text{ mph.}\end{aligned}$$

It is quite true that all this additional work is really unnecessary in answering the original question. However, Eq. (9.3) allows us to answer some other questions that are less immediate. It is undoubtedly the case that during that hour and a half the speedometer did not constantly read $58\frac{2}{3}$ mph. Variations in road conditions, curves, hills, and an occasional highway patrol car must have led to changes in the *instantaneous velocity*, which is the reading observed on the speedometer. What connection is there between average velocity and instantaneous velocity? At a particular time $t_0$, the instantaneous velocity may be quite different from the average velocity over a long period of time. However, if we shorten the elapsed time period $h$ to a few seconds, we see that the average velocity will be nearly the same as the instantaneous velocity, except when we are accelerating or decelerating very rapidly. And even when we are rapidly accelerating or decelerating we can still get a close approximation of the instantaneous velocity by using the average velocity if we are willing to shorten the time period $h$ to a few microseconds. In short, *instantaneous velocity is simply the limit of the average velocity as the elapsed time interval $h$ tends to zero.* If we let $v(t)$ be the instantaneous velocity at time $t$ we can express this statement as

$$v(t) = \lim_{h\to 0} \frac{s(t + h) - s(t)}{(t + h) - (t)},$$

or

$$v(t) = \lim_{h\to 0} \frac{s(t + h) - s(t)}{h}. \tag{9.4}$$

**2.** A ten-year-old spruce tree is observed to grow an additional two feet during a three-month period. How fast is it growing?

This question is really a trap, because there is insufficient information to give an adequate answer. The data that have been supplied are sufficient for determining only the *average* growth rate, given by the difference quotient

$$\text{average growth rate} = \frac{\text{total growth}}{\text{elapsed time}} = \frac{2}{1/4} = 8 \text{ ft/year.}$$

However, this is no indication of the actual (instantaneous) growth rate, since we do not know whether the tree has been sustaining a constant growth during the three-month period. It may well be that no growth rate occurred over the first two months due to cold weather or drought. Again, the *instantaneous growth rate* is the limit of the average growth rate as the elapsed time interval tends to zero.

**3.** In economics or in the business world, nearly all decisions are made in terms of adding (or subtracting) a few units of a particular item to the inventory or the production run. These extra units are generally referred to as the *margin*. An analysis of the actual cost of this margin, or *marginal cost*, determines whether or not the additional units will be produced or stocked. The main reason that such an analysis is necessary is that unchecked production, for example, will create such a surplus that items may eventually have to be sold at a loss. Thus, there is a point of diminishing returns beyond which a good manager will not wish to venture.

The marginal cost $MC$ is determined by the difference quotient

$$MC = \frac{\text{change in total cost}}{\text{change in output}}. \tag{9.5}$$

Thus $MC$ is the average cost per unit *of the margin*.

Let's examine Eq. (9.5) in more detail. Suppose that $x$ units had already been produced at a cost of $C(x)$ dollars, and that the margin consists of $h$ additional units. Then the change in the total cost is

$$C(x + h) - C(x),$$

while the change in output is $h = (x + h) - x$. Thus, the marginal cost is

$$MC = \frac{C(x + h) - C(x)}{(x + h) - x},$$

or

$$MC = \frac{C(x + h) - C(x)}{h}. \tag{9.6}$$

Note that we have again obtained the same kind of a difference quotient as in the previous two examples. This suggests that we again examine the limit of this difference quotient as $h \to 0$. We shall see later that this limit is the instantaneous rate of change of the total cost curve. When the total cost curve begins to rise at an increasing rate we are at the point of diminishing returns.

It is useful to have a visual interpretation of the difference quotient and an intuitive idea of the meaning of the limit of the difference quotient as $h$ tends to zero. The following example illustrates the abstract situation.

***Example 1*** Consider the graph of the function $y = x^2$ shown in Fig. 9.1. Let $x_0$ be any fixed value of $x$ and suppose temporarily that $h$ is positive. The change in $y$-values can be computed by substituting each $x$-value in the equation $y = x^2$ and subtracting the $y$-value at $x_0$ from the $y$-value at $x_0 + h$, that is,

$$\text{change in } y = (x_0 + h)^2 - x_0^2.$$

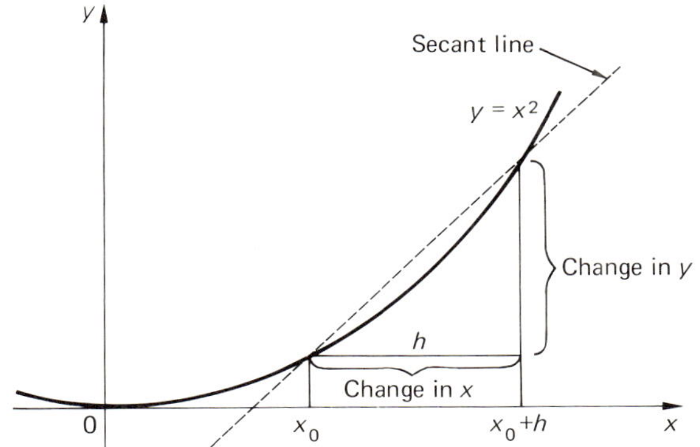


**Figure 9.1**

Thus, the difference quotient

$$\frac{\text{change in } y}{\text{change in } x} = \frac{(x_0 + h)^2 - x_0^2}{(x_0 + h) - x_0}$$

is simply the *slope of the straight line passing through the two points* $(x_0, x_0^2)$ *and* $(x_0 + h, (x_0 + h)^2)$ *on the curve* $y = x^2$. Such a line is called a *secant line*.

Let's do the arithmetic and find the slope of this secant line:

$$\text{Slope of secant line} = \frac{\text{change in } y}{\text{change in } x} = \frac{(x_0 + h)^2 - x_0^2}{(x_0 + h) - x_0}$$

$$= \frac{x_0^2 + 2x_0h + h^2 - x_0^2}{x_0 + h - x_0}$$

$$= \frac{2hx_0 + h^2}{h}.$$

We can factor an $h$ out of both terms in the numerator and cancel it with the $h$ in the denominator, obtaining

$$\text{slope of secant line} = \frac{h(2x_0 + h)}{h} = 2x_0 + h. \tag{9.7}$$

Thus, the limit of the slopes of the secant lines, as $h$ tends to zero, equals

$$\lim_{h \to 0} (2x_0 + h) = 2x_0.$$

The meaning of this limit may be grasped by looking at Fig. 9.2, where we have drawn three secant lines corresponding to the differences in $x$-values $h$, $h/2$, and $h/4$. Each secant line has been labeled by the respective change in $x$. You may notice that as $h$ gets closer to zero, the secant lines get closer to the line that is *tangent* to (just touches) the curve $y = x^2$ at the point $(x_0, x_0^2)$. Thus, *the slopes of the secant lines approach the slope of the tangent line as* $h \to 0$.

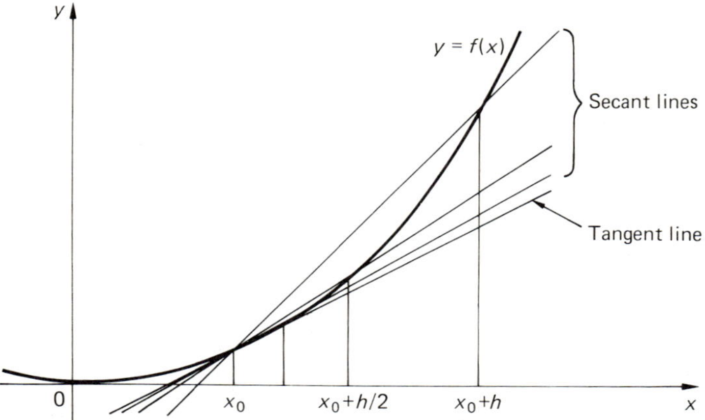


**Figure 9.2**

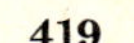

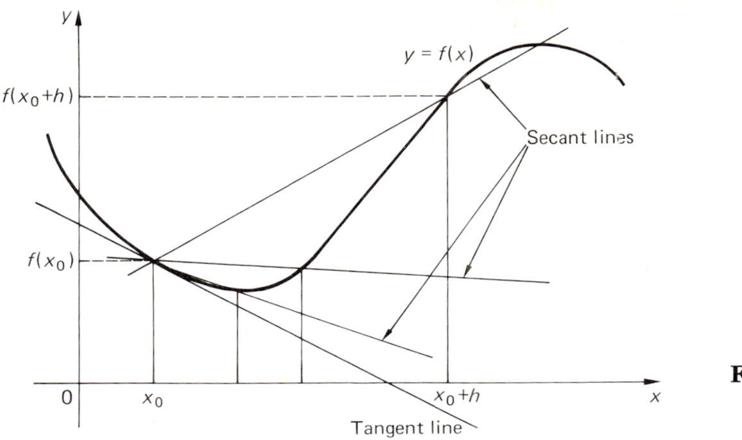


**Figure 9.3**

We now repeat this analysis for any function $y = f(x)$. The change in $y$ (see Fig. 9.3) may be expressed in functional notation by writing

$$\text{change in } y = f(x_0 + h) - f(x_0).$$

Hence, the slope of the secant line joining the point $(x_0, f(x_0))$ to the point $((x_0 + h), f(x_0 + h))$ is the difference quotient

$$\text{Slope of secant line} = \frac{f(x_0 + h) - f(x_0)}{h}, \tag{9.8}$$

that is, the change in $y$ divided by the change in $x$. Finally, the slope of the line tangent to the curve $y = f(x)$ at the point $(x_0, f(x_0))$ is the limit of the slopes of these secant lines as $h$ tends to zero, or

$$\text{Slope of tangent line} = \lim_{h \to 0} \frac{f(x_0 + h) - f(x_0)}{h}. \tag{9.9}$$

This last identity is *the* fundamental concept of differential calculus. It is extremely important that you understand it thoroughly.

The following two examples illustrate the procedure involved in finding the slope of the tangent line to a curve as stated in Eq. (9.9).

***Example 2*** Find the slope of the tangent line to the curve $y = f(x) = 3x^2$ at $x = 2$.

Substituting $x = 2$ in the equation $y = f(x) = 3x^2$, we obtain

$$f(2) = 3(2)^2 = 12,$$

so the curve $y = 3x^2$ passes through the point (2, 12). We wish to find the tangent line to the curve $y = 3x^2$ at the point (2, 12). Let the change in $x$ be $h$. Then the change in $y$ is the difference between the $y$-value at $x = 2 + h$ and the $y$-value at $x = 2$, or

$$\begin{aligned} f(2 + h) - f(2) &= 3(2 + h)^2 - 3(2)^2 \\ &= 3(4 + 4h + h^2) - 12 \\ &= 12h + 3h^2. \end{aligned}$$

Thus the slope of the secant line through these two points is the difference quotient

$$\frac{f(2 + h) - f(2)}{(2 + h) - (2)} = \frac{12h + 3h^2}{h} = 12 + 3h. \tag{9.10}$$

To find the slope of the tangent line to the curve $y = 3x^2$ at $x = 2$, we need only take the limit of Eq. (9.10) as $h \to 0$. Hence,

$$\text{slope of the tangent line} = \lim_{h \to 0} (12 + 3h) = 12.$$

If we wish, we can now write the equation of this tangent line, since we know its slope and the point (2, 12) through which it passes. We can use the point-slope formula given in Table 2.2 or simply recall that if $(x, y)$ is any other point on the tangent line, then

$$\frac{y - 12}{x - 2} = \text{slope of the tangent line} = 12,$$

or

$$\begin{aligned} y &= 12(x - 2) + 12 \\ &= 12x - 12. \end{aligned}$$

***Example 3*** Find the equation of the tangent line to the curve

$$y = f(x) = \frac{x - 1}{x}$$

at $x = 3$.

Here the difference quotient (the slope of the secant line) is

$$\begin{aligned} \frac{f(3 + h) - f(3)}{(3 + h) - 3} &= \frac{f(3 + h) - f(3)}{h} \\ &= \frac{\dfrac{(3 + h) - 1}{(3 + h)} - \dfrac{3 - 1}{3}}{h}. \end{aligned}$$

This expression looks very complicated, but is quite easy to simplify if we first find the common denominator of the two terms in the numerator. We have

$$\frac{f(3+h)-f(3)}{h} = \frac{\dfrac{3[(3+h)-1]-(3+h)(3-1)}{3(3+h)}}{h}$$

$$= \frac{6+3h-6-2h}{3(3+h)h}$$

$$= \frac{h}{3(3+h)h} = \frac{1}{3(3+h)}.$$

Now we take the limit as $h \to 0$, obtaining the slope of the tangent line

$$\lim_{h\to 0} \frac{f(3+h)-f(3)}{h} = \lim_{h\to 0} \frac{1}{3(3+h)} = \frac{1}{9}.$$

Finally, we substitute $x = 3$ in $y = f(x)$ to obtain the point on the curve $(3, f(3))$ through which the tangent line passes. Since

$$f(3) = \frac{3-1}{3} = \frac{2}{3},$$

we find that the equation of the tangent line satisfies

$$\frac{y-(2/3)}{x-3} = \text{slope of tangent line} = \frac{1}{9}$$

so that

$$y - (2/3) = (1/9)(x-3)$$

or

$$y = (1/9)x + (1/3).$$

---

**EXERCISES 9.1**

*In Exercises 1–20, find the equation of tangent line to the curve $y = f(x)$ at the given point.*

**1.** $f(x) = 2x^2,\ x = 1$

**2.** $f(x) = -5x^2,\ x = 2$

**3.** $f(x) = 3x^2 + x,\ x = -1$

**4.** $f(x) = 4x^2 - 2x,\ x = 2$

**5.** $f(x) = 5x^2 - 3x + 1,\ x = 0$

**6.** $f(x) = 5x^2 + x + 3,\ x = 1$

**7.** $f(x) = 2x^3,\ x = 2$

**8.** $f(x) = x^3 - x,\ x = -1$

**9.** $f(x) = x^4 + x^2 + 1,\ x = 0$

**10.** $f(x) = x^5 - x^2 + 4,\ x = 1$

**11.** $f(x) = \dfrac{2x - 1}{x},\ x = 1$

**12.** $f(x) = \dfrac{3x + 2}{4x},\ x = -1$

**13.** $f(x) = \dfrac{x}{x + 1},\ x = 1$

**14.** $f(x) = \dfrac{2x}{x - 3},\ x = 2$

**15.** $f(x) = \dfrac{x + 2}{x + 1},\ x = 0$

**16.** $f(x) = \dfrac{x + 1}{x^2},\ x = 1$

**17.** $f(x) = \dfrac{x^2}{x + 1},\ x = 0$

**18.** $f(x) = \dfrac{x^2 + 1}{x^2},\ x = -1$

**19.** $f(x) = \dfrac{x^2 + 1}{x + 1},\ x = 0$

**20.** $f(x) = \dfrac{x + 1}{x^2 + 1},\ x = 1$

**21.** The total cost of producing 100,000 refrigerators is \$28,950,000. If an extra 1000 refrigerators will raise the cost to \$29,250,000, what is the marginal cost?

**22.** A motorist leaves home at 9 A.M. and is 80 miles away from home at 10:30 A.M. By 12:15 P.M. he is 170 miles from home. What is his average velocity between 9 and 10:30 A.M.? What is his average velocity between 10:30 A.M. and 12:15 P.M.?

**23.** A rock is dropped from the top of a building 80 ft high. Suppose the height of the rock above the ground $t$ seconds after release is

$y(t) = 80 - 16\,t^2$.

a) Find the rock's average velocity over the first two seconds that it falls. (The velocity will be *negative* since the rock is *falling*.)
b) Find the average velocity of the rock over the time interval between $(2 - h)$ seconds and 2 seconds.
c) Use the result in part (b) to calculate the instantaneous velocity at time $t = 2$.

**24.** In Example 5 in Section 3.4 we obtained the formula

$$y = 310.82x^{-0.987}$$

relating the mean body weight $x$ (in grams) of birds to the percentage $y$ of body weight lost as water per day.
a) Find the average rate of change in the percentage of body weight loss between $x = 20$ g and $x = 25$ g.
b) Compare the result of part (a) with the average rate of change between $x = 24$ g and $x = 25$ g.
c) Repeat part (a) for $x = 24.9$ g and $x = 25$ g.
d) Repeat part (a) for $x = 24.99$ g and $x = 25$ g.
e) Estimate, based on the above calculations and any others you care to make, what the *instantaneous* change in the percentage of body weight loss will be at $x = 25$ g.

## 9.2 THE DERIVATIVE

Much of the material in the preceding four sections has been presented so that the reader will be able to understand the concept of derivatives. In Section 9.1 we learned that the average rate of change of a function $y = f(x)$, which we called the *difference quotient*, represented the slope of the secant line joining the two points $(x_0, f(x_0))$ and $(x_0 + h, f(x_0 + h))$, that is,

$$\text{slope of secant line} = \frac{f(x_0 + h) - f(x_0)}{h}. \qquad \textbf{(9.11)}$$

We saw also that the limit, as $h$ tends to zero, of the slopes of the secant lines determined by different values of $h$ tends to the slope of the line tangent to the curve $y = f(x)$ at the point $(x_0, f(x_0))$:

$$\text{slope of tangent line} = \lim_{h \to 0} \frac{f(x_0 + h) - f(x_0)}{h}. \qquad \textbf{(9.12)}$$

This slope represents the *instantaneous rate of change* of the function $y = f(x)$ at $x = x_0$. Now it is very awkward to continue calling the limit in Eq. (9.12) the "slope of the tangent line to the curve $y = f(x)$ at $x = x_0$," so instead we will call it the *derivative of f at $x_0$*, and we will express this definition symbolically by writing

$$f'(x_0) = \lim_{h \to 0} \frac{f(x_0 + h) - f(x_0)}{h}. \qquad \textbf{(9.13)}$$

From now on, whenever we see a prime it will denote the derivative of the preceding function. Thus, $g'(x)$ stands for the derivative of the function $y = g(x)$, while $(x^2 + 2x + 4)'$ represents the derivative of the function $y = x^2 + 2x + 4$. This notation for the derivative was invented by the great French mathematician Joseph Louis Lagrange (1736–1813).

There are several other notations that are frequently used for the derivative.* Each has certain advantages and disadvantages. In addition to Lagrange's "prime" notation above, we will make frequent use (especially in Chapter 10) of the notation introduced by Gottfried Leibniz (1646–1716):

$$\frac{dy}{dx} = \frac{df(x_0)}{dx} = \lim_{h \to 0} \frac{f(x_0 + h) - f(x_0)}{h}, \qquad \textbf{(9.14)}$$

---

* If $y = f(x)$, then

$$y',\ f'(x_0),\ D_x y,\ D_x f(x_0),\ \frac{dy}{dx},\ \frac{df(x_0)}{dx},\ \dot{y},\ \text{and}\ \dot{f}(x_0)$$

are all commonly used notations for the derivative.

which is read "$dy$ over $dx$" or "$df$ at $x_0$ over $dx$." This notation is particularly suggestive if we interpret the letter $d$ as an instantaneous "difference," that is, $dy/dx$ suggests the difference quotient

$$\frac{\text{difference in } y}{\text{difference in } x},$$

or

$$\frac{\text{difference in } f \text{ at } x_0}{\text{difference in } x} = \frac{f(x_0 + h) - f(x_0)}{(x_0 + h) - (x_0)} = \frac{f(x_0 + h) - f(x_0)}{h}. \quad \textbf{(9.15)}$$

Indeed, when $h$ is very close to zero we have the approximation

$$\frac{dy}{dx} = \frac{df}{dx}(x_0) \approx \frac{\text{difference in } y}{\text{difference in } x} = \frac{f(x_0 + h) - f(x_0)}{h}, \quad \textbf{(9.16)}$$

since the right side of Eq. (9.16) tends to the left side as $h$ approaches zero.

The following two examples will accustom us to the language of derivatives.

***Example 1*** Find the derivative of the function

$$y = f(x) = x^2 + 2x + 4 \quad \textbf{(9.17)}$$

at $x = 2$.

The problem asks us to determine $f'(2)$, the derivative of the function $y = f(x)$ at $x = 2$. We can do this by substituting $x = 2$ in Eq. (9.13), obtaining

$$f'(2) = \lim_{h \to 0} \frac{f(2 + h) - f(2)}{h}. \quad \textbf{(9.18)}$$

To find the actual value of $f'(2)$, we must replace the functional notation in Eq. (9.18) with the actual functions. That is, we must replace $f(2 + h)$ and $f(2)$ with what they equal. This is done by using Eq. (9.17). We replace each $x$ with $(2 + h)$ to obtain

$$f(2 + h) = (2 + h)^2 + 2(2 + h) + 4,$$

or

$$f(2 + h) = (4 + 4h + h^2) + (4 + 2h) + 4 = h^2 + 6h + 12,$$

and we replace each $x$ with 2 to determine

$$f(2) = (2)^2 + 2(2) + 4 = 12.$$

Substituting these two expressions into Eq. (9.18) yields

$$f'(2) = \lim_{h\to 0}\frac{(h^2 + 6h + 12) - (12)}{h} = \lim_{h\to 0}\frac{h^2 + 6h}{h}.$$

Factoring an $h$ from both terms in the numerator of the quotient and canceling this $h$ with the $h$ in the denominator, we have

$$f'(2) = \lim_{h\to 0}\frac{h(h + 6)}{h} = \lim_{h\to 0}(h + 6) = 6.$$

Thus, the derivative of $f$ at $x = 2$ is $f'(2) = 6$.

---

It is important to note that *the derivative of a function is also a function*, since it associates a real number with each number $x_0$ at which it is defined. In the following example we will determine the derivative of the function $f(x)$ at all values of $x$.

***Example 2*** Find $dy/dx$ when

$$y = f(x) = \frac{1}{x + 1}.$$

This problem calls for us to determine the derivative at any value of $x$. By Eq. (9.14) we have

$$\frac{dy}{dx} = \lim_{h\to 0}\frac{f(x + h) - f(x)}{h}$$

since no particular point $x_0$ has been specified. We must now determine the function $f(x + h)$. To do so we replace every $x$ in $f(x)$ by $(x + h)$ in the same way that we replaced every $x$ in Example 1 by $(2 + h)$:

$$f(x + h) = \frac{1}{(x + h) + 1} = \frac{1}{x + h + 1}.$$

Thus,

$$\frac{dy}{dx} = \lim_{h\to 0}\frac{\dfrac{1}{x + h + 1} - \dfrac{1}{x + 1}}{h},$$

and we can combine both terms in the numerator by using the common denominator $(x + 1)(x + h + 1)$, obtaining

$$\frac{dy}{dx} = \lim_{h\to 0}\frac{\dfrac{(x + 1) - (x + h + 1)}{(x + 1)(x + h + 1)}}{h}$$

$$= \lim_{h\to 0}\frac{-h}{h(x + 1)(x + h + 1)}.$$

Canceling the $h$ that multiplies the denominator with the $h$ in the numerator, we have

$$\frac{dy}{dx} = \lim_{h \to 0} \frac{-1}{(x+1)(x+h+1)}$$

$$= \frac{-1}{(x+1)(x+1)} = \frac{-1}{(x+1)^2}. \tag{9.19}$$

This is the derivative of $y = 1/(x+1)$ at any point $x$. If we wish to determine the derivative at $x = -3$, we merely substitute $-3$ for $x$ in Eq. (9.19) and obtain

$$\frac{df(-3)}{dx} = \frac{-1}{[(-3)+1]^2} = \frac{-1}{(-2)^2} = \frac{-1}{4}.$$

Equation (9.19) emphasizes that the derivative of the function

$$y = f(x) = \frac{1}{x+1}$$

is again a function

$$\frac{dy}{dx} = f'(x) = \frac{-1}{(x+1)^2}.$$

---

Since the derivative and the continuity of a function are both defined in terms of limits, it is natural to ask if there is any connection between the two concepts. Theorem 9.1 and the two examples that follow it show that

1. The function $f(x)$ is continuous at every point at which it has a derivative, but
2. The function $f(x)$ need not have a derivative at points at which it is continuous.

In other words, having a derivative implies continuity but a function can be continuous without having a derivative.

**THEOREM 9.1** **Suppose $f(x)$ has a derivative at $x_0$. Then $f$ is continuous at $x_0$.**

***Proof*** By the definition of continuity (see Eq. 8.60) we need to prove that

$$\lim_{x \to x_0} f(x) = f(x_0).$$

Let $x = x_0 + h$; then $x \to x_0$ if and only if $h \to 0$, so the equation above can be rewritten as

$$\lim_{h\to 0} f(x_0 + h) = f(x_0).$$

We must show that this equation is valid, so we begin with the left side and add $0 = -f(x_0) + f(x_0)$, obtaining

$$\lim_{h\to 0} f(x_0 + h) = \lim_{h\to 0} [f(x_0 + h) - f(x_0) + f(x_0)].$$

Since zero has been added, both sides are equal. Next we multiply the first two terms on the right by

$$1 = \frac{h}{h}$$

to obtain

$$\lim_{h\to 0} f(x_0 + h) = \lim_{h\to 0} \left\{\frac{[f(x_0 + h) - f(x_0)]h}{h} + f(x_0)\right\}.$$

Again equality is preserved since multiplication by 1 does not change what we have. Using the laws of limits in Theorem 8.1 we may rewrite the right side above as three limits:

$$\lim_{h\to 0} f(x_0 + h) = \left[\lim_{h\to 0} \frac{f(x_0 + h) - f(x_0)}{h}\right]\left[\lim_{h\to 0} h\right] + \left[\lim_{h\to 0} f(x_0)\right].$$

The first of these equals the derivative $f'(x_0)$, the second is zero, and the third is $f(x_0)$ since it does not depend on $h$. Hence we have

$$\lim_{h\to 0} f(x_0 + h) = f'(x_0) \cdot 0 + f(x_0) = f(x_0),$$

and the theorem is proved. ■

***Example 3*** Consider the function

$$y = f(x) = |x|.$$

The V-shaped graph of this function shows that $f$ is continuous at the bottom of the V:

$$\lim_{x\to 0} f(x) = \lim_{x\to 0} |x| = 0 = f(0).$$

However, we will see that the 90° corner at $x = 0$ precludes the existence of a derivative at this point since infinitely many lines just touch the graph at this point. Thus, no *one* line is tangent to the curve $y = x$ at this point. Consider the difference quotient

$$\frac{f(h) - f(0)}{h} = \frac{|h| - 0}{h} = \frac{|h|}{h}.$$

If $h$ is positive then $|h|/h = 1$, but if $h$ is negative $|h|/h = -1$. Thus the limit

$$\lim_{h\to 0} \frac{f(h) - f(0)}{h} = \lim_{h\to 0} \frac{|h|}{h}$$

does not exist, since *two* values are approached. Thus, $f(x)$ does not have a derivative at $x = 0$, even though it is continuous at that point.

---

The following example shows another situation in which continuity is present but the derivative does not exist.

***Example 4*** Consider the function $f(x) = \sqrt[3]{x}$. Since

$$\lim_{x\to 0} f(x) = \lim_{x\to 0} \sqrt[3]{x} = 0 = f(0),$$

the cube root function is definitely continuous at $x = 0$. However,

$$\lim_{h\to 0} \frac{f(h) - f(0)}{h} = \lim_{h\to 0} \frac{\sqrt[3]{h}}{h} = \lim_{h\to 0} \frac{1}{h^{2/3}}$$

does not exist, so there is no derivative at $x = 0$. Notice that we do have a derivative at all other points, since

$$\lim_{h\to 0} \frac{f(x + h) - f(x)}{h} = \lim_{h\to 0} \frac{\sqrt[3]{x + h} - \sqrt[3]{x}}{h}.$$

Multiplying numerator and denominator of the right side by

$$(\sqrt[3]{x + h})^2 + (\sqrt[3]{x + h})(\sqrt[3]{x}) + (\sqrt[3]{x})^2$$

we obtain (see Eq. 1.12)

$$\begin{aligned}\lim_{h\to 0} \frac{f(x + h) - f(x)}{h} &= \lim_{h\to 0} \frac{(\sqrt[3]{(x + h})^3 - (\sqrt[3]{x})^3}{h[(\sqrt[3]{x + h})^2 + (\sqrt[3]{x + h})(\sqrt[3]{x}) + (\sqrt[3]{x})^2]} \\ &= \lim_{h\to 0} \frac{x + h - x}{h[(\sqrt[3]{x + h})^2 + (\sqrt[3]{x + h})(\sqrt[3]{x}) + (\sqrt[3]{x})^2]}.\end{aligned}$$

Canceling the $x$'s in the numerator and dividing numerator and denominator by $h$, we have

$$\lim_{h\to 0}\frac{f(x+h)-f(x)}{h}=\lim_{h\to 0}\frac{1}{(\sqrt[3]{x+h})^2+(\sqrt[3]{x+h})(\sqrt[3]{x})+(\sqrt[3]{x})^2}.$$

Hence, the derivative of the cube root is

$$\frac{df(x)}{dx}=\frac{1}{3(\sqrt[3]{x})^2},\quad x\neq 0.$$

Notice that the denominator on the right vanishes when $x = 0$ and division by zero is prohibited.

---

The approximation suggested by Leibniz's notation in Eq. (9.16) has many useful applications. The following two examples illustrate some of its uses.

***Example 5*** Calculate an approximate value for $\sqrt[3]{7.9}$.

Rewriting Eq. (9.16) as

$$\frac{df(x)}{dx}\approx\frac{f(x+h)-f(x)}{h},$$

and observing that $\sqrt[3]{8}=2$, we let $f(x)=\sqrt[3]{x}$, $x = 8$, and $h = -0.1$. Using the derivative of $\sqrt[3]{x}$ that we determined in Example 4, we have

$$\frac{1}{3(\sqrt[3]{x})^2}\approx\frac{\sqrt[3]{x+h}-\sqrt[3]{x}}{h},$$

implying that

$$\frac{1}{3(\sqrt[3]{8})^2}\approx\frac{\sqrt[3]{8+(-0.1)}-\sqrt[3]{8}}{(-0.1)}$$

or

$$\frac{1}{3\cdot 4}\approx\frac{\sqrt[3]{7.9}-2}{-0.1}.$$

Hence,

$$\begin{aligned}\sqrt[3]{7.9}&\approx 2+(-0.1)\frac{1}{3\cdot 4}\\&\approx 2-\frac{0.1}{12}=2-0.0083=1.9917,\end{aligned}$$

which is very close to the actual cube root $\sqrt[3]{7.9} = 1.9916317$ (to seven decimal places).

***Example 6*** A bank purchases a square lot in mid-Manhattan, paying \$100 per square foot. The lot is described as being 223 ft long on each side, but careful remeasurement shows it is actually 222.47 ft long on each side. Approximately how much did the bank lose in the transaction?

We showed in Example 1 in Section 9.1 (see Eq. 9.7) that

$$(x^2)' = 2x.$$

Setting $x = 223$ and $h = -0.53$, we have by Eq. (9.16)

$$2x \approx \frac{(x+h)^2 - x^2}{h},$$

or

$$2(223) \approx \frac{(222.47)^2 - (223)^2}{-0.53}.$$

Hence, the difference in square footage is approximately

$$(223)^2 - (222.47)^2 \approx 2(223)(0.53) = 236.38 \text{ sq ft},$$

implying that the bank lost \$23,638 on the transaction. If we compute the exact answer we find that

$$(223)^2 - (222.47)^2 = 236.0991$$

so we see that the error is less than 0.3 sq ft.

---

## EXERCISES 9.2

*In Exercises 1–16, calculate the derivative of the given function $f(x)$ at the indicated values of $x$. If no value of $x$ is given, find the derivative for any $x$, that is, determine the function $f'(x)$.*

**1.** $f(x) = x^2 + 1, x = 1$

**2.** $f(x) = x^3 + 1$

**3.** $f(x) = x^2 + x + 1,\ x = 2$

**4.** $f(x) = 2x^3 + x,\ x = 3$

**5.** $f(x) = 5x^3 - 3x + 1,\ x = -1$

**6.** $f(x) = 2x^2 + 4x + 8$

**7.** $f(x) = x^4 - 1,\ x = -2$

**8.** $f(x) = 4x^4 + 1,\ x = 0, 1$

**9.** $f(x) = 3x^3 + 7,\ x = 1$

**10.** $f(x) = x^5 - 1$

**11.** $f(x) = x^6,\ x = 0, 1$

**12.** $f(x) = 2x^5,\ x = 1, 2$

**13.** $f(x) = \dfrac{x}{x^2 + 1}$

**14.** $f(x) = \dfrac{x}{x + 1},\ x = 1, 2, 3$

**15.** $f(x) = \dfrac{1+x}{1-x}, \; x = 2, 3$

**16.** $f(x) = \dfrac{2}{1+x^2}, \; x = 0, 1$

**17.** Show that $(\sqrt{x})' = 1/(2\sqrt{x})$. Use this result to approximate $\sqrt{9.2}$ and $\sqrt{15.93}$.

**18.** Suppose that the length of each side of the lot in Example 6 was 223.24 ft long. How much did the bank save?

**19.** A book publisher uses 1″ margins on each 6″ × 9″ page. Approximately how much more type can be set on each page (in square inches) if the publisher uses 0.88″ margins? If a 500-page book with 1″ margins is reset with 0.88″ margins, approximately how long will the book be? (*Hint:* Let $x$ be the margin size.)

**20.** Suppose the publisher in Exercise 19 decides to keep the 1″ margins but to use 7″ × 10″ pages. Approximately how many pages would the 500-page 6″ × 9″ book then occupy? (*Hint:* Let $x$ be the increase in each dimension.)

**21.** A rock dropped from the top of a building 80 ft high is

$$y(t) = 80 - 16t^2 \text{ ft}$$

above ground level $t$ seconds after it is released.

a) When does the rock reach the ground?

b) Find the velocity of the rock at any time $t$. (*Hint:* Remember that downward velocity is the instantaneous rate of change of height.)

c) Use the formula obtained in part (b) to calculate the velocity of the rock every 0.5 sec after it is released and at the moment of impact.

★ **22.** In Example 5 in Section 2.7 we described the indifference curve of a shopper in a supermarket. Suppose Mary's indifference curve with highest utility is given by the equation $xy = 60$. Find the price per pound of hamburger if potatoes cost 50¢/lb and she buys 12 lb of potatoes. How much money does Mary spend? (*Hint:* Find the equation of the tangent line to the indifference curve at $x = 12$.)

★ **23.** The procedure we will develop in this problem, called *Newton's method,* is an efficient tool for finding the roots of a function. Suppose we seek a root of the function $f(x)$. Let $x_0$ be any point and suppose a root of $f$ is at $x_0 + h$, that is, $f(x_0 + h) = 0$.

a) Use Eq. (9.16) to show that

$$x_0 + h \approx x_0 - \frac{f(x_0)}{f'(x_0)}.$$

b) Newton's method is given by the recurrence equation

$$x_{n+1} = x_n - \frac{f(x_n)}{f'(x_n)}.$$

Give a plausible argument suggesting that

$$\lim_{n \to 0} x_n = x_0 + h.$$

c) Show that if $f(x) = x^2 - a$, then Newton's method yields

$$x_{n+1} = \frac{1}{2}\left(x_n + \frac{a}{x_n}\right).$$

d) Set $a = 2$ and $x_0 = 1$, and show that the sequence $x_0$, $x_1$, $x_2$, $x_3, \ldots$ obtained by using part (c) tends to $\sqrt{2} \approx 1.414213562$. Why does this happen?

---

## 9.3 SIX RULES OF DIFFERENTIATION

In Sections 9.1 and 9.2 we found derivatives by taking the limit of a difference quotient as $h$ tends to zero. In this section we present six rules that will permit us to avoid the limit process. In particular, once we have learned these six rules and the chain rule discussed in Section 9.4, we will be able to find the derivatives of all expressions involving roots and powers of the independent variables without the direct use of limits. Although some of these rules look complicated, they represent an enormous saving in time and effort over what would be required in using the limit definition of a derivative directly.

First, however, we add an additional word to our mathematical vocabulary. The process of finding the derivative of a function is called *differentiation*. Thus, instead of saying "find the derivative of the function $f(x)$," we may simply say "differentiate $f(x)$."

In presenting each of these rules we will first consider a practical example illustrating that rule. One point must be emphasized: In the initial example and the proof of the rule we *must* use the limit definition of derivative. However, once the rule has been proved, we need never return to the limit definition in an identical situation. After each rule we provide several examples of how the rule is used in practice, and that will be the way we use it henceforth.

First, let's consider the constant function $y = 3$. If we plot this function in the $xy$-plane we obtain a horizontal line passing through the number 3 on the $y$-axis. Since the slope of a straight line is given by the ratio (rise/run) and this line has zero rise, it has slope zero. Thus, the difference quotient is always zero for any two points on this line, which implies that the derivative of the function $f(x) = 3$ is $f'(x) = 0$. This motivates our first rule.

***RULE 1*** The derivative of a constant function is zero, that is,

$$(c)' = 0$$

for any constant.

To prove this rule let $f(x) = c$ for all $x$. Then $f(x + h) = c$ and using the limit definition of the derivative (9.13) we have

$$f'(x) = \lim_{h\to 0}\frac{f(x + h) - f(x)}{h} = \lim_{h\to 0}\frac{c - c}{h} = \lim_{h\to 0}\frac{0}{h} = 0.$$

***Example 1*** Using Rule 1, we have

$$(17)' = 0, \qquad (2)' = 0, \qquad (0)' = 0, \qquad (-1/3)' = 0.$$

---

The next function we will consider is a power of $x$:

$$f(x) = x^4.$$

In this case we cannot conclude anything from the geometry of the graph of $y = f(x)$. Instead we must rely entirely on the limit definition of the derivative. We have

$$(x^4)' = f'(x) = \lim_{h\to 0}\frac{f(x+h) - f(x)}{h}$$

$$= \lim_{h\to 0}\frac{(x+h)^4 - x^4}{h}.$$

We now use Pascal's triangle (see Section 1.3) to expand $(x + h)^4$, and obtain

$$(x^4)' = \lim_{h\to 0}\frac{(x^4 + 4x^3h + 6x^2h^2 + 4xh^3 + h^4) - x^4}{h}.$$

Canceling the $x^4$-terms and factoring a multiple of $h$ from the remaining terms in the numerator, we get

$$(x^4)' = \lim_{h\to 0}\frac{h(4x^3 + 6x^2h + 4xh^2 + h^3)}{h}$$

$$= \lim_{h\to 0}(4x^3 + 6x^2h + 4xh^2 + h^3) = 4x^3.$$

---

We might wish to compare this result with Example 1 in Section 9.1, where we showed that $(x^2)' = 2x$. These two examples might lead us to suspect that when we differentiate a power of $x$, the power is reduced by one and the resulting function is multiplied by a constant equal to the previous power. For example,

$$(x^2)' = 2x^{2-1} \quad \text{and} \quad (x^4)' = 4x^{4-1}.$$

This leads to the following rule.

***RULE 2*** $(x^n)' = nx^{n-1}$, for *any* number $n$. **(9.20)**

Although Rule 2 is valid for all values of $n$, we will prove it here only for positive integers. A complete proof of Rule 2 is technically complicated.

To show that Eq. (9.20) holds, let $f(x) = x^n$ and apply the limit definition of the derivative. We obtain

$$(x^n) = f'(x) = \lim_{h\to 0}\frac{f(x+h) - f(x)}{h}$$
$$= \lim_{h\to 0}\frac{(x+h)^n - x^n}{h}. \quad (9.21)$$

From Pascal's triangle we know the first two terms of $(x+h)^n$ and that the remaining terms involve higher powers of $h$:

$$(x+h)^n = x^n + nx^{n-1}h + (\text{terms involving } h^k, \quad k \geq 2).$$

Substituting this expression into Eq. (9.21) yields

$$(x^n)' = \lim_{h\to 0}\frac{[x^n + nx^{n-1}h + h^2 \times (\text{other terms})] - x^n}{h}$$

so that

$$(x^n)' = \lim_{h\to 0}\frac{h[nx^{n-1} + h \times (\text{other terms})]}{h}.$$

Canceling the multiples of $h$ in the numerator and denominator, we have

$$(x^n)' = \lim_{h\to 0}[nx^{n-1} + h \times (\text{other terms})] = nx^{n-1}.$$

***Example 2*** Using Rule 2 we obtain

a) $(x^8)' = 8x^7$

b) $(x^{91})' = 91x^{90}$

c) $(x^{2.3})' = 2.3x^{1.3}$

d) $(x^{-4})' = -4x^{-5}$

e) $(\sqrt{x})' = (x^{1/2})' = (1/2)x^{-1/2}$

f) $\left(\dfrac{1}{x^{2/3}}\right)' = (x^{-2/3})' = (-2/3)x^{-5/3}$

g) $(x)' = (x^1)' = 1x^0 = 1$

h) $(x^0)' = 0x^{-1} = 0$

The result in (h) should not surprise us, since $x^0 = 1$ and we already know from Rule 1 that the derivative of a constant is zero.

---

Generally, however, we don't encounter powers of $x$ alone. Often they are multiplied by some constant, such as

$$f(x) = 5x^4.$$

In this situation, using the limit definition of a derivative, we have

$$(5x^4)' = f'(x) = \lim_{h\to 0}\frac{f(x+h)-f(x)}{h}$$
$$= \lim_{h\to 0}\frac{5(x+h)^4-5x^4}{h} \tag{9.22}$$

We can now factor the 5 from both terms in the numerator, and *since a constant does not affect the limiting process* (see Theorem 8.1d), we can carry the 5 outside the limit. Hence,

$$(5x^4)' = 5\lim_{h\to 0}\frac{(x+h)^4-x^4}{h}.$$

What remains inside the limit is exactly the same difference quotient that we had when we differentiated $x^4$. Thus,

$$(5x^4)' = 5(x^4)' = 5(4x^3) = 20x^3.$$

This leads to the following rule.

***RULE 3*** Constants that multiply a function are carried along under differentiation:

$$[cf(x)]' = c[f(x)]'. \tag{9.23}$$

To prove Eq. (9.23) we turn to the limit definition of the derivative (9.13) and observe that $c$ can be carried outside the limit:

$$[cf(x)]' = \lim_{h\to 0}\frac{cf(x+h)-cf(x)}{h}$$
$$= \lim_{h\to 0}\frac{c[f(x+h)-f(x)]}{h}$$
$$= c\lim_{h\to 0}\frac{f(x+h)-f(x)}{h} = cf'(x).$$

***Example 3*** Using Rule 3 we have the following:

a) $(4x^2)' = 4(2x) = 8x$

b) $(-3x^8)' = -3(8x^7) = -24x^7$

c) $\left(\frac{-2}{3}x^{1/2}\right)' = \frac{-2}{3}\left(\frac{1}{2}x^{-1/2}\right) = \frac{-1}{3}x^{-1/2}$

d) $(\sqrt{3}x^{-1.51})' = \sqrt{3}(-1.51x^{-2.51}) = -1.51\sqrt{3}x^{-2.51}$

e) $(5x)' = 5(x^1)' = 5(1 \cdot x^0) = 5$

f) $(17x^0)' = 17(0 \cdot x^{-1}) = 0$

Again (f) is no surprise if we remember that $x^0 = 1$. Then $17x^0 = 17$ and by Rule 1, the derivative of a constant is zero.

---

What happens if our function is the sum (or difference) of two or more powers of $x$? For example, suppose

$$f(x) = 2x^2 - 5x + 4.$$

The next rule proves that the derivative of this function,

$$f'(x) = 4x - 5,$$

is obtained by differentiating each term of the function $f(x)$ separately,

$$(2x^2)' = 2(x^2)' = 2(2x) = 4x,$$

$$(-5x)' = -5(x')' = -5(1 \cdot x^0) = -5,$$

and

$$(4)' = 0,$$

and then taking the sum of these derivatives.

***RULE 4*** The derivative of a sum of functions equals the sum of the derivatives of the functions:

$$[f(x) + g(x)]' = f'(x) + g'(x). \tag{9.24}$$

The proof of this rule is again a direct application of the limit definition of the derivative (9.13):

$$[f(x) + g(x)]' = \lim_{h \to 0} \frac{[f(x+h) + g(x+h)] - [f(x) + g(x)]}{h}$$

$$= \lim_{h \to 0} \frac{f(x+h) - f(x)}{h} + \frac{g(x+h) - g(x)}{h}$$

by rearranging the terms in the numerator. But the limit of each of these difference quotients, as $h$ tends to zero, will be the derivative of the respective function. Thus, Eq. (9.24) has been verified.

Although Rule 4 is stated expressly in terms of sums of functions, *it also applies to differences*, since subtraction is simply the addition of a negative term. Before proceeding to the remaining rules, it is advantageous to summarize the preceding four rules and see what implications can be drawn from them.

| |
|---|
| **1.** $(c)' = 0$, $c$ constant. |
| **2.** $(x^n)' = nx^{n-1}$, for any real number $n$. |
| **3.** $[cf(x)]' = cf'(x)$, $c$ constant. |
| **4.** $[f(x) + g(x)]' = f'(x) + g'(x)$. |

If we consider these four rules carefully, we note that with them we can find the derivative of any polynomial, since polynomials are simply finite sums of multiples of integer powers of $x$, as the following example demonstrates.

***Example 4*** Find the derivatives of the following polynomials

a) $p(x) = 7x^3 + 4x^2 - 3x + 8$;

b) $q(x) = \sqrt{2}x^4 - 17x^2 - \pi x + 1/2$.

To solve (a) we note first that $p(x)$ is the sum of four terms; thus, by Rule 4,

$$p'(x) = (7x^3)' + (4x^2)' + (-3x)' + (8)'.$$

Each term can be differentiated, so using Rules 1 and 3 we obtain

$$p'(x) = 7(x^3)' + 4(x^2)' + (-3)(x^1)' + (8)',$$

and by Rule 2 it follows that

$$p'(x) = 7(3x^2) + 4(2x^1) + (-3)(1x^0) + 0,$$

or

$$p'(x) = 21x^2 + 8x - 3.$$

Instead of belaboring the repeated use of the rules in solving (b), let's simply differentiate each term as we come to it, and then do the arithmetic to obtain the final answer:

$$q'(x) = \sqrt{2}(4x^3) - 17(2x^1) - \pi(1 \cdot x^0) + 0$$

or

$$q'(x) = 4\sqrt{2}x^3 - 34x - \pi.$$

---

Other expressions can also be solved in this fashion as is illustrated by the following example.

***Example 5*** Differentiate these functions:

a) $f(x) = \sqrt{x} - (4/\sqrt{x})$;

b) $g(x) = 3x^{\pi} - 2x^{1/3} + 7x^{-3/4} + 12$;

c) $h(x) = 2x^{7/9} - \frac{2}{3}x^{4/3} + 1.8x^{-0.6} - 5$.

Again we obtain each derivative by differentiating each term in the function as we come to it. First we change the roots to exponents:

$$f'(x) = (x^{1/2} - 4x^{-1/2})' = \frac{1}{2}x^{-1/2} + (-4)\left(\frac{-1}{2}x^{-3/2}\right)$$

or

$$f'(x) = \frac{1}{2}x^{-1/2} + 2x^{-3/2}.$$

Similarly,

$$g'(x) = 3(\pi x^{\pi-1}) + (-2)\left(\frac{1}{3}x^{-2/3}\right) + 7\left(\frac{-3}{4}x^{-7/4}\right) + 0$$

or

$$g'(x) = 3\pi x^{\pi-1} - \frac{2}{3}x^{-2/3} - \frac{21}{4}x^{-7/4},$$

and

$$h'(x) = 2\left(\frac{7}{9}x^{-2/9}\right) - \frac{2}{3}\left(\frac{4}{3}x^{1/3}\right) + 1.8(-0.6x^{-1.6}) + 0$$

or

$$h'(x) = \frac{14}{9}x^{-2/9} - \frac{8}{9}x^{1/3} - 1.08x^{-1.6}.$$

Example 5(a) shows that in dealing with derivatives of roots it is best to merely change the root into an exponential. For example,

$$(\sqrt[3]{x})' = (x^{1/3})' = \frac{1}{3}x^{-2/3},$$

$$(\sqrt[5]{x^3})' = (x^{3/5})' = \frac{3}{5}x^{-2/5},$$

and

$$\left(\frac{1}{\sqrt[4]{x^3}}\right)' = (x^{-3/4})' = \frac{-3}{4}x^{-7/4}.$$

---

Now that we have successfully dealt with the derivative of a sum or difference of functions, it is logical to consider derivatives of products and quotients of functions. It would be very nice if it were true that the derivative of a product is the product of the derivatives, *but this is simply not the case*! Neither is it true that the derivative of a quotient is the quotient of the derivatives! The following two examples should convince you that these simplistic rules do not work. We know that

$$x^4 \cdot x^3 = x^7,$$

but

$$(x^4)' \cdot (x^3)' = (4x^3) \cdot (3x^2) = 12x^5,$$

while

$$(x^7)' = 7x^6.$$

Similarly,

$$x^7/x^4 = x^3,$$

but

$$(x^7)'/(x^4)' = 7x^6/4x^3 = (7/4)x^3,$$

while

$$(x^3)' = 3x^2.$$

Thus rules for the derivatives of products or quotients of functions must be more complicated than we would assume at first. The two correct rules are stated below. We will demonstrate their application before indicating their proof.

***RULE 5 PRODUCT RULE*** The derivative of a product of two functions is the sum of the product of each function with the derivative of the other:

$$[f(x)g(x)]' = f(x)g'(x) + g(x)f'(x). \tag{9.25}$$

***RULE 6 QUOTIENT RULE*** The derivative of a quotient of two functions is given by the expression

$$\left(\frac{f(x)}{g(x)}\right)' = \frac{g(x)f'(x) - f(x)g'(x)}{g^2(x)}. \tag{9.26}$$

Let's see how well these two rules work in our previous examples. If we let $f(x) = x^4$ and $g(x) = x^3$ in Eq. (9.25), we have

$$\begin{aligned}(x^4 \cdot x^3)' &= x^4(x^3)' + x^3(x^4)' \\ &= x^4(3x^2) + x^3(4x^3) = 7x^6,\end{aligned}$$

which is exactly what we would get by differentiating $x^7$.

Now letting $f(x) = x^7$ and $g(x) = x^4$ in Eq. (9.26) yields

$$\left(\frac{x^7}{x^4}\right)' = \frac{x^4(x^7)' - x^7(x^4)'}{(x^4)^2}$$

$$= \frac{x^4(7x^6) - x^7(4x^3)}{x^8}$$

$$= \frac{7x^{10} - 4x^{10}}{x^8} = 3x^2 = (x^3)',$$

exactly the result we were supposed to get.

Success in finding the derivative of products and quotients depends *entirely* on remembering exactly what the two rules state. You should memorize Eqs. (9.25) and (9.26), since any deviation from their exact statement is almost certain to lead to errors.

***Example 6*** Find the derivatives of these expressions:

a) $(4x - 2)(6x^3 + 5x^2 + 7x + 1)$;

b) $\dfrac{1 - x}{3x + 2}$;

c) $\dfrac{(4x + x^2)(x - 3)}{x^3 + x + 2}$.

In (a) we let $f(x) = 4x - 2$ and $g(x) = 6x^3 + 5x^2 + 7x + 1$ and apply the product rule, obtaining

$$\begin{aligned}[f(x)g(x)]' &= (4x - 2)(6x^3 + 5x^2 + 7x + 1)' \\ &\quad + (6x^3 + 5x^2 + 7x + 1)(4x - 2)' \\ &= (4x - 2)(18x^2 + 10x + 7) + (6x^3 + 5x^2 + 7x + 1)(4).\end{aligned}$$

After we multiply and add like terms, we obtain

$$[(4x - 2)(6x^3 + 5x^2 + 7x + 1)]' = 96x^3 + 24x^2 + 36x - 10.$$

For (b) let $f(x) = 1 - x$ and $g(x) = 3x + 2$ and apply the quotient rule. Then

$$\left[\frac{1 - x}{3x + 2}\right]' = \frac{(3x + 2)(1 - x)' - (1 - x)(3x + 2)'}{(3x + 2)^2}$$

$$\left[\frac{1-x}{3x+2}\right]' = \frac{(3x+2)(-) - (1-x)(3)}{(3x+2)^2}$$

$$= \frac{-3x-2-3+3x}{(3x+2)^2} = \frac{-5}{(3x+2)^2}.$$

In (c) we must use both rules since we have a quotient whose numerator is a product. First we use the quotient rule:

$$\left[\frac{(4+x^2)(x-3)}{x^3+x+2}\right]'$$
$$= \frac{(x^3+x+2)[(4+x^2)(x-3)]' - [(4+x^2)(x-3)](x^3+x+2)'}{(x^3+x+2)^2}.$$

Now, using the product rule, we differentiate the first term in brackets in the numerator, obtaining

$$\left[\frac{(4+x^2)(x-3)}{x^3+x+2}\right]'$$
$$= \frac{(x^3+x+2)[(4+x^2)(x-3)' + (x-3)(4+x^2)'] - [(4+x^2)(x-3)](x^3+x+2)'}{(x^3+x+2)^2}$$

$$= \frac{(x^3+x+2)[(4+x^2)(1) + (x-3)(2x)] - [(4+x^2)(x-3)](3x^2+1)}{(x^3+x+2)^2}.$$

Simplifying by adding like terms, we have

$$\left[\frac{(4+x^2)(x-3)}{x^3+x+2}\right]'$$
$$= \frac{(x^3+x+2)(4-6x+3x^2) - (x^3-3x^2+4x-12)(3x^2+1)}{(x^3+x+2)^2}$$
$$= \frac{3x^4-6x^3+39x^2-12x+20}{(x^3+x+2)^2}$$

after a lengthy multiplication.

---

To close this section we now show how the product and quotient rules are obtained. We make use of the limit definition of the derivative and a very common trick in mathematical proofs: adding two terms whose sum is zero.

By the limit definition of the derivative we know that

$$[f(x)g(x)]' = \lim_{h\to 0} \frac{f(x+h)g(x+h) - f(x)g(x)}{h}, \tag{9.27}$$

that is, the limit of the difference quotient obtained by taking the difference of the products of the functions at $x + h$ minus the products at $x$ and dividing by $(x + h) - x = h$. Now insert the zero

$$-f(x+h)g(x) + f(x+h)g(x)$$

in the numerator of the right side of Eq. (9.27), obtaining

$$\begin{aligned}&[f(x)g(x)]' \\ &\quad = \lim_{h\to 0} \frac{f(x+h)g(x+h) - f(x+h)g(x) + f(x+h)g(x) - f(x)g(x)}{h}.\end{aligned}$$

The equality is preserved since we have added zero. Factoring the $f(x + h)$ term from the first two terms and $g(x)$ from the last two terms in the numerator, we obtain by separation

$$\begin{aligned}[f(x)g(x)]' = \lim_{h\to 0} \Bigg\{ f(x+h)\left[\frac{g(x+h)-g(x)}{h}\right] \\ + g(x)\left[\frac{f(x+h)-f(x)}{h}\right]\Bigg\}.\end{aligned} \tag{9.28}$$

Now, recalling that the limit of a sum or product is the sum or product of the limits, we can rewrite Eq. (9.28) as

$$\begin{aligned}[f(x)g(x)]' = \left[\lim_{h\to 0} f(x+h)\right]\left[\lim_{h\to 0}\frac{g(x+h)-g(x)}{h}\right] \\ + \left[\lim_{h\to 0} g(x)\right]\left[\lim_{h\to 0}\frac{f(x+h)-f(x)}{h}\right].\end{aligned}$$

The differentiability of $f(x)$ implies its continuity, so that the first limit equals $f(x)$, while the third equals $g(x)$ since it does not depend on $h$.

Hence

$$\begin{aligned}[f(x)g(x)]' &= \left[\lim_{h\to 0} f(x+h)\right]\left[\lim_{h\to 0}\frac{g(x+h)-g(x)}{h}\right] \\ &\quad + [g(x)]\left[\lim_{h\to 0}\frac{f(x+h)-f(x)}{h}\right] \\ &= f(x)g'(x) + g(x)f'(x).\end{aligned}$$

The quotient rule is also obtained by adding zero to the numerator. This time we have

$$\left[\frac{f(x)}{g(x)}\right]' = \lim_{h\to 0} \frac{\dfrac{f(x+h)}{g(x+h)} - \dfrac{f(x)}{g(x)}}{h},$$

so we add

$$\frac{-f(x)}{g(x+h)} + \frac{f(x)}{g(x+h)}$$

to obtain

$$\begin{aligned}\left[\frac{f(x)}{g(x)}\right]' &= \lim_{h\to 0} \frac{\dfrac{f(x+h)}{g(x+h)} - \dfrac{f(x)}{g(x+h)} + \dfrac{f(x)}{g(x+h)} - \dfrac{f(x)}{g(x)}}{h} \\ &= \lim_{h\to 0}\left\{\left[\frac{1}{g(x+h)}\right]\left[\frac{f(x+h)-f(x)}{h}\right] \right. \\ &\quad \left. + \left[\frac{f(x)}{h}\right]\left[\frac{1}{g(x+h)} - \frac{1}{g(x)}\right]\right\}. \end{aligned} \tag{9.29}$$

Simplifying the last term in Eq. (9.29) by finding the common denominator yields

$$\begin{aligned}\left[\frac{f(x)}{g(x)}\right]' &= \lim_{h\to 0}\left\{\left[\frac{1}{g(x+h)}\right]\left[\frac{f(x+h)-f(x)}{h}\right] \right. \\ &\quad \left. + \left[\frac{f(x)}{h}\right]\left[\frac{g(x)-g(x+h)}{g(x)g(x+h)}\right]\right\} \\ &= \lim_{h\to 0} \frac{1}{g(x+h)}\left[\frac{f(x+h)-f(x)}{h}\right] \\ &\quad - \frac{f(x)}{g(x)g(x+h)}\left[\frac{g(x+h)-g(x)}{h}\right]. \end{aligned}$$

Taking the limit of each term as $h$ tends to zero, we have

$$\begin{aligned}\left[\frac{f(x)}{g(x)}\right]' &= \frac{1}{g(x)} f'(x) - \frac{f(x)}{g(x)\cdot g(x)} g'(x) \\ &= \frac{g(x)f'(x) - f(x)g'(x)}{g^2(x)}, \end{aligned}$$

since the differentiability of $g(x)$ implies its continuity (see Theorem 9.1).

The six rules discussed in this section are summarized in abbreviated notation below. It will be easier to memorize this chart than each rule separately.

1. $c' = 0$, $c$ constant.
2. $(x^n)' = nx^{n-1}$, $n$ real.
3. $(cf)' = cf'$, $c$ constant.
4. $(f \pm g)' = f' \pm g'$.
5. $(fg)' = fg' + gf'$.
6. $\left(\frac{f}{g}\right)' = \frac{gf' - fg'}{g^2}$.

## EXERCISES 9.3

*In Exercises 1–30, use the six rules given in the above chart to differentiate the given functions. If a value of x is specified, calculate the derivative at that point and find the equation of the tangent line to the curve at that point.*

1. $f(x) = 18$
2. $f(x) = -5$, $x = 2$
3. $f(x) = 2/3$, $x = 1$
4. $f(x) = x^{12}$
5. $f(x) = x^{171}$, $x = 1$
6. $f(x) = x^{-4}$, $x = 2$
7. $f(x) = x^{1/2}$
8. $f(x) = x^{1/3}$, $x = 8$
9. $f(x) = 5x^0$, $x = 3$
10. $f(x) = 7x^{-1/3}$
11. $f(x) = 2x^3 + 3x - 4$
12. $f(x) = x^2 + 2x + 3$, $x = 0$
13. $f(x) = 17x^5 - 2x^2 - 3$, $x = 0$
14. $f(x) = 4x^{1/5} - 7\sqrt{x}$
15. $f(x) = 4\sqrt[3]{x} + 7/\sqrt{x}$, $x = 1$
16. $f(x) = 5\sqrt[3]{x^2} - 4\sqrt[4]{x^3}$
17. $(x + 1)(x - 2)$
18. $(x^2 + 1)(x + 4)$, $x = 0$
19. $(x^2 + 4)(\sqrt{x} - 1/\sqrt{x})$
20. $(x^2 + 3x + 1)(\sqrt{x} + \sqrt[3]{x^2})$
21. $(x^2 + 3x + 2)(x^2 + 1)$, $x = 1$
22. $(\sqrt{x} + 2\sqrt[3]{x})(\sqrt{x} - 4x)$, $x = 1$
23. $\frac{x + 1}{x - 1}$, $x = 0$
24. $\frac{x^2 - 1}{x^2 + 1}$, $x = 1$
25. $\frac{2x - 7}{3x^2 + 4}$, $x = 1$
26. $\frac{x^2 + 1}{x^2 + x + 1}$
27. $\frac{1 + 2x}{1 + 2x + 3x^2}$
28. $\frac{x^4 + x^2 + 1}{3x^3 + 2x + 2}$
29. $\frac{(x + 1)(x^2 + 1)}{3x^2 + 2x + 1}$
30. $\frac{4x^4 + 2x^2 + 1}{(x + 1)(2x - 3)}$, $x = 0$

**31.** Show using the rules of derivatives that

$$[f(x) + g(x) + h(x)]' = f'(x) + g'(x) + h'(x).$$

**32.** Find a formula that obtains the derivative

$$[f(x)g(x)h(x)]'.$$

(*Hint:* Apply the product rule repeatedly.)

**33.** Find a formula for the derivative

$$[1/f(x)]'.$$

**34.** Find the derivative

$$\left[\frac{f(x)g(x)}{h(x)j(x)}\right]'.$$

**35.** A rock is thrown upward from the top of a 96-ft high building at an initial velocity of 80 ft/sec. The height of the rock at any time $t$ until impact is given by the equation

$$h(t) = -16t^2 + 80t + 96.$$

a) When does the rock reach the ground?
b) Find the velocity of the rock at any time $t$. When is the velocity equal to zero?
c) What happens when the velocity of the rock is zero?

---

## 9.4 THE CHAIN RULE

The six rules of differentiation that we studied in Section 9.3 can be used to differentiate any product or quotient of polynomials. However, they do not apply to many other simple algebraic expressions, such as

$$y = \sqrt{x^2 + 1}.$$

Although Rule 2 tells us how to differentiate any power of $x$, it does not tell us how to differentiate powers of *other* expressions—in this case, $x^2 + 1$. Actually, what we really need is a rule that tells us how to differentiate *a function of another function*. For example, $y$ can be constructed from the two functions $f(t) = \sqrt{t}$ and $t(x) = x^2 + 1$, since by the functional notation

$$f[t(x)] = f(x^2 + 1) = \sqrt{x^2 + 1} = y.$$

In order to motivate this important rule, which is the theme of this section, we shall first obtain the derivative of $\sqrt{x^2 + 1}$ and see if it can be related *in any way* to the derivatives of $f(t)$ and $t(x)$.

Using the limit definition of the derivative (9.13), we have

$$\frac{dy}{dx} = \lim_{h \to 0} \frac{\sqrt{(x + h)^2 + 1} - \sqrt{x^2 + 1}}{h}. \tag{9.30}$$

The trick now is to multiply the numerator and denominator of Eq. (9.30) by

$$\sqrt{(x+h)^2+1}+\sqrt{x^2+1},$$

obtaining

$$\begin{aligned}\frac{dy}{dx} &= \lim_{h\to 0}\frac{[(x+h)^2+1]-[x^2+1]}{h[\sqrt{(x+h)^2+1}+\sqrt{x^2+1}]}\\ &= \lim_{h\to 0}\frac{h(2x+h)}{h[\sqrt{(x+h)^2+1}+\sqrt{x^2+1}]}.\end{aligned}$$

Canceling the multiple $h$'s in the numerator and denominator and letting $h$ tend to zero in the resulting expression yields

$$\frac{dy}{dx}=\frac{2x}{\sqrt{x^2+1}+\sqrt{x^2+1}}=\frac{2x}{2\sqrt{x^2+1}}. \tag{9.31}$$

Although we can cancel the twos in Eq. (9.31), we shall not do so temporarily, so that we can connect this answer with the derivatives of $f(t)=\sqrt{t}$ and $t(x)=x^2+1$.

As we saw in Section 9.3, if $t$ and $x$ are variables, then

$$\frac{df}{dt}=(t^{1/2})'=\tfrac{1}{2}t^{-1/2}=\frac{1}{2\sqrt{t}}$$

and

$$\frac{dt}{dx}=(x^2+1)'=2x.$$

Now if we *multiply* these two derivatives, we get

$$\frac{df}{dt}\frac{dt}{dx}=(\sqrt{t})'(x^2+1)'=\frac{2x}{2\sqrt{t}}. \tag{9.32}$$

Substituting $t=x^2+1$ into Eq. (9.32) we obtain exactly the same expression that we did on the right side of Eq. (9.31). Thus, to find the derivative of $\sqrt{x^2+1}$, we first take the derivative of the square root, *as if it were simply the square root of a single variable,* and multiply the result by the derivative of the expression inside the root. Note that in Leibniz's notation this is simply

$$\frac{dy}{dx}=\frac{df}{dt}\frac{dt}{dx},$$

suggesting that in "canceling" the "differences" in the intermediate variable $t$, we "chain" the two derivatives together. This explains the name of the following rule.

***CHAIN RULE*** Let $y(x) = f[t(x)]$ be the *composite function* obtained by replacing each $t$ in the function $f(t)$ by the function $t(x)$. Suppose the functions $f(t)$ and $t(x)$ are both differentiable. Then the derivative of the composite function $y$ with respect to $x$ is given by

$$\frac{dy}{dx} = \frac{df}{dt}\frac{dt}{dx}$$

or, in Lagrange's notation,

$$y'(x) = f'[t(x)]t'(x). \tag{9.33}$$

As Eq. (9.33) states, to find the derivative of a "function of another function," simply find the derivative of each function, multiply the results, and substitute the function $t(x)$ wherever the variable $t$ appears in $f'(t)$. The use of the chain rule is illustrated in the following three examples.

***Example 1*** Find $[\sqrt[3]{4x^3 - 7x + 1}]'$.

Here we have two functions: $f(t) = \sqrt[3]{t}$ and $t = 4x^3 - 7x + 1$. Since

$$(\sqrt[3]{t})' = (t^{1/3})' = \frac{1}{3}t^{-2/3} = \frac{1}{3t^{2/3}}$$

and

$$(4x^3 - 7x + 1)' = 4(3x^2) - 7(1) + 1(0),$$

we obtain by the chain rule

$$[\sqrt[3]{4x^3 - 7x + 1}]' = \frac{1}{3t^{2/3}} \cdot (12x^2 - 7) = \frac{12x^2 - 7}{3(4x^3 - 7x + 1)^{2/3}}.$$

It is not necessary to go through all the steps that we went through above. Moving from left to right in the expression

$$\sqrt[3]{4x^3 - 7x + 1},$$

we first encounter the cube root, $\sqrt[3]{\quad}$. Its derivative is

$$\frac{1}{3}(\quad)^{-2/3};$$

we fill in the blank with whatever function is inside the root—in this case, $4x^3 - 7x + 1$. Then we differentiate the function inside and multiply the two results; that is,

$$[\sqrt[3]{4x^3 - 7x + 1}]' = \frac{1}{3}(\quad)^{-2/3} \cdot (\quad)'$$

$$= \frac{1}{3}(4x^3 - 7x + 1)^{-2/3} \cdot (12x^2 - 7).$$

***Example 2*** Differentiate $\sqrt[4]{(2x^3 + 1)^3}$.

Rewriting the function as $(2x^3 + 1)^{3/4}$, we first differentiate $(\quad)^{3/4}$ and multiply the result (with the blanks filled in by $2x^3 + 1$) by the derivative of $2x^3 + 1$. Thus,

$$[(2x^3 + 1)^{3/4}]' = \frac{3}{4}(2x^3 + 1)^{-1/4} \cdot (2x^3 + 1)'$$

$$= \frac{3}{4}(2x^3 + 1)^{-1/4} \cdot (6x^2)$$

$$= \frac{9x^2}{2\sqrt[4]{2x^3 + 1}}.$$

---

We can apply the chain rule to compositions of more than two functions. This amounts to simply using the chain rule repeatedly.

***Example 3*** Differentiate $y = \sqrt{x + \sqrt[3]{x^2 + 1}}$.

Moving from left to right we first encounter the square root, so that

$$\frac{dy}{dx} = \frac{1}{2}(x + \sqrt[3]{x^2 + 1})^{-1/2} \cdot \frac{d}{dx}(x + \sqrt[3]{x^2 + 1}).$$

But

$$\frac{d}{dx}(x + \sqrt[3]{x^2 + 1}) = 1 + \frac{d}{dx}(\sqrt[3]{x^2 + 1})$$

$$= 1 + \frac{1}{3}(x^2 + 1)^{-2/3} \cdot \frac{d}{dx}(x^2 + 1)$$

$$= 1 + \frac{1}{3}(x^2 + 1)^{-2/3} \cdot (2x).$$

Thus, replacing this term in the first equation, we obtain

$$\frac{dy}{dx} = \frac{1}{2\sqrt{x + \sqrt[3]{x^2 + 1}}} \cdot \left[1 + \frac{2x}{3(x^2 + 1)^{2/3}}\right].$$

---

To prove the chain rule we rewrite the approximation formula in Eq. (9.16) as

$$f(t + k) - f(t) \approx \frac{df}{dt} \cdot k. \tag{9.34}$$

To obtain Eq. (9.34) we merely replaced $x$ with $t$ and $h$ with $k$ and then cross multiplied the $k$. Now consider the difference quotient

$$\frac{y(x + h) - y(x)}{h} = \frac{f[t(x + h)] - f[t(x)]}{h},$$

where $y(x) = f[t(x)]$ is the composite function we wish to differentiate. Since $t(x)$ is differentiable it must be continuous (see Theorem 9.1), so $t(x + h)$ is close to $t(x)$ when $h$ is near zero. Thus we can let

$$t(x + h) = t(x) + k \tag{9.35}$$

where $k$ is small, and the difference quotient becomes

$$\frac{y(x + h) - y(x)}{h} = \frac{f[t(x) + k] - f[t(x)]}{h}.$$

Replacing the numerator of the right side by its approximation in Eq. (9.34), we have

$$\frac{y(x + h) - y(x)}{h} \approx \frac{df}{dt} \cdot \frac{k}{h},$$

and using Eq. (9.35) to replace $k$, we obtain

$$\frac{y(x + h) - y(x)}{h} \approx \frac{df}{dt} \cdot \frac{t(x + h) - t(x)}{h}. \tag{9.36}$$

The approximate equality ($\approx$) becomes an equality as $k \to 0$ in Eq. (9.34) so the same holds true in Eq. (9.36). Hence, if we take the limit of both sides as $h \to 0$, it follows that $k = t(x + h) - t(x) \to 0$, since $t$ is continuous, and we obtain

$$\begin{aligned} \frac{dy}{dx} &= \lim_{h \to 0} \frac{y(x + h) - y(x)}{h} \\ &= \lim_{h \to 0} \frac{df}{dt} \cdot \frac{t(x + h) - t(x)}{h} = \frac{df}{dt} \cdot \frac{dt}{dx}. \end{aligned}$$

## EXERCISES 9.4

*Use the chain rule to differentiate each of the following functions.*

1. $\sqrt[3]{x^2 + 1}$
2. $\sqrt{x^2 - 4}$
3. $\sqrt{x^2 + x + 1}$
4. $\sqrt[3]{x^3 + 2x^2 + 3}$
5. $\sqrt{5x^2 - 4x}$
6. $\sqrt{6x^3 - \sqrt{x}}$
7. $\sqrt[4]{x^2 - \sqrt{x}}$
8. $\sqrt[4]{(2x^2 - \sqrt{x})^3}$
9. $\sqrt{(x^2 + x + 2)^5}$
10. $\sqrt[5]{(x^2 + x + 2)^2}$
11. $\sqrt{x + \sqrt{x + 1}}$
12. $\sqrt{x + \sqrt[3]{x + 1}}$
13. $\sqrt[3]{x + \sqrt{x + 1}}$
14. $\sqrt[3]{x + \sqrt[3]{x + 1}}$
15. $\sqrt{x^2 + \sqrt[3]{x^3 - x}}$
16. $\sqrt{x + \sqrt[3]{x^2 + \sqrt[4]{x^3 - 1}}}$
17. $\sqrt{(x + 1)\sqrt{x - 1}}$
18. $\sqrt[3]{(x + 1)/(x - 1)}$
19. $\sqrt{x + \sqrt{x - \sqrt{x + 1}}}$
20. $\sqrt{x\sqrt{x - 1}/(x - 7)}$
21. $(x + \sqrt{x})^{15}$
22. $(x^2 - 4x + 3)^8$
23. $(x^2 - 14)^7$
24. $[\sqrt{x} - (1/\sqrt{x})]^6$
25. $[x + (\sqrt{x} + 1)^2]^3$
26. $(x + \sqrt{x^2 + 1})^{15}$
27. $\{x + [x^2 + (x^3 + 1)^2]^3\}^4$
28. $\sqrt{x + (x^3 + \sqrt{x})^7}$
29. $\left(\dfrac{x + 1}{x}\right)^9$
30. $\left(\dfrac{x^2 + 1}{x - 1}\right)^{10}$

---

## 9.5 RATES OF CHANGE

In Sections 9.1 and 9.2 we presented several situations in which difference quotients provided a means of measuring the average change of distance, height, or cost over a period of time or over a production run. We saw that if the time interval, for example, tended to zero, then the average change became an instantaneous change in distance or height. In this section we will elaborate on these and other examples in which the derivative has practical applications.

### Acceleration and Velocity

**Free fall** We have already seen that velocity is the instantaneous rate of change of distance, while acceleration is the instantaneous rate of change of velocity. We will now discuss the formulas associated with free fall near the surface of the earth.

Suppose a ball is thrown with an initial velocity $v_0$ from the top of a building of height $h_0$. We will adopt the convention that $v_0$ is positive if the ball is thrown upward and is negative if the ball is thrown downward. Now, if we neglect air resistance, the only force acting on the ball is gravity, pulling the ball downward with an acceleration of $-g$. Thus the acceleration $a(t)$ of the ball at any time $t$ is the constant $-g$, or, in symbols,

$$a(t) = -g.$$

(The negative sign is in keeping with our convention of indicating downward motion by negative quantities.) Since the acceleration is the instantaneous rate of change of the velocity $v(t)$ of the ball at any time $t$, we can also express the acceleration as the derivative of the velocity. Thus, $a(t) = v'(t)$, so the previous equation can also be written as

$$v'(t) = -g. \tag{9.37}$$

The problem now is to find a function $v(t)$ whose derivative is the constant $-g$, where $t$ is the independent variable. It is clear that if we differentiate the function $-gt$ with respect to $t$ we will obtain $-g$ as the derivative. However, this is not the only such function, since any added constant $c$ will vanish upon differentiation. Thus, any function

$$v(t) = -gt + c$$

will satisfy Eq. (9.37) when differentiated with respect to $t$. (In Section 10.1 we will show that only functions of this form satisfy Eq. 9.37.)

The problem of finding the velocity would be very complicated if we had no way of determining which constant $c$ to add, since then there would be infinitely many possible velocities. In this situation we do have one additional bit of information, namely, the *initial velocity* $v_0$. Thus we know that if we set $t = 0$ the velocity must equal $v_0$, that is,

$$v_0 = v(0) = -g \cdot 0 + c = c.$$

Hence the constant $c$ equals the initial velocity $v_0$, so that the velocity at any time $t$ is given by the unique expression

$$v(t) = -gt + v_0. \tag{9.38}$$

Before proceeding with our development we will consider a practical application. Suppose the ball is thrown upward with an initial velocity $v_0 = 80$ ft/sec. Since the force of gravity near the surface of the earth is approximately $g \approx 32$ ft/sec$^2$, the velocity of this ball at any time $t$ is given by

$$v(t) = -32t + 80 \text{ (ft/sec)}$$

until it hits the ground. Thus two seconds after it is released the velocity is

$$v(2) = -32(2) + 80 = 16 \text{ ft/sec}$$

in an upward direction, while three seconds later it is

$$v(5) = -32(5) + 80 = -80 \text{ ft/sec}$$

downward. When $t = 2.5$ sec the velocity is

$$v(2.5) = -32(2.5) + 80 = 0 \text{ ft/sec}$$

so that the ball has reached the top of its trajectory.

Now, returning to Eq. (9.38), we recall that the velocity is the instantaneous rate of change of the height $h(t)$ of the ball at any time $t$. Hence $v(t) = h'(t)$, so we may rewrite Eq. (9.38) as

$$h'(t) = -gt + v_0. \qquad \textbf{(9.39)}$$

Our problem is similar to that we had before: to find a function $h(t)$ whose derivative satisfies Eq. (9.39). The $v_0$ constant term is easy since it is simply the derivative of $v_0 t$. To get $-gt$ we must differentiate a term involving $t^2$. Note that

$$\left(\frac{-gt^2}{2}\right)' = \frac{-g}{2} \cdot 2t = -gt$$

so we see that any expression of the form

$$h(t) = \frac{-gt^2}{2} + v_0 t + c$$

will yield the desired derivative since the constant $c$ vanishes under differentiation. To determine which constant $c$ is required, we use the fact that the initial height $h_0$ is given. Setting $t = 0$ we have

$$h_0 = h(0) = \frac{-g}{2}(0)^2 + v_0 \cdot 0 + c = c,$$

so the constant $c$ must be the initial height. Thus the height at any time $t$ is given by the equation

$$h(t) = -\tfrac{1}{2}gt^2 + v_0 t + h_0. \qquad \textbf{(9.40)}$$

***Example 1*** A ball is thrown upward at 80 ft/sec from a 96-ft tall building. The height at any time $t$ is

$$h(t) = \frac{-32}{2}t^2 + 80t + 96.$$

At the top of the ball's trajectory, which we saw occurs when $t = 2.5$ sec, the height is

$$h(2.5) = -16(2.5)^2 + 80(2.5) + 96 = 196 \text{ ft}.$$

After five seconds the ball is

$$h(5) = -16(5)^2 + 80(5) + 96 = 96 \text{ ft}$$

high, that is, even with the top of the building. If it misses the building it will hit the ground when the height is zero. Thus, setting $h(t) = 0$ we solve the equation

$$0 = -16t^2 + 80t + 96$$

for $t$. Dividing both sides by $-16$ we obtain

$$0 = t^2 - 5t - 6 = (t - 6)(t + 1).$$

Since $t = -1$ is impossible (we haven't thrown the ball yet), the ball hits the ground after six seconds.

---

**Velocity and geometry** We frequently need to use geometric considerations when solving problems involving related rates. The following examples illustrate three typical situations.

***Example 2*** Sir Galahad is trying to rescue Princess Ann, who has been cruelly imprisoned by her wicked stepmother in a turret of the castle. Sir Galahad has placed a 50-ft ladder against the outside wall so that the top of the ladder is even with the window in the turret, 40 ft above the ground (see Fig. 9.4). Just as he reaches the window, the bottom of the ladder starts sliding away from the tower at a velocity of 5 ft/sec. How fast is the top of the ladder (and Sir Galahad) falling?

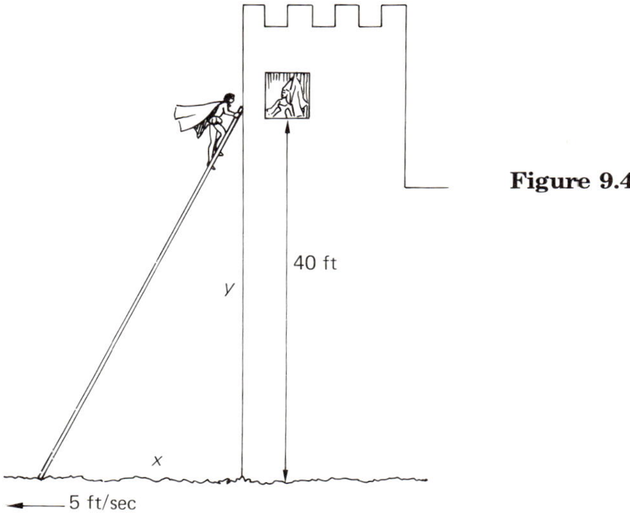


**Figure 9.4**

In this problem there are two velocities involved: the velocity of the base of the ladder horizontally away from the tower, and the velocity of the top of the ladder in a vertical direction. Let $x(t)$ and $y(t)$ be the horizontal distance between the base of the ladder and the tower and the height of the top of the ladder, respectively, at any time $t \geq 0$. Since the angle between the tower and the ground is assumed to be a right angle, we can use the Pythagorean theorem to express the length of the ladder in terms of the distances $x(t)$ and $y(t)$. We have

$$[x(t)]^2 + [y(t)]^2 = 2500 \text{ (or } 50^2). \tag{9.41}$$

Since $x(t)$ and $y(t)$ are both functions of $t$, we can differentiate Eq. (9.41) with respect to $t$, obtaining by the chain rule

$$2x(t)x'(t) + 2y(t)y'(t) = 0. \tag{9.42}$$

The function $x'(t)$ is the horizontal velocity with which the ladder is moving away from the tower, hence, $x'(t) = 5$ ft/sec. The function $y'(t)$ is the vertical velocity of the top of the ladder at any time $t$, which is the quantity we are trying to find. Solving Eq. (9.42) for $y'(t)$, we have

$$y'(t) = \frac{-2x(t)x'(t)}{2y(t)} = \frac{-x(t)x'(t)}{y(t)}. \tag{9.43}$$

Since we wish to know how fast Sir Galahad is falling when the top of the ladder is 40 ft high, we set $y(t) = 40$ ft and $x'(t) = 5$ ft/sec in Eq. (9.43). We have one unknown left, $x(t)$, which can be determined using Eq. (9.41) with $y(t) = 40$ ft. Thus,

$$\begin{aligned}[x(t)]^2 &= 2500 - [y(t)]^2 \\ &= 2500 - (40)^2 = 900,\end{aligned}$$

so that $x(t) = 30$ ft. Substituting this value also into Eq. (9.43) yields

$$y'(t) = \frac{-(30)(5)}{40} = \frac{-15}{4}\text{ ft/sec},$$

that is, Sir Galahad is falling (initially) at a velocity of $3\frac{3}{4}$ ft/sec.

We might also ask how fast Sir Galahad is falling two seconds later if the velocity of the bottom of the ladder away from the tower is constantly 5 ft/sec. The same equations (9.41)–(9.43) hold but the values of $x(t)$ and $y(t)$ have changed. Since $x(t)$ was initially 30 ft in distance, and the ladder is moving at a constant velocity $x'(t) = 5$ ft/sec,

$$\begin{aligned}x(2) &= x(0) + 2x'(t) \\ &= 30 + 2(5) = 40 \text{ ft}.\end{aligned}$$

Substituting this value into Eq. (9.41) yields

$$[y(2)]^2 = 2500 - [x(2)]^2$$
$$= 2500 - (40)^2 = 900$$

so that $y(2) = 30$ ft. Hence Eq. (9.43) implies that

$$y'(2) = \frac{-x(2)x'(2)}{y(2)} = \frac{-40(5)}{30}$$
$$= \frac{-20}{3}\text{ ft/sec.}$$

Thus, Sir Galahad is falling at a velocity of $6\frac{2}{3}$ ft/sec at time $t = 2$ sec.

***Example 3*** Young Douglas fir trees often grow at a prodigious rate when climatic and soil conditions are favorable. A certain tree is 10 ft high and the average diameter of its trunk is 3 in. Suppose the height is increasing at a rate of 5 ft/year, while the average increase of the diameter of the trunk is 1.5 in./year. How fast is the volume of the trunk changing?

Suppose we assume that the overall shape of the trunk is conical (see Fig. 9.5) with the base diameter being twice as big as the average diameter. The volume of a cone is given by the equation

$$V = \frac{\pi}{12} d^2 h. \tag{9.44}$$

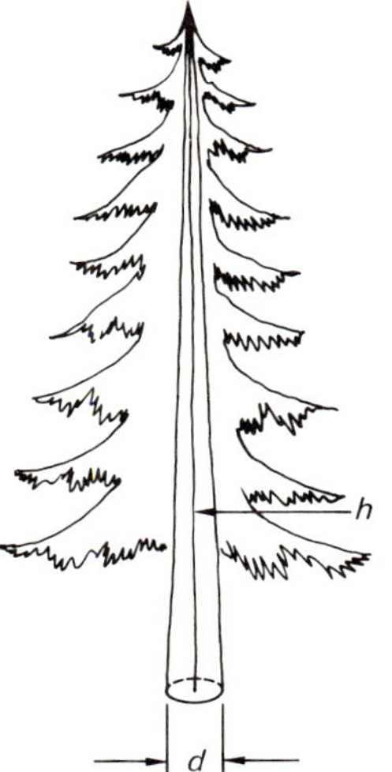


**Figure 9.5**

Here the volume $V$, base diameter $d$, and height $h$ are all functions of time $t$. Differentiating both sides of Eq. (9.44) with respect to $t$, we obtain (omitting the independent variable $t$ to simplify the equation)

$$V' = (\pi/12)[d^2h' + 2dd'h]. \tag{9.45}$$

Now, at the time in question, we have $d = \frac{1}{2}$ ft, $h = 10$ ft, $d' = \frac{1}{8}$ ft/yr, and $h' = 5$ ft/yr so that the change in volume is

$$V' = \frac{\pi}{12}\left[\left(\frac{1}{2}\right)^2(5) + (2)\left(\frac{1}{2}\right)\left(\frac{1}{8}\right)(10)\right] = \frac{5\pi}{24}\,\text{ft}^3/\text{yr}.$$

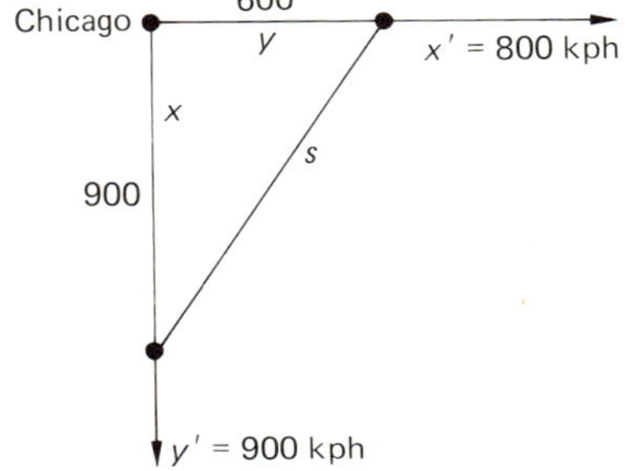


**Figure 9.6**

***Example 4*** At noon United Airlines flight 1 leaves Chicago flying south at 900 kph (kilometers per hour). Fifteen minutes later flight 2 leaves Chicago going east at 800 kph. How fast are the two flights separating at 1 P.M.?

The situation is illustrated in Fig. 9.6. The distance $s$ between the two planes, as well as the vertical and horizontal distances $x$ and $y$, respectively, that the planes have traveled since takeoff, are all functions that vary with time $t$. Since the triangle formed by $x$, $y$, and $s$ satisfies the Pythagorean formula, we have

$$s^2 = x^2 + y^2.$$

Differentiating both sides with respect to time $t$ yields

$$2ss' = 2xx' + 2yy'$$

so that

$$s' = \frac{xx' + yy'}{s}. \tag{9.46}$$

Now $x = 900$ km, $y = 600$ km, $x' = 900$ kph, $y' = 800$ kph, and $s$ is determined by the Pythagorean formula

$$s^2 = (900)^2 + (600)^2 = 1{,}170{,}000,$$

so that $s \approx 1081.67$ km. Thus,

$$s' \approx \frac{900(900) + 600(800)}{1081.67} = 1192.61\text{ kph}.$$

**Geometric Changes** As we have seen above, geometry plays a useful role in problems involving velocities. In those situations time is the independent variable. Other problems can be formulated in which time is not involved; instead the independent variable is one of the geometric parameters. We illustrate this situation with the following example.

***Example 5*** Mothballs shrink through surface evaporation. Since mothballs are spherical, they retain their spherical shape since the evaporation rate is constant at each point on their surface. Find the rate of change of the volume in terms of the radius of the mothball.

Since the volume of a sphere can be expressed entirely in terms of the radius, we need only differentiate

$$V(r) = \frac{4}{3}\pi r^3$$

with respect to $r$. Hence $V'(r) = 4\pi r^2$, that is, the change in volume with respect to the radius is $4\pi r^2$. It is worth noting that the surface area of a sphere is given by the formula $S(r) = 4\pi r^2$. Thus the rate of change in volume equals the surface area, a result that corresponds exactly with the evaporation process present in the mothball.

---

**Economics** In economics we frequently study the relations that exist between various goods and their prices. Often it is not possible to express these relations in functional notation, but it is generally the case that the *economic* conditions of the situation impose one or more *mathematical* properties. An example is provided by demand functions and curves. By a *demand function* we mean a continuous function $p = f(x)$, where $x$ is the number of units of the good and $p$ is the price per unit. Economic considerations dictate that as the price of the good increases, demand will fall *provided* there are no changes in the number of consumers, their preferences, their income, or the prices of all other goods. Thus, demand curves are generally shown as decreasing functions. Empirical evidence indicates that the following functions provide suitable models for actual demand curves:

**1.** $p = a - bx$,

**2.** $p = \sqrt{a - bx}$,

**3.** $p = a - bx \pm cx^2$,

**4.** $p = \dfrac{a}{x + b} - c$,

where $a$, $b$, and $c$ are all nonnegative constants. Clearly (1) is a downward sloping line, (2) and (3) are parabolas, and (4) is a hyperbola.

The *total revenue* derived from a demand of $x$ units sold at the price $p = f(x)$ is the function

$$R(x) = xp = xf(x).$$

If the form of the demand curve is known, then it is easy to determine that of the total revenue curve.

A third example of an economic function is the total cost $C$ arising in the production of $x$ units of a good. In this situation it is usually assumed that there are fixed overhead costs incurred even when no units are produced, and that costs increase with production. A reasonable model will, therefore, cross the vertical axis above the origin and rise continuously from left to right. Usually total *cost functions* are of the following types:

1. $C = a + bx,$
2. $C = ax^2 + bx + c,$
3. $C = \sqrt{a + bx} + c,$
4. $C = \dfrac{ax^2 + bx + c}{x + d},$

with all coefficients nonnegative.

The *total profit* function is found by subtracting total cost from total revenue:

$$P = R - C = xp - C.$$

We can now use derivatives to determine the *marginal revenue, cost,* or *profit,* that is, the instantaneous change in revenue, cost, or profit obtained by increasing the output of a good from the demand $x$.

***Example 6*** Pemex is the government-owned monopoly that controls all aspects of petroleum production and sales in Mexico. Suppose Mexico has a demand $x$ of three billion gallons of gasoline per month, and that the demand and total cost functions are

$$p = 0.7 - (4x)(10^{-11}) \text{ dollars/gal}$$

and

$$C = 10^4\sqrt{5x} + 2(10^8) \text{ dollars.}$$

Setting $x = 3(10^9)$, we obtain a price $p = 58¢/\text{gal}$, a total revenue of $R = xp = 1{,}740{,}000{,}000$ dollars, a total cost of $C = 1{,}424{,}744{,}871$ dollars, and a total profit of $P = R - C = \$315{,}255{,}129$ per month. What will be the effect of increasing production?

To find the marginal revenue, marginal cost, and marginal profit, we differentiate the total revenue and total cost functions:

$$R' = \frac{d}{dx}\{xp\} = \frac{d}{dx}\left\{0.7x - \frac{4x^2}{10^{11}}\right\} = 0.7 - \frac{8x}{10^{11}},$$

and

$$C' = \frac{5(10^4)}{2\sqrt{5x}}.$$

Setting $x = 3(10^9)$ in these equations we find that the marginal revenue = 46¢/gal and the marginal cost = 20.41¢/gal. Since $P' = R' - C'$, the marginal profit is 25.59¢/gal. This marginal profit is over twice as much as the *average profit* per gallon that was made on the first three billion gallons of gasoline,

$$\text{average profit} = \frac{\text{total profit}}{\text{demand}} = 10.51¢/\text{gal},$$

and would encourage Pemex to increase production.

---

## EXERCISES 9.5

*In Exercises 1–12, a ball is thrown from a building of height $h_0$ with initial velocity $v_0$. Find in each exercise:*

a) When will the ball reach the ground?
b) When did the ball attain its maximum height?
c) What was the maximum height attained by the ball?
d) When did the ball attain height $h_1$?

Use $g = 32\text{ ft/sec}^2 = 9.8\text{ m/sec}^2$.

**1.** $h_0 = 64$ ft, $v_0 = 80$ ft/sec, $h_1 = 96$ ft

**2.** $h_0 = 96$ ft, $v_0 = 64$ ft/sec, $h_1 = 80$ ft

**3.** $h_0 = 49$ m, $v_0 = 21$ m/sec, $h_1 = 70$ m

**4.** $h_0 = 49$ m, $v_0 = -21$ m/sec, $h_1 = 28$ m

**5.** $h_0 = 98$ m, $v_0 = 28$ m/sec, $h_1 = 70$ m

**6.** $h_0 = 21$ m, $v_0 = 28$ m/sec, $h_1 = 100$ m

**7.** $h_0 = 80$ ft, $v_0 = -96$ ft/sec, $h_1 = 10$ ft

**8.** $h_0 = 88$ ft, $v_0 = 80$ ft/sec, $h_1 = 200$ ft

**9.** $h_0 = 100$ ft, $v_0 = 90$ ft/sec, $h_1 = 200$ ft

**10.** $h_0 = 200$ ft, $v_0 = 75$ ft/sec, $h_1 = 100$ ft

**11.** $h_0 = 75$ m, $v_0 = 50$ m/sec, $h_1 = 100$ m

**12.** $h_0 = 125$ m, $v_0 = 30$ m/sec, $h_1 = 100$ m

**13.** A man 6 ft tall walks at a rate of 3 ft/sec directly away from a streetlamp 15 ft above the street. At what rate is the length of his shadow changing when he is 10 ft from the lamp?

**14.** A missile is rising vertically at a constant rate of 2000 ft/sec. How fast is the distance from the missile to a telemetry station, on level ground one mile away from the launch site, changing one second after liftoff? Two seconds after liftoff?

**15.** Atlanta is 750 miles by air southwest from New York City. At noon a plane leaves Atlanta flying north at 500 mph, and 20 minutes later a plane leaves New York flying west at 600 mph.

a) When are the two planes closest?
b) How close do the two planes come together?
c) How fast is the distance between the two planes changing at 1 P.M.?

**16.** Ship A is 100 kilometers south of ship B and is headed east at 10 kph. Ship B is headed north at 12 kph. How fast is the distance between the two ships changing two hours later?

**17.** A disk-shaped aspirin tablet dissolves in water in such a way that the core retains its shape until the tablet is fully dissolved. Find the rate at which the area of the circle is changing in terms of its radius.

**18.** Sand is falling at a rate of 2 $m^3$/sec in a conical pile whose height equals the radius of its base. How fast is the height changing when the pile is 4 m high?

**19.** The natives of the third planet of the star Daneb are shaped like hemispheres. Their body, composed of 99% liquid nitrogen, loses fluids during the day through pores on its surface, but not on the flat base of the hemisphere. Find the rate of change of their volume in terms of the radius of their base. What would happen if they were shaped like cones whose height equals the radius of the base?

**20.** A calculator manufacturer estimates that she can sell 200,000 programmable calculators at a price of \$160 apiece. Find the marginal profit if the demand and total cost functions are

$$p = \frac{10^8}{x + 3(10)^5} - 40 \quad \text{and} \quad C = 10^6 + 120x.$$

**21.** The manufacturer in Exercise 20 discovers that to produce that many calculators she will have to put part of the work force on overtime. This changes the total cost equation to

$$C = 10^6 + 120x + (10^{-9})x^2.$$

Find the marginal profit with this total cost function. Should she try to increase production?

★ **22.** A Japanese competitor to the calculator manufacturer in Exercise 20 introduces his own competing calculator at a unit price of \$140. The

Japanese manufacturer exports 80,000 of his calculators to the United States. Which strategy will produce the greatest profits for the U.S. manufacturer?

a) Lower the unit price to $140 and produce enough calculators to meet the *added* demand.
b) Maintain the present price of $160 and cut production to 120,000 units.
c) Drop the price to $150 and produce enough calculators to meet the new demand.

---

## 9.6 DERIVATIVES OF EXPONENTIAL AND LOGARITHMIC FUNCTIONS

In the preceding sections of this chapter we have developed seven rules that allow us to differentiate any *algebraic* function involving sums, products, quotients, and powers of a single variable $x$. In the next two sections we will provide the necessary tools for differentiating the exponential, logarithmic, and trigonometric functions. This section will deal only with the exponential and logarithmic functions.

The key to finding the derivatives of *all* exponential and logarithmic functions is to obtain the derivative of the function

$$f(x) = e^x. \tag{9.47}$$

It is important to note that one *cannot* use Rule 2, $(x^n)' = nx^{n-1}$, to differentiate the exponential function $e^x$. In the exponential the power is the variable, while in $x^n$ the base is the variable. Thus the two functions are totally different and consequently satisfy different rules of differentiation.

As we have done before whenever we wished to develop a new rule of differentiation, we turn to the limit definition of the derivative Eq. (9.13) and define

$$f'(x) = \lim_{h \to 0} \frac{f(x + h) - f(x)}{h}$$

so that for the function in Eq. (9.47) we have

$$(e^x)' = \lim_{h \to 0} \frac{e^{x+h} - e^x}{h}. \tag{9.48}$$

[Remember that the functional notation $f(x + h)$ merely requires that we substitute $x + h$ for every $x$ in $f(x)$.] Now $e^{x+h} = e^x e^h$, so we can factor an $e^x$-term from the numerator of the right side of Eq. (9.48), obtaining

$$(e^x)' = \lim_{h \to 0} e^x \left(\frac{e^h - 1}{h}\right).$$

Since $e^x$ does not depend on $h$, we can move the $e^x$-term outside the limit to obtain

$$(e^x)' = e^x\left(\lim_{h\to 0}\frac{e^h - 1}{h}\right). \tag{9.49}$$

Our problem now is to determine

$$\lim_{h\to 0}\frac{e^h - 1}{h}. \tag{9.50}$$

Although it is possible to obtain the limit (9.50) from purely theoretical considerations, the proof is too difficult to be presented at this level. Instead we will provide numerical evidence that will lead us to the correct solution. The use of a calculator is essential to the following discussion.

The difficulty in (9.50), as with all limits associated with derivatives, is that the difference quotient is undefined when $h = 0$. However, we have seen that we need not calculate the value of the difference quotient at $h = 0$, but merely to discover what the difference quotient approaches as $h$ tends to zero. Thus, to find the limit (9.50), we need only calculate the difference quotient for several values of $h$ that get closer and closer to zero. In Table 9.1 we have used a calculator to obtain the values of the difference quotient. You should be able to reproduce these values with your calculator if it has an $e^x$ key.

**Table 9.1**

| $h$ | $\dfrac{e^h - 1}{h}$ |
|---|---|
| 1.000 | 1.71828 |
| 0.100 | 1.05171 |
| 0.010 | 1.00502 |
| 0.001 | 1.00050 |
| −0.001 | 0.99950 |
| −0.010 | 0.99502 |
| −0.100 | 0.95163 |
| −1.000 | 0.63212 |

As we can see from Table 9.1, the closer we get to $h = 0$, the closer the difference quotient gets to 1. Hence we can surmise (correctly) that

$$\lim_{h\to 0}\frac{e^h - 1}{h} = 1. \tag{9.51}$$

Substituting this fact into Eq. (9.49) yields

$$(e^x)' = e^x, \tag{9.52}$$

that is, *the derivative of the exponential function $e^x$ is itself.* This is one rule you should not have any trouble remembering!

With this fact and the chain rule, we can now differentiate any exponential function, as the following examples illustrate.

***Example 1*** Find the derivatives of these functions:

a) $e^{7x}$,
b) $3e^{x^2}$,
c) $e^{e^x}$.

Note that in each of these problems the given function is an exponential function of the form $e^{g(x)}$, where $g(x)$ is another function. The chain rule indicates that we are to differentiate the exponential function first, treating $g(x)$ as if it were simply an $x$, and then to differentiate $g(x)$ and multiply the two derivatives together. Hence,

a) $(e^{7x})' = e^{7x} \cdot (7x)' = 7e^{7x}$;
b) $(3e^{x^2})' = 3e^{x^2} \cdot (x^2)' = 6xe^{x^2}$;
c) $(e^{e^x})' = e^{e^x} \cdot (e^x)' = e^{e^x}e^x$.

***Example 2*** In Example 3 in Section 3.2 we obtained the function

$$y = f(x) = 179{,}323{,}175e^{0.0142x}$$

as a predictor of the U.S. population $x$ years after 1960. Suppose we wish to find the rate that the population was increasing in 1965. Since the growth rate is simply the derivative, we need to calculate $f'(5)$, because 1965 is $x = 5$ years after 1960. Hence

$$\begin{aligned} f'(x) &= 179{,}323{,}175(e^{0.0142x})' \\ &= 179{,}323{,}175(0.0142)e^{0.0142x} \\ &= 2{,}546{,}389.1e^{0.0142x}. \end{aligned}$$

Now substituting $x = 5$ in this equation, we have

$$f'(5) = 2{,}546{,}389.1e^{0.071} = 2{,}733{,}755.5,$$

so the U.S. population in 1965 was growing at a rate of approximately 2,733,756 persons per year.

The limit in Eq. (9.51) has several useful applications. Since the quotient tends to 1 as $h \to 0$, we have the approximation

$$\frac{e^h - 1}{h} \approx 1$$

for $h$ close to zero. Multiplying both sides by $h$ and adding 1 to both sides yields $e^h \approx 1 + h$. Taking the $h$th root we have

$$e \approx (1 + h)^{1/h},$$

and setting $k = 1/h$, we finally obtain the approximation

$$e \approx \left(1 + \frac{1}{k}\right)^k, \tag{9.53}$$

valid for *large* values of $k$ since $k$ is the reciprocal of $h$. Equation (9.53) suggests (correctly) that as $k$ tends to infinity we have

$$e = \lim_{k \to \infty} \left(1 + \frac{1}{k}\right)^k. \tag{9.54}$$

This limit has interesting applications in finance, as the following example illustrates.

***Example 3*** In Example 6 in Section 8.1 we developed a formula for computing the amount $A$ in an account if a principal of $P$ is allowed to compound at $i\%$ annual interest for a period of $n$ years. The equation we obtained (see Eq. 8.15),

$$A = P(1 + i)^n,$$

must be modified if the interest payments are more frequent than once a year. For example, if the interest is paid quarterly then the amount in the account must be increased by $(i/4)\%$ every three months. The amount could now be calculated by the equation

$$A = P\left(1 + \frac{i}{4}\right)^{4n},$$

since there are four times as many payments as before but each percentage increase is only $\frac{1}{4}$ as much. Similarly, if the interest is compounded daily, we have

$$A = P\left(1 + \frac{i}{365}\right)^{365n},$$

since 365 payments of $(i/365)\%$ of the amount present are made each year.

The more frequent the payments, the greater the final amount, because each payment is a percentage of the amount then present *including all accumulated interest.* Thus, the sooner the interest is received, the quicker it will generate additional income. For this reason, banks that offer daily interest on savings accounts are preferable to those that pay interest less frequently, all other conditions being equal.

Some accounts pay interest even more frequently: The interest is *compounded continuously*. To see what this means, we first suppose that the number of payments annually is $m$. The amount in the account after $n$ years would then be

$$A = P\left(1 + \frac{i}{m}\right)^{nm},$$

since $m$ payments of $(i/m)\%$ of the amount present are made each year. If we let the number of payments $m$ tend to infinity, we obtain

$$\begin{aligned} A &= \lim_{m\to\infty} P\left(1 + \frac{i}{m}\right)^{nm} \\ &= P \lim_{m\to\infty}\left(1 + \frac{1}{(m/i)}\right)^{(m/i)(in)}. \end{aligned}$$

Setting $k = m/i$, and noting that $k$ tends to infinity if $m \to \infty$, we have by the laws of exponents and Eq. (9.54)

$$A = P\left[\lim_{k\to\infty}\left(1 + \frac{1}{k}\right)^k\right]^{in} = Pe^{in}. \tag{9.55}$$

**Table 9.2**

| $m$ (frequency) | $A = \begin{cases} P[1 + (i/m)]^{nm}, & m < \infty \\ Pe^{in}, & m = \infty \end{cases}$ | | |
|---|---|---|---|
| 1 (annually) | $1000(1 + 0.06)^{10}$ | $= 1000(1.06)^{10}$ | = \$1790.85 |
| 4 (quarterly) | $1000\left(1 + \frac{0.06}{4}\right)^{(4)10}$ | $= 1000(1.015)^{40}$ | = \$1814.02 |
| 365 (daily) | $1000\left(1 + \frac{0.06}{365}\right)^{(365)10}$ | $= 1000(1.000164)^{3650}$ | = \$1822.03 |
| $\infty$ (continuously) | $1000e^{(0.06)10}$ | $= 1000e^{0.6}$ | = \$1822.12 |

Table 9.2 shows the amount that would accumulate in $n = 10$ years if $P = \$1000$ is deposited at $i = 6\%$ interest compounded $m$ times a year.

We can also use Eq. (9.52) to find the derivative of exponential functions with other bases. Suppose $a$ is a positive constant. Then

$$(a^x)' = a^x \ln a, \tag{9.56}$$

since

$$(a^x)' = (e^{x \ln a})' = e^{x \ln a}(x \ln a)' = a^x \ln a.$$

It should be noticed that the differentiation formula (9.56) includes Eq. (9.52) since if $a = e$, then $\ln a = \ln e = 1$. In practice, it is generally more convenient to use Eq. (9.56) instead of replacing each $a$ by $e^{\ln a}$.

***Example 4*** Differentiate the following functions:

a) $7^x$
b) $8^{3^x}$
c) $6^{x^2}$.

By Eq. (9.56) and the chain rule we have

a) $(7^x)' = 7^x \ln 7$;
b) $(8^{3^x})' = 8^{3^x} \ln 8 \cdot (3^x)' = 8^{3^x} \ln 8 \cdot (3^x) \ln 3$;
c $(6^{x^2})' = 6^{x^2} \ln 6 \cdot (x^2)' = 6^{x^2} \ln 6 \cdot (2x)$.

---

We now consider the logarithmic functions. The key to finding the derivatives of the logarithmic functions is the identity

$$x = e^{\ln x}.$$

Differentiating both sides with respect to $x$ we have

$$\begin{aligned} 1 &= e^{\ln x}(\ln x)' \\ &= x(\ln x)'. \end{aligned}$$

Dividing both sides by $x$ we obtain

$$(\ln x)' = \frac{1}{x}. \tag{9.57}$$

For logarithms to other bases we can use Theorem 3.2 with $a = e$:

$$\log_b x = \frac{\ln x}{\ln b},$$

implying, since $\ln b$ is a constant, that

$$(\log_b x)' = \frac{1}{x \ln b}. \tag{9.58}$$

Note that Eq. (9.58) includes Eq. (9.57) when $b = e$, since $\ln e = 1$.

***Example 5*** Differentiate the following functions:

a) $\ln(4x^2)$;
b) $\log(\ln x)$;
c) $\log_2(3^x)$.

By Eq. (9.58) and the chain rule we obtain

a) $$[\ln(4x^2)]' = \frac{1}{4x^2}(4x^2)' = \frac{8x}{4x^2} = \frac{2}{x};$$

b) $$[\log(\ln x)]' = \frac{1}{(\ln x)\ln 10}(\ln x)' = \frac{1}{x(\ln x)\ln 10};$$

c) $$[\log_2(3^x)]' = \frac{1}{3^x \ln 2}(3^x)' = \frac{3^x \ln 3}{3^x \ln 2} = \frac{\ln 3}{\ln 2}.$$

The result in (c) might surprise you since we have differentiated a complicated function and obtained the constant ($\ln 3/\ln 2$). Note, however, that

$$\log_2(3^x) = x \log_2 3$$

since logarithms exponentiate by multiplying. That $\log_2 3 = \ln 3/\ln 2$ follows from Eq. (3.30).

***Example 6*** Suppose the demand function for a product is

$$p = \frac{1}{3}\log\frac{700}{x}.$$

Then the marginal revenue is given by the function

$$\begin{aligned}\frac{dR}{dx} &= (xp)' = \left(\frac{x}{3}\log\frac{700}{x}\right)' \\ &= \frac{1}{3}\log\frac{700}{x} + \frac{x}{3}\frac{1}{(700/x)\ln 10}\left(\frac{700}{x}\right)' \\ &= \frac{1}{3}\log\frac{700}{x} - \frac{1}{3\ln 10}.\end{aligned}$$

---

The following box summarizes the rules we have presented in this section. You should memorize these four formulas, since you will use them often in the rest of this book. A more complete list of the rules of differentiation is presented in Appendix 3.

| | |
|---|---|
| **1.** $(e^x)' = e^x$ | **2.** $(a^x)' = a^x \ln a$ |
| **3.** $(\ln x)' = 1/x$ | **4.** $(\log_a x)' = 1/(x \ln a)$ |

## EXERCISES 9.6

*Differentiate each function in Exercises* 1–20 *using the methods of this section.*

**1.** $e^{-5x}$

**2.** $e^{-3x^3}$

**3.** $7^{2x}$

**4.** $8^{-x^2}$

**5.** $8^{\ln x}$

**6.** $e^{\sqrt{1+x^2}}$

**7.** $3^{\sqrt{4-2x^2}}$

**8.** $e^{\log x}$

**9.** $x^{2x}$

**10.** $x^{\ln x}$

**11.** $x^{2^x}$

**12.** $2^{x^2}$

**13.** $\log \sqrt{4 + x^2}$

**14.** $\log_8 \{\log_5 [\log_2 (5^x)]\}$

**15.** $2^{\ln x}$

**16.** $3^{\log x}$

**17.** $e^{2^{3^{4^x}}}$

**18.** $2^{x^{3^x}}$

**19.** $3^{4^{\ln x}}$

**20.** $4^{x^x}$

**21.** In Example 2 (see p. 463), find the growth rate of the population in 1980.

**22.** In Example 3 in Section 3.4 we obtained the formula

$$y = 10{,}000 + 54{,}000(0.8865)^x$$

as the number of thousands of tons of anthracite produced $x$ years after 1945. Find the production rate in 1970.

*In Exercises* 23–28, *find the amount A that would accumulate in n years if a principal P were deposited at i% interest compounded continuously.*

**23.** $P = \$100$, $i = 6\%$, $n = 5$

**24.** $P = \$300$, $i = 8\%$, $n = 7$

**25.** $P = \$250$, $i = 12\%$, $n = 10$

**26.** $P = \$75$, $i = 9\%$, $n = 6$

**27.** $P = \$900$, $i = 7\%$, $n = 8$

**28.** $P = \$1200$, $i = 12\%$, $n = 20$

---

## ▶ 9.7 DERIVATIVES OF TRIGONOMETRIC FUNCTIONS*

In this section we will obtain the derivatives of the six trigonometric functions. Using these six rules for differentiation and the chain rule, we can find the derivative of any expression involving trigonometric functions.

* This section can be omitted with no loss of continuity. All problems involving trigonometric functions in the remainder of the book are indicated by ▶.

All the derivatives in this section depend on finding the derivative of one of the trigonometric functions, in this case, $\sin x$. We will provide proofs for the following derivatives after we have seen several examples of their use. The following box lists the important expressions discussed in this section.

| | |
|---|---|
| $(\sin x)' = \cos x$ | $(\cos x)' = -\sin x$ |
| $(\tan x)' = \sec^2 x$ | $(\cot x)' = -\csc^2 x$ |
| $(\sec x)' = \sec x \tan x$ | $(\csc x)' = -\csc x \cot x$ |

***Example 1*** Find the derivatives of the following functions:

a) $\sin(x^2)$; b) $\tan(2x^3 + 3x)$;
c) $\csc(e^x)$; d) $\cos\sqrt{x^2 - 1}$;
e) $\cot(x \sec x)$; f) $\sec(\ln \sin x)$.

Again we must make use of the chain rule and the other rules of differentiation. We have the following:

a) $[\sin(x^2)]' = \cos(x^2) \cdot (x^2)' = 2x \cos x^2$;

b) $$[\tan(2x^3 + 3x)]' = \sec^2(2x^3 + 3x) \cdot (2x^3 + 3x)'$$
$$= (6x^2 + 3)\sec^2(2x^3 + 3x);$$

c) $$[\csc(e^x)]' = -\csc(e^x)\cot(e^x) \cdot (e^x)'$$
$$= -e^x \csc(e^x)\cot(e^x);$$

d) $$(\cos\sqrt{x^2 - 1})' = -\sin\sqrt{x^2 - 1} \cdot (\sqrt{x^2 - 1})'$$
$$= \frac{-x}{\sqrt{x^2 - 1}} \sin\sqrt{x^2 - 1};$$

e) $$[\cot(x \sec x)]' = -\csc^2(x \sec x) \cdot (x \sec x)'$$
$$= -\csc^2(x \sec x) \cdot (\sec x + x \sec x \tan x)$$
$$= -\csc^2(x \sec x) \sec x \cdot (1 + x \tan x);$$

f) $$\begin{aligned}[\sec(\ln\sin x)]' &= \sec(\ln\sin x)\tan(\ln\sin x)\cdot(\ln\sin x)' \\ &= \sec(\ln\sin x)\tan(\ln\sin x)\frac{1}{\sin x}(\sin x)' \\ &= \frac{\cos x}{\sin x}\sec(\ln\sin x)\tan(\ln\sin x) \\ &= \cot x\sec(\ln\sin x)\tan(\ln\sin x).\end{aligned}$$

***Example 2*** A lighthouse is situated 2 miles off a straight shore. The light revolves at 2 rpm. How fast does the light sweep along the shore at (a) its nearest point; (b) a point a mile away along the shore from the nearest point?

This situation is illustrated in Fig. 9.7. Let $x = 0$ denote the closest point to the lighthouse along the shore. Suppose the light is shining at an angle of $\theta$ from the line joining the lighthouse to the nearest point on shore. The distance $x$ along the shore from the nearest point to the point where the light is shining is

$$x = 2\tan\theta,$$

since the tangent is the quotient of the opposite side divided by the adjacent side of the triangle. Now, both $x$ and $\theta$ are functions of time $t$, since the angle and the distance on shore are always changing. Thus we can differentiate both sides of $x = 2\tan\theta$ with respect to $t$, obtaining

$$x'(t) = 2\sec^2\theta(t)\cdot\theta'(t) \tag{9.59}$$

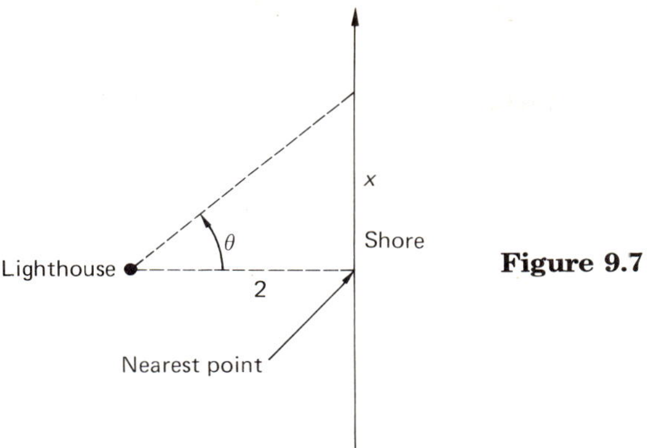


**Figure 9.7**

by the chain rule. Equation (9.59) provides the answer to both questions. In (a) we are assuming that the angle $\theta = 0$ radians. Since $\theta'(t)$ is the velocity at which the angle is changing,

$$\begin{aligned}\theta'(t) &= 2 \text{ rpm} = 2 \cdot 2\pi \text{ radians/min}\\ &= 240\pi \text{ radians/hr.}\end{aligned}$$

Substituting these two values into Eq. (9.59) we obtain

$$x'(t) = 2 \sec^2(0) \cdot 240\pi = 480\pi \text{ miles/hr.}$$

(b) Here the angle $\theta$ satisfies $\tan\theta = 1/2$. Since

$$\sec^2\theta = 1 + \tan^2\theta = 1 + \frac{1}{4} = \frac{5}{4},$$

Eq. (9.59) yields

$$x'(t) = 2\left(\frac{5}{4}\right)240\pi = 600\pi \text{ miles/hr.}$$

It is important always to use radian measure when making such calculations, since the differentiation formulas for trigonometric functions are based on measuring the angles in radians. Failure to do so will result in incorrect answers.

---

We now turn to the proofs of the differentiation formulas given on p. 468. We first find the derivative of $\sin x$ using the limit definition of the derivative (9.13):

$$(\sin x)' = \lim_{h\to 0}\frac{\sin(x+h) - \sin x}{h}.$$

Using the sum of angles formula (9) from Appendix 2 we have

$$\sin(x + h) = \sin x \cos h + \cos x \sin h,$$

so we may rewrite the derivative in the form

$$\begin{aligned}(\sin x)' &= \lim_{h\to 0}\frac{\cos x \sin h + \sin x(\cos h - 1)}{h}\\ &= \cos x \cdot \lim_{h\to 0}\frac{\sin h}{h} + \sin x \cdot \lim_{h\to 0}\frac{\cos h - 1}{h}.\end{aligned} \tag{9.60}$$

The problem now is to find the two limits in Eq. (9.60). We can again use a calculator to determine these limits; the results of these calculations are shown in Table 9.3. Remember to set your calculator to radian measure when checking the entries in this table.

**Table 9.3**

| $h$ | $(\sin h)/h$ | $(\cos h - 1)/h$ |
|---|---|---|
| 0.1 | 0.998334166 | −0.049958348 |
| 0.01 | 0.999983332 | −0.004999950 |
| 0.001 | 0.999999800 | −0.000500100 |
| 0.0001 | 0.999999995 | −0.000051000 |
| −0.001 | 0.999999985 | 0.000499700 |
| −0.01 | 0.999983347 | 0.004999920 |
| −0.1 | 0.998334169 | 0.049958344 |

It is apparent from Table 9.3 that

$$\lim_{h\to 0}\frac{\sin h}{h} = 1, \qquad \lim_{h\to 0}\frac{\cos h - 1}{h} = 0,$$

so that $(\sin x)' = \cos x$.

It is also possible to prove this fact using geometry. Consider Fig. 9.8, in which a portion of the unit circle is illustrated. Then we have

$$\text{area}\,\triangle OAD \le \text{area sector}\ OBD \le \text{area}\,\triangle OBC. \tag{9.61}$$

Since the lines $OB$ and $OD$ have length 1, we have

$$AD = \frac{AD}{OD} = \sin h,$$

$$OA = \frac{OA}{OD} = \cos h,$$

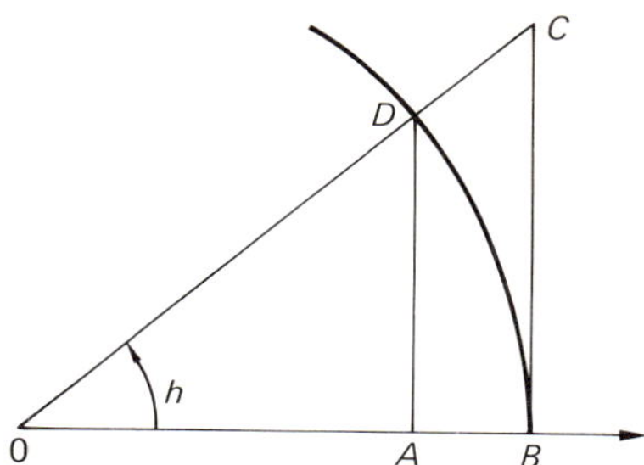


**Figure 9.8**

and

$$CB = \frac{CB}{OB} = \tan h = \frac{\sin h}{\cos h}.$$

The area of a sector of any circle is found by multiplying half the angle (in radians) by the radius squared. This fact is a consequence of the area of the circle,

$$\text{Area} = \pi r^2 = \frac{1}{2}(2\pi)r^2,$$

where the angle $h = 2\pi$ radians $= 360°$. Hence Eq. (9.61) becomes

$$\frac{1}{2}\sin h \cos h \le \frac{1}{2}h \le \frac{1}{2}\frac{\sin h}{\cos h}, \tag{9.62}$$

since the radius is 1. Multiplying each term by $2/(\sin h)$ yields

$$\cos h \le \frac{h}{\sin h} \le \frac{1}{\cos h} \tag{9.63}$$

and both ends of this inequality tend to 1 as $h \to 0$. If $h$ is negative, the inequalities in Eq. (9.63) are reversed, but the same conclusion is attained (see Exercise 17 on p. 475).

Thus,

$$\lim_{h\to 0}\frac{h}{\sin h} = 1,$$

so the reciprocal also has limit equal to 1. To show that the second limit vanishes, multiply top and bottom by $\cos h + 1$, obtaining

$$\lim_{h\to 0}\frac{\cos h - 1}{h} = \lim_{h\to 0}\frac{(\cos h - 1)(\cos h + 1)}{h(\cos h + 1)}$$

$$= \lim_{h\to 0}\frac{\cos^2 h - 1}{h(\cos h + 1)}.$$

Since $\sin^2 h = 1 - \cos^2 h$, we can rewrite this last limit as a product, and we have

$$\begin{aligned}\lim_{h\to 0}\frac{\cos h - 1}{h} &= \lim_{h\to 0}\left(\frac{\sin h}{h}\right)\left(\frac{-\sin h}{\cos h + 1}\right)\\ &= \lim_{h\to 0}\frac{\sin h}{h}\cdot\lim_{h\to 0}\frac{-\sin h}{\cos h + 1} = 1\cdot\left(\frac{0}{2}\right) = 0.\end{aligned}$$

Once we have proved that $(\sin x)' = \cos x$, it is easy to prove the other rules of differentiation using appropriate trigonometric identities. First, $\cos x = \sin[(\pi/2) - x]$ and $\sin x = \cos[(\pi/2) - x]$, so that

$$(\cos x)' = \left[\sin\left(\frac{\pi}{2} - x\right)\right]' = \cos\left(\frac{\pi}{2} - x\right)\cdot\left(\frac{\pi}{2} - x\right)' = -\sin x.$$

For the tangent and cotangent we use the identities

$$\tan x = \frac{\sin x}{\cos x}, \qquad \cot x = \frac{\cos x}{\sin x},$$

and the quotient rule to obtain

$$\begin{aligned}(\tan x)' &= \left(\frac{\sin x}{\cos x}\right)' = \frac{\cos x\,(\sin x)' - \sin x\,(\cos x)'}{\cos^2 x}\\ &= \frac{\cos^2 x + \sin^2 x}{\cos^2 x} = \frac{1}{\cos^2 x} = \sec^2 x,\end{aligned}$$

and

$$\begin{aligned}(\cot x)' &= \left(\frac{\cos x}{\sin x}\right)' = \frac{\sin x\,(\cos x)' - \cos x\,(\sin x)'}{\sin^2 x}\\ &= \frac{-\sin^2 x - \cos^2 x}{\sin^2 x} = \frac{-1}{\sin^2 x} = -\csc^2 x.\end{aligned}$$

Similarly,

$$(\sec x)' = \left(\frac{1}{\cos x}\right)' = \frac{\cos x\cdot(1)' - 1\,(\cos x)'}{\cos^2 x} = \frac{\sin x}{\cos^2 x} = \sec x\tan x,$$

and

$$\begin{aligned}(\csc x)' &= \left(\frac{1}{\sin x}\right)' = \frac{\sin x\cdot(1)' - 1\,(\sin x)'}{\sin^2 x}\\ &= \frac{-\cos x}{\sin^2 x} = -\csc x\cot x.\end{aligned}$$

## EXERCISES 9.7

*In Exercises 1–16, find the derivative of the function.*

1. $\sin(3x + 7)$
2. $\cos(x^2 + 2)$
3. $\cos\sqrt{1 + x^2}$
4. $\sin(3x^2 + x + 1)$
5. $x \sin x$
6. $\cos(\ln x)$
7. $\tan x^3$
8. $\sec(7^x)$
9. $x^2 \cot x$
10. $\tan^2(\sin x)$
11. $\csc^2(e^x)$
12. $\sec^3(\ln x)$
13. $\tan(\csc\sqrt{1 + x})$
14. $\cot(x \log x)$
15. $\sec(x \cot x)$
16. $\csc(x^x)$
17. Show that if $h < 0$ the inequalities in Eq. (9.62) are reversed. Divide again by $2/(\sin h)$ to obtain Eq. (9.63) and prove that

$$\lim_{h \to 0} \frac{\sin h}{h} = 1.$$

18. An airport searchlight is shining on the bottom of a uniform cloud layer 2000 ft above the ground. The searchlight is moving along a straight line at an angular velocity of 1 radian/sec. How fast is the light moving on the cloud when the beam is directly overhead?
19. A parachute is descending at 10 ft/sec. An observer 500 ft away from the parachute notices the parachute when it is still 300 ft above the ground. How fast is the angle of elevation between the observer and the parachute changing?
20. An airplane is flying at 30,000 ft above sea level. An observer directly below the path of the airplane hears the noise of its jets when the angle of elevation is 75°. If the observer is at sea level and the plane is flying at 700 mph away from the observer, how fast is the angle of elevation changing?
21. A dock is 1 m above water at high tide and 5 m above water at low tide. Assuming the motion of the tides has a period of 12 hr 25 min and that low tide occurred at 2:00 P.M., how fast is the water level rising at 4:00 P.M.?
22. A robin is most active at dawn and dusk. Assuming its body heat production has a 12-hr period and ranges from a minimum of 0.4 cal/hr to a maximum of 1.6 cal/hr during periods of greatest activity, find its heat production at 10:00 A.M. if dawn occurs at 6:00 A.M. What is the rate at which heat production is changing at 10:00 A.M.?

---

## 9.8 OPTIMIZATION: THE TWO DERIVATIVE TESTS

A farmer has 1000 ft of 3-strand barbed-wire fencing and wishes to build a rectangular-shaped enclosure for a cow. One edge of the enclosure is bounded by a stream, the other three by the fencing (see Fig. 9.9a). How long should the horizontal sides $x$ be so as to maximize the area of the enclosure?

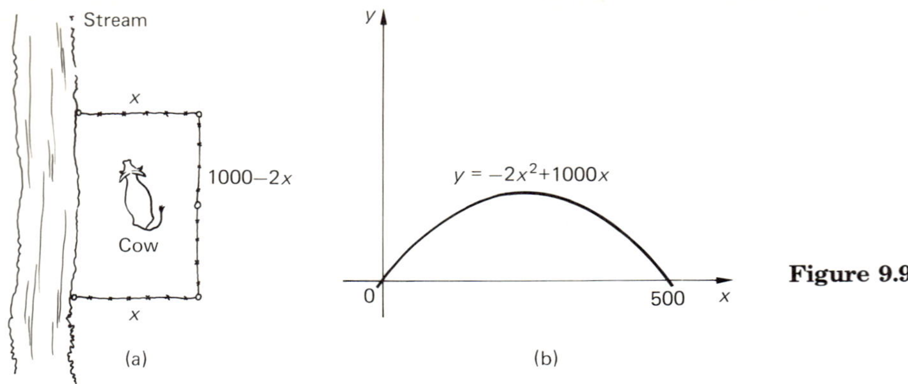


**Figure 9.9**

Since the two horizontal sides have length $x$, the one vertical side that must be built from fencing has length $1000 - 2x$. Thus, the area $y$ is given by the product of the lengths of the sides, that is,

$$y = x(1000 - 2x),$$

or

$$y = -2x^2 + 1000x. \tag{9.64}$$

A graph reveals that the function is a parabola (see Fig. 9.9b). The maximum area corresponds to the top of the curve where the slope of the tangent to the curve will be zero. Thus, to find the value of $x$ at which the maximum occurs, we must differentiate Eq. (9.64), set the derivative equal to zero, and solve for $x$. Hence,

$$y' = -4x + 1000 = 0,$$

so that $x = 250$ ft. Therefore, the horizontal sides of the enclosure must be 250 ft long, while the vertical side is 500 ft long. The maximum area will then be $250(500) = 125{,}000$ sq ft.

This example illustrates the main concept discussed in this section: that derivatives can be used to find maximum and minimum values of functions.

Consider the graph $y = f(x)$ that is illustrated in Fig. 9.10. The point $x_1$ is called a *local maximum* of $y = f(x)$, since $f(x_1) \geq f(x)$ for all $x$ in an interval $|x - x_1| < c_1$ containing the point $x_1$. Similarly, $x_4$ is a local maximum. The point $x_2$ is a *local minimum* of $y = f(x)$, since $f(x_2) \leq f(x)$ for all $x$ in an interval $|x - x_2| < c_2$ about $x_2$, as is the point $x_5$. If the function $y = f(x)$ is differentiable, then the derivative at each of the points $x_1$, $x_2$, $x_3$, $x_4$, and $x_5$ will vanish, since the tangent to the curve is horizontal at each of these points. Any point $x$ at which $f'(x) = 0$ is called a *critical point* of $y = f(x)$. Note, however, that the point

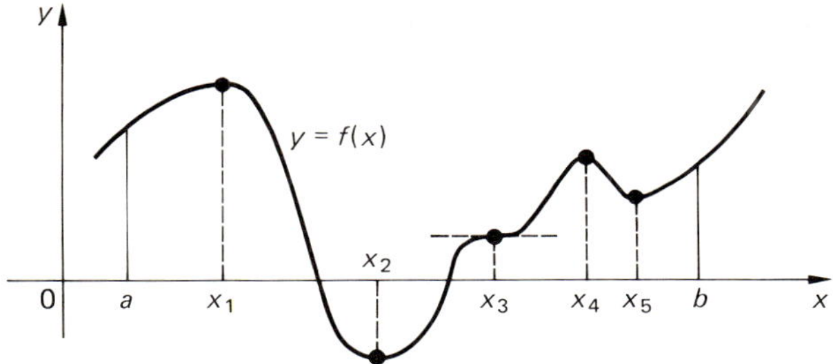


**Fig. 9.10**
Critical points of $y = f(x)$.

$x_3$ is critical *without* being a local maximum or minimum.

Suppose we are interested in finding the maximum or minimum values of a function $y = f(x)$ in an interval $a \le x \le b$. *If the function is differentiable we need only consider the critical points and the endpoints a and b of the interval.* These are the only points that matter, because if the maximum or minimum occurs inside the interval it will have a zero derivative and be a critical point, and if it does not occur inside the interval it must be an endpoint. The following example illustrates this process.

***Example 1*** Where does the function

$$y = \frac{x^4}{4} - x^3 + x^2 + 7 \tag{9.65}$$

assume its maximum value in the interval $-1 \le x \le 5/2$?

Differentiating this function we have

$$y' = x^3 - 3x^2 + 2x = x(x - 1)(x - 2), \tag{9.66}$$

which has the roots $x = 0, 1, 2$. These are the critical points, so if we calculate $y$ at these points *and* at the endpoints $x = -1$, and $5/2$ of the interval, the point with the greatest value will be the maximum of the function in the interval $-1 \le x \le 5/2$. Substituting the values $x = -1, 0,$ 1, 2, and 5/2 into Eq. (9.65), we have

$$y(-1) = 9\tfrac{1}{4}, \qquad y(0) = 7, \qquad y(1) = 7\tfrac{1}{4}, \qquad y(2) = 7, \qquad y(5/2) = 7\tfrac{25}{64}.$$

Thus the maximum occurs at the endpoint $x = -1$ where Eq. (9.65) equals $9\frac{1}{4}$. We also see that the minimum, $y = 7$, occurs at two places:

$x = 0$ and $x = 2$. Thus in this example the maximum occurred at an endpoint and the minimums occurred inside the interval.

---

The procedure we used in this example is very useful in determining the maximum or minimum value of a differentiable function or an interval, but it does not reveal whether a given critical point is a local maximum, local minimum, or neither. It is not hard to distinguish between these alternatives; Fig. 9.10 provides the necessary clues.

> The curve *rises until it reaches the local maximum, and then drops* after passing by. The situation reverses for a local minimum, the curve *dropping until it reaches the local minimum and rising afterward.*

How does this translate into the language of derivatives? We know that when the derivative is positive the curve must be rising, while when the derivative is negative the curve is dropping. The four possible situations are shown in Fig. 9.11: In each part, $y'(x_1) = 0$ and $x_1$ lies between the nearby points $x_0 < x_1 < x_2$. Assuming the derivative is continuous, we consider the derivative $y'$ at the points $x_0$ and $x_1$. Since $x_0 < x_1 < x_2$, we have:

1. A local maximum at $x_1$ if $y'(x_0) > 0 > y'(x_2)$;
2. A local minimum at $x_1$ if $y'(x_0) < 0 < y'(x_2)$;
3. Neither a local maximum nor a local minimum at $x_1$ otherwise.

It is important to *choose $x_0$ and $x_2$ closer to $x_1$ than any other point at which the derivative is zero;* otherwise, you may have erroneous results.

Returning to Example 1, we determine which of the points $x = 0, 1$, and 2 are local maxima or local minima. Using Eq. (9.66) we find the derivative at the points $x = \pm 1/2$. Now

$$y'(-1/2) = -15/8 \quad \text{and} \quad y'(1/2) = 3/8,$$

so the curve is dropping at $x = -1/2$ and rising at $x = 1/2$. By Fig. 9.11(b) we have a local minimum at $x = 0$. To check $x = 1$, we consider $x = 1/2$ and $x = 3/2$. As before,

$$y'(1/2) = 3/8 \quad \text{and} \quad y'(3/2) = -3/8,$$

so that $x = 1$ is a local maximum. Finally, to test $x = 2$ we consider $x = 3/2$ and $x = 9/4$. Then $y'(3/2) = -3/8$ and $y'(9/4) = 45/64$, so that $x = 2$ is a local minimum.

Gathering all the facts we have discussed, we obtain the following result.

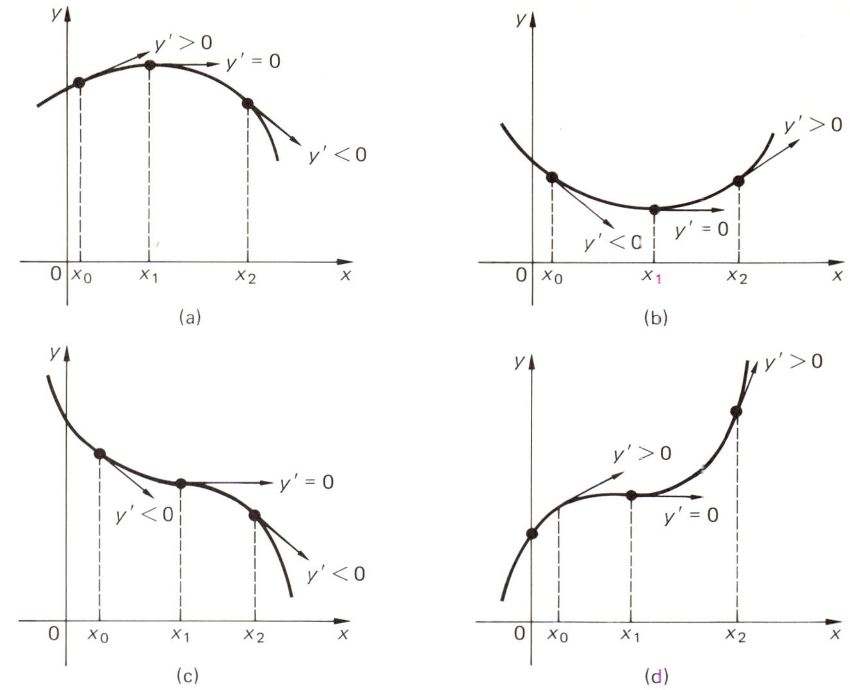


**Fig. 9.11**
(a) Local maximum; (b) local minimum; (c) neither; (d) neither.

**First Derivative Test** **The local maxima and minima of a function $y = f(x)$ having a continuous derivative are determined by testing those points $x_1$ at which $f'(x_1) = 0$. To test $x_1$, select nearby values $x_0 < x_1 < x_2$ and examine the derivatives at these points. Then**

**(1) $x_1$ is a local maximum if $f'(x_0) > 0 > f'(x_2)$,**

**or**

**(2) $x_1$ is a local minimum if $f'(x_0) < 0 < f'(x_2)$.**

**If neither situation occurs, $x_1$ is not a local maximum or minimum.**

***Example 2*** An orange juice packer orders ten million frozen orange juice cans from a manufacturer. Each cylindrical can is to hold 10 fluid ounces ($\approx 295.74\ \text{cm}^3$) of frozen concentrate, and have cardboard sides and metal ends. The cardboard for the sides costs 4c per m$^2$, while the metals cost 8¢ per m$^2$. What dimensions should the cans have to minimize the cost of each can?

Let $r$ be the radius of the base and $h$ the height of each can. The volume of the can is found by multiplying the area of the base ($\pi r^2$) by the height, that is,

$$V = \pi r^2 h.$$

In this example we are given the volume $V = 295.74$, so dividing by $\pi r^2$ we obtain

$$h = \frac{295.74}{\pi r^2}. \tag{9.67}$$

We shall make use of this identity later on.

Since each end is metal and has an area of $\pi r^2$, the cost of the two ends (assuming there is no waste of metal in making the ends) is found by multiplying the total area by the cost of the metal per $\text{cm}^2$. Thus,

$$\text{cost of ends} = 2\pi r^2 \cdot \frac{8}{(100)^2} = \frac{\pi r^2}{625}\text{ cents.}$$

To find the cost of the sides, we multiply the perimeter of the base $2\pi r$ by the height $h$, obtaining the area, and multiply this by the cost of the cardboard per $\text{cm}^2$, or

$$\text{cost of sides} = 2\pi rh \cdot \frac{4}{(100)^2} = \frac{\pi rh}{1250}\text{ cents.}$$

Thus, the total cost of the materials in each can is

$$C = \frac{\pi r^2}{625} + \frac{\pi rh}{1250}. \tag{9.68}$$

Equation (9.68) determines cost in terms of the radius and height of the can. Before we can use the first derivative test, we must eliminate one of the variables $r$ or $h$, since the techniques in this section only apply to the situation in which there is one independent variable and one dependent variable. This is where we use Eq. (9.67) again: We replace $h$ in Eq. (9.70) by Eq. (9.69), obtaining

$$\begin{aligned} C &= \frac{\pi r^2}{625} + \frac{\pi r}{1250}\left(\frac{295.74}{\pi r^2}\right). \\ &= \frac{\pi r^2}{625} + \frac{0.236592}{r} \end{aligned} \tag{9.69}$$

for all $r \geq 0$ since the radius of the can cannot be negative. Differentiating Eq. (9.69) with respect to $r$, we obtain

$$C'(r) = \frac{2\pi r}{625} - \frac{0.236592}{r^2}. \tag{9.70}$$

Setting Eq. (9.70) equal to zero, we solve for $r$, obtaining

$$r^3 = 0.236592\left(\frac{625}{2\pi}\right) = \frac{73.935}{\pi},$$

or

$$r \approx 2.86572 \text{ cm}. \tag{9.71}$$

We now have three values to test: that in Eq. (9.71) and the endpoints of the interval $r \geq 0$. Note that Eq. (9.69) becomes arbitrarily large as $r$ decreases to zero or increases to infinity, so neither of the endpoints provides a minimum. Thus, we must check that Eq. (9.71) yields a local minimum. We can do so by substituting $r = 2$ and $r = 3$ into the derivative (9.70). But

$$C'(2) = \frac{4\pi}{625} - \frac{0.236592}{4} \approx -0.03904$$

and

$$C'(3) = \frac{6\pi}{625} - \frac{0.236592}{9} \approx 0.00387,$$

so $r \approx 2.86572$ cm is a local minimum. Substituting this value into Eqs. (9.67) and (9.69) we obtain a height of $h \approx 11.46285$ cm and a total cost of $C(2.86572) \approx 0.12384$¢ for each can.

---

There is an alternative procedure for determining whether a critical point is a local maximum or minimum. To understand this procedure you must remember that a derivative is a function of the independent variable. Thus, in many cases it is possible to take the derivative of $f'(x)$, obtaining the *second derivative* of the original function $y = f(x)$. Indeed, it may be possible to continue taking derivatives indefinitely! We use the notations $y''$ and $f''(x)$ for the second derivative, but other notations are common:

$$\text{Second derivatives:} \quad \frac{d^2y}{dx^2}, \quad \frac{d^2f(x)}{dx^2}, \quad \ddot{y}, \quad D_x^2y;$$

$$n\text{th derivatives:} \quad \frac{d^ny}{dx^n}, \quad \frac{d^nf(x)}{dx^n}, \quad y^{(n)}, \quad f^{(n)}(x), \quad D_x^ny.$$

As we have seen, a local maximum occurs whenever the slope *decreases* from positive to negative values. Thus, the graph of the curve $y = f'(x)$ is decreasing and must have negative slope. Hence, the second derivative $f''(x)$ must be negative, since it is the derivative of the curve $y = f'(x)$. Similarly, a local minimum is identified by the fact that the slope of $y = f(x)$ increases from negative to positive values. Therefore, the curve $y = f'(x)$ must be increasing, implying that $f''(x)$ is positive. We gather all this information in the following result.

**Second Derivative Test** **The local maxima and minima of a function $y = f(x)$ having a second derivative at each point are determined by testing those points $x_1$ at which $f'(x_1) = 0$. The point $x_1$ is**

**(1) a local maximum if $f''(x_1) > 0$,**

**or**

**(2) a local minimum if $f''(x_1) < 0$.**

If $f''(x_1) = 0$, the second derivative test fails. The point $x_1$ may be either a local maximum or minimum or neither. *No information is conveyed*! When this occurs, we must use the first derivative test to resolve the question.

Because of this failure of the second derivative test, it should only be used when it is easy to find the second derivative. At all other times, the first derivative test is preferred. The following examples illustrate the use of the second derivative test.

***Example 2 (continued)*** Suppose we had decided to use the second derivative test to determine whether $r \approx 2.86572$ is a local maximum or minimum of Eq. (9.69). Differentiating Eq. (9.70) once again, we obtain

$$C''(r) = \frac{2\pi}{625} + \frac{2(0.236592)}{r^3},$$

and substituting $r \approx 2.86572$ into this equation yields

$$C''(2.86572) \approx 0.3016,$$

which is positive, implying that we have a local minimum.

***Example 3*** A barber can give 120 haircuts a week if he charges \$3 per haircut. For each 50¢ increase in price he loses an average of 10 customers per week. How much should he charge to maximize his revenue?

The weekly revenue that the barber receives is the product of the number of haircuts and the price per haircut. Let $x$ be the number of 50¢ increases in price. Then the price per haircut is

$$3.00 + 0.50x,$$

while the number of haircuts given is

$$120 - 10x.$$

Thus, the revenue received by the barber is

$$\begin{aligned} R(x) &= (3.00 + 0.50x)(120 - 10x) \\ &= 360 + 30x - 5x^2. \end{aligned} \qquad \textbf{(9.72)}$$

To optimize this function we take the derivative and set it equal to zero:

$$R'(x) = 30 - 10x = 0.$$

Thus, $x = 3$ is the only critical point for this problem. Since $R''(x) = -10$, the second derivative test implies that $x = 3$ is a local maximum. However, is it a maximum? We know that $x \geq -6$ since barbers do not pay their customers, and $x \leq 12$ since there cannot be a negative number of customers. Thus, to be sure that $x = 3$ is a maximum we should compare $R(3)$ with the value of $R(x)$ at the endpoints $x = -6$ and 12. Since $R(3) = \$405$ and $R(-6) = R(12) = 0$, it is clear that $x = 3$ is *the* maximum.

---

The next example will show that it *is* important to consider the endpoints.

***Example 4*** A semitrailer traveling at $x$ mph on an expressway having 40 mph minimum and 55 mph maximum speed limits uses

$$y = 9 + \frac{x}{10} + \frac{x^2}{225}$$

gallons of diesel fuel per hour. Diesel fuel costs 50¢ per gallon. The driver is paid \$8 per hour, and fixed costs for operating the truck average \$1 per mile. Find the legal velocity that will make the cost per mile a minimum.

Since $y(\text{gal/hr}) \div x(\text{miles/hr}) = y/x(\text{gal/mile})$ is the amount of fuel consumed per mile, the total cost per mile for operating the truck at $x$ mph is

$$C(x) = 0.5\left(\frac{y}{x}\right) + \frac{8}{x} + 1,$$

or

$$C(x) = \frac{4.5}{x} + \frac{1}{20} + \frac{x}{450} + \frac{8}{x} + 1. \tag{9.73}$$

Differentiating with respect to $x$ and setting the derivative equal to zero, we have

$$C'(x) = \frac{-4.5}{x^2} + \frac{1}{450} - \frac{8}{x^2} = 0, \tag{9.74}$$

from which we obtain

$$\frac{12.5}{x^2} = \frac{1}{450} \quad \text{or} \quad x^2 = 5625.$$

Hence $x = 75$ mph is the only critical point for the function $C(x)$. But 75 mph is not a legal velocity, so there are no critical points in the interval $40 \leq x \leq 55$. Thus, only the endpoints need be tested. Since

$C(40) = 1.451388\ldots$ and $C(55) = 1.3994949\ldots$, the minimum cost per mile is attained at a velocity of 55 mph.

---

## EXERCISES 9.8

*In Exercises 1–10, find the maximum and minimum of the given function over the indicated interval, using the first derivative test.*

1. $y = 5x^2 + 3x + 7, -1 \leq x \leq 1$
2. $y = 6x^2 - 7x + 1, -7 \leq x \leq 5$
3. $y = 1 - x - x^2, -1 \leq x \leq 1$
4. $y = 6x^2 + x + 18, 0 \leq x \leq 5$
5. $y = xe^{-x^2}, |x| < \infty$
6. $y = (2 - x) \ln x, 1 \leq x \leq 2$
7. $y = 2x^3 + 3x^2 - 12x + 5, -3 \leq x \leq 3$
8. $y = x^x, 1/2 \leq x \leq 2$
9. $y = x/(1 + x^2), |x| < \infty$
10. $y = \sin x/(1 + x^2), |x| < \infty$

*In Exercises 11–20, find the maximum and minimum of the given function over the interval indicated, using the second derivative test.*

11. $y = 4x^2 + 5x + 1, -1 \leq x \leq 1$
12. $y = 5x^2 + 7x + 6, -3 \leq x \leq 4$
13. $y = 12 + x - 7x^2, -3 \leq x \leq 1$
14. $y = 9 - 8x - 7x^2, -3 \leq x \leq 3$
15. $y = 2x^3 + 15x^2 + 6x + 9, -5 \leq x \leq 0$
16. $y = e^{-x} \ln x, 1 \leq x < \infty$
17. $y = \ln x/x, 1 \leq x < \infty$
18. $y = e^{-x^2}, |x| < \infty$
19. $y = x \sin x, 0 \leq x \leq 2\pi$
20. $y = e^x \sin x, -2\pi \leq x \leq 2\pi$
21. Fred's Taco Shop can sell 300 tacos a day at a price of 50¢ apiece. Each 10¢ increase in the price cuts the sales by 50.
    a) How much should be charged in order to maximize the revenue?
    b) If the cost of making each taco is 20¢, how much should be charged to maximize profits?
22. A butcher can sell 200 lb of sirloin every day if he charges \$1.89 per lb. If he raises the price to \$2.39 per lb., he only sells 75 lb. His costs are \$1.50 per

lb. What should he charge to maximize profits? Assume the demand curve is linear with price.

23. Suppose a farmer wishes to build two adjacent identical rectangular enclosures, each having one side along the stream. How should 1000 ft of barbed-wire fencing be used to maximize the area of each enclosure?

24. A dog breeder wishes to build a row of 10 identical kennels. If each kennel is to have an area of 60 sq ft, find the minimum length of chain link fencing that will be required. Assume no top is required and that the gates are included in the total length.

25. Suppose the dog breeder in Exercise 24 can use an existing brick wall as the back of each kennel. If we assume the same conditions as before, what is the minimum length of chain link fencing that now is required?

26. A long metal pipe is to be moved down an L-shaped corridor. The corridor is 8 ft wide on the horizontal side and 6 ft wide on the vertical side. If we neglect the width of the pipe, what is the length of the longest pipe that can be carried around the corner?

27. In Example 2, assume the ends of the cans are cut from square pieces of metal producing some waste metal in the process. What should the dimension of the can be in this situation so that the manufacturer minimizes the cost of materials?

28. A farmer wishes to construct a V-shaped trough from a piece of metal 4 m long and 1 m wide (see Fig. 9.12). The ends of the trough will be made of wood. How wide should the opening at the top be made in order to maximize the volume of the trough?

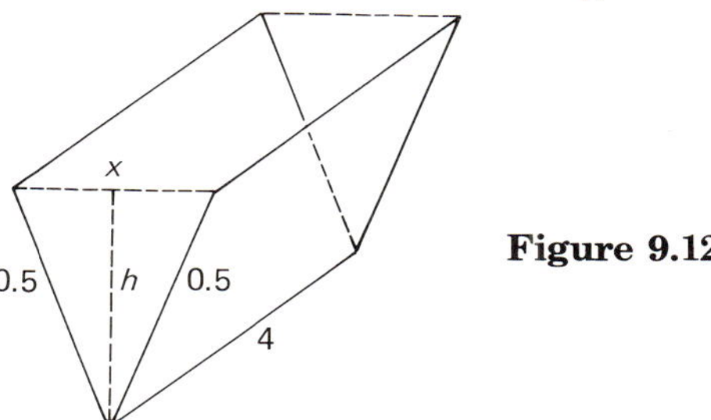


**Figure 9.12**

29. A farmer wishes to construct the trough shown in Fig. 9.13 from a piece of metal 10 ft long and 6 ft wide. The ends are to be made of wood. How wide should the opening at the top be made in order to maximize the volume of the trough?

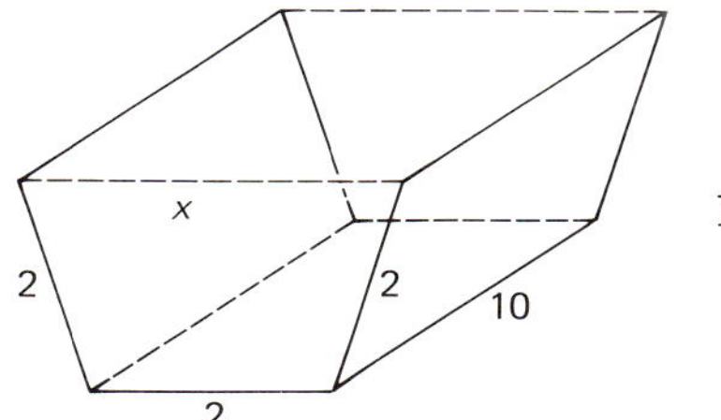


**Figure 9.13**

**30.** A piece of wire 12″ long is to be cut to construct a circle and a square. What are the dimensions of the circle and square having the minimum total area? Maximum total area?

---

## 9.9 PARTIAL DERIVATIVES

At this point we have discussed only derivatives of functions of one independent variable. However, most real-life problems involve several independent variables. The methods and applications of derivatives may be extended to such situations by using *partial derivatives*, the basic idea of which is quite simple: *Differentiate with respect to one independent variable, treating all other independent variables as constants.* For example, if

$$y = 2x^2 - 5xt + 7t^2,$$

then the partial derivative of $y$ with respect to $x$ is

$$\frac{\partial y}{\partial x} = 4x - 5t,$$

while the partial derivative of $y$ with respect to $t$ is

$$\frac{\partial y}{\partial t} = -5x + 14t.$$

In terms of limits we have the following.

**DEFINITION** **Suppose $y = f(x, t)$. Then the partial derivatives of $y$ with respect to $x$ or $t$ are given by the limits**

$$\frac{\partial y}{\partial x} = \lim_{h \to 0} \frac{f(x + h, t) - f(x, t)}{h}$$

**and**

$$\frac{\partial y}{\partial t} = \lim_{k \to 0} \frac{f(x, t + k) - f(x, t)}{k}.$$

Other commonly used notations for these partial derivatives are

$$\frac{\partial y}{\partial x} = y_x = \frac{\partial f}{\partial x} = f_x = D_1 f \quad \text{and} \quad \frac{\partial y}{\partial t} = y_t = \frac{\partial f}{\partial t} = f_t = D_2 f.$$

It is easy to interpret the partial derivatives geometrically once we graph the function $y = f(x, t)$ in three-dimensional $xty$-space. As the following example shows, the graph of $y = f(x, t)$ is a surface in three-space.

***Example 1*** Consider the function

$$y = x^2 + t^2. \tag{9.75}$$

Although our graph must necessarily be two-dimensional, we will use perspective to simulate three-dimensional space. In Fig. 9.14 we visualize the positive $t$-axis as emerging outward perpendicular to the $xy$-plane. Since $x^2 + t^2$ cannot be negative, $y$ must be nonnegative. Consider any circle in the $xt$-plane centered at the origin. If the radius of the circle is $r$, its equation is

$$x^2 + t^2 = r^2$$

implying, by Eq. (9.75), that the $y$-value associated with the points on this circle must be $r^2$. Pictorially, this tells us that all points on the surface $y = x^2 + t^2$ lying above the circle $x^2 + y^2 = r^2$ are at a "$y$-height" of $r^2$ units. Continuing in this fashion we obtain the surface of all points of the form $(x, t, x^2 + t^2)$ shown in Fig. 9.14. This surface may be visualized by revolving the parabola $y = x^2$ about the $y$-axis. For this reason, this surface is often called a *paraboloid of revolution.*

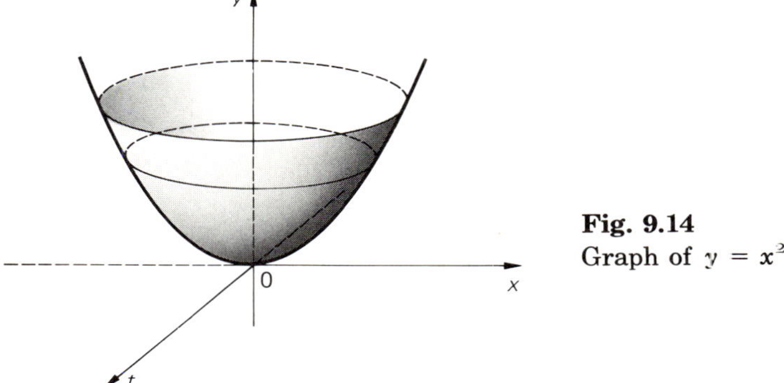


**Fig. 9.14**
Graph of $y = x^2 + t^2$

Generally, the equation $y = f(x, t)$ produces a surface consisting of all points whose coordinates are of the form $(x, t, f(x, t))$. Since $\partial y/\partial x$ and $\partial y/\partial t$ are *also* functions of $x$ and $t$, they can also be thought of as surfaces in three-space. Thus, we may select a point $(x_0, t_0)$ in the $xt$-plane and determine the "vertical" height of either of these surfaces above that point. But what connection do these partials *at the point* $(x_0, t_0)$,

$$\left.\frac{\partial y}{\partial x}\right|_{(x_0, t_0)} \quad \text{and} \quad \left.\frac{\partial y}{\partial t}\right|_{(x_0, t_0)},$$

have with the original surface $y = f(x, t)$? To answer this question we begin by graphing the surface $y = f(x, t)$. Then the partial derivative of $y$ with respect to $x$ *at any point* $(x_0, t_0)$,

$$\left.\frac{\partial y}{\partial x}\right|_{(x_0, t_0)}, \tag{9.76}$$

is simply the slope, at the point $(x_0, t_0)$, of the tangent line to the curve $C$ obtained by intersecting the surface $y = f(x, t)$ with the plane $t = t_0$ (see Fig. 9.15). Since all $t$-values are constant in the plane $t = t_0$, the values of $y$ along the curve $C$ depend entirely on the independent variable $x$. This explains why we treat all other independent variables as constants when taking partial derivatives with respect to one independent variable. Similarly, the partial derivative

$$\left.\frac{\partial y}{\partial t}\right|_{(x_0, t_0)}$$

is the slope at $(x_0, t_0)$ of the tangent line to the curve $C'$ obtained by intersecting the surface $y = f(x, t)$ with the plane $x = x_0$.

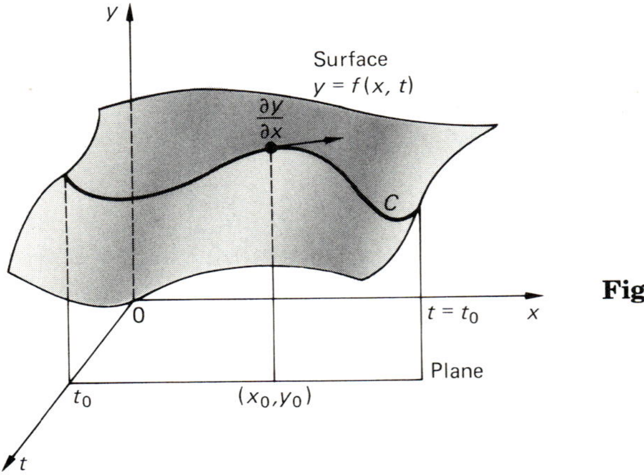


**Figure 9.15**

***Example 2*** Find the partial derivatives of

$$y = x^2 + t^2 \tag{9.77}$$

at the point $(x_0, t_0) = (1, 0)$.

Since

$$\frac{\partial y}{\partial x} = 2x \quad \text{and} \quad \frac{\partial y}{\partial x} = 2t,$$

we have

$$\left.\frac{\partial y}{\partial x}\right|_{(1,0)} = 2 \cdot 1 = 2 \quad \text{and} \quad \left.\frac{\partial y}{\partial t}\right|_{(1,0)} = 2 \cdot 0 = 0.$$

To see what is happening, first "slice" the paraboloid in Fig. 9.14 along the $xy$-plane ($t = 0$). The curve we get is simply the parabola $y = x^2$ whose derivative is $y' = 2x$. Note that this agrees with the computed partial derivative $\partial y/\partial x$. The slope of the tangent line to the parabola at $x = 1$ is 2, using either $y'$ or $\partial y/\partial x$. Now slice the paraboloid along the plane $x = 1$ parallel to the $ty$-plane. The parabola $y = t^2 + 1$ is obtained (set $x = 1$ in Eq. 9.77), with derivative $y' = 2t = \partial y/\partial t$. The minimum of this parabola occurs when $t = 0$, at which point the slope of the tangent to the curve is zero.

Since the partial derivatives $\partial y/\partial x$ and $\partial y/\partial t$ are again functions of $x$ and $t$, it is possible to take partial derivatives of these surfaces. These second partials, denoted by

$$\frac{\partial}{\partial x}\left(\frac{\partial y}{\partial x}\right) = \frac{\partial^2 y}{\partial x^2}, \qquad \frac{\partial}{\partial x}\left(\frac{\partial y}{\partial t}\right) = \frac{\partial^2 y}{\partial x\, \partial t},$$

$$\frac{\partial}{\partial t}\left(\frac{\partial y}{\partial x}\right) = \frac{\partial^2 y}{\partial t\, \partial x}, \qquad \frac{\partial}{\partial t}\left(\frac{\partial y}{\partial t}\right) = \frac{\partial^2 y}{\partial t^2},$$

are simply a generalization of second derivatives.* If the function $y = f(x, t)$ is sufficiently "smooth," it may be possible to continue this process indefinitely, generating third partials, fourth partials, etc.

***Example 3*** Find the first and second partials of the function

$$y = e^{2x} \ln t.$$

We have

$$\frac{\partial y}{\partial x} = 2e^{2x} \ln t, \qquad \frac{\partial y}{\partial t} = \frac{e^{2x}}{t},$$

so that

$$\frac{\partial^2 y}{\partial x^2} = \frac{\partial}{\partial x}\left(\frac{\partial y}{\partial x}\right) = \frac{\partial}{\partial x}(2e^{2x} \ln t) = 4e^{2x} \ln t,$$

$$\frac{\partial^2 y}{\partial x\, \partial t} = \frac{\partial}{\partial x}\left(\frac{\partial y}{\partial t}\right) = \frac{\partial}{\partial x}\left(\frac{e^{2x}}{t}\right) = \frac{2e^{2x}}{t},$$

---

* Other common notations are:

$$\partial^2 y/\partial x^2 = y_{xx}, \quad \partial^2 y/\partial x\, \partial t = y_{xt}, \quad \partial^2 y/\partial t\, \partial x = y_{tx}, \quad \partial^2 y/\partial t^2 = y_{tt}.$$

$$\frac{\partial^2 y}{\partial t\,\partial x} = \frac{\partial}{\partial t}\left(\frac{\partial y}{\partial x}\right) = \frac{\partial}{\partial t}(2e^{2x}\ln t) = \frac{2e^{2x}}{t},$$

$$\frac{\partial^2 y}{\partial t^2} = \frac{\partial}{\partial t}\left(\frac{\partial y}{\partial t}\right) = \frac{\partial}{\partial t}\left(\frac{e^{2x}}{t}\right) = \frac{-e^{2x}}{t^2}.$$

Notice that the two second-order *mixed* partials are equal:

$$\frac{\partial^2 y}{\partial t\,\partial x} = \frac{\partial^2 y}{\partial x\,\partial t}. \tag{9.78}$$

This identity will hold whenever the second-order mixed partials are continuous in both variables, which is usually the situation in practice.

---

Just as derivatives are useful in determining maximum and minimum values of a function of a single independent variable, partial derivatives can be used to maximize or minimize a function of several independent variables. The following theorem is the equivalent of the second derivative test for function of two variables. Just as in the second derivative test, in some situations the test below will fail; when this happens other more advanced techniques can be used. The proof of this theorem can be found in most advanced calculus texts.

**THEOREM 9.2** **Let $y = f(x, t)$ be a function having continuous second partial derivatives. The local maxima and minima of $y = f(x, t)$ can be determined by testing the *critical points* $(x_1, t_1)$ *at which both* $\partial f/\partial x$ and $\partial f/\partial t$ *are zero.***

**Suppose we wish to test the critical point $(x_1, t_1)$. Let**

$$\Delta = \left(\frac{\partial^2 f}{\partial x^2}\bigg|_{(x_1,t_1)}\right)\cdot\left(\frac{\partial^2 f}{\partial t^2}\bigg|_{(x_1,t_1)}\right) - \left(\frac{\partial^2 f}{\partial x\,\partial t}\bigg|_{(x_1,t_1)}\right)^2;$$

**then $(x_1, t_1)$ is**

**a) A *local maximum* if $\Delta > 0$ and $\dfrac{\partial^2 f}{\partial x^2}\bigg|_{(x_1,t_1)} < 0$;**

**b) A *local minimum* if $\Delta > 0$ and $\dfrac{\partial^2 f}{\partial x^2}\bigg|_{(x_1,t_1)} > 0$;**

**c) Neither a local maximum or a minimum if $\Delta < 0$.**

**Otherwise the test fails and no conclusion can be reached.**

***Example 4*** Find the local extrema of the function

$$y = xt.$$

First we need to find the critical points of this function. Setting the first partials equal to zero, we have

$$\frac{\partial y}{\partial x} = t = 0 \quad \text{and} \quad \frac{\partial y}{\partial t} = x = 0,$$

so the only critical point is at the origin (0, 0). The second partials are

$$\frac{\partial^2 y}{\partial x^2} = \frac{\partial^2 y}{\partial t^2} = 0 \quad \text{and} \quad \frac{\partial^2 y}{\partial x\, \partial t} = 1,$$

so $\Delta = -1$. By Theorem 9.2(c), the critical point (0, 0) is neither a local maximum nor a local minimum. We can discover what is happening at (0, 0) if we check the sign in each quadrant of the $xt$-plane. Note that in the first and third quadrants $y = xt \geq 0$, while in the second and fourth quadrants $y = xt \leq 0$. A graph of the function $y = xt$ in three dimensional space is shown in Fig. 9.16. For obvious reasons the point (0, 0, 0) is called a *saddle point.*

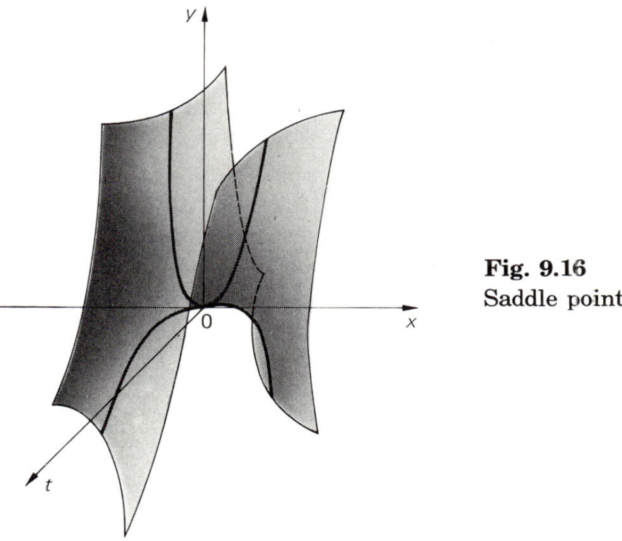


**Fig. 9.16** Saddle point.

***Example 5*** Find the local maxima and minima of the function

$$y = x^3 + t^3 + 4xt.$$

To locate the critical points we set the first partials equal to zero:

$$\frac{\partial y}{\partial x} = 3x^2 + 4t = 0 \quad \text{and} \quad \frac{\partial y}{\partial t} = 3t^2 + 4x = 0.$$

To solve these equations simultaneously, we plot the two curves in the $xt$-plane (see Fig. 9.17), noting that they are both parabolas and meet at $(0, 0)$ and at some point on the line $x = t$ in the third quadrant. Setting $x = t$ in either equation, we get

$$3t^2 + 4t = 0,$$

or $t(3t + 4) = 0$, implying that $t = 0$ or $t = -4/3$. The solution $x = t = 0$ yields the critical point $(0, 0)$, while $x = t = -4/3$ yields the critical point $(-4/3, -4/3)$.

Taking the second partials, we have

$$\frac{\partial^2 y}{\partial x^2} = 6x, \qquad \frac{\partial^2 y}{\partial t^2} = 6t, \qquad \frac{\partial^2 y}{\partial x\, \partial t} = 4$$

so

$$\Delta = 36xt - 16.$$

Substituting $(0, 0)$ and $(-4/3, -4/3)$ into $\Delta$, we obtain

$$\Delta(0, 0) = -16 \quad \text{and} \quad \Delta(-4/3, -4/3) = 48.$$

Thus $(0, 0)$ is neither a maximum nor a minimum (it again is a saddle point), while $(-4/3, -4/3)$ is a local maximum since

$$\frac{\partial^2 y}{\partial x^2}\left(\frac{-4}{3}, \frac{-4}{3}\right) = 6x\bigg|_{(-4/3,\,-4/3)} = -8.$$

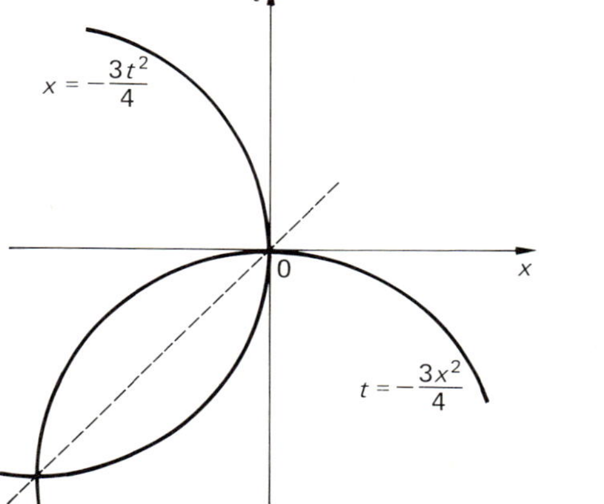


**Figure 9.17**

***Example 6*** In Section 2.8 we discussed the uses of the least squares theorem when we are fitting a straight line through a set of data points

$(x_1, y_1), (x_2, y_2), \ldots, (x_n, y_n)$. The theorem states that the straight line

$$y = m(x - \bar{x}) + b, \tag{9.79}$$

where the constants $m$ and $b$ are given by

$$m = \frac{(x_1y_1 + x_2y_2 + \cdots + x_ny_n) - n\bar{x}\bar{y}}{(x_1^2 + x_2^2 + \cdots + x_n^2) - n(\bar{x})^2}, \quad b = \bar{y}, \tag{9.80}$$

*minimizes* the squares of the errors between predicted and observed $y$-values:

$$\begin{aligned} E &= \{y_1 - [m(x_1 - \bar{x}) + b]\}^2 \\ &\quad + \{y_2 - [m(x_2 - \bar{x}) + b]\}^2 + \cdots + \{y_n - [m(x_n - \bar{x}) + b]\}^2. \end{aligned} \tag{9.81}$$

The symbols $\bar{x}$ and $\bar{y}$ stand for the averages

$$\bar{x} = \frac{1}{n}(x_1 + x_2 + \cdots + x_n),$$

$$\bar{y} = \frac{1}{n}(y_1 + y_2 + \cdots + y_n).$$

To prove the least squares theorem we will find the minimum of the nonnegative function (9.81). Observe that the data points are all fixed so that the only independent variables are the coefficients $m$ and $b$, implying that $E = E(m, b)$ is a function of $m$ and $b$. We begin by setting the first partials of (9.81) with respect to $m$ and $b$ equal to zero, so that we can locate the critical points:

$$\begin{aligned} \frac{\partial E}{\partial m} &= -2\{y_1 - [m(x_1 - \bar{x}) + b]\}(x_1 - \bar{x}) \\ &\quad - \cdots -2\{y_n - [m(x_n - \bar{x}) + b]\}(x_n - \bar{x}) \\ &= 2\{m[(x_1 - \bar{x})^2 + \cdots + (x_n - \bar{x})^2] \\ &\quad + b[(x_1 - \bar{x}) + \cdots + (x_n - \bar{x})] \\ &\quad - [y_1(x_1 - \bar{x}) + \cdots + y_n(x_n - \bar{x})]\} \\ &= 0, \end{aligned}$$

and

$$\begin{aligned} \frac{\partial E}{\partial b} &= -2\{y_1 - [m(x_1 - \bar{x}) + b]\} - \cdots -2\{y_n - [m(x_n - \bar{x}) + b]\} \\ &= 2\{m[(x_1 - \bar{x}) + \cdots + (x_n - \bar{x})] + nb - [y_1 + \cdots + y_n]\} \\ &= 0. \end{aligned}$$

Now

$$(x_1 - \bar{x}) + \cdots + (x_n - \bar{x}) = (x_1 + \cdots + x_n) - n\bar{x} = 0$$

and

$$\begin{aligned}(x_1 - \bar{x})^2 &+ \cdots + (x_n - \bar{x})^2 \\ &= (x_1^2 - 2x_1\bar{x} + \bar{x}^2) + \cdots + (x_n^2 - 2x_n\bar{x} + \bar{x}^2) \\ &= (x_1^2 + \cdots + x_n^2) - 2\bar{x}(x_1 + \cdots + x_n) + n\bar{x}^2 \\ &= (x_1^2 + \cdots + x_n^2) - 2\bar{x}(n\bar{x}) + n\bar{x}^2 \\ &= (x_1^2 + \cdots + x_n^2) - n\bar{x}^2.\end{aligned}$$

Hence

$$\begin{aligned}\frac{\partial E}{\partial m} &= 2\{m[(x_1^2 + \cdots + x_n^2) - n\bar{x}^2] \\ &\qquad - [(y_1x_1 + \cdots + y_nx_n) - \bar{x}(y_1 + \cdots + y_n)]\} \\ &= 2\{m[(x_1^2 + \cdots + x_n^2) - n\bar{x}^2] - [(y_1x_1 + \cdots + y_nx_n) - n\bar{x}\bar{y}]\} \\ &= 0 \end{aligned} \tag{9.82}$$

and

$$\frac{\partial E}{\partial b} = 2\{nb - n\bar{y}\} = 0.$$

Solving for $m$ and $b$, we obtain the values in Eq. (9.80)

$$m = \frac{(y_1x_1 + \cdots + y_nx_n) - n\bar{x}\bar{y}}{(x_1^2 + \cdots + x_n^2) - n\bar{x}^2}, \quad b = \bar{y},$$

as the coordinates of the critical point. We must now check that this critical point is a minimum. Differentiating Eq. (9.82) once more with respect to $m$ and $b$, we obtain the second partials

$$\frac{\partial^2 E}{\partial m^2} = 2[(x_1^2 + \cdots + x_n^2) - n\bar{x}^2], \qquad \frac{\partial^2 E}{\partial b^2} = 2n, \qquad \frac{\partial^2 E}{\partial m\, \partial b} = 0,$$

so that

$$\begin{aligned}\Delta &= 4n[(x_1^2 + \cdots + x_n^2) - n\bar{x}^2] \\ &= 4n[(x_1 - \bar{x})^2 + \cdots + (x_n - \bar{x})^2].\end{aligned}$$

But this sum of squares is always positive unless $x_1 = x_2 = \cdots = x_n = \bar{x}$ (in which case $m$ is undefined), so the critical point is a local minimum.

Since this is the *only* critical point, the local minimum must be a minimum. This completes the proof of the least squares theorem.

***Example 7*** The total number of cars in use in the United States from 1960 to 1974 is given in the table below:

| $x$ = year | $y$ = number of cars |
|---|---|
| 1960 | 56,935,000 |
| 1965 | 68,935,000 |
| 1970 | 80,427,000 |
| 1973 | 89,781,000 |
| 1974 | 92,583,000 |

Using the expressions for $m$ and $b$ in Eq. (9.80) we obtain, after a brief calculation (see Section 2.8) using $\bar{x} = 1968.4$ and $b = \bar{y} = 77{,}732{,}200$,

$$m = \frac{(7.6539 \times 10^{11}) - 5(1968.4)(77{,}732{,}200)}{19{,}373{,}130 - 5(1968.4)^2} = \frac{342{,}507{,}500}{137.2}$$

$$= 2{,}532{,}853.5,$$

implying that the least squares line (9.79) is

$$y = 2{,}532{,}853.5(x - 1968.4) + 77{,}732{,}200. \tag{9.83}$$

If we wish to use Eq. (9.83) to predict the number of cars in use in 1985, we obtain

$$y = 2{,}532{,}853.5(1985 - 1968.4) + 77{,}732{,}200 = 119{,}777{,}568.$$

***Example 8*** A foundry produces cast-iron stoves as a sideline. It sells an average of 20,000 stoves per year. Since stoves are only a small sideline, the foundry produces $y$ units on the first day of each inventory cycle of length $t$ (years). This production run yields an inventory of $x$ units, the other $(y - x)$ units being shipped to meet backorders. The production cost per run is \$4000 since retooling is necessary but the raw material is cheap, the inventory (storage) cost of each unit is \$10 per year, and the shortage cost (due to increased office processing costs) is \$8 per unit per year. Determine $t$, $x$, and $y$ so that the total cost $C$ is minimized.

Figure 9.18 illustrates the inventory cycle. Using similar triangles, we see that

$$\frac{x}{t_0} = \frac{y}{t} = \frac{y - x}{t - t_0}. \tag{9.84}$$

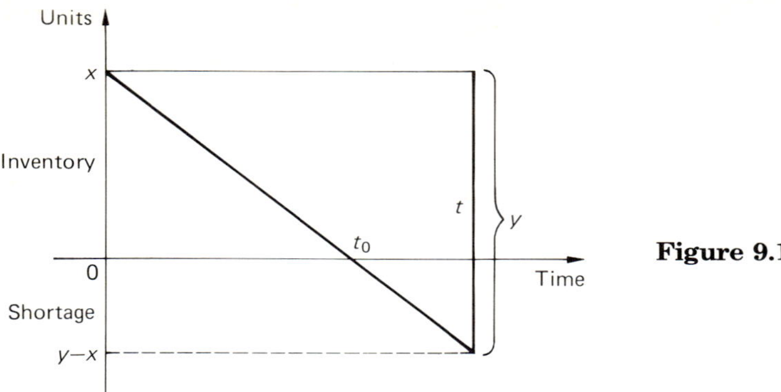


**Figure 9.18**

The cycle length $t$ is generally a fraction of a year. Since 20,000 stoves are sold each year and $y$ are produced on the first day of each cycle,

$$\frac{20{,}000}{y} = \frac{1}{t} \tag{9.85}$$

cycles per year are required to fill all orders.

Floor space in the warehouse is occupied $t_0/t$ of the time with an average of $x/2$ stoves in stock at that time. Thus inventory cost is

$$10\left(\frac{xt_0}{2t}\right) = \frac{5xt_0}{t} \quad \text{dollars.}$$

By Eq. (9.84) $t_0/t = x/y$, so inventory cost can be rewritten as $5x^2/y$ dollars per year.

Shortage costs may be found by multiplying the average number of short units $(y - x)/2$ by the fraction of the year that a shortage exists $(t - t_0)/t$ *and* the annual unit shortage cost of \$8. By Eq. (9.84) we see that the shortage costs are

$$\frac{8(y - x)}{2}\left(\frac{t - t_0}{t}\right) = \frac{4(y - x)^2}{y} \quad \text{dollars per year.}$$

Finally, total production costs are

$$4000\,\frac{\text{dollars}}{\text{cycle}} \cdot \frac{1}{t}\,\frac{\text{cycles}}{\text{year}} = \frac{80{,}000{,}000}{y}\,\frac{\text{dollars}}{\text{year}}.$$

Hence, total costs per year equal

$$C = \frac{5x^2}{y} + 4\,\frac{(y - x)^2}{y} + \frac{80{,}000{,}000}{y} \text{ (dollars).} \tag{9.86}$$

We can now apply Theorem 9.2. First we find the first partials of $C$:

$$\frac{\partial C}{\partial x} = \frac{10x}{y} - \frac{8(y-x)}{y}$$

and

$$\frac{\partial C}{\partial y} = \frac{-5x^2}{y^2} + 4 - \frac{4x^2}{y^2} - \frac{80{,}000{,}000}{y^2},$$

or

$$\frac{\partial C}{\partial x} = \frac{18x}{y} - 8 \quad \text{and} \quad \frac{\partial C}{\partial y} = 4 - \frac{(9x^2 + 80{,}000{,}000)}{y^2}. \tag{9.87}$$

Setting both equations equal to zero, we have

$$\frac{x}{y} = \frac{4}{9} \quad \text{and} \quad 9\left(\frac{x}{y}\right)^2 + \frac{80{,}000{,}000}{y^2} = 4. \tag{9.88}$$

Substituting the first equation into the second yields

$$\frac{80{,}000{,}000}{y^2} = 4 - 9\left(\frac{4}{9}\right)^2 = 4 - \frac{16}{9} = \frac{20}{9},$$

or

$$y^2 = 36{,}000{,}000 \quad \text{and} \quad y = 6000 \text{ units.}$$

This implies by the first equation in (9.88) that

$$x = \tfrac{4}{9}y = 2666\tfrac{2}{3} \text{ units.}$$

To see if this is a local minimum we must calculate

$$\frac{\partial^2 C}{\partial x^2} = \frac{18}{y}, \qquad \frac{\partial^2 C}{\partial x\,\partial y} = \frac{-18x}{y^2}, \qquad \frac{\partial^2 C}{\partial y^2} = \frac{18x^2 + 160{,}000{,}000}{y^3},$$

so that at the critical point $(2666\frac{2}{3}, 6000)$, $\Delta > 0$ and $(\partial^2 C/\partial x^2) > 0$. Hence a local minimum is attained. In short, since $t = y/20{,}000$, we find that the foundry should cast $y = 6000$ stoves every $t = 0.3$ years (or $109\frac{1}{2}$ days), and stock $2666\frac{2}{3}$ stoves at the beginning of each inventory cycle. Although it is impossible to stock a fraction of a stove, the fraction indicates that on an average $2666\frac{2}{3}$ units should be kept as initial inventory.

---

It is also possible to optimize a function of several variables subject to one or more constraints. The technique that we now describe is called the *method of Lagrange multipliers*. We will describe the process for a function of two independent variables and a single constraint. The following example illustrates how the method is applied. Notice as you go through this example that the method of Lagrange multipliers is very similar to the technique of linear programming, but is used for *nonlinear* functions and constraints.

***Example 9*** A supermarket shopper has \$12.00 with which to buy $x$ lb of ground beef and $y$ lb of potatoes. The shopper's indifference curve (see Example 5 in Section 2.7) is given by the function

$$u = x^2y,$$

and the prices of ground beef and potatoes are \$1.00/lb and 20¢/lb, respectively. What combination of ground beef and potatoes will maximize the utility function $u$, that is, will be the most desirable combination in the mind of the shopper?

In this problem we wish to maximize the utility function

$$u = x^2y, \tag{9.89}$$

subject to the budgetary constraint function

$$1.00x + 0.20y = 12.00. \tag{9.90}$$

The first step when using the method of Lagrange multipliers is to construct a new function $F(x, y, k)$ consisting of the function we wish to maximize less $k$ times the constraint, where the parameter $k$, called the *Lagrange multiplier*, is unknown. Hence we have

$$F(x, y, k) = x^2y - k\left(x + \frac{y}{5} - 12\right). \tag{9.91}$$

The next step is to find the critical points of this function. To do so we set the first partials of $F$ with respect to $x$, $y$, and $k$ equal to zero and solve for $x$, $y$, and $k$. Thus we have

$$\frac{\partial F}{\partial x} = 2xy - k = 0,$$

$$\frac{\partial F}{\partial y} = x^2 - \frac{k}{5} = 0, \tag{9.92}$$

$$\frac{\partial F}{\partial k} = -x - \frac{y}{5} + 12 = 0.$$

Solving the first two equations for $k$ we have

$$k = 2xy = 5x^2. \tag{9.93}$$

Thus $2y = 5x$, or $y = 5x/2$, which we substitute in the constraint function (9.90), obtaining

$$x + \frac{1}{5}\left(\frac{5x}{2}\right) - 12 = 0,$$

or

$$\frac{3x}{2} - 12 = 0.$$

Thus $x = 8$, so $y = 5x/2 = 20$, implying that the maximal utility is obtained by buying 8 lb of ground beef and 20 lb of potatoes.

---

We summarize the steps in the method of Lagrange multipliers: Suppose we wish to optimize the function $u = u(x, y)$ subject to the constraint $g(x, y) = 0$.

1. Construct the function

   $$F(x, y, k) = u(x, y) - kg(x, y).$$

2. Find the critical points of the function $F$ by setting the first partials equal to zero,

   $$\frac{\partial F}{\partial x} = 0, \qquad \frac{\partial F}{\partial y} = 0, \qquad \frac{\partial F}{\partial k} = 0,$$

   and solving for $x$, $y$, and $k$.

3. Test the points so obtained to make sure they fulfill the conditions we are seeking. In this brief presentation we will avoid this step, since it requires a second derivative test for functions of *three* independent variables.

***Example 10*** Arteries and veins resemble long cylinders of radius $R$ (cm) and length $L$ (cm). The blood that flows in these vessels has an inner friction called its viscosity $n$, which is measured in *poise* (g cm$^{-1}$ sec$^{-1}$), honoring J. L. Poiseuille (1799–1869), a French physiologist. The walls of the vessels also provide friction, slowing the flow of the blood. If $P$ is the pressure difference (g cm$^{-1}$ sec$^{-2}$) between two ends of the vessel, then the velocity of a laminar flow $r$ cm away from the axis is given by the empirical formula

$$v = \frac{P}{4nL}(R^2 - r^2).$$

The velocity reaches its maximum along the axis ($r = 0$).

Suppose we wish to find the dimensions of the blood vessel having maximum axial velocity subject to the condition

$$R^2 - L^2 - 10 = 0.$$

We assume that the pressure and viscosity are the same for each possible blood vessel.

By the method of Lagrange multipliers we construct the function

$$F = F(R, L, k) = \frac{PR^2}{4nL} - k(R^2 - L^2 - 10),$$

since we are interested only in axial flow velocities ($r = 0$). Taking the first partials with respect to $R$ and $P$, we have

$$\frac{\partial F}{\partial R} = \frac{2PR}{4nL} - 2Rk = 0,$$

$$\frac{\partial F}{\partial L} = \frac{-PR^2}{4nL^2} + 2kL = 0.$$

Solving for $k$, we obtain

$$k = \frac{P}{4nL} = \frac{PR^2}{8nL^3},$$

which implies by cross multiplication that $R^2 = 2L^2$. Thus by the constraint equation, $L^2 = 10$, so $L = \sqrt{10}$ and $R = \sqrt{20}$.

---

**EXERCISES 9.9**

*In Exercises 1–12, find the first and second partials of each of the given functions.*

**1.** $y = xt$

**2.** $y = x^2t$

**3.** $y = e^{xt}$

**4.** $y = \ln(x + t)$

**5.** $y = e^x \ln t$

**6.** $y = e^x\sqrt{1 + t^2}$

**7.** $y = e^{x/t}$

**8.** $y = \sqrt{x^2 + t}$

**9.** $y = te^x$

**10.** $y = x^t$

▶ **11.** $y = e^x \sin t$

▶ **12.** $y = \cos(xt)$

*In Exercises 13–18, find the relative extrema of the given functions (if any).*

**13.** $y = x^2t$

**14.** $y = x^2 + t^2 + 4xt$

**15.** $y = x^3 + t^3 - 9xt$

**16.** $y = e^x + t^3 - x^2t$

▶ **17.** $y = e^x \sin t$

▶ **18.** $y = \cos x + t^2$

*Use the least squares theorem to fit a straight line to each set of data in Exercises 19–22.*

**19.**

| $x$ | $y$ |
|---|---|
| 1 | 4 |
| 2 | 8 |
| 3 | 12 |
| 7 | 22 |
| 12 | 39 |

**20.**

| $x$ | $y$ |
|---|---|
| −2 | −6.1 |
| −1 | −1.9 |
| 0 | 2.1 |
| 1 | 6.3 |
| 2 | 9.9 |

**21.**

| $x$ | $y$ |
|---|---|
| 0 | −3.4 |
| 5 | −1.0 |
| 10 | 1.4 |
| 15 | 4.0 |
| 20 | 6.6 |

**22.**

| $x$ | $y$ |
|---|---|
| 1.0 | −4 |
| 3.5 | −7 |
| 5.0 | −8 |
| 7.5 | −11 |
| 9.0 | −12 |

*Use Lagrange multipliers to maximize the given utility function in the purchase of ground beef and potatoes, where the budget and the prices of ground beef and potatoes are as indicated.*

**23.** $u = xy$, ground beef at 90¢/lb, potatoes at 30¢/lb, budget = \$10

**24.** $u = xy^2$, ground beef at 80¢/lb, potatoes at 20¢/lb, budget = \$12

**25.** $u = x^2y^2$, ground beef at \$1.00/lb, potatoes at 10¢/lb, budget = \$15

**26.** $u = xe^y$, ground beef at \$1.00/lb, potatoes at 20¢/lb, budget = \$20

▶ **27.** $u = \tan(xy)$, ground beef at 80¢/lb, potatoes at 20¢/lb, budget = \$25

**28.** Repeat Example 8 (see p. 495) if the factory can sell 40,000 stoves per year and shortage costs are only \$4 per unit per year.

**29.** Repeat Example 8 if inventory costs per unit per year are \$12, all other facts remaining constant.

## CHAPTER VOCABULARY

**chain rule** $\frac{dy}{dx} = \frac{dy}{dt}\frac{dt}{dx}$

**composite function** $f[t(x)]$

**continuous compounding** $A = Pe^{in}$

**cost function** $C(x)$

**critical point**

**demand function** $p(x)$

**derivative** $f'(x) = \frac{df}{dx} = \frac{dy}{dx}$

**difference quotient** $\frac{f(x_0 + h) - f(x_0)}{h}$

**differentiation**

**endpoint (of interval)**

**first derivative test**

**free fall**

**growth rate**

**initial velocity** $v_0$

**instantaneous**

**Lagrange multipliers**

**least squares line**

**least squares theorem**

**local maximum**
**local minimum**
**margin**
**marginal change**
**marginal cost**
**marginal profit**
**marginal revenue**
**mixed partials**
**Newton's method** $x_{n+1} = x_n - f(x_n)/f'(x_n)$
**partial derivative** $\dfrac{\partial y}{\partial x}, \dfrac{\partial y}{\partial t}$
**product rule** $(fg)' = f'g + fg'$
**quotient rule** $\left(\dfrac{f}{g}\right)' = \dfrac{gf' - fg'}{g^2}$
**saddle point**
**secant line**
**second derivative test**
**tangent line**
**total profit** $P = R - C$
**total revenue** $R(x) = xp(x)$
**velocity** $v$

# CHAPTER 10 INTEGRAL CALCULUS

## 10.1 THE ANTIDERIVATIVE

An antiderivative of a function $f(x)$ is simply another function $F(x)$ whose derivative is $f(x)$. In symbols we can express this concept by saying that $F(x)$ is an antiderivative of $f(x)$ if

$$F'(x) = f(x). \tag{10.1}$$

We saw several instances in Chapter 9 in which the antiderivative served a useful purpose. The most important application we met was in Section 9.5 when we discussed free fall. We saw then that acceleration is the derivative of velocity, which in turn implies that velocity is an antiderivative of acceleration. Specifically, if the acceleration is given by the constant function

$$a(t) = -g,$$

each of the velocity functions

$$v(t) = -gt + c,$$

with $c$ an arbitrary constant, is an antiderivative of $a(t)$, because

$$\frac{d}{dt}(-gt + c) = -g.$$

This simple example illustrates an extremely important property: *If a function $f(x)$ has an antiderivative, then it has infinitely many antiderivatives, but each differs from the others only by a constant.* To see why this is so, suppose that both $F(x)$ and $G(x)$ are antiderivatives of the function $f(x)$. Then

$$F'(x) = f(x) \quad \text{and} \quad G'(x) = f(x),$$

so the function $H(x) = F(x) - G(x)$ has the derivative

$$H'(x) = F'(x) - G'(x) = f(x) - f(x) = 0.$$

What functions have the properties that their derivative vanishes? Select any point on the graph of $H(x)$. Then since the slope $H'(x) = 0$, the graph of $H(x)$ must be a horizontal line. It follows that $H(x)$ must equal some constant $c$. Hence

$$F(x) - G(x) = H(x) = c,$$

proving that any two antiderivatives differ only by a constant.

Returning to the velocity $v(t)$, we see that unless some additional information has been given, such as the initial velocity, there is no way to determine the value of the constant $c$. If we do not assign a specific value to $c$, we simply refer to it as an *arbitrary constant.*

It is somewhat cumbersome to keep saying "the antiderivative of $f(x)$ is $F(x)$," so we use the special mathematical notation

$$\int f(x)\,dx = F(x) + c, \tag{10.2}$$

where an arbitrary constant $c$ has been added on the right side to emphasize the fact that infinitely many antiderivatives exist, each differing from the others by a constant.

Each symbol on the left side of Eq. (10.2) has a special meaning: The $\int$ instructs us to *integrate* (find the antiderivative of) the function $f(x)$, and the $dx$ reminds us that the *integration* (antidifferentiation) is in terms of the variable $x$. We usually call the left side of Eq. (10.2) the *indefinite integral* of $f(x)$.

In terms of this notation, and remembering that integration is merely reverse differentiation, we already know a number of indefinite integrals. For example,

$$\int x^n\,dx = \frac{x^{n+1}}{n+1} + c, \quad n \neq -1, \tag{10.3}$$

since

$$\left[\frac{x^{n+1}}{n+1} + c\right]' = \frac{(n+1)x^n}{n+1} = x^n.$$

Because constants that multiply a function are *carried along* under differentiation, the same must be true for the reverse process of integration, so that

$$\int af(x)\,dx = a\int f(x)\,dx. \tag{10.4}$$

Since the derivative of a sum is the sum of the derivatives, the same must be true for indefinite integrals, that is,

$$\int f(x) + g(x)\,dx = \int f(x)\,dx + \int g(x)\,dx. \tag{10.5}$$

Equations (10.3)–(10.5) and the remarks about arbitrary constants are the integral equivalents of the first four rules of differentiation that we learned in Section 9.3. We found that these rules enabled us to differentiate any polynomial function, so Eqs. (10.3)–(10.5) should provide the necessary tools for integrating any polynomial. The following two examples illustrate the procedure.

***Example 1*** Complete the integration

$$\int (5x^4 + 3x^2 - 7x + 2)\,dx.$$

By Eqs. (10.4) and (10.5) we can separate the integral into four separate integrals and bring the constants outside of each integral:

$$\int (5x^4 + 3x^2 - 7x + 2)\,dx = 5\int x^4\,dx + 3\int x^2\,dx - 7\int x\,dx + 2\int x^0\,dx.$$

Using Eq. (10.3), we can integrate each power of $x$, obtaining

$$\int (5x^4 + 3x^2 - 7x + 2)\,dx = 5\frac{x^5}{5} + 3\frac{x^3}{3} - 7\frac{x^2}{2} + \frac{2x}{1} + c$$

$$= x^5 + x^3 - \frac{7}{2}x^2 + 2x + c.$$

It is not necessary to repeatedly use the rules (10.4) and (10.5); instead we should learn to integrate each term in a sum when we get to it.

***Example 2*** Integrate the following functions:

a) $f(x) = \sqrt{x} - 4/\sqrt{x}$,

b) $g(x) = 3x^{\pi} - 2x^{1/3} + 7x^{-3/4} + 12$;

c) $h(x) = 2x^{7/9} - (2/3)x^{4/3} + 1.8x^{-0.6} - 5$.

Converting the roots into exponents we have

a) $$\int f(x)\,dx = \int (x^{1/2} - 4x^{-1/2})\,dx = \frac{x^{3/2}}{(3/2)} - 4\left[\frac{x^{1/2}}{(1/2)}\right] + c = \frac{2}{3}x^{3/2} - 8x^{1/2} + c;$$

b) $$\int g(x)\,dx = \int (3x^{\pi} - 2x^{1/3} + 7x^{-3/4} + 12)\,dx$$

$$= 3\left[\frac{x^{n+1}}{\pi + 1}\right] - 2\left[\frac{x^{4/3}}{(4/3)}\right] + 7\left[\frac{x^{1/4}}{(1/4)}\right] + 12x + c$$

$$= \frac{3x^{\pi+1}}{\pi + 1} - \frac{3}{2}x^{4/3} + 28x^{1/4} + 12x + c;$$

c) $$\int h(x)\,dx = \int [2x^{7/9} - (2/3)x^{4/3} + 1.8x^{-0.6} - 5]\,dx$$

$$= 2\left[\frac{x^{16/9}}{(16/9)}\right] - \frac{2}{3}\left[\frac{x^{7/3}}{(7/3)}\right] + 1.8\left[\frac{x^{0.4}}{(0.4)}\right] - 5x + c$$

$$= \frac{9}{8}x^{16/9} - \frac{2}{7}x^{7/3} + 0.45x^{0.4} - 5x + c.$$

---

Notice that Eq. (10.3) is undefined when $n = -1$ because the integral of $x^{-1}$ is not a power of $x$. Instead, remembering that $(\ln x)' = 1/x$, we have

$$\int \frac{1}{x}\,dx = \ln|x| + c. \tag{10.6}$$

Since $(e^x)' = e^x$, it follows that

$$\int e^x\,dx = e^x + c. \tag{10.7}$$

Similarly, $(a^x)' = a^x \ln a$ for any given positive constant $a \neq 1$, yielding

$$\int a^x\,dx = \frac{a^x}{\ln a} + c. \tag{10.8}$$

***Example 3*** Integrate the following functions:

a) $f(x) = 7e^x - (3/x)$;

b) $g(x) = 5^x + (x + 1)/x$;

c) $h(x) = 5x^2 - 2(6)^x$.

Using Eqs. (10.6)–(10.8) we have

a) $$\int \left(7e^x - \frac{3}{x}\right) dx = 7e^x - 3\ln x + c;$$

b) $$\int \left(5^x + \frac{x+1}{x}\right) dx = \int \left(5^x + 1 + \frac{1}{x}\right) dx$$
$$= \frac{5^x}{\ln 5} + x + \ln x + c;$$

c) $$\int [5x^2 - 2(6)^x]\,dx = 5\left[\frac{x^3}{3}\right] - 2\left[\frac{6^x}{\ln 6}\right] + c$$
$$= \frac{5}{3}x^3 - \frac{2}{\ln 6}(6)^x + c.$$

---

For convenience we summarize the integrals we have determined in this section in Table 10.1. We shall indicate two useful methods for determining indefinite integrals of a variety of functions in the following two sections. A more complete table of integrals can be found in Appendix 3.

**Table 10.1**

$$\int cf(x)\,dx = c\int f(x)\,dx$$

$$\int f(x) + g(x)\,dx = \int f(x)\,dx + \int g(x)\,dx$$

$$\int x^n\,dx = \begin{cases} \dfrac{x^{n+1}}{n+1} + c, & \text{if } n \neq 1 \\ \ln x + c, & \text{if } n = -1 \end{cases}$$

$$\int e^x\,dx = e^x + c$$

$$\int a^x\,dx = \frac{a^x}{\ln a} + c, \quad a > 0, a \neq 1$$

## EXERCISES 10.1

*Use Table 10.1 to integrate the functions in Exercises 1–16.*

**1.** $f(x) = x^2 + 2x - 1$

**2.** $f(x) = x^3 - 1$

**3.** $f(x) = 3x^2 - 4x + 7$

**4.** $f(x) = (5x^2 - 1)/3$

**5.** $f(x) = 3x^7 - 4x^2 + 1$

**6.** $g(x) = 9x^{18} + 7x^9 - 3$

**7.** $g(x) = 4x^3 - 2x$

**8.** $h(x) = 9 - 5x + 7x^3$

**9.** $g(x) = \sqrt{x} - \sqrt[3]{x}$

**10.** $h(x) = \sqrt[3]{x^2} - 4\sqrt[5]{x}$

**11.** $f(x) = 7e^x - \dfrac{4}{x}$

**12.** $g(x) = \dfrac{x^2 - 1}{x^2}$

**13.** $f(x) = (15)^x - (14)^x$

**14.** $h(x) = 6e^x - 5(7)^x$

**15.** $g(x) = \dfrac{\sqrt{x} - x}{\sqrt{x}}$

**16.** $h(x) = 1.8x^{-1.8} + 3e^x - e(3)^x$

*Complete the integrations in Exercises 17–24.*

**17.** $\displaystyle\int e^t\,dt$

**18.** $\displaystyle\int \frac{7}{s}\,ds$

**19.** $\displaystyle\int 6^z\,dz$

**20.** $\displaystyle\int (e^t - 7t^{1/2})\,dt$

**21.** $\displaystyle\int (4e^m - \sqrt[3]{m})\,dm$

**22.** $\displaystyle\int \frac{q - \sqrt{q}}{\sqrt{q}}\,dq$

**23.** $\displaystyle\int \frac{\sqrt{t} - 4}{t}\,dt$

**24.** $\displaystyle\int (4^m + m^4)\,dm$

▶ *Since* $(\sin x)' = \cos x$ *and* $(\cos x)' = -\sin x$, *we also have the indefinite integrals*

$$\int \sin x\,dx = -\cos x + c, \qquad \int \cos x\,dx = \sin x + c.$$

*Use these two integrals to solve Exercises* 25–28.

**25.** $\int 4\cos x\,dx$

**26.** $\int (3\sin x - 2\cos x)\,dx$

**27.** $\int 14\cos z\,dz$

**28.** $\int 5\sin t\,dt$

▶ **29.** Verify, using derivatives, that $\int \sec^2 x\,dx = \tan x + c$.

▶ **30.** Show that $\int \csc^2 x\,dx = -\cot x + c$.

▶ **31.** Show that $\int \sec x \tan x\,dx = \sec x + c$.

▶ **32.** Find $\int \csc x \cot x\,dx$.

---

## 10.2 SUBSTITUTION TECHNIQUES IN INTEGRATION

It is frequently the case that the indefinite integrals we derived in Section 10.1 cannot be used immediately in solving a particular problem. For example,

$$\int xe^{x^2}\,dx \tag{10.9}$$

certainly does not resemble any of the integrals we computed. When we encounter an indefinite integral that we do not recognize, we should always try to use the following substitution techniques to transform the integral into one for which a solution is known.

To see how the method works, we will transform (10.9) into the indefinite integral (10.7) by a *change of variables*. Some of the steps in the method will seem mysterious and purely mechanical at first. Don't let this bother you, since there is a good reason for each step, which will be explained later on. Instead, concentrate solely on the mechanical procedure.

Set $u = x^2$ and differentiate this function with respect to $x$, obtaining

$$\frac{du}{dx} = 2x,$$

or

$$x = \frac{1}{2}\frac{du}{dx}. \tag{10.10}$$

Notice that we have written the derivative in the form $du/dx$ rather than $u'(x)$ as has been our habit. Substitute $u$ for the exponent of the exponential term and Eq. (10.10) for the $x$ in (10.9) to obtain

$$\int e^u \frac{1}{2}\frac{du}{dx}\,dx. \tag{10.11}$$

Next we "cancel" the $dx$-terms, obtaining the integral

$$\frac{1}{2}\int e^u\,du \tag{10.12}$$

in terms of only the $u$-variable. This "cancellation" is the reason we use $du/dx$ rather than $u'(x)$. Now Eq. (10.12) is easily solved:

$$\frac{1}{2}\int e^u\,du = \frac{1}{2}e^u + c, \tag{10.13}$$

and reversing the substitution by replacing each $u$ in (10.13) with $x^2$ we have the solution

$$\int xe^{x^2}\,dx = \frac{1}{2}e^{x^2} + c. \tag{10.14}$$

If we differentiate the right side of Eq. (10.14) we obtain the function

$$\left(\frac{1}{2}e^{x^2} + c\right)' = \frac{1}{2}e^{x^2}\cdot 2x = xe^{x^2},$$

so we have indeed obtained the solution of the indefinite integral (10.9).

Before discussing the method let's solve another, quite different, indefinite integral so that the similarities in the two procedures will be more apparent. Consider the integral

$$\int \frac{1}{x\ln x}\,dx. \tag{10.15}$$

Let $u = \ln x$; then

$$\frac{du}{dx} = \frac{1}{x}$$

so (10.15) can be rewritten as

$$\int \frac{1}{u}\frac{du}{dx}\,dx$$

by substituting $u$ for $\ln x$ and $du/dx$ for $1/x$. "Canceling" the $dx$-terms yields

$$\int \frac{1}{u}\,du = \ln u + c,$$

and replacing each $u$ with $\ln x$, we have

$$\int \frac{1}{x \ln x}\, dx = \ln(\ln x) + c.$$

To check, we differentiate the solution, obtaining

$$[\ln(\ln x) + c]' = \frac{1}{\ln x} \cdot \frac{1}{x} = \frac{1}{x \ln x}.$$

If we compare these two solutions, we might ask the following questions:

**1.** How do we know what substitution $u = f(x)$ to use?

**2.** Why do we substitute $du/dx$ for some terms and not others?

**3.** Why can the $dx$-terms be canceled?

At this point you might not be too happy with the answer to the first question: experience. After you have solved a good many such problems you will usually be able to decide immediately what substitutions are likely to lead to a solution. There is no set rule other than the answer to the second question: It is necessary that exactly one $(du/dx)$-term be present and that *all* other terms be expressed as functions of $u$. In Section 10.4 we will show that changing variables is essentially a change of coordinate systems and the $(du/dx)$-term is simply the instantaneous change of scale between the two coordinate systems. The analogy you should use in trying to understand the procedure is that of converting a length of $x$ feet into $u$ meters by the formula $u = kx$. In that case, the $(du/dx)$-term is simply the conversion coefficient $k = 0.3048$ m/ft, and

$$u(m) = k\left(\frac{\text{m}}{\cancel{\text{ft}}}\right) \cdot x\,(\cancel{\text{ft}}).$$

In particular, this analogy also explains why the $dx$-terms are canceled since the only role they serve is that of reminding us of the units or variable of integration.

***Example 1*** Find the solution of the indefinite integral

$$\int e^{5x}\, dx.$$

Clearly we wish to transform the integral into a multiple of Eq. (10.7), so the obvious choice of a substitution is $u = 5x$. Then $du/dx = 5$ so $(1/5)\, du/dx = 1$ and

$$\int e^{5x}\, dx = \int e^{u} \frac{1}{5} \frac{du}{dx}\, dx = \frac{1}{5} \int e^{u}\, du$$

$$= \frac{1}{5} e^{u} + c = \frac{1}{5} e^{5x} + c.$$

***Example 2*** Solve the integral

$$\int \frac{dx}{x - 5}.$$

This integral is very similar to Eq. (10.6), suggesting the substitution $u = x - 5$. Then $du/dx = 1$ so

$$\int \frac{dx}{x - 5} = \int \frac{1}{u} \frac{du}{dx}\, dx = \int \frac{1}{u}\, du$$
$$= \ln u + c = \ln |x - 5| + c.$$

We have taken the absolute value of $x - 5$ to emphasize that logarithms of negative numbers are not allowed.

***Example 3*** Solve the integral

$$\int \frac{x\, dx}{\sqrt{x^2 + 4}}. \tag{10.16}$$

This integral is slightly more complicated than the previous ones in that the choice of substitution is not as evident. One possible choice might be to set $u = x^2 + 4$, since this substitution may transform the integral to a power of $u$. Now $du/dx = 2x$ so $x = (1/2)\, du/dx$ and

$$\int \frac{x\, dx}{\sqrt{x^2 + 4}} = \int \frac{1}{\sqrt{u}} \frac{1}{2} \frac{du}{dx}\, dx = \frac{1}{2} \int u^{-1/2}\, du$$
$$= \frac{1}{2} \frac{u^{1/2}}{(1/2)} + c = \sqrt{x^2 + 4} + c.$$

It is interesting to note that another substitution would also have worked: $w = \sqrt{x^2 + 4}$. Then

$$\frac{dw}{dx} = \frac{1}{2}(x^2 + 4)^{-1/2} \cdot 2x = \frac{x}{\sqrt{x^2 + 4}},$$

which is the function we wish to integrate so that Eq. (10.16) becomes

$$\int \frac{x\, dx}{\sqrt{x^2 + 4}} = \int \frac{dw}{dx}\, dx = \int dw = \int w^0\, dw$$
$$= w + c = \sqrt{x^2 + 4} + c.$$

This example demonstrates that there may often be several possible ways of obtaining the solution.

***Example 4*** Solve the integral

$$\int \frac{dx}{x^2 - 1}. \tag{10.17}$$

This integral is somewhat difficult to obtain because the most immediate substitutions fail. For example, if we try $u = x^2 - 1$ then $du/dx = 2x$, but there is no $x$-term in the numerator. Since a $(du/dx)$-term is necessary, we might multiply and divide by $x$ and use the fact that $x^2 = u + 1$ to obtain

$$\begin{aligned} \int \frac{dx}{x^2 - 1} &= \int \frac{x\,dx}{x(x^2 - 1)} \\ &= \int \frac{1}{u\sqrt{u + 1}} \frac{1}{2} \frac{du}{dx}\,dx = \frac{1}{2} \int \frac{du}{u\sqrt{u + 1}}, \end{aligned}$$

but this integral is even more complicated than (10.17). The key to solving this integral is to notice that

$$x^2 - 1 = (x - 1)(x + 1).$$

This suggests that it might be possible to express the fraction

$$\frac{1}{x^2 - 1} = \frac{1}{(x - 1)(x + 1)}$$

as a sum of two simpler fractions

$$\frac{1}{(x - 1)(x + 1)} = \frac{A}{x - 1} + \frac{B}{x + 1}, \tag{10.18}$$

since we always find the common denominator on the right side of Eq. (10.18) by multiplying the two denominators together. Adding the two fractions on the right side of Eq. (10.18) together, we have

$$\frac{1}{(x - 1)(x + 1)} = \frac{A(x + 1) + B(x - 1)}{(x - 1)(x + 1)}.$$

For equality to hold the numerators must be equal, so if we gather the terms involving $x$ together we have

$$1 = (A + B)x + (A - B).$$

The left side has no $x$-term, implying that $A + B$ must be zero. Then $A - B$ must equal 1, so we obtain the system of equations

$$A + B = 0,$$

$$A - B = 1.$$

Solving this system simultaneously, we obtain $A = \frac{1}{2}$ and $B = -\frac{1}{2}$. Thus,

$$\frac{1}{x^2 - 1} = \frac{1}{2}\left[\frac{1}{x - 1} - \frac{1}{x + 1}\right]$$

so that

$$\int \frac{dx}{x^2 - 1} = \frac{1}{2}\int \frac{dx}{x - 1} - \frac{1}{2}\int \frac{dx}{x + 1}.$$

We now solve each integral on the right side separately. Set $u = x - 1$ and $v = x + 1$; then since

$$\frac{du}{dx} = 1 \quad \text{and} \quad \frac{dv}{dx} = 1,$$

we have the integrals

$$\frac{1}{2}\int \frac{dx}{x - 1} = \frac{1}{2}\int \frac{1}{u}\frac{du}{dx}\,dx = \frac{1}{2}\int \frac{du}{u} = \frac{1}{2}\ln|u| + c,$$

$$\frac{1}{2}\int \frac{dx}{x + 1} = \frac{1}{2}\int \frac{1}{v}\frac{dv}{dx}\,dx = \frac{1}{2}\int \frac{dv}{v} = \frac{1}{2}\ln|v| + c.$$

Thus,

$$\begin{aligned}\int \frac{dx}{x^2 - 1} &= \frac{1}{2}\ln|u| - \frac{1}{2}\ln|v| + c \\ &= \frac{1}{2}\ln|x - 1| - \frac{1}{2}\ln|x + 1| + c \\ &= \frac{1}{2}\ln\left|\frac{x - 1}{x + 1}\right| + c.\end{aligned}$$

---

The technique of separating the function we wish to integrate into a sum or difference of simpler expressions is called the method of *partial fractions*. This method can always be used when the function we wish to integrate is a quotient of polynomials with a denominator that can be factored.

The success we have had in solving these indefinite integrals might lead us to the false assumption that every indefinite integral has a solution consisting of combinations of the elementary functions. This is certainly not the case. For example, the indefinite integrals

$$\int e^{x^2}\,dx \quad \text{and} \quad \int \frac{dx}{\sqrt[3]{1 + x^7}}$$

do not have elementary solutions. Appendix 3 contains a short table of the most common indefinite integrals. A much longer table of integrals may be found in the *CRC Standard Mathematical Tables* (Cleveland, Ohio: CRC Press, 1975).

---

**▶ Trigonometric Substitutions** This brief subsection is for those readers who have studied Section 9.7 and remember the derivatives of the trigonometric functions. Since

$$(\sin x)' = \cos x \quad \text{and} \quad (\cos x)' = -\sin x,$$

it follows that

$$\int \sin x\, dx = -\cos x + c \quad \text{and} \quad \int \cos x\, dx = \sin x + c. \qquad \textbf{(10.19)}$$

The following three examples illustrate the use of the substitution method in solving indefinite integrals of trigonometric functions.

**▶ *Example 5*** Find the solution of the integral

$$\int \sin^2 x \cos^3 x\, dx.$$

The trick here is to use the identity

$$\cos^2 x = 1 - \sin^2 x$$

in order to obtain an integral with a single ($\cos x$)-term. Then

$$\begin{aligned}\int \sin^2 x \cos^3 x\, dx &= \int \sin^2 x \cdot (1 - \sin^2 x) \cos x\, dx \\ &= \int \sin^2 x \cos x\, dx - \int \sin^4 x \cos x\, dx.\end{aligned}$$

We now have two integrals to solve. Let $u = \sin x$; then $du/dx = \cos x$, so

$$\begin{aligned}\int \sin^2 x \cos x\, dx &= \int u^2 \frac{du}{dx}\, dx = \int u^2\, du \\ &= \frac{u^3}{3} + c = \frac{\sin^3 x}{3} + c\end{aligned}$$

and

$$\begin{aligned}\int \sin^4 x \cos x\, dx &= \int u^4 \frac{du}{dx}\, dx = \int u^4\, du \\ &= \frac{u^5}{5} + c = \frac{\sin^5 x}{5} + c.\end{aligned}$$

Hence, subtracting the two results we have

$$\int \sin^2 x \cos^3 x \, dx = \frac{1}{3} \sin^3 x - \frac{1}{5} \sin^5 x + c.$$

Had the sines and cosines been reversed we would have changed all but one sine to cosines and used the substitution $u = \cos x$.

▶ ***Example 6*** Solve the integral

$$\int \tan x \, dx.$$

Remembering that $\tan x = \sin x/\cos x$, we let $u = \cos x$. Then $du/dx = -\sin x$, so that

$$\begin{aligned} \int \tan x \, dx &= \int \frac{\sin x}{\cos x} dx = -\int \frac{1}{u} \frac{du}{dx} dx \\ &= -\int \frac{du}{u} = -\ln |u| + c \\ &= -\ln |\cos x| + c. \end{aligned}$$

▶ ***Example 7*** Find

$$\int \sec x \, dx.$$

This integral requires a farfetched substitution. Let $u = \sec x + \tan x$; then

$$\frac{du}{dx} = \sec x \tan x + \sec^2 x = \sec x (\sec x + \tan x).$$

Multiplying both numerator and denominator of the integral by $\sec x + \tan x$, we have

$$\begin{aligned} \int \sec x \, dx &= \int \frac{\sec x (\sec x + \tan x)}{\sec x + \tan x} dx = \int \frac{du}{u} \\ &= \ln |u| + c \\ &= \ln |\sec x + \tan x| + c. \end{aligned}$$

---

Trigonometry can also be used when solving integrals involving square roots of algebraic quantities. This fact is based on the observation that if we let $x = \sin u$, then

$$\sqrt{1 - x^2} = \sqrt{1 - \sin^2 u} = \sqrt{\cos^2 u} = \cos u,$$

which eliminates the square root. The following example illustrates this method.

---

▶ ***Example 8*** Solve the integral

$$\int \frac{dx}{\sqrt{4 - x^2}}. \tag{10.20}$$

Here we must also eliminate the 4 from within the root. To do so we let $x = 2 \sin u$. Then

$$\sqrt{4 - x^2} = \sqrt{4 - 4 \sin^2 u} = \sqrt{4 \cos^2 u} = 2 \cos u.$$

Now we differentiate both sides of $x = 2 \sin u$ with respect to $x$, obtaining

$$1 = 2 \cos u \frac{du}{dx},$$

or

$$\frac{du}{dx} = \frac{1}{2 \cos u}.$$

Therefore,

$$\int \frac{dx}{\sqrt{4 - x^2}} = \int \frac{1}{2 \cos u}\, dx = \int \frac{du}{dx}\, dx$$

$$= \int du = u + c.$$

To complete the solution we must replace $u$ by a function of $x$. Since $\sin u = x/2$, the function $u$ *is that angle (in radians) whose sine equals* $x/2$. We denote this function $u$ by the notation $u = \sin^{-1}(x/2)$, where the *inverse sine function*, $\sin^{-1}$, reverses the effect of the sine function, and vice versa:

$$\text{angle (in radians)} \underset{\sin^{-1}}{\overset{\sin}{\rightleftarrows}} \text{number.}$$

Thus $u = \sin^{-1}(x/2)$, yielding

$$\int \frac{dx}{\sqrt{4 - x^2}} = \sin^{-1}(x/2) + c.$$

---

## EXERCISES 10.2

*Use the substitution method to solve Exercises* 1–10.

1. $\int e^{7x}\, dx$

2. $\int \frac{\ln x}{x}\, dx$

3. $\int \frac{dx}{4 - x}$

4. $\int \frac{x\,dx}{4 - x^2}$

5. $\int \frac{(\ln x)^2}{x}\,dx$

6. $\int \frac{e^x\,dx}{1 + e^x}$

7. $\int \frac{x\,dx}{\sqrt{9 + x^2}}$

8. $\int \frac{x\,dx}{4 + x^2}$

9. $\int \frac{x\,dx}{\sqrt{9 - x^2}}$

10. $\int \frac{x\,dx}{\sqrt{9x^2 - 4}}$

*Use partial fractions to solve Exercises* 11–16.

11. $\int \frac{dx}{x^2 - 4}$

12. $\int \frac{dx}{25x^2 - 9}$

13. $\int \frac{dx}{x^2 - 3}$

14. $\int \frac{dx}{2x^2 - 7}$

15. $\int \frac{dx}{x + x^2}$

16. $\int \frac{dx}{4x - x^2}$

▶ *Use trigonometric substitutions to solve Exercises* 17–28.

17. $\int \sin x \cos x\,dx$

18. $\int \sin^2 x \cos x\,dx$

19. $\int \cos^2 x \sin x\,dx$

20. $\int \cos^3 x \sin x\,dx$

21. $\int \tan 2x\,dx$

22. $\int x \tan (x^2)\,dx$

23. $\int \csc x\,dx$

24. $\int \cot x\,dx$

25. $\int \frac{dx}{\sqrt{9 - x^2}}$

26. $\int \frac{dx}{4 + x^2}$ (*Hint:* Let $x = 2 \tan u$.)

27. $\int \frac{dx}{\sqrt{4 - 9x^2}}$

28. $\int \frac{dx}{9 + 4x^2}$

---

## 10.3 INTEGRATION BY PARTS

The second most commonly used method of determining indefinite integrals is based on the product formula of differentiation. Before we develop the procedure it is useful to examine the following example. Consider the differentiation of $x \ln x$ using the product formula:

$$(x \ln x)' = x\frac{1}{x} + \ln x \cdot (1),$$

or

$$(x \ln x)' = 1 + \ln x.$$

If we integrate both sides of this equation with respect to $x$, the left side will simply be the product $x \ln x$, since integration reverses the process of differentiation. Thus we have

$$x \ln x = \int (x \ln x)' \, dx = \int 1 \, dx + \int \ln x \, dx$$

so that

$$\int \ln x \, dx = x \ln x - \int 1 \, dx = x \ln x - x + c. \tag{10.21}$$

Thus, we have determined the indefinite integral of $\ln x$ by integrating the product formula of differentiation. Of course, we cannot rely on accidentally determining useful integrals. Instead we should analyze the procedure and see if there is a systematic way of exploiting the method to obtain those indefinite integrals whose solution we seek.

To discover what benefits may be obtained from this procedure we integrate the product formula of differentiation

$$\frac{d}{dx}(uv) = u\frac{dv}{dx} + v\frac{du}{dx}$$

with respect to $x$, again remembering that integration reverses the process of differentiation. Thus, we obtain

$$uv = \int u\frac{dv}{dx}\, dx + \int v\frac{du}{dx}\, dx,$$

or subtracting the second term on the right from both sides,

$$\int u\frac{dv}{dx}\, dx = uv - \int v\frac{du}{dx}\, dx. \tag{10.22}$$

Equation (10.22) may be abbreviated by "canceling" the $dx$-terms (for the same reasons as those in Section 10.2), yielding the usual form of the formula for *integration by parts:*

$$\int u \, dv = uv - \int v \, du. \tag{10.23}$$

These two formulas are very useful. They tell us that if we divide the function we wish to integrate into *two parts, $u$ and $dv/dx$*, then the integral of $u\,(dv/dx)$ is the difference between the function $uv$ and the integral $\int v\,(du/dx)\,dx$. If the latter integral is one for which a solution is known, then a solution of the original integral is obtained. The following six examples illustrate this process.

***Example 1*** Use the method of integration by parts to solve the integral

$$\int xe^{3x}\,dx. \tag{10.24}$$

The basic idea in such problems is to choose $u$ and $dv/dx$ in such a way that the integral

$$\int v\frac{du}{dx}\,dx$$

can be solved. Although there is no infallible way of deciding which part to call $u$ and which to call $dv/dx$, the acronym LATE is frequently effective. The letters,

L = logarithmic function,
A = algebraic function,
T = trigonometric function,
E = exponential function,

list the priorities of the various possible terms of the function we wish to integrate. The idea is to assign to the function $u$ the highest priority term, and let the rest of the function be $dv/dx$. In this case, we have an algebraic function $x$ and an exponential function $e^{3x}$, so we let

$$u = x \quad \text{and} \quad \frac{dv}{dx} = e^{3x}.$$

Then* $v = e^{3x}/3$ and $du/dx = 1$, so that

$$\int xe^{3x}\,dx = x\left(\frac{e^{3x}}{3}\right) - \int\frac{e^{3x}}{3}(1)\,dx. \tag{10.25}$$

This last integral may be solved by the substitution $u = 3x$ (see Example 1 in Section 10.2), yielding

$$\int xe^{3x}\,dx = \frac{xe^{3x}}{3} - \frac{e^{3x}}{9} + c.$$

---

Had we searched the table of integrals in Appendix 3 we would have found that Formula 41 (see p. 596) becomes Eq. (10.24) if we let $n = 1$ and $a = 3$. Substituting these values into the formula would have yielded Eq. (10.25), and Formula 4 in Appendix 3 would have

* It is unnecessary to include the arbitrary constant $c$ at this point. Instead, we find an antiderivative of Eq. (10.24) by eliminating the integral on the right side of Eq. (10.25) and *then* we add the arbitrary constant. This procedure will always be followed.

produced the final solution. It is interesting to note that this and several other formulas in the table have an integral as part of their solution. Generally, the new integral resembles the original integral with changes in the values of the exponents. Such expressions are called *reduction formulas*, the basic idea being to use them repeatedly until a final solution can be obtained. For example, if we wished to solve $\int x^2e^{4x}\,dx$, we let $n = 2$ and $a = 4$ and use Formula 41:

$$\int x^2e^{4x}\,dx = \frac{x^2e^{4x}}{4} - \frac{2}{4}\int xe^{4x}\,dx.$$

Now let $n = 1$ and $a = 4$, obtaining from Formula 41 again,

$$\int x^2e^{4x}\,dx = \frac{x^2e^{4x}}{4} - \frac{1}{2}\left[\frac{xe^{4x}}{4} - \frac{1}{4}\int e^{4x}\,dx\right];$$

this last integral may now be solved by substitution (or Formula 4), yielding

$$\int x^2e^{4x}\,dx = \frac{x^2e^{4x}}{4} - \frac{xe^{4x}}{8} + \frac{e^{4x}}{32} + c.$$

It should be noticed that the reduction formula 41 is an easy consequence of the method of integration by parts. Using the acronym LATE we set $u = x^n$ and $dv/dx = e^{ax}$. Then

$$\frac{du}{dx} = nx^{n-1} \quad \text{and} \quad v = \frac{e^{ax}}{a}$$

so that

$$\int x^ne^{ax}\,dx = \frac{x^ne^{ax}}{a} - \frac{n}{a}\int x^{n-1}e^{ax}\,dx,$$

which is the desired reduction formula.

***Example 2*** Find the solution of the indefinite integral

$$\int x\sqrt{4 + x}\,dx.$$

The acronym LATE cannot be used in this case since all terms are algebraic. However, Example 1 provides a clue since we were able to find a solution by eliminating the $x$-term by setting $u = x$. Then $dv/dx = \sqrt{4 + x}$ so that

$$\frac{du}{dx} = 1 \quad \text{and} \quad v = \frac{2}{3}(4 + x)^{3/2}.$$

Integrating by parts, we obtain

$$\int x\sqrt{4+x}\,dx = \frac{2x}{3}(4+x)^{3/2} - \frac{2}{3}\int (4+x)^{3/2}\,dx,$$

and this last integral is easily solved by substitution, yielding

$$\begin{aligned}\int x\sqrt{4+x}\,dx &= \frac{2x}{3}(4+x)^{3/2} - \frac{2}{3}\frac{(4+x)^{5/2}}{5} + c \\ &= \frac{2x}{3}(4+x)^{3/2} - \frac{2}{15}(4+x)^{5/2} + c.\end{aligned}$$

***Example 3*** Solve the integral

$$\int x(\ln x)^2\,dx.$$

Here we can apply LATE, so we set $u = (\ln x)^2$ and $dv/dx = x$. Then

$$\frac{du}{dx} = \frac{2\ln x}{x} \quad \text{and} \quad v = \frac{x^2}{2},$$

so integrating by parts we have

$$\begin{aligned}\int x(\ln x)^2\,dx &= \frac{x^2(\ln x)^2}{2} - \int \frac{x^2}{2}\frac{2\ln x}{x}\,dx \\ &= \frac{(x\ln x)^2}{2} - \int x\ln x\,dx.\end{aligned}$$

Integrating the last integral by parts with $u = \ln x$ and $dv/dx = x$, we have

$$\frac{du}{dx} = \frac{1}{x} \quad \text{and} \quad v = \frac{x^2}{2}$$

and

$$\begin{aligned}\int x(\ln x)^2\,dx &= \frac{(x\ln x)^2}{2} - \left[\frac{x^2}{2}\ln x - \int \frac{x^2}{2}\frac{1}{x}\,dx\right] \\ &= \frac{(x\ln x)^2}{2} - \frac{x^2\ln x}{2} + \frac{x^2}{4} + c.\end{aligned}$$

▶ ***Example 4*** Find the solution of the indefinite integral

$$\int x\sin x\,dx.$$

Using LATE, we set $u = x$ and $dv/dx = \sin x$. Then $du/dx = 1$ and $v = -\cos x$, so that

$$\int x \sin x \, dx = x(-\cos x) - \int (-\cos x)\, dx$$
$$= -x \cos x + \sin x + c.$$

▶ ***Example 5*** Solve

$$\int e^{2x} \sin x \, dx.$$

Using LATE we set $u = \sin x$ and $dv/dx = e^{2x}$. Then $du/dx = \cos x$ and $v = e^{2x}/2$, so integrating by parts we have

$$\int e^{2x} \sin x \, dx = \frac{e^{2x} \sin x}{2} - \frac{1}{2}\int e^{2x} \cos x \, dx.$$

Once again we integrate by parts with $u = \cos x$ and $dv/dx = e^{2x}$, so that $du/dx = -\sin x$, $v = e^{2x}/2$, and

$$\int e^{2x} \sin x \, dx = \frac{e^{2x}}{2} \sin x - \frac{1}{2}\left[\frac{e^{2x}}{2} \cos x - \frac{1}{2}\int e^{2x}(-\sin x)\, dx\right]$$
$$= \frac{e^{2x}}{2} \sin x - \frac{e^{2x}}{4} \cos x - \frac{1}{4}\int e^{2x} \sin x \, dx.$$

But this last integral is exactly the same as the original integral. Carrying both integrals to the left side, we have

$$\frac{5}{4}\int e^{2x} \sin x \, dx = \frac{e^{2x}}{4}(2 \sin x - \cos x) + c,$$

so that

$$\int e^{2x} \sin x \, dx = \frac{e^{2x}}{5}(2 \sin x - \cos x) + c.$$

This example illustrates the fact that you should not give up if your first try fails to solve the problem immediately. Often it may be necessary to try several different times before you obtain a solution. It may also be necessary to combine several methods.

▶ ***Example 6*** Find the solution of the integral

$$\int \sin^2 x \, dx.$$

Here using LATE is of no help. Instead, let $u = \sin x$ and $dv/dx = \sin x$, so that

$$\frac{du}{dx} = \cos x \quad \text{and} \quad v = -\cos x.$$

Integrating by parts we have

$$\int \sin^2 x\, dx = -\sin x \cos x + \int \cos^2 x\, dx.$$

We could integrate by parts again, but unfortunately this would not help. (Try it!) Instead we use the fact that $\cos^2 x = 1 - \sin^2 x$. Substituting this identity into the rightmost integral yields

$$\begin{aligned}\int \sin^2 x\, dx &= -\sin x \cos x + \int (1 - \sin^2 x)\, dx \\ &= -\sin x \cos x + \int dx - \int \sin^2 x\, dx.\end{aligned}$$

Combining the two integrals involving $\sin^2 x$, we have

$$2\int \sin^2 x\, dx = x - \sin x \cos x + c,$$

or

$$\int \sin^2 x\, dx = \frac{x - \sin x \cos x}{2} + c.$$

---

## EXERCISES 10.3

*Solve Exercises 1–24 using the method of integration by parts.*

**1.** $\int x^2 e^{3x}\, dx$

**2.** $\int x e^{7x}\, dx$

**3.** $\int x^3 e^{2x^2}\, dx$

**4.** $\int x^5 e^{x^3}\, dx$

**5.** $\int (4 + x)^2 e^x\, dx$

**6.** $\int x\sqrt{1 + x}\, dx$

**7.** $\int \frac{x}{\sqrt{4 + x}}\, dx$

**8.** $\int x^2\sqrt{1 - x}\, dx$

9. $\int \frac{x}{\sqrt{1+x}}\,dx$

10. $\int \frac{x^3}{\sqrt{1-x}}\,dx$

11. $\int x \ln x\,dx$

12. $\int x^2 \ln x\,dx$

13. $\int x^2 (\ln x)^2\,dx$

14. $\int x (\ln x)^3\,dx$

15. $\int (1+x)e^x \ln x\,dx$

16. $\int \ln (x^2+1)\,dx$

▶ 17. $\int x^3 \sin (2x)\,dx$

▶ 18. $\int \sin^2 (4x)\,dx$

▶ 19. $\int x^2 \cos x\,dx$

▶ 20. $\int x \sin 2x\,dx$

▶ 21. $\int e^x \cos x\,dx$

▶ 22. $\int e^x \sin 4x\,dx$

▶ 23. $\int \cos^2 (3x)\,dx$

▶ 24. $\int \cos^2 x\,dx$

---

## 10.4 THE FUNDAMENTAL THEOREM OF CALCULUS

Up to now, integration has been merely a game of reversing the differentiation process. In this and the next four sections we will see that integration has many practical applications. For example, an important result of this section (Theorem 10.1) shows that integrals can be used to determine areas of irregularly shaped regions.

Before stating the main results, we need to define the *definite integral:*

$$\int_a^b f(x)\,dx. \tag{10.26}$$

(Remember that all previous integrals were called indefinite integrals.) Although the notation in (10.26) is similar to that of the indefinite integral

$$\int f(x)\,dx,$$

two additional symbols have been added: the *limits of integration a* and *b* at either end of the integral sign $\int$. The presence or absence of these values determines whether the integral is definite or indefinite.

So far we have discussed only how to *recognize* a definite integral. We now show how to *evaluate* a definite integral. Suppose a solution—

that is, an antiderivative—is known for the indefinite integral

$$\int f(x)\,dx = F(x) + c. \tag{10.27}$$

Then the definite integral (10.26) is simply the difference of the values of the solution (10.27) at $b$ and at $a$:

$$\begin{aligned}\int_a^b f(x)\,dx &= [F(b) + c] - [F(a) + c]\\ &= F(b) - F(a).\end{aligned}$$

*The arbitrary constant $c$ cancels out since it is both added and subtracted.* Since only the values of $F(x)$ at $x = a$ and $x = b$ are involved in the solution, we will often use the notation

$$\int_a^b f(x)\,dx = F(x)\Big|_a^b,$$

which completely omits the arbitrary constant $c$ (since it cancels out), and indicates that the bottom value $F(a)$ is subtracted from the top value $F(b)$:

$$F(x)\big|_a^b = F(b) - F(a).$$

***Example 1*** Solve the definite integral

$$\int_2^e \ln x\,dx.$$

We have already found the indefinite integral in Eq. (10.21):

$$\int \ln x\,dx = x \ln x - x + c.$$

Thus, letting $F(x) = x \ln x - x$, we have

$$\begin{aligned}\int_2^e \ln x\,dx &= x \ln x - x\Big|_2^e\\ &= (e \ln e - e) - (2 \ln 2 - 2)\\ &= 2 - 2 \ln 2,\end{aligned}$$

since $\ln e = 1$.

It is not necessary that the limits of integration $a$ and $b$ be numbers. They can also be variables.

**Example 2** Determine the definite integral

$$\int_{2x}^{x^2} te^t\,dt.$$

Integrating by parts we see that

$$\int te^t\,dt = te^t - \int e^t\,dt = e^t(t-1) + c.$$

Setting $F(t) = e^t(t-1)$, we have

$$\int_{2x}^{x^2} te^t\,dt = e^t(t-1)\Big|_{2x}^{x^2}$$

$$= e^{x^2}(x^2-1) - e^{2x}(2x-1).$$

---

At the beginning of this section we mentioned that integrals can be used in determining area. Let's see what connection there is between the definite integral and the areas of rectangles and triangles. Consider the curves shown in Fig. 10.1. The area of the rectangle is the product of its base $b$ by its height $h$, or $bh$, while the area of the triangle is $\frac{1}{2}bh$. Let $y = f(x) = h$ be the upper edge of the rectangle in Fig. 10.1(a). Then

$$\int_0^b f(x)\,dx = \int_0^b h\,dx = hx\Big|_0^b = hb - 0 = hb,$$

so the definite integral of the horizontal line $y = f(x) = h$ between (the vertical lines) $x = 0$ and $x = b$ agrees with the area of the rectangle bounded by these lines and the $x$-axis. Similarly, the line $y = f(x) = (h/b)x$ agrees with the hypotenuse of the triangle in Fig. 10.1(b) between $x = 0$ and $x = b$. Integrating this function between 0 and $b$ we

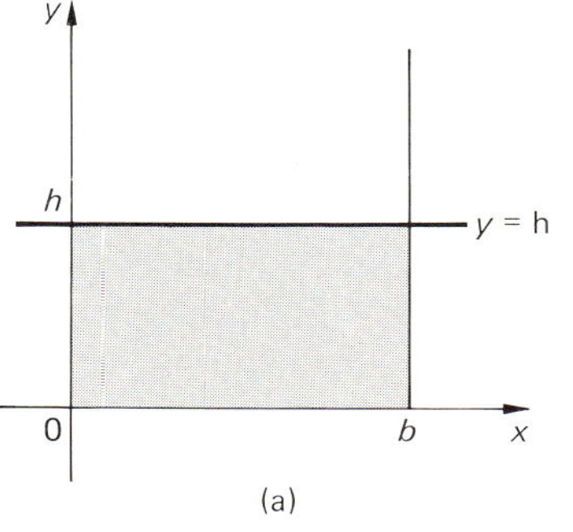


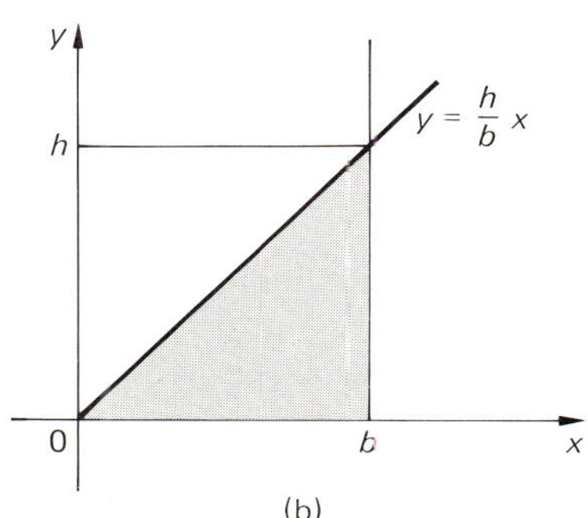


**Figure 10.1**

have

$$\int_0^b f(x)\,dx = \int_0^b (h/b)x\,dx$$
$$= \frac{h}{b}\frac{x^2}{2}\bigg|_0^b = \frac{h}{b}\left(\frac{b^2}{2} - 0\right) = \frac{hb}{2},$$

which again is the area of the region bounded by the curves $y = f(x)$, the $x$-axis, and the vertical lines $x = 0$ and $x = b$—in this case, the triangle in Fig. 10.1(b). Thus, in these two cases we have demonstrated that the definite integral

$$\int_a^b f(x)\,dx$$

yields the value of the area of the figure bounded by the curve $y = f(x)$, the $x$-axis, and the vertical lines $x = a$ and $x = b$ (see Fig. 10.2). We now prove that this result is valid for a very large class of irregularly shaped regions.

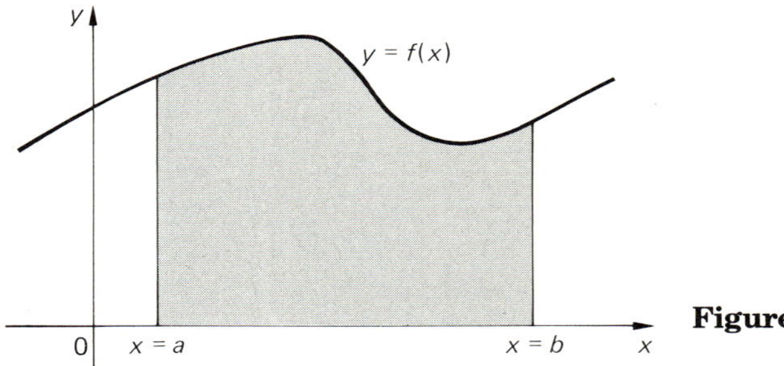


**Figure 10.2**

**THEOREM 10.1** **Let $f(x)$ be any nonnegative continuous function. Then the area of the region bounded by the curve $y = f(x)$, the vertical lines $x = a$ and $x = b$, and the $x$-axis (see the shaded region in Fig. 10.2) is given by the definite integral**

$$\text{Area} = \int_a^b f(x)\,dx. \tag{10.28}$$

***Proof*** Denote the area of the lighter shaded region in Fig. 10.3 by $A(x)$. It is evident (although we will not prove it) that as we gradually increase the value of $x$ in the interval $a \leq x \leq b$, the "area" function $A(x)$ increases continuously, since the right edge of the region is being moved farther to the right. The difference

$$A(x + h) - A(x)$$

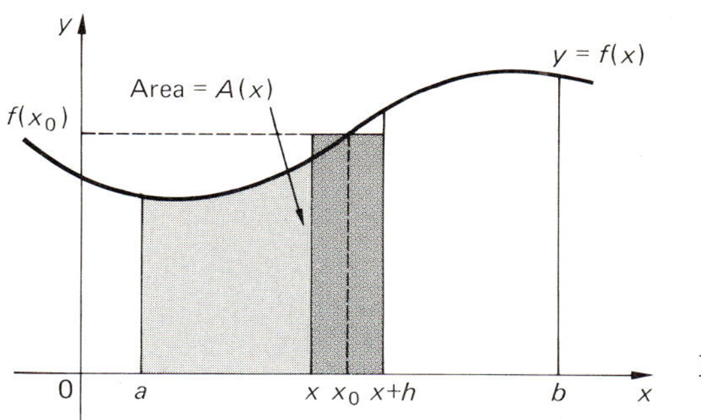


Figure 10.3

is simply the area of the darker shaded region bounded by the curve $y = f(x)$, the $x$-axis, and the vertical lines passing through $x$ and $x + h$.

A second look at Fig. 10.3 shows (although we will not prove this fact either) that the darker shaded region has the same area as the rectangle with identical base and height $f(x_0)$, determined by observing where the top edge of the rectangle intersects the curve $y = f(x)$. This is one reason we required $f(x)$ to be continuous, as it might not be possible to find a point of intersection between the top edge of the rectangle and the curve $y = f(x)$ if $f(x)$ were not continuous. Thus, equating the two areas, we have

$$A(x + h) - A(x) = f(x_0)h, \tag{10.29}$$

where $x_0$ lies in the interval $x \le x_0 \le x + h$. If we divide by $h$ on both sides of Eq. (10.29), we obtain the familiar difference quotient of differentiation on the left side:

$$\frac{A(x + h) - A(x)}{h} = f(x_0).$$

This suggests we might wish to take the limit of both sides as $h$ tends to zero, obtaining the derivative

$$A'(x) = \lim_{h \to 0} \frac{A(x + h) - A(x)}{h} = \lim_{h \to 0} f(x_0). \tag{10.30}$$

Since $x_0$ lies in the interval $x \le x_0 \le x + h$, it must tend to $x$ as $h \to 0$. Again using the fact that $f$ is continuous, we see that the rightmost term in Eq. (10.30) has $f(x)$ as its limit so that

$$A'(x) = f(x). \tag{10.31}$$

Now, Eq. (10.31) says that the "area" function $A(x)$ is an antiderivative of $f(x)$, implying that

$$\int f(x)\,dx = A(x) + c.$$

But then the definite integral

$$\int_a^b f(x)\,dx = A(b) - A(a) \tag{10.32}$$

and $A(a) = 0$ since there is zero area in the region between the line $x = a$ *and itself*. Since $A(b)$ denotes the area bounded by the curve $y = f(x)$, the $x$-axis, and the vertical lines $x = a$ and $x = b$, the theorem is proved. ■

We have been very particular about requiring that the curve $y = f(x)$ stay on or above the $x$-axis. The reason we have done so is that *when the curve lies below the $x$-axis the definite integral is negative.* This occurs because the function value $f(x_0)$ in Eq. (10.29), where the rectangle meets the curve $y = f(x)$, will be negative whenever $y = f(x)$ lies under the $x$-axis. In this case, Eq. (10.29) yields the *negative* of the value of the area bounded by the curve $y = f(x)$, the $x$-axis, and the vertical lines through $x$ and $x + h$. Actually, we can still find this area by merely changing the sign of the solution of the definite integral.

***Example 3*** Find the area of the shaded region shown in Fig. 10.4.

First we calculate the definite integral

$$\int_0^1 (x^2 - 1)\,dx = \frac{x^3}{3} - x\Big|_0^1 = \frac{1}{3} - 1 - (0) = \frac{-2}{3}.$$

Changing the sign, we obtain an area of 2/3.

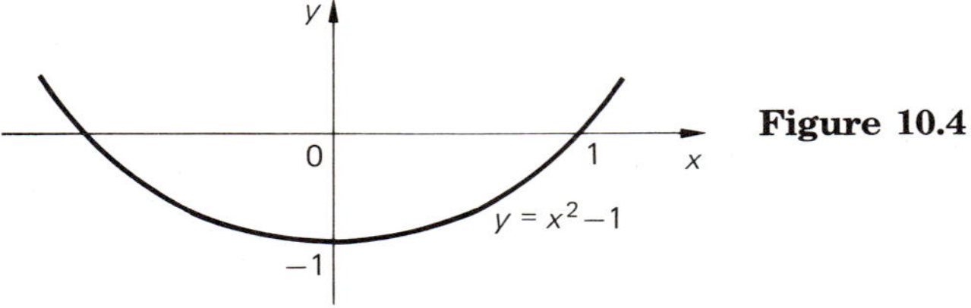


**Figure 10.4**

---

If part of the curve lies above the $x$-axis and part lies below, we simply divide the region into separate parts by finding the zeros of the function and integrating each part separately. If we then change the sign of all the negative definite integrals we obtain the total area.

***Example 4*** Find the area of the two shaded regions in Fig. 10.5 bounded by the $x$-axis and the curve $y = x(x^2 - 1)$. Here

$$\int_{-1}^{0} x(x^2 - 1)\, dx = \int_{-1}^{0} (x^3 - x)\, dx = \left.\frac{x^4}{4} - \frac{x^2}{2}\right|_{-1}^{0}$$

$$= 0 - \left(\frac{1}{4} - \frac{1}{2}\right) = \frac{1}{4},$$

but

$$\int_{0}^{1} x(x^2 - 1)\, dx = \int_{0}^{1} (x^3 - x)\, dx = \left.\frac{x^4}{4} - \frac{x^2}{2}\right|_{0}^{1}$$

$$= \left(\frac{1}{4} - \frac{1}{2}\right) - 0 = -\frac{1}{4}.$$

Changing the sign of this last term we get a total area of $\frac{1}{2}$.

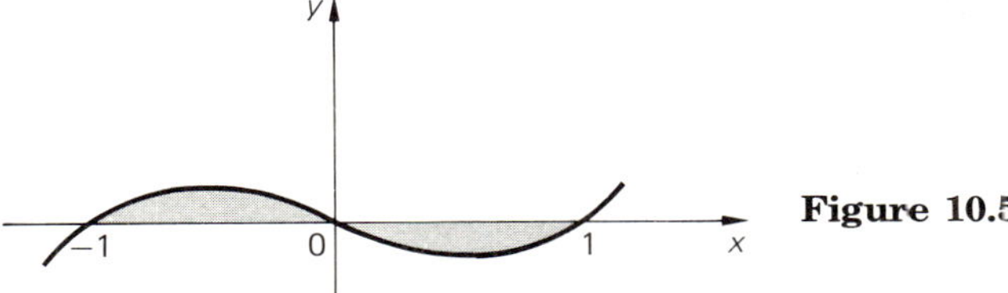


**Figure 10.5**

---

Returning to the proof of Theorem 10.1, we see that if we allow the use of negative values of area in the definition of the "area" function $A(x)$, we are able to repeat all the steps in this proof for an *arbitrary* (instead of for only a positive) continuous function $f(x)$. This proves that *any* continuous function $f(x)$ has an antiderivative given by the "area" function $A(x)$. Since all antiderivatives of $f(x)$ differ only by a constant (see Section 10.1), if $F(x)$ is any antiderivative of $f(x)$, we have $F(x) = A(x) + c$. Then

$$F(b) - F(a) = A(b) - A(a) = A(b),$$

since $A(a) = 0$, and Theorem 10.1 yields the following result.

**FUNDAMENTAL THEOREM OF CALCULUS** **If a function $f(x)$ is continuous in an interval $a \le x \le b$, then the function possesses antiderivatives in that interval. If $F(x)$ is any antiderivative of $f(x)$ on that interval, then**

$$\int_a^b f(x)\, dx = F(b) - F(a).$$

As a final test we will prove that a circle of radius $r$ has area $\pi r^2$. This familiar result is rarely proved, since calculus-free proofs require complicated geometrical constructions. Here, however, we plan to make full use of the techniques of calculus.

***Example 5*** We will determine the area of the circle

$$x^2 + y^2 = r^2 \tag{10.33}$$

lying in the first quadrant (see the shaded region in Fig. 10.6). Solving Eq. (10.33) for $y$, we have

$$y^2 = r^2 - x^2,$$

so that the equation of the circle in the first quadrant is equivalent to the function

$$y = f(x) = \sqrt{r^2 - x^2}.$$

By Theorem 10.1 it follows that the area of this quarter circle is

$$A = \int_0^r \sqrt{r^2 - x^2}\, dx. \tag{10.34}$$

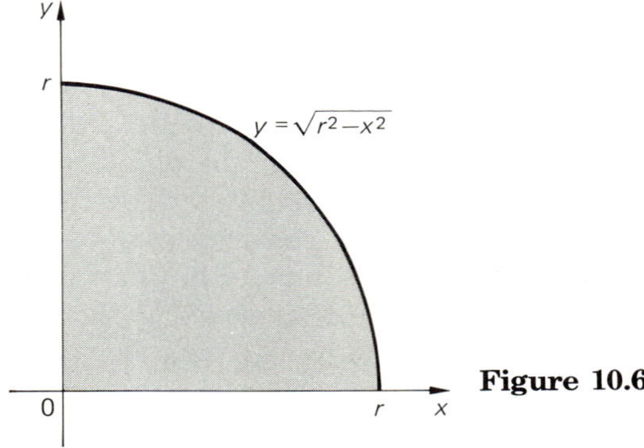


**Figure 10.6**

If we wish to find an antiderivative for the indefinite integral

$$\int \sqrt{r^2 - x^2}\, dx, \tag{10.35}$$

we must use trigonometric substitutions. (If you have skipped the sections involving trigonometry, simply use Formula 14 from Appendix 3 and proceed directly to Eq. 10.38.) Setting $x = r \cos t$, we have

$$\frac{dx}{dt} = -r \sin t \quad \text{and} \quad \sqrt{r^2 - x^2} = \sqrt{r^2(1 - \cos^2 t)} = r \sin t,$$

so that with this change of variable (10.35) becomes

$$\int \sqrt{r^2 - x^2}\, dx = -r^2 \int \sin^2 t\, dt. \tag{10.36}$$

We have already found an antiderivative for the integral on the right side of Eq. (10.36) in Example 6 in Section 10.3. Hence

$$-r^2 \int \sin^2 t\, dt = r^2\left[\frac{\sin t \cos t - t}{2}\right] + c. \tag{10.37}$$

Reversing the substitution $x = r\cos t$, we see that

$$\cos t = \frac{x}{r}, \qquad t = \cos^{-1}\left(\frac{x}{r}\right),$$

and

$$\sin t = \sqrt{1 - \cos^2 t} = \sqrt{1 - (x/r)^2} = \frac{\sqrt{r^2 - x^2}}{r}.$$

Combining Eqs. (10.36) and (10.37) and substituting the values above in the right side of (10.37), we have

$$\int \sqrt{r^2 - x^2}\, dx = \frac{x}{2}\sqrt{r^2 - x^2} - \frac{r^2}{2}\cos^{-1}\left(\frac{x}{r}\right) + c. \tag{10.38}$$

Substituting $x = r$ and $x = 0$ into this antiderivative, we obtain the value of the definite integral

$$\begin{aligned}\int_0^r \sqrt{r^2 - x^2}\, dx &= \left[\frac{r}{2}\sqrt{r^2 - r^2} - \frac{r^2}{2}\cos^{-1}\left(\frac{r}{r}\right)\right] \\ &\quad - \left[\frac{0}{2}\sqrt{r^2 - 0} - \frac{r^2}{2}\cos^{-1}\left(\frac{0}{r}\right)\right] \\ &= \frac{r^2}{2}[\cos^{-1}(0) - \cos^{-1}(1)].\end{aligned}$$

Since $\cos 0 = 1$ and $\cos(\pi/2) = 0$, the inverse function satisfies

$$\cos^{-1}(1) = 0 \quad \text{and} \quad \cos^{-1}(0) = \pi/2,$$

implying that

$$\int_0^r \sqrt{r^2 - x^2}\, dx = \frac{r^2}{2}\left(\frac{\pi}{2} - 0\right) = \frac{\pi r^2}{4}. \tag{10.39}$$

If we multiply this value by 4, we obtain the familiar expression for the area of a circle with radius $r$.

---

## EXERCISES 10.4

*In Exercises 1–12, compute the given definite integrals.*

**1.** $\int_1^8 \frac{dt}{\sqrt[3]{t^2}}$

**2.** $\int_{-1}^1 (3x + 4)^3\, dx$

**3.** $\int_0^2 x\sqrt{1 + 2x^2}\, dx$

**4.** $\int_0^1 x\sqrt[3]{1 - x^2}\, dx$

**5.** $\int_1^{\ln 3} e^{3x}\, dx$

**6.** $\int_0^5 xe^{x^2}\, dx$

**7.** $\int_1^e x \ln x\, dx$

**8.** $\int_3^8 x\sqrt{1 + x}\, dx$

▶ **9.** $\int_{-2}^2 \sin (x + 2)\pi\, dx$

▶ **10.** $\int_{-\pi/4}^{\pi/4} \tan x\, dx$

▶ **11.** $\int_0^{\pi/2} x \sin x\, dx$

▶ **12.** $\int_0^{\pi} e^x \sin (2x)\, dx$

*Find the area of each of the regions indicated in Exercises 13–22 bounded by the following lines:*

**13.** $y = x^2$, $y = 0$, and $x = 1$

**14.** $y = x^3$, $y = 0$, $x = 1$, and $x = 2$

**15.** $y = \ln x$, $y = 0$, and $x = 2$

**16.** $y = e^x$, $y = 0$, $x = -1$, and $x = 1$

**17.** $y = 1/x$, $y = 0$, $x = 1$, and $x = 2$

**18.** $y = 5 - x$ and $y = 6/x$. (*Hint:* Find where the curves intersect and subtract the smaller area from the larger area.)

▶ **19.** One link between $y = \sin x$ and $y = \cos x$

▶ **20.** First arch of $e^{-x} \sin 2x$

▶ **21.** $y = \sin x$, $y = 0$, and $x = 3\pi/4$

▶ **22.** $y = \frac{\sin x}{x}$, $x = 0$, $x = \pi/2$, and $y = 0$

*The limits of integration need not be finite. If either is infinite, we define the resulting* **improper integral** *by the limit*

$$\int_a^{\infty} f(x)\, dx = \lim_{b\to\infty} \int_a^b f(x)\, dx,$$

*or*

$$\int_{-\infty}^b f(x)\, dx = \lim_{a\to-\infty} \int_a^b f(x)\, dx,$$

*if it exists. If the limit does not exist we say that the improper integral* **diverges**.

*Calculate the following improper integrals if they exist.*

**23.** $\int_1^{\infty} e^{-x}\, dx$

**24.** $\int_0^{\infty} xe^{-4x^2}\, dx$

**25.** $\int_{-2}^{\infty} \frac{1}{(x+3)^2}\, dx$

**26.** $\int_{-\infty}^{0} e^{2x}\, dx$

**27.** $\int_{-\infty}^{2} \frac{1}{x-3}\, dx$

▶ **28.** $\int_{-\infty}^{\infty} \frac{dx}{1+x^2}$

**29.** Show that it only requires a finite amount of paint to paint the area between the curve $y = xe^{-x^2}$ and the $x$-axis but that an infinite amount of paint is required to outline the edges of this region.

*There are other kinds of improper integrals. For example, a and b may both be finite but f(x) may become infinite at a or b. Then*

$$\int_a^b f(x)\, dx = \lim_{h \to 0} \int_a^{b-h} f(x)\, dx,$$

*if f becomes infinite at x = b, and*

$$\int_a^b f(x)\, dx = \lim_{h \to 0} \int_{a+h}^{b} f(x)\, dx,$$

*if f becomes infinite at x = a, if the limits exist. Otherwise the integrals diverge.*
*Find the following improper integrals, if they exist:*

**30.** $\int_0^1 \frac{dx}{\sqrt{x}}$

**31.** $\int_0^1 \frac{dx}{\sqrt{1-x}}$

**32.** $\int_0^1 \frac{dx}{x}$

▶ **33.** $\int_0^1 \frac{dx}{\sqrt{x-x^2}}$ (*Hint:* Break this integral in two.)

---

## 10.5 APPLICATIONS OF INTEGRAL CALCULUS

Definite integrals can be used for many different purposes besides the determination of areas of irregularly shaped regions. In this section and in Sections 10.7 and 10.8, we will see a number of ways in which integrals are used in business and science. The applications we give are not a comprehensive list, but merely a sampling of the many possible uses of integration.

***Example 1*** In Section 9.1 we defined the velocity at which an object travels as the (instantaneous) rate of change, or *derivative*, of the distance traveled with respect to the time $t$. Thus, we may reverse the process and obtain the distance traveled by *integrating* the velocity function $v(t)$:

$$\text{Distance traveled} = \int v(t)\, dt. \qquad \textbf{(10.40)}$$

Although we did not say so, Eq. (10.40) was used in determining the height above the ground $h(t)$ of an object falling under the effect of gravity in the absence of wind resistance (see Section 9.5). In that case the velocity function is

$$v(t) = -gt + v_0,$$

where $v_0$ is the velocity at time $t = 0$, and the height at any time $t$ is

$$h(t) = \int v(t)\,dt = \frac{-gt^2}{2} + v_0 t + h_0,$$

with $h_0$ the height at time $t = 0$.

If the initial velocity $v_0 \leq 0$ (the object is at rest or falling) we can calculate the (downward) distance traveled by the object between the times $t = t_0$ and $t = t_1$ by using the definite integral:

$$\begin{aligned}\text{Distance traveled} &= h(t_0) - h(t_1)\\ &= -\int_{t_0}^{t_1} v(t)\,dt\\ &= \int_{t_0}^{t_1} (gt - v_0)\,dt.\end{aligned}$$

For example, if the object was originally at rest ($v_0 = 0$) it falls

$$\int_1^4 gt\,dt = \frac{gt^2}{2}\bigg|_1^4 = \frac{g}{2}(16 - 1) = \frac{15g}{2} = 240 \text{ ft}$$

($g = 32$ ft/sec$^2$), between the times $t = 1$ sec and $t = 4$ sec. Similarly, if the velocity of an object is given by the function

$$v(t) = t^3 + 9t + 7,$$

the distance it travels from its original location in $t$ seconds is

$$\begin{aligned}\text{Distance traveled} &= \int_0^t v(t)\,dt\\ &= \frac{t^4}{4} + \frac{9t^2}{2} + 7t\bigg|_0^t\\ &= \frac{t^4}{4} + \frac{9}{2}t^2 + 7t.\end{aligned}$$

***Example 2*** The concentration of suspended particles or a gas in the atmosphere may be measured by finding their mass per unit volume. The concentration of a pollutant in the atmosphere will vary with time due to climatological effects, chemical decomposition and production of the pollutant, deposition, absorption by plants, etc. The concentration $C(t)$ of a given pollutant is often a useful index of its effects, particularly when a living organism reacts quickly to a certain level of concentration. Odor is a good example of this situation: If the threshold concentration is exceeded for one second, the odor is detected.

Suppose the concentration function $C(t)$ for some gas has been determined over a certain period of time. We plot this function as shown in Fig. 10.7. The shaded area under the curve between the times $t = t_0$ and $t = t_1$ is defined as the *dosage* $D$ of the pollutant measured in mass $\times$ time units per unit volume. Hence, we compute the definite integral

$$D = \int_{t_0}^{t_1} C(t)\, dt$$

to find the dosage of the pollutant in the time interval $t_0 \leq t \leq t_1$. Dosage may be a more important index than concentration, especially when the living organism reacts slowly to the pollutant. For example, pine needles are able to withstand very brief exposures to relatively high concentrations of sulfur dioxide $SO_2$ in the atmosphere, but will die under prolonged low-level concentrations.

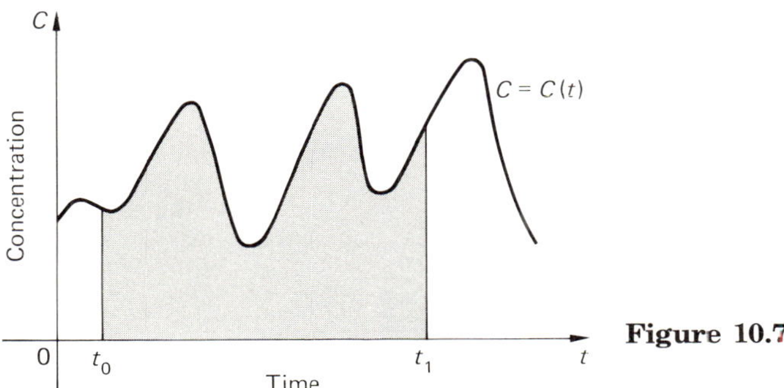


**Figure 10.7**

To calculate the actual exposure of a living organism to an airborne pollutant, we multiply the *inspiration rate* $r$ measured in units of volume per unit of time by the dosage $D$. This yields the mass $rD$ of the pollutant that has been inspired since

$$\frac{\text{mass} \times \text{time}}{\text{volume}} \cdot \frac{\text{volume}}{\text{time}} = \text{mass}.$$

The concentration function $C(t)$ is usually not expressed in terms of simple functions of $t$. Most frequently we only have strip charts of the concentrations, and occasionally only a table of the concentration levels at fixed equally spaced values of the time variable $t$. In these situations, mechanical and numerical inaccuracies prevent an *exact* determination of the dosage. However, it is very easy to obtain a good *approximation* of the dosage by the use of numerical integration techniques. One such method, called the *trapezoidal rule*, will be discussed in Section 10.6. This rule allows us to find a good approximation of the area under a curve by using only the function values at equally spaced values of time. At that time we will approximate the dosage in a real situation.

***Example 3*** The *demand equation* for a given product is a function $p = f(x)$ that specifies the relation between the price $p$ of a given product and the number of units $x$ of the product that are available for consumers to purchase. In a free market we observe that when the number of units $x$ is small enough to create a shortage, the consumers are willing to pay an increased price for the product. Conversely, when the number of units $x$ is so large that it exceeds the demand for that product, the producers lower the price of the product to encourage additional sales. The graph of a typical demand equation is shown in Fig. 10.8. The shape of such a curve can be found by the following procedure.

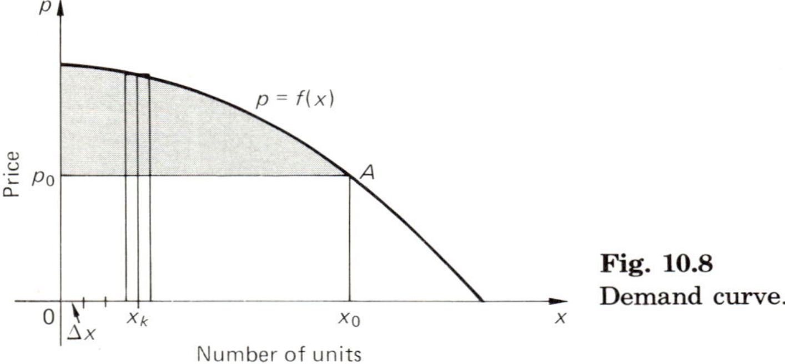


**Fig. 10.8**
Demand curve.

A monopolist produces batches of equal size $\Delta x$ of the product. The batch size could be as small as one, in the case of a very sophisticated computer system, or millions, in the case of pins or needles. The first batch $\Delta x$ is sold to the highest bidder for the maximal price of $p_1$ dollars per unit, and the value $p_1$ is associated with a point $x_1$ in the interval $0 \leq x \leq \Delta x$. The next batch is sold to the second highest bidder at a unit price $p_2\,(\leq p_1)$ and that value is assigned to a point $x_2$ in the interval $\Delta x \leq x \leq 2\Delta x$. The procedure is repeated until no buyers can be found, at which time the unit price is set equal to zero. This process is called *perfect discrimination*. If $n$ batches are sold, the total revenue obtained by the monopolist under perfect discrimination is

$$S_n = f(x_1)\,\Delta x + f(x_2)\,\Delta x + \cdots + f(x_n)\,\Delta x. \tag{10.41}$$

Note that the revenue obtained from the $k$th batch corresponds to the area of a rectangle of base $\Delta x$ and height $f(x_k)$ (see Fig. 10.8).

If the size of each batch is made smaller and smaller and the number of batches increases without bound, we obtain a continuous demand curve such as the one shown in Fig. 10.8. If the monopolist were now able to sell $x_0$ units of the product under conditions of perfect discrimination, the revenue would equal the area under the curve $p = f(x)$ between $x = 0$ and $x = x_0$:

$$\begin{Bmatrix} \text{Revenue for } x_0 \text{ sales under} \\ \text{perfect discrimination} \end{Bmatrix} = \int_0^{x_0} f(x)\, dx. \tag{10.42}$$

However, perfect discrimination is rarely possible in a free market. Instead, several producers compete to produce a continuous supply of a product, varying the output in order to meet a fixed demand $x_0$ at a market price of $p_0$ dollars per unit. The revenue the producers obtain is $R_0 = p_0 x_0$ given by the area of the rectangle $0x_0Ap_0$ in Fig. 10.8. The shaded region in Fig. 10.8 represents the difference between the revenue the monopolist would receive under perfect discrimination and that received in a free market at the market price $p_0$. The area of the shaded region is thus a savings (in dollars) for the consumers and is called the *consumer's surplus* corresponding to the market price $p_0$. Thus,

$$\begin{Bmatrix} \text{Consumer's surplus} \\ \text{at market price } p_0 \end{Bmatrix} = \int_0^{x_0} f(x)\, dx - p_0 x_0. \tag{10.43}$$

For example, suppose the demand curve for a certain type of calculator is

$$p = 200 - 10^{-8}x^2$$

dollars per calculator. Then if three different companies compete to produce $x_0 = 100{,}000$ calculators, the consumer's surplus is

$$\{\text{Consumer's surplus}\} = \int_0^{100{,}000} \left(200 - \frac{x^2}{10^8}\right) dx - x_0 p_0,$$

where $p_0 = 200 - 10^{-8}(10^5)^2 = 200 - 100 = \$100/\text{unit}$. Hence

$$\{\text{Consumer's surplus}\} = 200x - \frac{x^3}{3 \cdot 10^8}\Big|_0^{10^5} - 10^5(10^2)$$

$$= 2 \cdot 10^7 - \frac{1}{3} \cdot 10^7 - 10^7 = \$6{,}666{,}666.67.$$

In other words, had the producer been a monopolist and able to auction off the calculators, he or she would have received

$$\int_0^{10^5} \left(200 - \frac{x^2}{10^8}\right) dx = \$16{,}666{,}666.67$$

in revenue, but the free market produced a revenue of \$10,000,000 for the producers, saving \$6,666,666.67 for the consumers.

---

It is worth noticing in the example above that we have suggested an analogy between sums and integrals (see Eqs. 10.41 and 10.42). Since the definite integral

$$\int_a^b f(x)\, dx$$

yields the area under the curve $y = f(x)$ between $x = a$ and $x = b$, we can approximate this area by drawing a collection of $n$ side-by-side rectangles of fixed base length $\Delta x = (b - a)/n$ and height equal to the value of $f(x)$ at some point in the interval, such as along the right side of each rectangle (see Fig. 10.9). To calculate the area of the $k$th rectangle, note that the right edge of the rectangle lies on the line $x = x_k = a + k\Delta x$, where $x_0 = a$ and $x_n = b$. Thus the height of this rectangle is $f(x_k)$ and the length of the base is $\Delta x$, so that the area of the shaded rectangle is $f(x_k)\,\Delta x$. Adding these areas together, we obtain

$$f(x_1)\,\Delta x + f(x_2)\,\Delta x + f(x_3)\,\Delta x + \cdots + f(x_n)\,\Delta x.$$

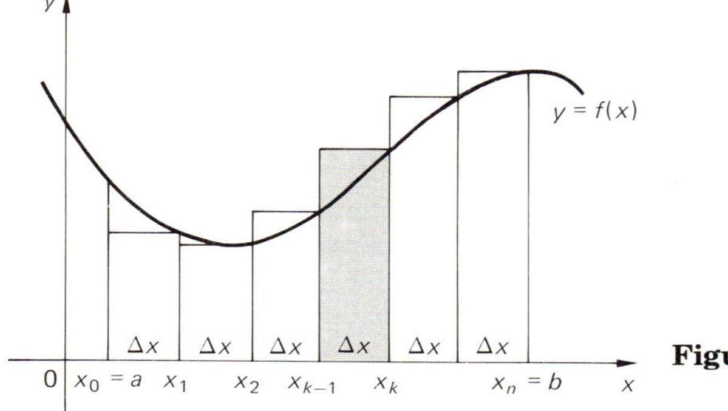


**Figure 10.9**

Notice that each term in this sum looks the same except for the subscript, which changes from term to term. There is a useful summation notation, sometimes called the "sigma notation," for abbreviating sums of this sort. We write

$$\sum_{k=1}^{n} f(x_k)\,\Delta x \tag{10.44}$$

to indicate that the terms $f(x_k)\,\Delta x$ are to be summed together with the subscripts ranging *by integers* from $k = 1$ to $k = n$. Hence, we have the approximation

$$\int_a^b f(x)\,dx \approx \sum_{k=1}^{n} f(x_k)\,\Delta x. \tag{10.45}$$

Notice the resemblance between the integral and sum signs, the functions $f(x)$ and $f(x_k)$, and the symbols $dx$ and $\Delta x$. This resemblance is not just a coincidence; instead it is a useful similarity that can be used to obtain a wide range of useful applications of definite integrals. Indeed, it can be proved that

$$\int_a^b f(x)\,dx = \lim_{n\to\infty} \sum_{k=1}^{n} f(x_k)\,\Delta x.$$

This proof is complicated and will not be done here. It can be found in most books on advanced calculus. We now illustrate two applications of Eq. (10.45).

***Example 4*** Suppose we revolve the shaded area shown in Fig. 10.10 around the $x$-axis, obtaining the solid of revolution illustrated. What is the volume of this solid?

Again we divide the interval $a \le x \le b$ into $n$ equal subintervals of length $\Delta x = (b - a)/n$ and consider the $n$ side-by-side rectangles with which we previously approximated the area. If we revolve the $k$th rectangle around the $x$-axis, we will obtain a cylinder of height $\Delta x$ and radius $f(x_k)$, where $x_k = a + k\Delta x$. If you look at Fig. 10.10 you will see that these cylinders resemble a bunch of coins or disks stacked next to each other.

Since the volume of a cylinder is simply the product of its height and the area of its base, we obtain

$$\pi f^2(x_k)\,\Delta x$$

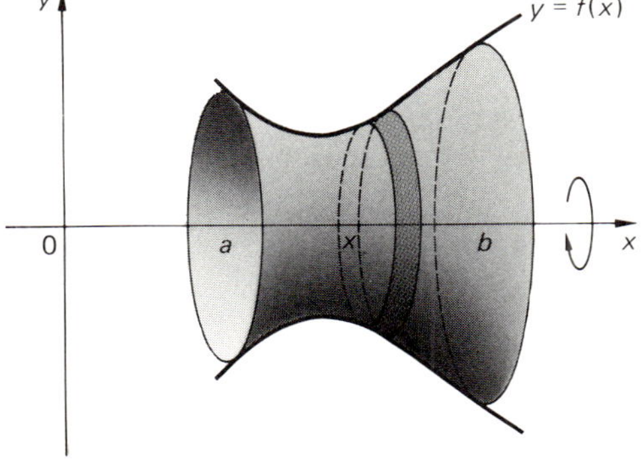


**Fig. 10.10**
Solid of revolution.

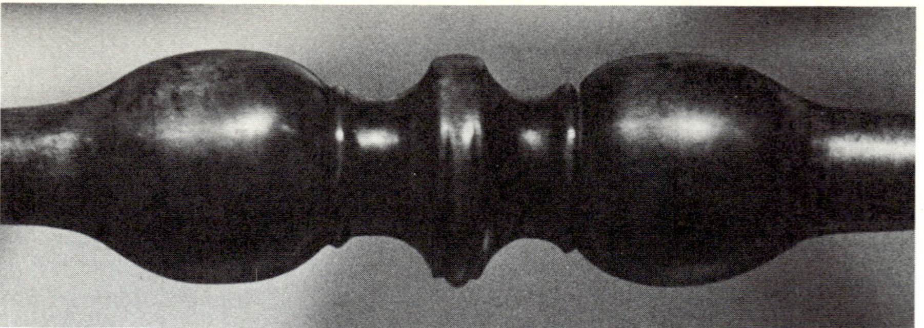

as the volume of the $k$th cylinder. Adding all these volumes we obtain an approximation to the volume of the solid of revolution:

$$\text{Approximate volume} = \sum_{k=1}^{n} \pi f^2(x_k)\,\Delta x. \tag{10.46}$$

We now use the similarity we discussed above to obtain a definite integral that will yield the exact volume. Replacing the sum with an integral, $f(x_k)$ with $f(x)$, and $\Delta x$ with $dx$, we have the *disk formula for a volume of revolution:*

$$\text{Volume} = \int_a^b \pi f^2(x)\,dx. \tag{10.47}$$

We shall now exhibit some uses of this equation. Suppose we wish to find the volume of a sphere of radius $r$. If we revolve the quarter circle $y = \sqrt{r^2 - x^2}, 0 \le x \le r$, shown in Fig. 10.11 around the $x$-axis, we obtain the volume of the right hemisphere. That volume is half the volume of a sphere.

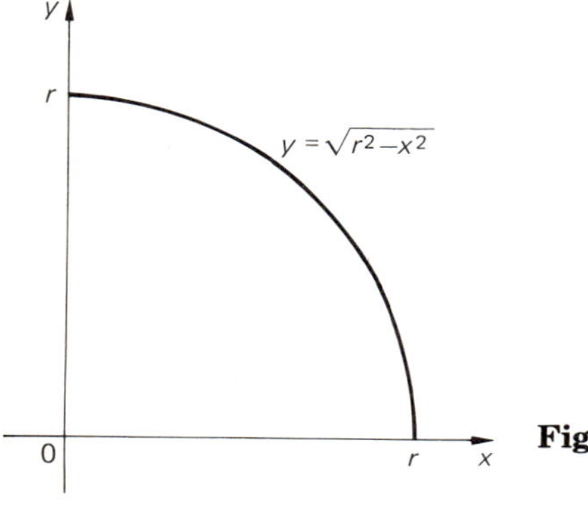


**Figure 10.11**

Here $y = f(x) = \sqrt{r^2 - x^2}$, so Eq. (10.47) yields

$$\text{Volume of hemisphere} = \int_0^r \pi(\sqrt{r^2 - x^2})^2\,dx$$

$$= \pi \int_0^r (r^2 - x^2)\,dx$$

$$= \pi\left(r^2x - \frac{x^3}{3}\right)\Bigg|_0^r = \frac{2\pi}{3}r^3.$$

Thus, the volume of a sphere is $(4/3)\pi r^3$ (cubic units).

Revolving the shaded area in Fig. 10.12 around the $x$-axis will yield a truncated cone-shaped solid of revolution. Since $y = f(x) = x^2$, the solid of revolution has volume (in cubic units)

$$\text{Volume} = \int_{1/2}^1 \pi(x^2)^2\,dx = \pi\int_{1/2}^1 x^4\,dx$$

$$= \pi\frac{x^5}{5}\Bigg|_{1/2}^1 = \frac{\pi}{5}\left(1 - \frac{1}{32}\right) = \frac{31}{160}\pi.$$

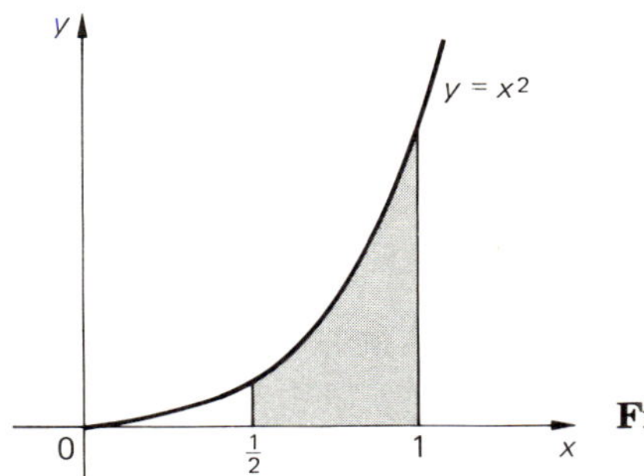


Figure 10.12

***Example 5*** Suppose we graph the curve $y = f(x)$ in the $xy$-plane and wish to find the *arc length of the curve* between $x = a$ and $x = b$ (see Fig. 10.13). Again we divide the interval $a \le x \le b$ into $n$ equal subintervals of length $\Delta x = (b - a)/n$ and construct a collection of right triangles, with sides parallel to the axes, such that the hypotenuse of the $k$th right triangle is the line segment joining $(x_{k-1}, f(x_{k-1}))$ to $(x_k, f(x_k))$, where $x_k = a + k\,\Delta x$. The length of this hypotenuse can be obtained by using the Pythagorean theorem, since the length of the horizontal side of the triangle is $\Delta x$, while the length of the vertical side is $f(x_k) - f(x_{k-1})$. Thus,

$$\text{Length of } k\text{th hypotenuse} = \sqrt{(\Delta x)^2 + [f(x_k) - f(x_{k-1})]^2}.$$

Now, if we sum all the lengths of these hypotenuses together, we will

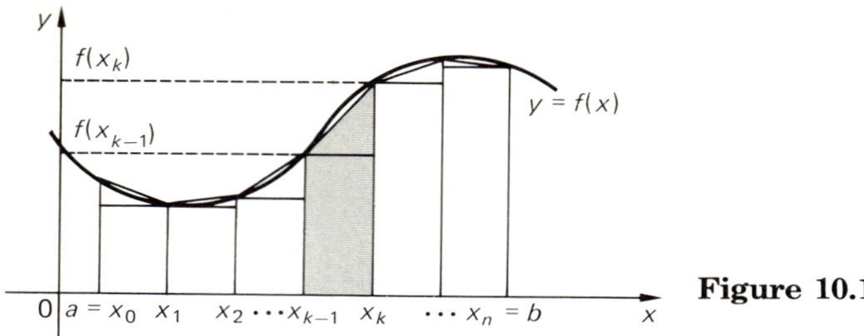


**Figure 10.13**

obtain an approximation of the arc length of the curve over the interval $a \le x \le b$:

$$\begin{aligned}\text{Approximate arc length} &= \sum_{k=1}^{n} \sqrt{(\Delta x)^2 + [f(x_k) - f(x_{k-1})]^2} \\ &= \sum_{k=1}^{n} \sqrt{1 + \left[\frac{f(x_k) - f(x_{k-1})}{\Delta x}\right]^2}\,\Delta x.\end{aligned} \tag{10.48}$$

To use the similarity previously discussed, we must approximate the difference quotient:

$$\frac{f(x_k) - f(x_{k-1})}{\Delta x} = \frac{f(x_k) - f(x_{k-1})}{x_k - x_{k-1}}. \tag{10.49}$$

If we take the limit of the difference quotient (10.49) as $\Delta x$ tends to zero, we obtain the derivative

$$f'(x_k) = \lim_{x_{k-1} \to x_k} \frac{f(x_k) - f(x_{k-1})}{x_k - x_{k-1}}. \tag{10.50}$$

Replacing the difference quotient in Eq. (10.48) with the derivative $f'(x_k)$, we have

$$\text{Approximate arc length} = \sum_{k=1}^{n} \sqrt{1 + [f'(x_k)]^2}\,\Delta x. \tag{10.51}$$

Replacing the sum with a definite integral, $f'(x_k)$ with $f'(x)$, and $\Delta x$ with $dx$, we have the *arc length formula:*

$$\text{Arc length} = \int_a^b \sqrt{1 + [f'(x)]^2}\,dx. \tag{10.52}$$

We illustrate the use of this formula by finding the arc length of the curve $y = x^2$ in the interval $1/2 \le x \le 1$ (see Fig. 10.12). Since

$y = f(x) = x^2$, we have $f'(x) = 2x$ so that Eq. (10.51) yields

$$\text{Arc length} = \int_{1/2}^{1} \sqrt{1 + (2x)^2}\, dx$$

$$= \int_{1/2}^{1} \sqrt{1 + 4x^2}\, dx. \tag{10.53}$$

The substitution $x = t/2$ yields $dx/dt = 1/2$, so that Eq. (10.53) becomes

$$\int_{1/2}^{1} \sqrt{1 + 4x^2}\, dx = \int_{1}^{2} \sqrt{1 + 4\left(\frac{t}{2}\right)^2} \frac{dx}{dt}\, dt$$

$$= \frac{1}{2}\int_{1}^{2} \sqrt{1 + t^2}\, dt.$$

Using Formula 13 of Appendix 3, we find that the integral equals

$$\frac{1}{2}\int_{1}^{2} \sqrt{t^2 + 1}\, dt = \frac{1}{4}[t\sqrt{t^2 + 1} + \ln|t + \sqrt{t^2 + 1}|]\Big|_{1}^{2}$$

$$= \frac{1}{4}\{[2\sqrt{5} + \ln(2 + \sqrt{5})] - [\sqrt{2} + \ln(1 + \sqrt{2})]\}$$

$$= \frac{1}{4}\left[2\sqrt{5} - \sqrt{2} + \ln\frac{2 + \sqrt{5}}{1 + \sqrt{2}}\right] \approx 0.9050.$$

Thus, the curve is approximately 0.9050 units long.

---

## EXERCISES 10.5

*In Exercises 1–8, find the distance traveled by an object falling with velocity $v(t)$ in the time interval $1 \le t \le 5$ sec.*

**1.** $v(t) = -32t - 10$ ft/sec

**2.** $v(t) = -980t$ cm/sec

**3.** $v(t) = -32t + 10$ ft/sec

**4.** $v(t) = -980t + 500$ cm/sec

**5.** $v(t) = t^2 - 32t + 9$ ft/sec

**6.** $v(t) = 5t^3 - 980t + 100$ cm/sec

**7.** $v(t) = -40t + 15$ miles/sec

**8.** $v(t) = -270t - 500$ m/sec

*In Exercises 9–12, find the dosage over a 24-hour period given the concentrations $C(t)$. If an animal breathes in 10 liters of air per minute, how much pollutant is inspired each day?*

**9.** $C(t) = 1000 + t(24 - t)\, \mu g/m^3$

**10.** $C(t) = 200 + t^2(24 - t)\, \mu g/m^3$

▶ **11.** $C(t) = 10 + \cos(\pi t/12)\ \mu\text{g/m}^3$

▶ **12.** $C(t) = 12 + 3\sin(\pi t/6)\ \mu\text{g/m}^3$

*In Exercises 13–16, use Eq. (10.43) to find the consumer's surplus for the given demand equation $p = f(x)$ at the given market price $p_0$ or demand $x_0$.*

**13.** $p = 1800 - 3x^2,\ p_0 = 1500$

**14.** $p = 150 - 5x - x^2,\ x_0 = 5$

**15.** $p = 4800 - 3x^3,\ p_0 = 1800$

▶ **16.** $p = 1000\cos\left(\dfrac{\pi x}{1000}\right),\ x_0 = 400$

*In Exercises 17–24, find the volume of the solid of revolution obtained by revolving the given region about the x-axis. [Hint: Use the disk formula (10.47).]*

**17.** $y = \sqrt{x},\ x = 0,\ x = 4$, and the $x$-axis

**18.** $y = 3x - x^2,\ x = 0,\ x = 3$, and the $x$-axis

**19.** $y = (x - 1)(2 - x),\ x = 1,\ x = 2$, and the $x$-axis

**20.** $y = x^2$ and $y = 4x - x^2$ (*Hint:* Find where the two curves cross.)

**21.** $y = (r/h)x,\ x = 0,\ x = h$, and the $x$-axis (volume of a cone)

▶ **22.** $y = \sin x,\ x = 0,\ x = \pi$, and the $x$-axis

▶ **23.** $y = e^x \sin x,\ x = 0,\ x = \pi$, and the $x$-axis

▶ **24.** $y = \tan x,\ x = \dfrac{\pi}{3},\ x = \dfrac{\pi}{2}$, and the $x$-axis

*In Exercises 25–30, find the arc length of the given curve $y = f(x)$ over the interval $a \le x \le b$. (Hint: Use Eq. 10.52.)*

**25.** $y = 2x^{3/2}$ over $0 \le x \le 4$

**26.** $y = \ln(1 - x^2)$ over $1/4 \le x \le 1/2$

**27.** $y = \frac{1}{2}(e^x + e^{-x})$ over $0 \le x \le 1$

▶ **28.** $y = \sqrt{r^2 - x^2}$ over $0 \le x \le r$ (quarter circle)

▶ **29.** $y = \ln\cos x$ over $\pi/6 \le x \le \pi/4$

▶ **30.** $y = \ln\sin x$ over $\pi/4 \le x \le \pi/2$

**31.** Suppose we revolve the shaded region in Fig. 10.10 about the $y$-axis (instead of the $x$-axis). We now obtain a hollow cylinder whose volume may be approximated by the sum of the volumes of the "shells" obtained by revolving the approximating rectangles around the $y$-axis.

a) Show that the volume of the $k$th shell is exactly

$$\pi[x_k^2 - x_{k-1}^2]f(x_k).$$

b) Show that $x_k^2 - x_{k-1}^2$ is approximately equal to $2x_k\,\Delta x$.

c) Use the approximation to obtain the *shell formula for a volume of revolution:*

$$\text{Volume} = 2\pi \int_a^b x f(x)\,dx.$$

d) Use the shell formula above to find the volumes of revolution obtained by revolving the given regions about the $y$-axis.

(i) $y = mx$, $y = h$, and the $y$-axis (cone)

(ii) $y = \sqrt{r^2 - x^2}$ in the first quadrant (hemisphere)

(iii) $y = x \ln x$, $x = 1$, $x = 2$, and the $x$-axis

▶ (iv) $y = \sin x$, $x = 0$, $x = \pi/2$, and the $x$-axis

▶ (v) $y = e^x \cos x$, $x = 0$, $x = \pi/2$, and the $x$-axis

---

## 10.6 APPROXIMATION: THE TRAPEZOIDAL RULE

In this section we will describe one of the simplest methods for approximating the value of a definite integral. This technique, called the *trapezoidal rule*, is one of a great many procedures in numerical analysis for the approximate evaluation of integrals. These methods, called *quadrature formulas*, are very useful whenever:

1. The indefinite integral cannot be expressed in terms of elementary functions, or
2. The function that is being integrated is given only as a table of values.

When either of these situations occurs, we are unable to obtain the value of the definite integral by the usual procedure. Hence, we use these numerical methods in order to obtain an approximate value of the definite integral.

The formula for the trapezoidal rule is very easy to derive. Consider Fig. 10.14, in which the interval $a \le x \le b$ has been divided into $n$ intervals of length

$$\Delta x = \frac{b - a}{n}.$$

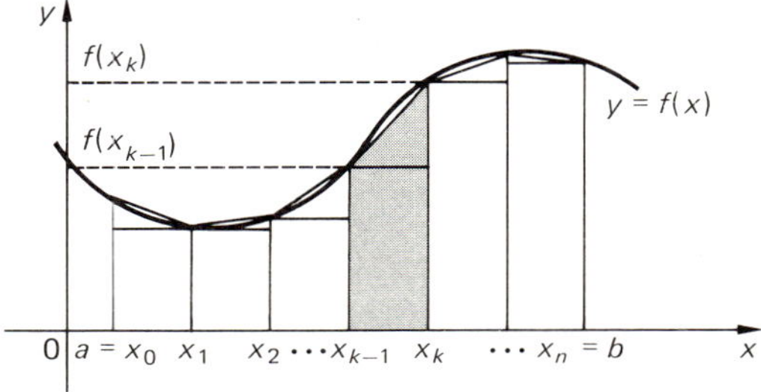


**Figure 10.14**

The area under the curve in the $k$th interval is closely approximated by the area of the shaded trapezoid formed by joining together the points $(x_{k-1}, f(x_{k-1}))$ and $(x_k, f(x_k))$ with a straight-line segment. Hence, if we add the areas of all these trapezoids, we will approximate the area under the curve.

Now, to find the area of each trapezoid, note that the $k$th trapezoid consists of a rectangle of height $f(x_{k-1})$ and base $\Delta x$ plus a right triangle with sides $\Delta x$ and $f(x_k) - f(x_{k-1})$. Hence,

$$\text{Area of } k\text{th trapezoid} = f(x_{k-1})\,\Delta x + \frac{1}{2}[f(x_k) - f(x_{k-1})]\,\Delta x$$

$$= \frac{f(x_k) + f(x_{k-1})}{2}\,\Delta x. \tag{10.54}$$

Adding all these areas together we have

$$\int_a^b f(x)\,dx \approx \frac{f(x_1) + f(x_0)}{2}\,\Delta x + \frac{f(x_2) + f(x_1)}{2}\,\Delta x + \frac{f(x_3) + f(x_2)}{2}\,\Delta x + \cdots + \frac{f(x_n) + f(x_{n-1})}{2}\,\Delta x,$$

or, factoring $\Delta x/2$ from each term,

$$\int_a^b f(x)\,dx \approx \frac{\Delta x}{2}\{[f(x_1) + f(x_0)] + [f(x_2) + f(x_1)] + [f(x_3) + f(x_2)] + \cdots + [f(x_n) + f(x_{n-1})]\}. \tag{10.55}$$

Now observe that each term $f(x_k)$ appears twice except for the first and last terms, $f(x_0)$ and $f(x_n)$. Hence Eq. (10.55) can be rewritten as

$$\int_a^b f(x)\,dx \approx \frac{\Delta x}{2}[f(x_0) + 2f(x_1) + \cdots + 2f(x_{n-1}) + f(x_n)]. \tag{10.56}$$

This formula is called the *trapezoidal rule*. The following three examples will illustrate its use.

***Example 1*** Use the trapezoidal rule with ten intervals ($n = 10$) to evaluate the definite integral

$$\int_0^1 \frac{dx}{1 + x^2}.$$

Since $a = 0$ and $b = 1$, we see that

$$\Delta x = \frac{b - a}{n} = \frac{1}{10} = 0.1;$$

thus we must evaluate the function at

$$x = 0.0, 0.1, 0.2, 0.3, 0.4, 0.5, 0.6, 0.7, 0.8, 0.9, 1.0.$$

It is useful to arrange the information in a table (see Table 10.2). The first column consists of the values of $x$, while the second column consists of the values of the corresponding values of $f(x) = 1/(1 + x^2)$. (We have rounded each answer to four decimal places.) Column three contains the factor with which each function value is multiplied in Eq. (10.56). Thus the first and last factors are one and all the rest are two. The final column contains the product of the second and third columns. The entries in the fourth column are added and then multiplied by $\Delta x/2 = 0.05$ to obtain the result of the trapezoidal rule. This yields the approximate value

$$\int_0^1 \frac{dx}{1 + x^2} \approx 0.7850. \qquad \textbf{(10.57)}$$

**Table 10.2**

| $x$ | $f(x) = 1/(1 + x^2)$ | Factor | Product |
|---|---|---|---|
| 0.0 | 1.0000 | 1 | 1.0000 |
| 0.1 | 0.9901 | 2 | 1.9802 |
| 0.2 | 0.9615 | 2 | 1.9230 |
| 0.3 | 0.9174 | 2 | 1.8348 |
| 0.4 | 0.8621 | 2 | 1.7242 |
| 0.5 | 0.8000 | 2 | 1.6000 |
| 0.6 | 0.7353 | 2 | 1.4706 |
| 0.7 | 0.6711 | 2 | 1.3422 |
| 0.8 | 0.6098 | 2 | 1.2196 |
| 0.9 | 0.5525 | 2 | 1.1050 |
| 1.0 | 0.5000 | 1 | 0.5000 |
| | | Sum = | 15.6996 |
| | Multiplied by $\Delta x/2$ (= 0.05) = | | 0.7850 |

In this case we can obtain the exact value of the definite integral by using Formula 6 from the integral tables in Appendix 3. We have

$$\int_0^1 \frac{dx}{1+x^2} = \tan^{-1} x \Big|_0^1 = \tan^{-1} 1 - \tan^{-1} 0$$

$$= \frac{\pi}{4} - 0 \approx 0.78539816. \quad \textbf{(10.58)}$$

We see that the approximate value (10.57) only differs from the actual value (10.58) by 0.00039816, which illustrates the effectiveness of this procedure.

***Example 2*** Use the trapezoidal rule with $n = 8$ to evaluate the integral

$$\int_0^2 e^{-x^2}\,dx.$$

Again we form a table as we did in Example 1. The results of the preliminary calculations are shown in Table 10.3 (correct to four decimal places). Thus the trapezoidal rule yields

$$\int_0^2 e^{-x^2}\,dx \approx 0.8817.$$

**Table 10.3**

| $x$ | $f(x) = e^{-x^2}$ | Factor | Product |
|---|---|---|---|
| 0.00 | 1.0000 | 1 | 1.0000 |
| 0.25 | 0.9394 | 2 | 1.8788 |
| 0.50 | 0.7788 | 2 | 1.5576 |
| 0.75 | 0.5698 | 2 | 1.1396 |
| 1.00 | 0.3679 | 2 | 0.7358 |
| 1.25 | 0.2096 | 2 | 0.4192 |
| 1.50 | 0.1054 | 2 | 0.2108 |
| 1.75 | 0.0468 | 2 | 0.0936 |
| 2.00 | 0.0183 | 1 | 0.0183 |
| | | | Sum = 7.0537 |
| | | | Multiplied by $\Delta x/2$ (= 0.125) = 0.8817 |

Tables for this integral have been evaluated* and the correct value for the definite integral is 0.88208138, indicating that the error in our approximation is 0.00038038, an error of less than 0.05%.

* M. Abramowitz and L. A. Stegun, *Handbook of Mathematical Functions*, National Bureau of Standards, Applied Mathematics Series, Vol. 55 (1964), p. 311.

***Example 3*** In an urban airshed photochemical study conducted for the Environmental Protection Agency,* the carbon monoxide (CO) concentrations at the corner of Long Beach Boulevard and 36th Street in Long Beach, California, were recorded. On September 29, 1969, the following readings were obtained:

| Time (A.M.) | 5 | 6 | 7 | 8 | 9 | 10 | 11 |
|---|---|---|---|---|---|---|---|
| Concentration (ppm) | 1.50 | 3.75 | 4.75 | 3.00 | 1.75 | 1.50 | 2.00 |
| (1 ppm = 1 part per million = 0.0224 $\mu g/m^3$) | | | | | | | |

Calculate the dosage of CO present at that corner over this six-hour period.

Since dosage is simply the integral

$$D = \int_5^{11} C(t)\, dt$$

(see Example 2 in Section 10.5), we need only apply the trapezoidal rule to the data above. Collecting the data in Table 10.4, we obtain $D \approx$ 16.5 ppm · hr.

**Table 10.4**

| $t$ = Time (A.M.) | $C(t)$ = Concentration (ppm) | Factor | Product |
|---|---|---|---|
| 5 | 1.50 | 1 | 1.50 |
| 6 | 3.75 | 2 | 7.50 |
| 7 | 4.75 | 2 | 9.50 |
| 8 | 3.00 | 2 | 6.00 |
| 9 | 1.75 | 2 | 3.50 |
| 10 | 1.50 | 2 | 3.00 |
| 11 | 2.00 | 1 | 2.00 |
| | | | Sum = 33.00 |
| | | | Multiplied by $\Delta x/2$ (= 1/2) = 16.5 |

* M. K. Lui and P. M. Roth, Urban Airshed Photochemical Simulation Model Study, EPA-R4-73-030d (July 1973).

## EXERCISES 10.6

*In Exercises 1–12, calculate an approximation to the given integral using the trapezoidal rule with n intervals. When possible, compare your answer to the exact value.*

1. $\int_1^4 x^2\,dx,\ n = 12$
2. $\int_2^5 x^3\,dx,\ n = 9$

▶ 3. $\int_0^1 \sqrt{1 - x^2}\,dx,\ n = 10$
4. $\int_0^{1/2} \frac{x}{\sqrt{1 - x^2}}\,dx,\ n = 5$

5. $\int_1^2 \frac{e^x}{x}\,dx,\ n = 10$
6. $\int_1^{10} \frac{\ln x}{x}\,dx,\ n = 9$

▶ 7. $\int_0^1 \sin x\,dx,\ n = 10$
▶ 8. $\int_0^{1/2} \cos(\pi x)\,dx,\ n = 8$

9. $\int_1^2 e^{-x^2}\,dx,\ n = 10$
10. $\int_1^2 x \ln x\,dx,\ n = 10$

▶ 11. $\int_0^{\pi} e^{x^2} \sin x\,dx,\ n = 8$
▶ 12. $\int_0^{\pi} \frac{\sin x}{x}\,dx,\ n = 12$

---

## 10.7 THE DIFFERENTIAL EQUATION $y' = ay + b$

A *differential equation* is simply any equation that involves at least one derivative. For example,

$$y' = y, \qquad yy'' = \sqrt{x + y^2}, \qquad y = \tan(y') + x$$

are all differential equations. In this section we will study some of the properties of the linear differential equation

$$y' = ay + b, \tag{10.59}$$

where $y' = dy/dx$ and $a$ and $b$ are functions of the independent variable $x$. Before dealing with the general case (10.59), it is useful to consider some special situations. Suppose $a \equiv 1$ and $b \equiv 0$, that is, $a$ and $b$ are constants, the first identically equal to 1 and the second to 0. Then we have the differential equation

$$y' = y, \tag{10.60}$$

We could ask ourselves what function $y = y(x)$ equals its own derivative, since that is the meaning of Eq. (10.60). (Do you remember?) If you don't remember, there is an easy method, called *separation of variables*, for finding the function $y(x)$. We rewrite (10.60) in the form

$$\frac{dy}{dx} = y, \tag{10.61}$$

and collect all the $y$-terms on one side and all the $x$-terms on the other:

$$\frac{dy}{y} = dx. \tag{10.62}$$

Now we take the indefinite integral of both sides of (10.62), obtaining

$$\ln y = \int \frac{dy}{y} = \int dx = x + c, \tag{10.63}$$

so that

$$y = e^{\ln y} = e^{x+c}$$

or

$$y = Ce^x, \tag{10.64}$$

where $C = e^c$. Since the constant $c$ is arbitrary, so is the constant $C$; thus any function $y(x) = Ce^x$ is a *solution* of the differential equation (10.60). For this reason Eq. (10.64) is called the *general solution* of the differential equation.

You may have noticed that the technique of separation of variables is very similar to the change of variables method we used in Section 10.2. This is not a coincidence; the two methods are essentially identical. To understand this claim observe that Eq. (10.61) is equivalent to the identity

$$\frac{1}{y}\frac{dy}{dx} = 1. \tag{10.65}$$

Now each term on the left side of Eq. (10.65) is a function of $x$, so we can integrate both sides of the equation with respect to $x$, obtaining

$$\int \frac{1}{y}\frac{dy}{dx}\,dx = \int dx.$$

But then the change of variables theorem gives us the center equality of (10.63). Thus, separating the variables initially is only a method for saving a step in the procedure.

We have shown that there are infinitely many solutions $y = Ce^x$ to the differential equation $y' = y$, one for each real value of the constant

$C$. Therefore, we need additional information if we are to single out a particular solution from this infinite collection. Suppose we are told that the solution must satisfy the ***condition***

$$y(4) = 2. \tag{10.66}$$

Then we can determine the constant $C$, since only one value of $C$ will satisfy Eq. (10.66). Setting $x = 4$ in Eq. (10.64), we have

$$2 = y(4) = Ce^4,$$

so that $C = 2e^{-4} \approx 0.0366$. The differential equation (10.60), together with the condition (10.66)

$$y' = y, \qquad y(4) = 2,$$

is called an *initial value problem*, and it is a fundamental theorem of differential equations that every initial value problem

$$y' = a(x)y + b(x), \qquad y(c) = k$$

has a unique solution.*

Linear differential equations have many applications in the biological and physical sciences, as the following three examples illustrate.

***Example 1*** When supplied with a favorable environment and ample quantities of nutrients, bacteria populations generally grow at a rate proportional to the size of the population. (Indeed, most organisms will satisfy this population growth law!) Let's translate this statement into a mathematical equation.

Let $P(t)$ be the number of bacteria (population) present at time $t$. The instantaneous rate at which the population is growing is simply the derivative $P'(t)$. Hence the growth law states that $P'(t)$ is proportional to $P(t)$, or in symbols,

$$P'(t) = kP(t), \tag{10.67}$$

where $k$ is some constant of proportionality. Rewriting Eq. (10.67) in the form

$$\frac{dP}{dt} = kP,$$

---

* The proof of this theorem is beyond the scope of this book. The interested reader may find the proof in almost every book on differential equations. See, for example, W. R. Derrick and S. I. Grossman, *Elementary Differential Equations with Applications* (Reading, Mass.: Addison-Wesley, 1976, p. 485).

we separate variables, obtaining

$$\frac{dP}{p} = k\,dt,$$

and integrate both sides:

$$\ln P = \int \frac{dP}{P} = \int k\,dt = kt + c.$$

Thus, we have

$$P(t) = e^{\ln P} = e^{kt+c},$$

or

$$P(t) = Ce^{kt}, \tag{10.68}$$

where $C = e^c$. This is the equation that we used for the population problems in Section 3.2 (see Example 3 in Section 3.2), with $C =$ 179,323,175 and $k = 0.0142$. We saw there that $k$ was the growth rate per individual, which agrees with Eq. (10.67), since $P'$ is the growth rate and $P$ is the number of individual bacteria.

In this case, we need to know the growth rate per individual $k$ and the population in a certain year in order to completely determine a formula for the population at any time $t$. Of course, no such equation will give an exact value of the population at all times, since the census figures $C$ and growth rate per individual $k$ are only approximate.

It is also possible to use the census figures for two different years in determining the coefficients $C$ and $k$. For example, the U.S. census figure for 1960 was 179,323,175, while in 1970 it was 200,251,326. Setting $t = 0$ for the year 1960, we have

$$P(1960) = Ce^{k\cdot 0} = 179{,}323{,}175,$$

$$P(1970) = Ce^{k\cdot 10} = 200{,}251{,}326.$$

Hence $C = 179{,}323{,}175$, so that

$$e^{10k} = \frac{200{,}251{,}326}{179{,}323{,}175} \approx 1.1167,$$

and taking the natural logarithm of both sides yields

$$10k = \ln(1.1167) \approx 0.1104,$$

or

$$k \approx 0.01104.$$

This suggests that the population formula for the United States should be

$$P(t) = 179{,}323{,}175e^{0.01104t},$$

where $t$ is the number of years after 1960.

***Example 2*** Newton's Law of Cooling states that the rate of change of the temperature difference between an object and its surrounding medium is proportional to the temperature difference. A skillet at 400°F is removed from a burner and is at 150°F two minutes later. If the kitchen temperature is 70°F, when will the skillet be at 100°F?

Let $T(t)$ be the temperature of the skillet at any time $t \geq 0$. Then the difference between the temperature of the skillet and its surrounding medium is

$$f(t) = T(t) - 70°\text{F}.$$

Newton's Law of Cooling is equivalent to the equation

$$\frac{df}{dt} = f'(t) = kf(t),$$

so we separate variables to obtain

$$\int \frac{df}{f} = \int k\,dt,$$

or

$$\ln f = kt + c.$$

Taking the exponential of both sides yields the general solution

$$f(t) = Ce^{kt},$$

or

$$T(t) = 70° + Ce^{kt}. \tag{10.69}$$

We now use the two known facts (or conditions) about the temperature of the skillet:

$$T(0) = 400°\text{F} \quad \text{and} \quad T(2) = 150°\text{F}.$$

Substituting these values into Eq. (10.69) yields

$$400° = 70° + Ce^{0\cdot k},$$

$$150° = 70° + Ce^{2\cdot k}.$$

The first equation implies that $C = 330°\text{F}$ and the second can be used to find $k$, since

$$e^{2k} = \frac{150 - 70}{C} = \frac{80}{330},$$

so that

$$2k = \ln(8/33) \quad \text{or} \quad k \approx -0.7085.$$

Now that we know the values of $C$ and $k$, we substitute them into Eq. (10.69), obtaining

$$T(t) = 70 + 330e^{-0.7085t}\ °\text{F}, \tag{10.70}$$

a formula for the temperature of the skillet at any time $t$. To find when the temperature will be 100°F we need only substitute $T = 100$ into Eq. (10.70) and solve for $t$,

$$100 = 70 + 330e^{-0.7085t},$$

so that

$$e^{-0.7085t} = \frac{30}{330} = \frac{1}{11},$$

implying that

$$-0.7085t = \ln(1/11) \quad \text{or} \quad t \approx 3.3843 \text{ min}.$$

***Example 3*** Infusion of glucose into the bloodstream is a common practice for the feeding of unconscious hospital patients. To study how the process works, let $G(t)$ be the amount of glucose (or blood sugar) in the bloodstream of a patient at any time $t$. Suppose that glucose is infused into the bloodstream at the constant rate of $A$ grams per minute. At the same time the patient's body metabolizes the glucose, removing it from the bloodstream at a rate proportional to the amount of glucose present. Thus, the difference between the intake and the removal of glucose from the bloodstream must equal the rate of change:

$$\text{Rate of change} = \text{Intake} - \text{Removal}. \tag{10.71}$$

In symbols, Eq. (10.71) translates into the differential equation

$$\frac{dG}{dt} = A - kG(t). \tag{10.72}$$

since $A$ g/min of glucose are being infused while an amount *proportional* to the glucose present, $G(t)$, is being metabolized. Our problem now is to solve the differential equation (10.72).

This equation is somewhat different from the ones we encountered in the preceding two examples, but separation of variables will again yield a solution. Cross multiplying we have

$$\frac{dG}{A - kG} = dt,$$

and integrating this identity yields

$$\frac{-1}{k}\ln(A - kG) = \int \frac{dG}{A - kG} = \int dt = t + c. \tag{10.73}$$

Thus,

$$\ln(A - kg) = -k(t + c),$$

so that

$$A - kG = e^{-kt-kc} = Ce^{-kt},$$

where $C$ is an arbitrary constant. Solving for $G(t)$ we have the general solution

$$G(t) = \frac{1}{k}[A - Ce^{-kt}]. \tag{10.74}$$

Since the constant of proportionality $k$ is positive, note that as the time increases, the $(Ce^{-kt})$-term tends to zero (why?), implying that the blood sugar approaches an equilibrium value of $A/k$. If we know the rate at which the patient metabolizes glucose, we can determine the value of the constant $k$. Then by carefully controlling the amount $A$ infused into the patient, the hospital can adjust the amount of glucose in the bloodstream so that the patient's body is adequately supplied with the nutrients it requires at all times.

---

This example illustrates the procedure for solving Eq. (10.59), when both coefficients $a$ and $b$ are constant. Now let's suppose that $a = a(x)$ is a nonconstant function of $x$ and that $b = 0$. Equation (10.59) becomes

$$\frac{dy}{dx} = a(x)y, \tag{10.75}$$

which may be separated into

$$\frac{dy}{y} = a(x)\,dx.$$

Integrating both sides of this equation yields

$$\ln y = \int a(x)\,dx + c$$

or

$$y = Ce^{\int a(x)\,dx}. \tag{10.76}$$

The functions (10.76) constitute an infinite set of solutions of the differential equation (10.75). If we can perform the integration $\int a(x)\,dx$, we will be able to obtain an elementary expression for the general solution.

***Example 4*** The economist Vilfredo Pareto (1848–1923) discovered an interesting law concerning the wealthy in stable economies. Let $y$ be the number of people in a stable economy having an income of at least $x$ dollars. Then the rate at which the population $y$ decreases is directly proportional to the population $y$ and inversely proportional to their income $x$, that is,

$$\frac{dy}{dx} = -k\frac{y}{x}.$$

Further, he found by empirical studies for various populations over a 400-year time period that the constants of proportionality cluster around the value $k = 1.5$.

Separating variables, we have the integrals

$$\int \frac{dy}{y} = -1.5 \int \frac{dx}{x}$$

or

$$\ln y = 1.5 \ln x + c = \ln x^{-1.5} + c.$$

Exponentiating, we obtain *Pareto's Law:*

$$y = cx^{-1.5}.$$

A different solution technique is required when both $a$ and $b$ are functions of $x$ in Eq. (10.59). Rewriting Eq. (10.59) as

$$y' - a(x)y = b(x),$$

we multiply both sides by $e^{-\int a(x)dx}$:

$$e^{-\int a(x)dx}[y' - a(x)y] = e^{-\int a(x)dx}b(x). \tag{10.77}$$

Now recalling that differentiation and integration are inverse processes, we notice that

$$\frac{d}{dx}\int a(x)\,dx = a(x),$$

which implies that

$$\frac{d}{dx}e^{-\int a(x)dx} = e^{-\int a(x)dx}\frac{d}{dx}\left[-\int a(x)\,dx\right] = -a(x)e^{-\int a(x)dx}.$$

Thus the left side of Eq. (10.77) is equal to

$$\begin{aligned}(ye^{-\int a(x)dx})' &= y'e^{-\int a(x)dx} + y(e^{-\int a(x)dx})' \\ &= e^{-\int a(x)dx}[y' - a(x)y].\end{aligned}$$

Thus, Eq. (10.77) may be rewritten in the form

$$(ye^{-\int a(x)dx})' = e^{-\int a(x)dx}b(x) \tag{10.78}$$

and we can eliminate the derivative on the left side of Eq. (10.78) by integrating both sides:

$$ye^{-\int a(x)dx} = \int e^{-\int a(x)dx}b(x)\,dx + c. \tag{10.79}$$

If each of the integrals in Eq. (10.79) can be solved, we obtain an elementary expression for the general solution of Eq. (10.59). Although Eq. (10.79) can be used as a "recipe" for obtaining the solution to the differential equation

$$y' = a(x)y + b(x),$$

it is generally easier to remember the procedure we have followed, rather than to memorize the complicated expression in Eq. (10.79). The following two examples illustrate the technique that we discussed above.

***Example 5*** Find the infinite set of solutions of the differential equation

$$y' = -\frac{2}{x}y + 5x^2. \tag{10.80}$$

Rewriting (10.80) as

$$y' + \frac{2}{x}y = 5x^2,$$

we multiply both sides by

$$e^{-\int(-2/x)\,dx} = e^{2\int dx/x} = e^{2\ln x} = x^2,$$

obtaining

$$x^2y' + 2xy = 5x^4.$$

The left side is equal to $(x^2y)'$, so integrating both sides we have

$$x^2y = \int 5x^4\,dx + c = x^5 + c.$$

Dividing both sides by $x^2$ yields the general solution

$$y = x^3 + cx^{-2}.$$

***Example 6*** Solve the initial value problem

$$y' - 2y = x^2e^{2x}, \qquad y(1) = e^2.$$

Multiplying both sides by

$$e^{-\int 2\,dx} = e^{-2x}$$

yields

$$(ye^{-2x})' = x^2.$$

Integrating both sides we have

$$ye^{-2x} = \int x^2\,dx = \frac{x^3}{3} + c,$$

so that the general solution is

$$y = e^{2x}\left[\frac{x^3}{3} + c\right].$$

Setting $x = 1$ we have

$$e^2 = y(1) = e^2\left(\frac{1}{3} + c\right)$$

so $c = 2/3$. Thus the solution of the initial value problem is

$$y = \frac{e^{2x}}{3}(x^3 + 2).$$

---

### EXERCISES 10.7

*In Exercises 1–16, find the general solution of the given differential equation. If an initial condition is given, solve the initial value problem.*

**1.** $y' = 2y$

**2.** $y' = 7y,\ y(1) = 2$

**3.** $y' = -3y$

**4.** $y' = -5y,\ y(2) = -3$

**5.** $y' = 2y + 5$

**6.** $y' = -3y + 5,\ y(-2) = 4$

**7.** $y' = xy$

**8.** $y' = e^x y,\ y(0) = 1$

**9.** $y' = y \ln x + x^x$

**10.** $y' - \dfrac{3y}{x} = x^3,\ y(1) = 4$

**11.** $y' + y = xe^{-x} + 1$

**12.** $y' = x^3 - 2xy,\ y(1) = 1$

▶ **13.** $y' = \dfrac{y}{1 + x^2}$

▶ **14.** $y' + y = \sin x,\ y(0) = 0$

▶ **15.** $y' + \dfrac{2y}{x} = \dfrac{\cos x}{x^2},\ y(\pi) = 0$

▶ **16.** $y' + y \cot x = 2x \csc x$

*In Exercises 17–22, assume that the population of the given country is growing at a rate proportional to the size of the population. Calculate each country's population in the year 2000, given the two census figures shown.*

**17.** Australia: 1968 census = 12,099,100; 1973 census = 13,268,000

**18.** Austria: 1968 census = 7,349,000; 1973 census = 7,520,000

**19.** Colombia: 1968 census = 19,825,000; 1973 census = 23,210,000

**20.** Czechoslovakia: 1968 census = 14,362,000; 1973 census = 14,580,000

**21.** India: 1968 census = 523,893,000; 1973 census = 574,220,000

**22.** Kenya: 1968 census = 10,209,000; 1973 census = 12,480,000

*In Exercises 23–28, use Newton's Law of Cooling to determine how long to bake a cake, at the given oven temperature, assuming that it takes exactly 30 minutes to change 80°F dough into a 170°F cake in a 350°F oven.*

**23.** 250°F

**24.** 300°F

**25.** 425°F

**26.** 400°F

**27.** 275°F

**28.** 375°F

*In Exercises 29–36, use separation of variables to find the general solution of the given differential equation.*

**29.** $y' = e^x/(2y)$

**30.** $xy' = 3y$

**31.** $y' = xy^2$

**32.** $y' = y^3\sqrt{1-x}$

**33.** $y' + y = y(xe^x + 1)$

**34.** $y' = xy(3 - 5y)$

▶ **35.** $y' = \dfrac{\cos x}{y}$

▶ **36.** $y' = \sqrt{1 - y^2}$

★★ **37.** On a certain day it began to snow early in the morning and the snow continued to fall at a constant rate. The velocity at which a snowplow can clear a road is inversely proportional to the height of the accumulated snow. The snowplow started at 9:00 A.M. and cleared four miles by noon. By 2:00 P.M. it had cleared another two miles. When did it start snowing?

★★ **38.** The king of Transylvania always adds cold cream to his hot coffee immediately upon being served. The queen waits to add the cold cream until just before she drinks her coffee. If both the king and queen are served at the same time and take their first sip simultaneously, who drinks the hotter coffee?

★ **39.** Show, using Pareto's Law, that the number of people having an income of $x$ dollars is approximately

$$N(x) = \frac{1.5C}{x^{2.5}}.$$

Assume that 10,000 individuals in the United States have incomes exceeding $200,000 per year. What is their *average* income? (*Hint:* Use an improper integral.)

---

## 10.8 SOME ECOLOGICAL MODELS

In this section we will present four situations in which the use of differential equations yields a solution to a real problem. These case studies are but a few ways in which differential equations can be applied. Similar models can be found in economics, sociology, and the physical sciences.

***Example 1*** In Example 6 in Section 3.2 we considered a population model in which the growth was restricted by the resources or food available. In that example, we used an S-shaped curve, called the *logistic* (or *sigmoidal*) curve, to model the population growth. We now see how that formula (Eq. 3.21) is derived.

Let $p(t)$ be the number of individuals in the population at any time $t$. Then $p'(t)$ is the growth rate of the population, so that

$$\frac{1}{p}\frac{dp}{dt} \tag{10.81}$$

is the *growth rate per individual* of the population. We now assume that the growth rate per individual is a linear function of the population: that it tends to $k$ as the population tends to zero, and it tends to zero as the population approaches the carrying capacity $A$. We can plot these points on a graph (see Fig. 10.15), where the horizontal axis is the population, while the vertical axis is the growth rate per individual $(1/p)(dp/dt)$. Using the slope-intercept formula for a line (see Section 2.2), we observe that the $y$-intercept is $k$ and the slope is

$$m = \frac{\text{rise}}{\text{run}} = \frac{-k}{A},$$

so that the line is given by

$$\frac{1}{p}\frac{dp}{dt} = \frac{-k}{A}p + k. \tag{10.82}$$

This is the differential equation that governs *logistic growth.*

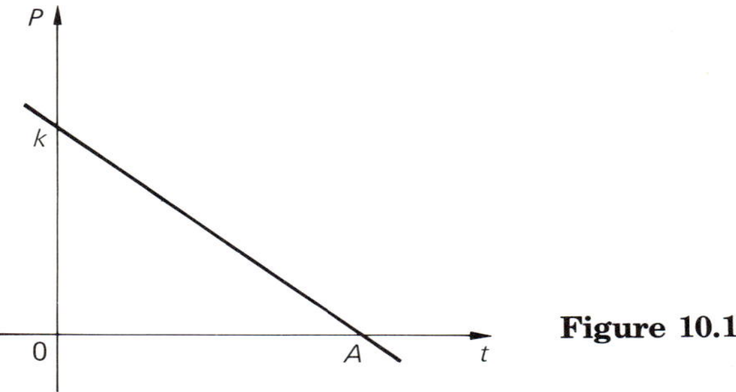


**Figure 10.15**

Rewriting Eq. (10.82) in the form

$$\frac{1}{p}\frac{dp}{dt} = \frac{k}{A}(A - p),$$

we separate variables to obtain the integral

$$\int \frac{A\,dp}{p(A - p)} = \int k\,dt. \tag{10.83}$$

The trick now is to notice that

$$\frac{1}{p} + \frac{1}{A - p} = \frac{A}{p(A - p)},$$

so that Eq. (10.83) becomes

$$\int \frac{1}{p} + \frac{1}{A - P}\, dp = \int k\, dt.$$

Integrating we have

$$\ln p - \ln (A - p) = kt + c,$$

or

$$\ln \frac{p}{A - p} = kt + c. \tag{10.84}$$

Exponentiating both sides we have

$$\frac{p(t)}{A - p(t)} = Ce^{kt}. \tag{10.85}$$

Now suppose the population at time $t = 0$ was $a$. Then if we substitute $t = 0$ into Eq. (10.85) we obtain

$$\frac{a}{A - a} = C.$$

Thus we may rewrite Eq. (10.85) as

$$\frac{p(t)}{A - p(t)} = \frac{a}{A - a} e^{kt}.$$

We now solve for $p(t)$. Cross multiplying and gathering all the $p(t)$-terms on the left side, we have

$$p(t)\left[1 + \frac{a}{A - a} e^{kt}\right] = \frac{Aa}{A - a} e^{kt}$$

or

$$p(t) = \frac{(Aa/(A - a))e^{kt}}{1 + (a/(A - a))e^{kt}}.$$

Multiplying the numerator and denominator of the right side by $(A - a)e^{-kt}$ yields

$$p(t) = \frac{Aa}{a + (A - a)e^{-kt}}. \tag{10.86}$$

This is the same equation as (3.21). Note that $p(0) = Aa/A = a$, while

$$\lim_{t\to\infty} p(t) = \lim_{t\to\infty} \frac{Aa}{a + (A - a)e^{-kt}} = \frac{Aa}{a} = A.$$

---

The following example illustrates another way in which logistic curves arise.

***Example 2*** We will now formulate a very simple model for the spread of a highly contagious, but nonfatal, disease. Let $s(t)$ be the number of susceptible individuals and $i(t)$ the number of infected individuals in the population. To simplify the model we will assume that once an individual is infected he or she remains infected and that the population size remains constantly equal to $n$. These assumptions are not unrealistic if the spread of the disease is extremely fast. Then

$$s(t) + i(t) = n \tag{10.87}$$

for any time $t$.

Now $i'(t)$ is the growth rate of infected individuals in the population. Since the growth of the epidemic depends on the number of susceptible individuals who come in contact with infected individuals, it is plausible to assume that the growth rate of infected individuals is jointly proportional to the number of susceptibles and to the number of infected individuals, or

$$\frac{di}{dt} = ksi. \tag{10.88}$$

This equation produces the largest growth rate when $s = i = n/2$ and decreases to zero as $s$ or $i$ tend to zero. (Prove this!) The constant $k$ is called the *specific infection rate* of the epidemic.

We now use Eq. (10.87) to rewrite Eq. (10.88) in terms of $i$ only:

$$\frac{di}{dt} = ki(n - i)$$

or

$$\frac{1}{i}\frac{di}{dt} = k(n - i). \tag{10.89}$$

Notice that Eq. (10.89) resembles Eq. (10.82), so we may use the same technique to find the general solution. Rewriting Eq. (10.89) as

$$\frac{n}{i(n - 1)}\,di = nk\,dt,$$

we have

$$\ln \frac{i}{n - i} = nkt + c$$

or

$$\frac{i(t)}{n - i(t)} = Ce^{nkt}.$$

Solving for $C$ in terms of $i(0)$, and then for $i(t)$, yields

$$i(t) = \frac{ni(0)}{i(0) + [n - i(0)]e^{-nkt}} . \tag{10.90}$$

Thus, if the constants $n$, $k$, and $i(0)$ are known, we can predict the number of infected (and hence the number of susceptible) individuals at any time $t$.

***Example 3*** The presence of temperature inversions and low wind speeds will often trap air pollutants in a mountain valley for an extended period of time. Gaseous sulfur compounds are often a significant problem and their study is complicated by rapid oxidation. For example, hydrogen sulfide $H_2S$ oxidizes into sulfur dioxide $SO_2$, which in turn oxidizes into a sulfate. The residence time (half-life) of these processes is approximately eight hours for the former and two days for the latter. Suppose a pulp mill is producing 12,000 lb/day of $H_2S$ from its settling ponds and 2400 lb/day of $SO_2$ from its stack. How much of each pollutant will there be in the valley at any time $0 \leq t \leq 72$ if all the pollutants are trapped in the valley for three days?

Let $x(t)$ be the number of pounds of $H_2S$ and $y(t)$ the number of pounds of $SO_2$ in the valley at time $t$. Then the rate of change of each pollutant depends on the difference between the input and removal of the pollutant (see Example 3 in Section 10.7):

Rate of change = Input − Removal.

First let's consider the removal of $H_2S$ from the atmosphere. If a fixed amount $A$ of $H_2S$ is introduced into the atmosphere at $t = 0$ and no more is added, the rate of change of $H_2S$ is proportional to the amount present, that is,

$$\frac{dx}{dt} = -kx, \tag{10.91}$$

where the constant of proportionality $k$ depends on the residence time of $H_2S$. This is the same differential equation that we considered in

Section 10.7. The solution of Eq. (10.91) is

$$x(t) = Ae^{-kt},$$

and since the residence time is 8 hr, we have $x(8) = A/2$, so that

$$\frac{A}{2} = Ae^{-8k},$$

or, canceling the $A$-terms and taking the natural logarithm,

$$-8k = \ln(1/2) = -\ln 2.$$

Thus $k = (\ln 2)/8$, that is, *the constant of proportionality is equal to* $\ln 2$ *divided by the residence time.*

Using hours as our unit of time and dividing the daily input of $H_2S$ by 24, we obtain

$$\frac{dx}{dt} = 500 - \frac{\ln 2}{8}x \tag{10.92}$$

as the differential equation governing the amount of $H_2S$ in the valley at any time $t$. We can construct a similar differential equation for $SO_2$. The input consists of the stack gases (100 lb/hr) and the $SO_2$ formed by conversion of $H_2S$ ($x \ln 2/8$). The removal is by an exponential decay process having residence time of 48 hours. Hence we obtain the differential equation

$$\frac{dy}{dt} = \left(100 + \frac{\ln 2}{8}x\right) - \frac{\ln 2}{48}y. \tag{10.93}$$

The two equations (10.92) and (10.93) form a *system of differential equations:*

$$\begin{aligned} \frac{dx}{dt} &= 500 - \frac{\ln 2}{8}x, \\ \frac{dy}{dt} &= 100 + \frac{\ln 2}{8}x - \frac{\ln 2}{48}y. \end{aligned} \tag{10.94}$$

We now solve this system under the assumption that clean air existed at time $t = 0$.

In this case the system (10.94) is easy to solve since the first equation is a linear differential equation in only the $x$-variable. Thus we can separate variables to solve Eq. (10.92):

$$\int \frac{dx}{500 - (\ln 2/8)\,x} = \int dt.$$

Integrating this equation we have

$$\frac{-8}{\ln 2}\ln\left(500 - \frac{\ln 2}{8}x\right) = t + c,$$

so that

$$\begin{aligned} 500 - \frac{\ln 2}{8}x &= Ce^{-(\ln 2)t/8} \\ &= C \cdot 2^{-t/8}. \end{aligned}$$

Hence

$$x(t) = \frac{8}{\ln 2}(500 - C \cdot 2^{-t/8}),$$

and we can determine $C$ by setting $t = 0$:

$$0 = x(0) = \frac{8}{\ln 2}(500 - C).$$

Thus $C = 500$ and we obtain

$$x(t) = \frac{4000}{\ln 2}(1 - 2^{-t/8}). \qquad \textbf{(10.95)}$$

We now substitute this equation into Eq. (10.93) to obtain a differential equation involving only the variables $y$ and $t$:

$$\frac{dy}{dt} = 100 + \frac{\ln 2}{8}\frac{4000}{\ln 2}(1 - 2^{-t/8}) - \frac{\ln 2}{48}y,$$

or

$$\frac{dy}{dt} + \frac{\ln 2}{48}y = 600 - 500 \cdot 2^{-t/8}.$$

Here we must multiply both sides by

$$e^{\int (\ln 2/48)\,dt} = e^{(\ln 2)t/48} = 2^{t/48},$$

obtaining

$$(y \cdot 2^{t/48})' = 600 \cdot 2^{t/48} - 500 \cdot 2^{-5t/48}.$$

Integrating both sides we have

$$y \cdot 2^{t/48} = 600\left(\frac{48}{\ln 2}\right)(2^{t/48}) + 500\left(\frac{48}{5\ln 2}\right)(2^{-5t/48}) + c$$

or

$$y(t) = \frac{4800}{\ln 2}(6 + 2^{-t/8}) + c \cdot 2^{-t/48}.$$

Since $y(0) = 0$, we have

$$0 = \frac{4800}{\ln 2}(7) + c,$$

hence

$$y(t) = \frac{4800}{\ln 2}(6 + 2^{-t/8} - 7 \cdot 2^{-t/48}). \tag{10.96}$$

***Example 4*** Let $P(t)$ denote the population of a certain country and $Q(t)$ the amount of air pollution at time $t$. The growth rate of the population is proportional to the population, but the death rate is influenced by the amount of air pollution. Hence, we have

$$\frac{dP}{dt} = kP - jQ. \tag{10.97}$$

Similarly, people cause pollution, so the growth rate of pollution is proportional to the population. However, most air pollutants decay in a variety of ways, and the rate of decay is proportional to the amount of pollutant present. Thus we obtain

$$\frac{dQ}{dt} = hP - iQ. \tag{10.98}$$

Equations (10.97) and (10.98) constitute a system of differential equations. In this case we cannot solve one of them first, since each equation involves both variables. However, all is not lost, as the following procedure will demonstrate.

Differentiating Eq. (10.97), we have

$$P'' = kP' - jQ',$$

and substituting Eq. (10.98) for $Q'$ in this equation yields

$$P'' = kP' - j(hP - iQ)$$

or

$$P'' = kP' - jhP + ijQ. \tag{10.99}$$

But by Eq. (10.97)

$$jQ = kP - P',$$

so we may replace the last term in Eq. (10.99) to obtain

$$P'' = kP' - jhP + i(kP - P')$$

or

$$P'' = (k - i)P' - (jh - ik)P.$$

Thus the system (10.97) and (10.98) yields the differential equation

$$\frac{d^2P}{dt^2} + (i - k)\frac{dP}{dt} + (jh - ik)P = 0.$$

To see how a differential equation of this type may be solved, suppose $i - k = -5$ and $jh - ik = 6$. We then have

$$\frac{d^2P}{dt^2} - 5\frac{dP}{dt} + 6P = 0,$$

or

$$\left[\left(\frac{d}{dt}\right)^2 - 5\frac{d}{dt} + 6\right]P = 0.$$

We have written the equation in this way to point out that we can treat the operation of taking a derivative as if it were a variable. So long as all the coefficients involved are constants, we can factor the expression in brackets, obtaining

$$\left(\frac{d}{dt} - 3\right)\left(\frac{d}{dt} - 2\right)P = 0. \qquad \textbf{(10.100)}$$

Now let

$$R = \left(\frac{d}{dt} - 2\right)P = \frac{dP}{dt} - 2P. \qquad \textbf{(10.101)}$$

Then Eq. (10.100) becomes

$$\left(\frac{d}{dt} - 3\right)R = \frac{dR}{dt} - 3R = 0$$

or

$$\frac{dR}{dt} = 3R.$$

This equation has the general solution (see Example 1 in Section 10.7)

$$R = C_1e^{3t}.$$

Substituting this function for the left side of Eq. (10.101) yields

$$\frac{dP}{dt} - 2P = C_1 e^{3t}.$$

Multiplying both sides by $e^{-2t}$ yields

$$(Pe^{-2t})' = e^{-2t}(P' - 2P) = C_1 e^{3t-2t} = C_1 e^t,$$

so that an integration gives us

$$Pe^{-2t} = C_1 e^t + C_2$$

or

$$P(t) = C_1 e^{3t} + C_2 e^{2t}. \qquad \textbf{(10.102)}$$

Note that the coefficients of the exponentials are the "roots" of the terms in parentheses in Eq. (10.100). This is no accident, but a direct consequence of Eq. (10.100). In general, if

$$\left(\frac{d}{dx} - a\right)\left(\frac{d}{dx} - b\right)y = 0, \quad a \neq b, \qquad \textbf{(10.103)}$$

then the general solution of Eq. (10.103) is

$$y = c_1 e^{ax} + c_2 e^{bx},$$

where $c_1$ and $c_2$ are arbitrary constants.

---

So far we have successfully found the solution of every differential equation we have encountered. This might lead you to the erroneous conclusion that all differential equations are easy to solve. *Nothing could be farther from the truth*! Remember that indefinite integrals are simply another way of writing a differential equation:

$$y' = f(x) \quad \text{is equivalent to} \quad y = \int f(x)\, dx,$$

and many integrals cannot be solved. Similarly, there are many other differential equations for which no method of solution is known. For example,

$$y' = e^{xy}$$

has no known solution. Indeed, the differential equations for which a solution can be found are an extremely small subset of the set of all differential equations. In this book we have tried to shield you from this unhappy fact by carefully selecting examples and exercises that can be solved by the techniques we have illustrated.

## EXERCISES 10.8

*In Exercises 1–8, use the method of Example 1 (see p. 563) to solve the given differential equation. If an initial condition is given, solve the initial value problem.*

**1.** $y' = y + y^2$

**2.** $y' = y - y^2, y(0) = 2$

**3.** $y' = 3y - 4y^2$

**4.** $y' = 3y^2 - 4y, y(1) = 5$

**5.** $P' = 4P^2 - 4P$

**6.** $z' = -3z - 5z^2, z(-1) = 1$

**7.** $y' = xy(1 - y)$

**8.** $y' = \dfrac{y - y^2}{x}, y(1) = 2$

*Use the methods of Examples 3 and 4 (see pp. 567–572) to solve the following systems of differential equations.*

**9.** $$\frac{dx}{dt} = -3x$$ $$\frac{dy}{dt} = 3x - 4$$

**10.** $$\frac{dx}{dt} = 2x - y + 3$$ $$\frac{dy}{dt} = 3y - 4$$

**11.** $$\frac{dx}{dt} = 4 - 2x$$ $$\frac{dy}{dt} = x + y - 5$$

**12.** $$\frac{dx}{dt} = x - 2y$$ $$\frac{dy}{dt} = x + 4y$$

**13.** $$\frac{dx}{dt} = -4x - y$$ $$\frac{dy}{dt} = x - 2y$$

**14.** $$\frac{dx}{dt} = 8x - y$$ $$\frac{dy}{dt} = 4x + 12y$$

*Solve the following initial value problems.*

**15.** $$\frac{dx}{dt} = 3x + 2y, x(0) = 1$$ $$\frac{dy}{dt} = x + 2y, y(0) = 4$$

**16.** $$\frac{dx}{dt} = -4x - y, x(0) = 2$$ $$\frac{dy}{dt} = x - 2y, y(0) = 0$$

**17.** Tank X contains 100 gallons of water in which 100 lb of salt are dissolved and tank Y contains 100 gallons of water. The brine flows from tank X into tank Y at 2 gal/min and is immediately stirred up. Suppose we add water to tank X at the rate of 2 gal/min and remove 2 gal/min of the mixture in tank Y.

a) When will tank Y contain a maximum amount of salt?
b) How much salt is in tanks X and Y at that time?
c) Find the amount of salt in tank Y after 20 minutes.
d) Does tank Y contain more salt than tank X after 30 minutes?

**18.** Suppose in Exercise 17 that only one gal/min of the mixture is removed from tank Y, and that tank Y can hold 1000 gal.
a) When will tank Y contain a maximum amount of salt?
b) How much salt is in tank Y after 20 minutes?
c) Which tank contains the most salt after 50 minutes?

**19.** Tank X contains 100 gal of water in which 100 lb of salt is dissolved and tank Y contains 100 gal of water. Water flows into tank X at a rate of 3 gal/min and the mixture flows from tank X into tank Y at a rate of 4 gal/min. From Y we pump 1 gal/min back into X while 3 gal/min are flushed away. Find the amount of salt in each tank at all times $t$.

**20.** Suppose an island contains only two types of mammals: rabbits and foxes. There is ample food so the birth rate of rabbits is proportional to the rabbit population, while their death rate depends only on the number of foxes. The birth rate for foxes depends on the number of rabbits available, while their death rate is proportional to the number of foxes.
a) Show that this situation yields the system

$$r' = ar - bf,$$
$$f' = cr - df,$$

where $r$ is the number of rabbits and $f$ is the number of foxes at any time $t$.

b) If $r(0) = 10{,}000$, $f(0) = 100$, $a = 4$, $b = 34$, $c = 0.1$, and $d = 0.4$, find a formula for the number of rabbits $r$ and foxes $f$ at any time $t$.

## CHAPTER VOCABULARY

**acceleration** $a(t)$

**antiderivative** (= integral)

**arbitrary constant** $c$

**arc length formula** $\int_a^b \sqrt{1+(f')^2(x)}\,dx$

**"area" function** $A(x)$

**change of variables**

**consumer's surplus**

**definite integral** $\int_a^b f(x)\,dx$

**demand equation**

**differential equation**

**disk formula for volume** $\pi\int_a^b f^2(x)\,dx$

**dosage**

**Fundamental Theorem of Calculus** $\int_a^b f(x)\,dx = F(b) - F(a)$

**general solution of a differential equation**

**growth rate per individual** $p'/p$

**half-life**

**improper integral**

**indefinite integral** $\int f(x)\,dx$

**initial condition**

**initial value problem** $y' = f(x, y), y(a) = b$

**inspiration rate**

**integral diverges**

**integration**

**integration by parts** $\int u\,dv = uv - \int v\,du$

**LATE**

**limits of integration**

**logistic curve** (= sigmoidal curve)

**logistic growth**

**Newton's Law of Cooling**

**Pareto's Law** $y = cx^{-1.5}$

**partial fractions**

**perfect discrimination**

**quadrature formulas**

**reduction formulas**

**residence time (pollution)**

**separation of variables**

**shell formula for volume** $2\pi \int_a^b xf(x)\,dx$

**sigma notation** $\sum_{k=1}^{n} f(x_k)$

**specific infection rate**

**substitution techniques**

**system of differential equations**

**trapezoidal rule** $\int_a^b f(x)\,dx$

$$\approx \frac{\Delta x}{2}\,[f(x_0) + 2f(x_1) + \cdots + 2f(x_{n-1}) + f(x_n)]$$

**velocity** $v(t)$

# APPENDIXES

Appendix 1 **A SUMMARY OF MATHEMATICS OF FINANCE TOPICS**

Appendix 2 **TRIGONOMETRY FORMULAS**

Appendix 3 **TABLE OF DERIVATIVES AND INTEGRALS**

Appendix 4 **COMMON LOGARITHMS**

Appendix 5 **NATURAL LOGARITHMS**

## *APPENDIX* 1 A SUMMARY OF MATHEMATICS OF FINANCE TOPICS

In this appendix we summarize the principal mathematics of finance topics that have been treated in this text. The reference numbers listed with each topic heading indicate other places in the text where these concepts are treated.

For uniformity, we use the following symbols:

$A$ = the amount (dollars),

$P$ = the principal (dollars),

$i$ = the annual interest rate (expressed as a decimal),

$n$ = the number of years involved,

$m$ = the number of payments each year.

A calculator will greatly simplify the arithmetic of these calculations.

---

### A1.1 COMPOUND INTEREST [3.2, 8.1, 9.6]

Suppose the principal $P$ is deposited in an account that one year later increases the funds on deposit by a fixed percentage. If we express this percentage as a decimal, we obtain the annual interest rate $i$. At the end of the first year we will have the original principal $P$ *plus* the percentage $iP$ earned as interest, for a total of

$$P + iP = P(1 + i). \tag{A1.1}$$

Equation (A1.1) illustrates the basic principle: *Whatever is in the account during a given year is increased by i times that amount at the end of that year.* Thus, the amount in the account after two years will be $P(1 + i)^2$, after three years, $P(1 + i)^3$, and at the end of $n$ years, the total amount will be

$$A = P(1 + i)^n. \tag{A1.2}$$

***Example 1*** Suppose \$100 is deposited in a bank that pays 5% interest compounded annually, and is left on deposit for 10 years. Here $P = 100$, $i = 0.05$, and $n = 10$, so the amount in the account at the end of that time is

$$A = 100(1 + 0.05)^{10} = 100(1.05)^{10} = 162.8895,$$

or approximately \$162.89.

Most banks pay interest more frequently than once a year. Many banks pay interest semiannually ($m = 2$), quarterly ($m = 4$), or even daily ($m = 365$). When interest is paid semiannually, only one half the annual interest is paid each time. Similarly, if interest is paid quarterly, only one fourth the annual interest is paid each time. Hence, if interest is paid $m$ times a year, the interest rate that is paid each time is $i/m$. Thus, Eq. (A1.2) must be modified by changing the interest rate to $i/m$ and increasing the number of payments to $nm$, since that many payments will be made in $n$ years. Therefore, we obtain the expression

$$A = P\left(1 + \frac{i}{m}\right)^{nm} \tag{A1.3}$$

as the amount after $n$ years when the principal $P$ is deposited in an account paying an annual interest rate $i$ compounded $m$ times per year.

***Example 2*** Suppose \$100 is deposited in a bank that pays 5% annual interest compounded quarterly, and is left on deposit for 10 years. Here $P = 100$, $i = 0.05$, $m = 4$, and $n = 10$, hence

$$\begin{aligned} A &= 100\left(1 + \frac{0.05}{4}\right)^{10(4)} = 100(1.0125)^{40} \\ &= \$164.3619, \end{aligned}$$

or approximately \$164.36. Notice that this is more than the amount obtained if the interest is compounded annually. In general:

> The amount in the account will increase faster if the interest payments are made more often.

---

Equation (A1.3) can also be used to calculate the amount in the account when the funds are left on deposit a fractional part of a year. Simply replace the exponent $nm$ with the number $q$ of times interest is accrued to the account:

$$A = P\left(1 + \frac{i}{m}\right)^{q}. \tag{A1.4}$$

Some banks compound their interest infinitely often. To calculate the amount in such an account after $n$ years (or even decimal parts of a year) we rewrite Eq. (A1.3) as

$$A = P\left(1 + \frac{i}{m}\right)^{(m/i)(in)}, \tag{A1.5}$$

and take the limit of Eq. (A1.5) as $m \to \infty$:

$$A = P\left[\lim_{m \to \infty} \left(1 + \frac{i}{m}\right)^{m/i}\right]^{in}. \tag{A1.6}$$

Substituting $k = m/i$, we can use Eq. (9.54) to replace the limit in Eq. (A1.6) by $e(= 2.718281828\ldots)$, obtaining

$$A = Pe^{in}. \tag{A1.7}$$

***Example 3*** Find the amount in the account after 10 years if \$100 is deposited at 5% annual interest compounded continuously. Here $P = 100$, $i = 0.05$, and $n = 10$, implying by Eq. (A1.7) that

$$A = 100e^{(0.05)10} = 100e^{0.5} = 164.8721,$$

or approximately \$164.87.

---

When discussing compound interest, we can ask how much we need to deposit *now*, at an annual compound interest rate $i$, so that in $n$ years we have an amount $A$ in that account. Since

$$A = P(1 + i)^n$$

from Eq. (A1.1), we need to solve for $P$ in terms of $A$, $i$, and $n$. Dividing both sides by $(1 + i)^n$, we obtain

$$P = A(1 + i)^{-n}. \tag{A1.8}$$

The principal we need to deposit now in order to have $A$ (dollars) in the account in $n$ years is called the *present value*. Equation (A1.8) is frequently used to *discount* promissory notes. Viewed from this perspective, we can think of $P$ as the *present value* and $A$ as the *future value* of a note or account.

***Example 4*** A merchant sells a three-year \$5000 promissory note to an investor. By doing so, the merchant is obliged to pay \$5000 when the note matures in three years. If the investor wishes to earn 9% annual compound interest, how much should he or she pay the merchant for the note?

Here $A = 5000$, $i = 0.09$, and $n = 3$ so that by Eq. (A1.8),

$$P = 5000(1 + 0.09)^{-3} = 5000(1.09)^{-3} = 3860.9174,$$

that is, the present value of the note is \$3860.92. This is the amount the investor should pay.

---

## EXERCISES A1.1

*In Exercises 1–12, calculate the amount A in each account when the principal P is deposited at an annual compound interest rate of i for n years.*

1. $P = \$1000$, $i = 0.04$ (4%), $n = 5$, compounded quarterly
2. $P = \$1000$, $i = 0.06$ (6%), $n = 4$, compounded daily
3. $P = \$1200$, $i = 0.08$ (8%), $n = 6$, compounded semiannually
4. $P = \$1200$, $i = 0.06$ (6%), $n = 7$, compounded daily
5. $P = \$1250$, $i = 0.05$ (5%), $n = 11$, compounded quarterly
6. $P = \$1250$, $i = 0.055$ ($5\frac{1}{2}$%), $n = 10\frac{1}{2}$, compounded monthly
7. $P = \$2500$, $i = 9\%$, $n = 5\frac{1}{4}$, compounded quarterly
8. $P = \$5200$, $i = 5\frac{1}{4}\%$, $n = 6$, compounded continuously
9. $P = \$18{,}000$, $i = 6\frac{1}{2}\%$, $n = 4$, compounded daily
10. $P = \$18{,}000$, $i = 6\frac{1}{2}\%$, $n = 4$, compounded continuously
11. $P = \$212{,}500$, $i = 7\frac{1}{2}\%$, $n = 3\frac{1}{3}$, compounded quarterly
12. $P = \$212{,}500$, $i = 7\frac{1}{4}\%$, $n = 3\frac{1}{3}$, compounded continuously
13. If the investor in Example 4 wished to earn 12% annual compound interest, how much should he or she pay the merchant for a three-year \$5000 promissory note?
14. An industrialist offers to sell a five-year \$100,000 promissory note for \$74,000. What annual interest rate would an investor obtain for his or her money?
15. Answer Exercise 14 if the length of time involved is only four years.
16. Find the present value of a ten-year \$100,000 promissory note at $8\frac{3}{4}\%$ annual compound interest.

---

## A1.2 ANNUITIES [8.1]

Annuities initially served as a way of providing for an individual's needs after retirement. Many present pension plans, including IRA accounts, are annuity plans, but annuities can also be used for many other purposes. An *ordinary annuity* operates in the following fashion: *A series of equal payments is made at equal time intervals coinciding with the period required for compounding the interest.* For example, in an IRA account an individual will typically deposit \$1500 annually, no later than 45 days after the end of the tax year, in an account paying an annual interest rate of 6%. How much will the account be worth in 20 years?

To see how we arrive at the solution to this problem, imagine that each deposit had been in a separate 6% annual interest account. If we add up the total in all these accounts, we obtain the sum

$$1500 + 1500(1.06) + 1500(1.06)^2 + \cdots + 1500(1.06)^{19},$$

where the first 1500 is the amount just deposited, 1500(1.06) is the amount from the previous year's deposit, and so forth. The exponent for the last term in this sum (which corresponds to the initial \$1500 deposit when the plan began) is 19, since it takes one year before the first compounding. Generally, if $P$ dollars are deposited annually in an account paying $i$% annual interest for $n$ years, the amount accrued is

$$A = P + P(1 + i) + P(1 + i)^2 + \cdots + P(1 + i)^{n-1}. \tag{A1.9}$$

Instead of using Eq. (A1.9), it is useful to obtain a more compact expression for the amount $A$. To do so, we multiply each term on both sides of Eq. (A1.9) by $(1 + i)$, obtaining

$$(1 + i)A = P(1 + i) + P(1 + i)^2 + P(1 + i)^3 + \cdots + P(1 + i)^n. \tag{A1.10}$$

Subtracting Eq. (A1.9) from Eq. (A1.10), we have

$$(1 + i)A - A = P(1 + i)^n - P, \tag{A1.11}$$

since all the terms $P(1 + i)^k$ with $1 \le k \le n - 1$ appear in both equations but with opposite signs. Since $(1 + i)A = A + iA$, Eq. (A1.11) yields

$$iA = P[(1 + i)^n - 1],$$

or

$$A = P\left[\frac{(1 + i)^n - 1}{i}\right]. \tag{A1.12}$$

We can use Eq. (A1.12) to compute the future value of the IRA account that we originally described. Here $P = 1500$, $i = 0.06$, and $n = 20$, so that

$$A = 1500\left[\frac{(1.06)^{20} - 1}{0.06}\right] = 1500\left[\frac{3.2071 - 1}{0.06}\right] = \$55{,}178.39.$$

If the deposits and compounding of interest occur $m$ times a year, Eq. (A1.12) becomes

$$A = \frac{P}{i}\left[\left(1 + \frac{i}{m}\right)^{nm} - 1\right], \tag{A1.13}$$

where $P$ is the total annual payment and $i$ is the annual interest rate.

### EXERCISES A1.2

*In Exercises 1–8, determine the future value of an annuity in which P dollars are deposited annually at i% annual interest for n years.*

1. $P = 50, \quad i = 0.05, \; n = 10$
2. $P = 500, \; i = 0.05, \; n = 20$
3. $P = 600, \quad i = 0.05, \; n = 30$
4. $P = 1000, \; i = 0.05, \; n = 30$
5. $P = 1500, \; i = 5\frac{1}{4}\%, \; n = 30$
6. $P = 1500, \; i = 5\frac{3}{4}\%, \; n = 30$
7. $P = 1500, \; i = 6\%, \quad n = 30$
8. $P = 1500, \; i = 6\frac{1}{2}\%, \; n = 30$
9. Derive an expression for an annuity in which monthly payments are made and the amount in the account is compounded each month.

★10. a) Suppose a total annual payment of $P$ is made on a daily basis into an annuity paying daily interest. Derive an expression for the amount in the account after $n$ years.
b) Use the equation you derived in part (a) to obtain the value of an annuity having a continuous cash flow and continuous compounding of the interest.

11. Justify Eq. (A1.13). (*Hint*: Each deposit is $P/m$.)

---

## A1.3 AMORTIZATION [8.2]

Suppose you obtain a five-year automobile loan for \$4000 at 12% annual interest from a credit union. The credit union arranges for you to make equal monthly payments $p$, which will repay the principal and interest over five years. This is the usual method for repaying loans and mortgages. How is the amount $p$ of the monthly payment determined?

To understand how the amount $p$ is determined, we must notice that the monthly payment is essentially an annuity account that in five years will equal the amount obtained by depositing $P = \$4000$ at $i = 12\%$ annual interest compounded monthly. Setting $n = 5$ and $m = 12$ and equating the expressions in (A1.3) and (A1.13), we have

$$\underbrace{\frac{mp}{i}\left[\left(1+\frac{i}{m}\right)^{nm}-1\right]}_{\text{Annuity}} = \underbrace{P\left(1+\frac{i}{m}\right)^{nm}}_{\text{Compound interest}}, \tag{A1.14}$$

where $m$ payments of $p$ dollars are made each year for a total annual payment of $mp$. Solving for $p$, we obtain by cross multiplication

$$p = \frac{P(i/m)[1+(i/m)]^{nm}}{[1+(i/m)]^{nm}-1},$$

and dividing numerator and denominator by $[1 + (i/m)]^{nm}$, we have

$$p = \frac{P(i/m)}{1-\left(1+\dfrac{i}{m}\right)^{-nm}}. \tag{A1.15}$$

Thus, in our example,

$$p = \frac{4000(0.12/12)}{1 - \left(1 + \dfrac{0.12}{12}\right)^{-12(5)}} = \frac{4000(0.01)}{1 - (1.01)^{-60}} = \frac{40}{1 - 0.5504} = \$88.98.$$

***Example 1*** A family borrows \$50,000 for 30 years at 9% annual interest compounded monthly to finance their home. How much will their monthly mortgage payment be, not including their payments for property taxes and insurance?

Here the principal of the loan $P = \$50{,}000$, $i = 0.09$, $n = 30$, and $m = 12$. Hence the monthly repayment of principal and interest is

$$p = \frac{50{,}000(0.09/12)}{1 - \left(1 + \dfrac{0.09}{12}\right)^{-30(12)}} = \frac{50{,}000(0.0075)}{1 - (1.0075)^{-360}}$$

$$= \frac{375}{1 - (0.0679)} = \$402.31.$$

The same equation may be used for other purposes.

***Example 2*** An individual invested \$1500 annually in an IRA account at 6% annual interest for 20 years. As was shown following Eq. (A1.12), the amount in the account at retirement is \$55,178.39. Instead of taking a lump-sum payment (and having to pay income tax on the entire amount if the deposits were tax-sheltered), the individual elects to receive a guaranteed annual payment for 15 years. Assuming the bank agrees to continue the 6% annual interest rate, how much will the annual payment be?

In this problem the principal $P = \$55{,}178.39$, $i = 0.06$, $n = 15$, and $m = 1$. Hence, by Eq. (A1.15) the annual payment is

$$p = \frac{(55{,}178.39)(0.06)}{1 - (1 + 0.06)^{-15}} = \$5681.32.$$

---

**EXERCISES A1.3**

*A family borrows P dollars at i% annual interest compounded monthly for n years. Find the amount of the monthly payment p.*

**1.** $P = 50{,}000$, $i = 9\%$, $n = 20$ **2.** $P = 50{,}000$, $i = 9\%$, $n = 25$

**3.** $P = 50{,}000$, $i = 8\frac{3}{4}\%$, $n = 25$ **4.** $P = 50{,}000$, $i = 8\frac{1}{2}\%$, $n = 30$

**5.** $P = 50{,}000,\ i = 8\frac{1}{2}\%,\ n = 25$

**6.** $P = 80{,}000,\ i = 9\%,\ n = 30$

**7.** $P = 80{,}000,\ i = 8\frac{1}{2}\%,\ n = 30$

**8.** $P = 80{,}000,\ i = 8\frac{1}{4}\%,\ n = 30$

**9.** A retiree deposits his life savings of $60,000 in an account that pays 5% annual interest compounded quarterly. How much can he withdraw from his account each quarter if he wishes the money to last him for 20 years?

**10.** Answer Exercise 9 if he decides to spend it all in 10 years.

---

## A1.4 EQUITY AND OTHER TOPICS [8.2]

The procedure we used to calculate the monthly payment in an amortization problem can also be used to determine the equity a borrower has in a certain property. By *equity* we mean that part of the principal that the borrower has repaid.

Assume that you obtained a five-year $4000 automobile loan at 12% annual interest compounded monthly and have made 20 payments. What is your remaining debt on the automobile? This debt will be the *difference* between the amount the principal would now be worth after compounding for 20 months, and the value of the "annuity" obtained by making 20 monthly payments of $88.98, as we computed following Eq. (A1.15):

$$R = 4000\left(1 + \frac{0.12}{12}\right)^{20} - \frac{12(88.98)}{0.12}\left[\left(1 + \frac{0.12}{12}\right)^{20} - 1\right] \qquad \textbf{(A1.16)}$$

$$= 4880.76 - 1959.25 = \$2921.51.$$

Thus, the equity you have in the car is 4000 − 2921.51 = $1078.49. Of course, depreciation may further reduce the value of your investment.

In general, if $q$ monthly payments of $p$ (dollars) have been made repaying an $n$-year $P$ (dollar) loan at $i\%$ annual interest compounded monthly, the remaining debt is

$$R = P\left(1 + \frac{i}{12}\right)^{q} - \frac{12p}{i}\left[\left(1 + \frac{i}{12}\right)^{q} - 1\right]. \qquad \textbf{(A1.17)}$$

***Example 1*** Suppose the family in Example 1 in Section A1.3 moves after living in their house for 76 months. How much do they still owe on their original $50,000 loan?

Here $P = 50{,}000$, $i = 0.09$, $q = 76$, and $p = 402.31$ (see Example 1 in Section A1.3), so that by Eq. (A1.17)

$$R = 50{,}000\left(1 + \frac{0.09}{12}\right)^{76} - \frac{12(402.31)}{0.09}\left[\left(1 + \frac{0.09}{12}\right)^{76} - 1\right]$$

$$= 88{,}225.50 - 41{,}009.43 = \$47{,}216.16.$$

Thus, although they have made payments for six years, they still owe over 94% of the principal.

---

It is interesting to see how much a homeowner pays over the life of a mortgage. For instance, over the 30 years of the mortgage in Example 1 in Section A1.3, the family would pay

$$402.31 \times 30 \times 12 = \$144{,}831.60,$$

or almost 290% of the original loan. What is the effect of paying off this loan faster? The following example studies this situation.

***Example 2*** Suppose the family in Example 1 in Section A1.3 decides to investigate the effect on the monthly payment of taking their \$50,000 loan at 9% annual interest (compounded monthly) for 20 or 25 years rather than 30. When $n = 20$, we find that the monthly payment is

$$p = \frac{50{,}000(0.09/12)}{1 - \left(1 + \dfrac{0.09}{12}\right)^{-20(12)}} = \$449.86,$$

while $n = 25$ changes the result to

$$p = \frac{50{,}000(0.09/12)}{1 - \left(1 + \dfrac{0.09}{12}\right)^{-25(12)}} = \$419.60.$$

Thus, \$17.29 more a month will reduce the length of time by five years, while \$47.55 more a month will reduce it by 10 years! Over the life of the loan, the family will pay

$$449.86 \times 20 \times 12 = \$107{,}966.40,$$

or 216% of the principal by taking 20 years to repay the loan. Thus, if the extra \$47.55 per month are available, a savings of \$36,865.20 can be made over the life of the two loans.

Our final application is somewhat more complicated.

***Example 3*** The family in Example 1 in Section A1.3 receives a large Christmas bonus one year and decides to use \$1000 of this bonus to reduce the amount they owe on their mortgage. Suppose they have already made 76 monthly payments so that their present remaining debt is \$47,216.16 (see Example 1). The \$1000 payment reduces their debt to \$46,216.16. How many more monthly payments of \$402.31 need they make?

This is basically the same problem as an amortization, except that the *number* of payments is required rather than the monthly payment. Using Eq. (A1.14) with $P = 46{,}216.16$, $i = 0.09$, $n = 12$, $p = 402.31$, and $nm = q$ *unknown*, we have

$$\frac{12(402.31)}{0.09}\left[\left(1 + \frac{0.09}{12}\right)^q - 1\right] = 46{,}216.16\left(1 + \frac{0.09}{12}\right)^q$$

or

$$(53{,}641.33 - 46{,}216.16)(1.0075)^q = 53{,}641.33,$$

Carrying the terms involving $(1.0075)^q$ to the left side, we have

$$(53{,}641.33 - 46{,}216.16)(1.0075)^q = 53{,}641.33,$$

or

$$(1.0075)^q = \frac{53{,}641.33}{7425.17} = 7.2243.$$

Taking common logarithms of both sides, we obtain

$$q \log 1.0075 = \log 7.2243,$$

or

$$q = \frac{\log 7.2243}{\log 1.0075} = \frac{0.8588}{0.0032} = 264.65,$$

or approximately 265 more payments. Adding the 76 payments already made, we see that about 341 payments would be made over the life of the loan. Hence the \$1000 payment at this time saves you from making 19 payments of \$402.31 later, giving a saving of

$$402.31 \times 19 - 1000 = 7643.89 - 1000 = \$6643.89.$$

---

A general formula for calculating the number of remaining payments is easily derived from Eq. (A1.14) by setting $q = nm$ and solving as we did above:

$$\frac{mp}{i}\left[\left(1 + \frac{i}{m}\right)^q - 1\right] = P\left(1 + \frac{i}{m}\right)^q$$

yields

$$\left(\frac{mp}{i} - P\right)\left(1 + \frac{i}{m}\right)^q = \frac{mp}{i},$$

or

$$\left(1 + \frac{i}{m}\right)^q = \frac{mp}{mp - Pi} = \frac{1}{1 - (Pi/mp)}.$$

Hence

$$q = \frac{-\log[1 - (Pi/mp)]}{\log[1 + (i/m)]} \tag{A1.18}$$

is the number of additional payments required to repay a debt of $P$ (dollars) at an annual interest rate $i$ compounded $m$ times a year by making $m$ payments of $p$ (dollars) per year. If $m = 12$ (the usual situation) Eq. (A1.18) reduces to

$$q = \frac{-\log[1 - (Pi/12p)]}{\log[1 + (i/12)]}. \tag{A1.19}$$

## EXERCISES A1.4

*In Exercises 1–8, find the remaining debt after q months of making payments on an n-year P-dollar loan at i% annual interest compounded monthly.*

1. $P = \$6000$, $i = 12\%$, $n = 4$, $q = 30$
2. $P = \$6000$, $i = 12\%$, $n = 5$, $q = 25$
3. $P = \$2000$, $i = 12\%$, $n = 2$, $q = 18$
4. $P = \$2000$, $i = 12\%$, $n = 3$, $q = 18$
5. $P = \$50{,}000$, $i = 9\%$, $n = 20$, $q = 100$
6. $P = \$50{,}000$, $i = 9\%$, $n = 30$, $q = 100$
7. $P = \$50{,}000$, $i = 8\frac{1}{2}\%$, $n = 20$, $q = 100$
8. $P = \$50{,}000$, $i = 8\frac{1}{2}\%$, $n = 30$, $q = 100$
9. How much more must one pay monthly to repay a 20-year \$40,000 mortgage at $8\frac{1}{2}\%$ annual interest (compounded monthly) than an identical 30-year mortgage?
10. Answer Exercise 9 if the annual interest rate is $8\frac{3}{4}\%$.
11. Answer Exercise 9 if the annual interest rate is $8\frac{1}{4}\%$.
12. A family has borrowed \$40,000 at $8\frac{1}{2}\%$ annual interest (compounded monthly) for 30 years. After 28 payments, they win \$2500 in a quiz show and decide to use all the money to reduce their debt. If they continue making their usual monthly payments, approximately how many fewer payments will they have to make? How much do they save by paying the \$2500 now?

**13.** Answer Exercise 12 assuming that the family uses only $2000 of their winnings to prepay their loan.

**14.** Answer Exercise 12 assuming that original principal was $50,000.

# TRIGONOMETRY FORMULAS

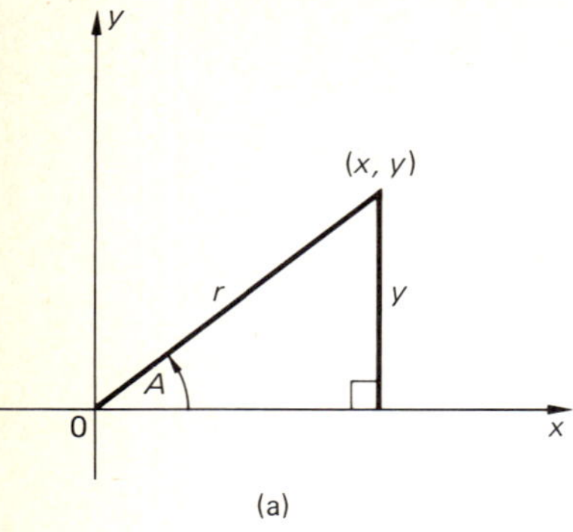


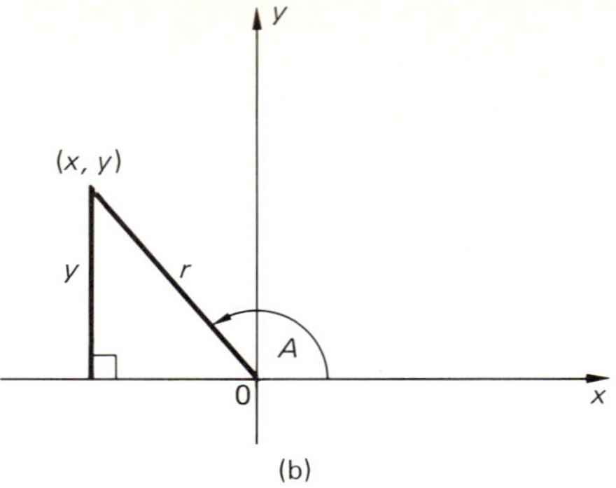


**Fig. A2.1**
a) Acute angle; (b) arbitrary angle.

**A2.1 PYTHAGOREAN THEOREM**

$r^2 = x^2 + y^2$

**A2.2 DEFINITION OF TRIGONOMETRIC FUNCTIONS**

sine $A$ = $\sin A = y/r$    cosine $A$ = $\cos A = x/r$

tangent $A$ = $\tan A = y/x$    cotangent $A$ = $\cot A = x/y$

secant $A$ = $\sec A = r/x$    cosecant $A$ = $\csc A = r/y$

**A2.3 TRIGONOMETRIC FUNCTIONS OF SOME SPECIAL ANGLES**

| Angle | sin | cos | tan | cot | sec | csc |
|---|---|---|---|---|---|---|
| 30° | $1/2$ | $\sqrt{3}/2$ | $1/\sqrt{3}$ | $\sqrt{3}$ | $2/\sqrt{3}$ | 2 |
| 45° | $1/\sqrt{2}$ | $1/\sqrt{2}$ | 1 | 1 | $\sqrt{2}$ | $\sqrt{2}$ |
| 60° | $\sqrt{3}/2$ | $1/2$ | $\sqrt{3}$ | $1/\sqrt{3}$ | 2 | $2/\sqrt{3}$ |
| 90° | 1 | 0 | $\infty$ | 0 | $\infty$ | 1 |
| 180° | 0 | −1 | 0 | $\infty$ | −1 | $\infty$ |
| 270° | −1 | 0 | $\infty$ | 0 | $\infty$ | −1 |

**A2.4 CONVERSION OF RADIAN MEASURE TO DEGREES AND VICE VERSA**

$180° = \pi$ radians,

$1° = \dfrac{\pi}{180}$ radians,

$1 \text{ radian} = \dfrac{180}{\pi}$ degrees

**A2.5 REDUCTION FORMULAS**

$\sin(-A) = -\sin A, \quad \cos(-A) = \cos A$

$\sin(90° - A) = \sin(90° + A) = \cos A$

$\cos(90° - A) = -\cos(90° + A) = \sin A$

$\sin(180° - A) = -\sin(180° + A) = \sin A$

$\cos(180° - A) = \cos(180° + A) = -\cos A$

$\sin(270° - A) = \sin(270° + A) = -\cos A$

$\cos(270° - A) = \sin(270° + A) = -\sin A$

$\sin(A + 360°) = \sin A, \quad \cos(A + 360°) = \cos A$

**A2.6 TRIGONOMETRIC IDENTITIES**

**Reciprocal relations**

1. $\sin A = 1/\csc A$
2. $\cos A = 1/\sec A$
3. $\tan A = 1/\cot A$

**Quotient relations**

4. $\tan A = \sin A/\cos A$
5. $\cot A = \cos A/\sin A$

**Pythagorean relations**

6. $\sin^2 A + \cos^2 A = 1$
7. $1 + \tan^2 A = \sec^2 A$
8. $1 + \cot^2 A = \csc^2 A$

**Sum and difference of angles formulas**

9. $\sin(A + B) = \sin A \cos B + \cos A \sin B$
10. $\sin(A - B) = \sin A \cos B - \cos A \sin B$
11. $\cos(A + B) = \cos A \cos B - \sin A \sin B$
12. $\cos(A - B) = \cos A \cos B + \sin A \sin B$
13. $\tan(A + B) = \dfrac{\tan A + \tan B}{1 - \tan A \tan B}$
14. $\tan(A - B) = \dfrac{\tan A - \tan B}{1 + \tan A \tan B}$

**Double-angle relations**

15. $\sin 2A = 2 \sin A \cos A$
16. $\cos 2A = \cos^2 A - \sin^2 A = 2\cos^2 A - 1 = 1 - 2\sin^2 A$
17. $\tan 2A = \dfrac{2 \tan A}{1 - \tan^2 A}$

**Half-angle relations**

18. $\sin \dfrac{A}{2} = \pm\sqrt{\dfrac{1 - \cos A}{2}}$

**19.** $\cos \frac{A}{2} = \pm\sqrt{\frac{1 + \cos A}{2}}$

**20.** $\tan \frac{A}{2} = \pm\sqrt{\frac{1 - \cos A}{1 + \cos A}} = \frac{1 - \cos A}{\sin A} = \frac{\sin A}{1 + \cos A}$

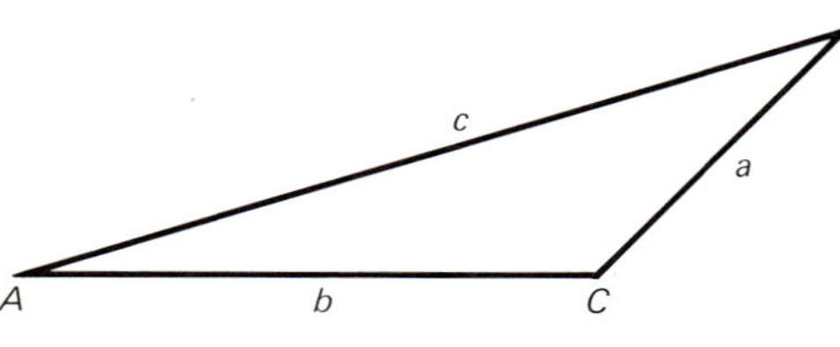


**Fig. A2.2** Oblique triangle.

**A2.7 OBLIQUE TRIANGLE FORMULAS**

Law of Sines: $\frac{\sin A}{a} = \frac{\sin B}{b} = \frac{\sin C}{c}$

Law of Cosines:

$$a^2 = b^2 + c^2 - 2bc \cos A$$
$$b^2 = a^2 + c^2 - 2ac \cos B$$
$$c^2 = a^2 + b^2 - 2ab \cos C$$

**DIFFERENTIATION FORMULAS**

Let $u$ and $v$ be functions of $x$ and let $a$ be a constant.

1. $\dfrac{d}{dx}(au) = a\dfrac{du}{dx}$

2. $\dfrac{d}{dx}(u^n) = nu^{n-1}\dfrac{du}{dx}$

3. $\dfrac{d}{dx}(u \pm v) = \dfrac{du}{dx} \pm \dfrac{dv}{dx}$

4. $\dfrac{d}{dx}(uv) = u\dfrac{dv}{dx} + v\dfrac{du}{dx}$

5. $\dfrac{d}{dx}\left(\dfrac{u}{v}\right) = \dfrac{v\dfrac{du}{dx} - u\dfrac{dv}{dx}}{v^2}$

6. $\dfrac{d}{dx}[u(v)] = \dfrac{du}{dv}\cdot\dfrac{dv}{dx}$

7. $\dfrac{d}{dx}(e^u) = e^u\dfrac{du}{dx}$

8. $\dfrac{d}{dx}(a^u) = a^u \ln a\dfrac{du}{dx}$

9. $\dfrac{d}{dx}(\ln u) = \dfrac{1}{u}\dfrac{du}{dx}$

10. $\dfrac{d}{dx}(\log_a u) = \dfrac{1}{u \ln a}\cdot\dfrac{du}{dx}$

11. $\dfrac{d}{dx}(\sin u) = \cos u\dfrac{du}{dx}$

12. $\dfrac{d}{dx}(\cos u) = -\sin u\dfrac{du}{dx}$

13. $\dfrac{d}{dx}(\tan u) = \sec^2 u\dfrac{du}{dx}$

14. $\dfrac{d}{dx}(\cot u) = -\csc^2 u\dfrac{du}{dx}$

15. $\dfrac{d}{dx}(\sec u) = \sec u \tan u\dfrac{du}{dx}$

16. $\dfrac{d}{dx}(\csc u) = -\csc u \cot u\dfrac{du}{dx}$

**INTEGRAL TABLES**

1. $\displaystyle\int u\,dv = uv - \int v\,du$

2. $\displaystyle\int x^n\,dx = \frac{x^{n+1}}{n+1} + c, \quad n \neq -1$

3. $\displaystyle\int \frac{dx}{x} = \ln|x| + c$

4. $\displaystyle\int e^{ax}\,dx = \frac{e^{ax}}{a} + c$

5. $\displaystyle\int \ln x\,dx = x\ln x - x + c$

6. $\displaystyle\int \frac{dx}{a^2 + x^2} = \frac{1}{a}\tan^{-1}\left(\frac{x}{a}\right) + c$

**7.** $\displaystyle\int \frac{dx}{a^2 - x^2} = \frac{1}{2a} \ln \left| \frac{a + x}{a - x} \right| + c$

**8.** $\displaystyle\int \frac{dx}{x^2 - a^2} = \frac{1}{2a} \ln \left| \frac{x - a}{x + a} \right| + c$

**9.** $\displaystyle\int \frac{dx}{\sqrt{x^2 \pm a^2}} = \ln \left| x + \sqrt{x^2 \pm a^2} \right| + c$

**10.** $\displaystyle\int \frac{dx}{\sqrt{a^2 - x^2}} = \sin^{-1} \left( \frac{x}{a} \right) + c$

**11.** $\displaystyle\int \frac{x\,dx}{\sqrt{x^2 \pm a^2}} = \sqrt{x^2 \pm a^2} + c$

**12.** $\displaystyle\int \frac{x\,dx}{\sqrt{a^2 - x^2}} = -\sqrt{a^2 - x^2} + c$

**13.** $\displaystyle\int \sqrt{x^2 \pm a^2}\,dx = \frac{x}{2} \sqrt{x^2 \pm a^2} \pm \frac{a^2}{2} \ln \left| x + \sqrt{x^2 \pm a^2} \right| + c$

**14.** $$\begin{aligned}\int \sqrt{a^2 - x^2}\,dx &= \frac{x}{2} \sqrt{a^2 - x^2} + \frac{a^2}{2} \sin^{-1} \left( \frac{x}{a} \right) + c \\ &= \frac{x}{2} \sqrt{a^2 - x^2} - \frac{a^2}{2} \cos^{-1} \left( \frac{x}{a} \right) + c\end{aligned}$$

**15.** $\displaystyle\int \frac{\sqrt{x^2 - a^2}}{x}\,dx = \sqrt{x^2 - a^2} - |a| \sec^{-1} \left( \frac{x}{a} \right) + c$

**16.** $\displaystyle\int \frac{\sqrt{a^2 \pm x^2}}{x}\,dx = \sqrt{a^2 \pm x^2} - a \ln \left| \frac{a + \sqrt{a^2 \pm x^2}}{x} \right| + c$

**17.** $\displaystyle\int \sin x\,dx = -\cos x + c$

**18.** $\displaystyle\int \sin^n x\,dx = -\frac{1}{n} \sin^{n-1} x \cos x + \frac{n - 1}{n} \int \sin^{n-2} x\,dx$

**19.** $\displaystyle\int \sin mx \sin nx\,dx = \frac{\sin (m - n)x}{2(m - n)} - \frac{\sin (m + n)x}{2(m + n)} + c, \quad m^2 \neq n^2$

**20.** $\displaystyle\int x^n \sin x\,dx = -x^n \cos x + n \int x^{n-1} \cos x\,dx$

**21.** $\displaystyle\int \cos x\,dx = \sin x + c$

**22.** $\int \cos^n x\,dx = \frac{1}{n}\cos^{n-1} x \sin x + \frac{n-1}{n}\int \cos^{n-2} x\,dx$

**23.** $\int \cos mx \cos nx\,dx = \frac{\sin(m-n)x}{2(m-n)} + \frac{\sin(m+n)x}{2(m+n)} + c, \quad m^2 \neq n^2$

**24.** $\int x^n \cos x\,dx = x^n \sin x - n\int x^{n-1} \sin x\,dx$

**25.** $\int \sin mx \cos nx\,dx = \frac{\cos(n-m)x}{2(n-m)} - \frac{\cos(n+m)x}{2(n+m)} + c, \quad m^2 \neq n^2$

**26.** $\int \sin x \cos x\,dx = \frac{1}{2}\sin^2 x + c$

**27.** $\int \tan x\,dx = -\ln|\cos x| + c$

**28.** $\int \tan^n x\,dx = \frac{\tan^{n-1} x}{n-1} - \int \tan^{n-2} x\,dx, \quad n \neq 1$

**29.** $\int \cot x\,dx = \ln|\sin x| + c$

**30.** $\int \cot^n x\,dx = -\frac{\cot^{n-1} x}{n-1} - \int \cot^{n-2} x\,dx, \quad n \neq 1$

**31.** $\int \sec x\,dx = \ln|\sec x + \tan x| + c$

**32.** $\int \sec^2 x\,dx = \tan x + c$

**33.** $\int \sec^n x\,dx = \frac{1}{n-1}\tan x \sec^{n-2} x + \frac{n-2}{n-1}\int \sec^{n-2} x\,dx$

**34.** $\int \csc x\,dx = \ln|\csc x - \cot x| + c$

**35.** $\int \csc^2 x\,dx = -\cot x + c$

**36.** $\int \csc^n x\,dx = -\frac{1}{n-1}\cot x \csc^{n-2} x + \frac{n-2}{n-1}\int \csc^{n-2} x\,dx$

**37.** $\int x^n \ln(ax)\,dx = \frac{x^{n+1}}{n+1}\left[\ln(ax) - \frac{1}{n+1}\right] + c, \quad n \neq -1$

**38.** $\int (\ln x)^n \, dx = x(\ln x)^n - n \int (\ln x)^{n-1} \, dx, \quad n \neq -1$

**39.** $\int \frac{(\ln x)^n}{x} \, dx = \frac{(\ln x)^{n+1}}{n+1} + c$

**40.** $\int \frac{dx}{x \ln x} = \ln (\ln x) + c$

**41.** $\int x^n e^{ax} \, dx = \frac{x^n e^{ax}}{a} - \frac{n}{a} \int x^{n-1} e^{ax} \, dx$

**42.** $\int e^{ax} \sin bx \, dx = \frac{e^{ax}}{a^2 + b^2} (a \sin bx - b \cos bx) + c$

**43.** $\int e^{ax} \cos bx \, dx = \frac{e^{ax}}{a^2 + b^2} (a \cos bx + b \sin bx) + c$

## Table of Common Logarithms

| *N* | 0 | 1 | 2 | 3 | 4 | 5 | 6 | 7 | 8 | 9 |
|---|---|---|---|---|---|---|---|---|---|---|
| 10 | .0000 | .0043 | .0086 | .0128 | .0170 | .0212 | .0253 | .0294 | .0334 | .0374 |
| 11 | .0414 | .0453 | .0492 | .0531 | .0569 | .0607 | .0645 | .0682 | .0719 | .0755 |
| 12 | .0792 | .0828 | .0864 | .0899 | .0934 | .0969 | .1004 | .1038 | .1072 | .1106 |
| 13 | .1139 | .1173 | .1206 | .1239 | .1271 | .1303 | .1335 | .1367 | .1399 | .1430 |
| 14 | .1461 | .1492 | .1523 | .1553 | .1584 | .1614 | .1644 | .1673 | .1703 | .1732 |
| 15 | .1761 | .1790 | .1818 | .1847 | .1875 | .1903 | .1931 | .1959 | .1987 | .2014 |
| 16 | .2041 | .2068 | .2095 | .2122 | .2148 | .2175 | .2201 | .2227 | .2253 | .2279 |
| 17 | .2304 | .2330 | .2355 | .2380 | .2405 | .2430 | .2455 | .2480 | .2504 | .2529 |
| 18 | .2553 | .2577 | .2601 | .2625 | .2648 | .2672 | .2695 | .2718 | .2742 | .2765 |
| 19 | .2788 | .2810 | .2833 | .2856 | .2878 | .2900 | .2923 | .2945 | .2967 | .2989 |
| 20 | .3010 | .3032 | .3054 | .3075 | .3096 | .3118 | .3139 | .3160 | .3181 | .3201 |
| 21 | .3222 | .3243 | .3263 | .3284 | .3304 | .3324 | .3345 | .3365 | .3385 | .3404 |
| 22 | .3424 | .3444 | .3464 | .3483 | .3502 | .3522 | .3541 | .3560 | .3579 | .3598 |
| 23 | .3617 | .3636 | .3655 | .3674 | .3692 | .3711 | .3729 | .3747 | .3766 | .3784 |
| 24 | .3802 | .3820 | .3838 | .3856 | .3874 | .3892 | .3909 | .3927 | .3945 | .3962 |
| 25 | .3979 | .3997 | .4014 | .4031 | .4048 | .4065 | .4082 | .4099 | .4116 | .4133 |
| 26 | .4150 | .4166 | .4183 | .4200 | .4216 | .4232 | .4249 | .4265 | .4281 | .4298 |
| 27 | .4314 | .4330 | .4346 | .4362 | .4378 | .4393 | .4409 | .4425 | .4440 | .4456 |
| 28 | .4472 | .4487 | .4502 | .4518 | .4533 | .4548 | .4564 | .4579 | .4594 | .4609 |
| 29 | .4624 | .4639 | .4654 | .4669 | .4683 | .4698 | .4713 | .4728 | .4742 | .4757 |
| 30 | .4771 | .4786 | .4800 | .4814 | .4829 | .4843 | .4857 | .4871 | .4886 | .4900 |
| 31 | .4914 | .4928 | .4942 | .4955 | .4969 | .4983 | .4997 | .5011 | .5024 | .5038 |
| 32 | .5051 | .5065 | .5079 | .5092 | .5105 | .5119 | .5132 | .5145 | .5159 | .5172 |
| 33 | .5185 | .5198 | .5211 | .5224 | .5237 | .5250 | .5263 | .5276 | .5289 | .5302 |
| 34 | .5315 | .5328 | .5340 | .5353 | .5366 | .5378 | .5391 | .5403 | .5416 | .5428 |
| 35 | .5441 | .5453 | .5465 | .5478 | .5490 | .5502 | .5514 | .5527 | .5539 | .5551 |
| 36 | .5563 | .5575 | .5587 | .5599 | .5611 | .5623 | .5635 | .5647 | .5658 | .5670 |
| 37 | .5682 | .5694 | .5705 | .5717 | .5729 | .5740 | .5752 | .5763 | .5775 | .5786 |
| 38 | .5798 | .5809 | .5821 | .5832 | .5843 | .5855 | .5866 | .5877 | .5888 | .5899 |
| 39 | .5911 | .5922 | .5933 | .5944 | .5955 | .5966 | .5977 | .5988 | .5999 | .6010 |
| 40 | .6021 | .6031 | .6042 | .6053 | .6064 | .6075 | .6085 | .6096 | .6107 | .6117 |
| 41 | .6128 | .6138 | .6149 | .6160 | .6170 | .6180 | .6191 | .6201 | .6212 | .6222 |
| 42 | .6232 | .6243 | .6253 | .6263 | .6274 | .6284 | .6294 | .6304 | .6314 | .6325 |
| 43 | .6335 | .6345 | .6355 | .6365 | .6375 | .6385 | .6395 | .6405 | .6415 | .6425 |
| 44 | .6435 | .6444 | .6454 | .6464 | .6474 | .6484 | .6493 | .6503 | .6513 | .6522 |
| 45 | .6532 | .6542 | .6551 | .6561 | .6571 | .6580 | .6590 | .6599 | .6609 | .6618 |
| 46 | .6628 | .6637 | .6646 | .6656 | .6665 | .6675 | .6684 | .6693 | .6702 | .6712 |
| 47 | .6721 | .6730 | .6739 | .6749 | .6758 | .6767 | .6776 | .6785 | .6794 | .6803 |
| 48 | .6812 | .6821 | .6830 | .6839 | .6848 | .6857 | .6866 | .6875 | .6884 | .6893 |
| 49 | .6902 | .6911 | .6920 | .6928 | .6937 | .6946 | .6955 | .6964 | .6972 | .6981 |
| 50 | .6990 | .6998 | .7007 | .7016 | .7024 | .7033 | .7042 | .7050 | .7059 | .7067 |
| 51 | .7076 | .7084 | .7093 | .7101 | .7110 | .7118 | .7126 | .7135 | .7143 | .7152 |
| 52 | .7160 | .7168 | .7177 | .7185 | .7193 | .7202 | .7210 | .7218 | .7226 | .7235 |
| 53 | .7243 | .7251 | .7259 | .7267 | .7275 | .7284 | .7292 | .7300 | .7308 | .7316 |
| 54 | .7324 | .7332 | .7340 | .7348 | .7356 | .7364 | .7372 | .7380 | .7388 | .7396 |

## Table of Common Logarithms (*Continued*)

| *N* | 0 | 1 | 2 | 3 | 4 | 5 | 6 | 7 | 8 | 9 |
|---|---|---|---|---|---|---|---|---|---|---|
| 55 | .7404 | .7412 | .7419 | .7427 | .7435 | .7443 | .7451 | .7459 | .7466 | .7474 |
| 56 | .7482 | .7490 | .7497 | .7505 | .7513 | .7520 | .7528 | .7536 | .7543 | .7551 |
| 57 | .7559 | .7566 | .7574 | .7582 | .7589 | .7597 | .7604 | .7612 | .7619 | .7627 |
| 58 | .7634 | .7642 | .7649 | .7657 | .7664 | .7672 | .7679 | .7686 | .7694 | .7701 |
| 59 | .7709 | .7716 | .7723 | .7731 | .7738 | .7745 | .7752 | .7760 | .7767 | .7774 |
| 60 | .7782 | .7789 | .7796 | .7803 | .7810 | .7818 | .7825 | .7832 | .7839 | .7846 |
| 61 | .7853 | .7860 | .7868 | .7875 | .7882 | .7889 | .7896 | .7903 | .7910 | .7917 |
| 62 | .7924 | .7931 | .7938 | .7945 | .7952 | .7959 | .7966 | .7973 | .7980 | .7987 |
| 63 | .7993 | .8000 | .8007 | .8014 | .8021 | .8028 | .8035 | .8041 | .8048 | .8055 |
| 64 | .8062 | .8069 | .8075 | .8082 | .8089 | .8096 | .8102 | .8109 | .8116 | .8122 |
| 65 | .8129 | .8136 | .8142 | .8149 | .8156 | .8162 | .8169 | .8176 | .8182 | .8189 |
| 66 | .8195 | .8202 | .8209 | .8215 | .8222 | .8228 | .8235 | .8241 | .8248 | .8254 |
| 67 | .8261 | .8267 | .8274 | .8280 | .8287 | .8293 | .8299 | .8306 | .8312 | .8319 |
| 68 | .8325 | .8331 | .8338 | .8344 | .8351 | .8357 | .8363 | .8370 | .8376 | .8382 |
| 69 | .8388 | .8395 | .8401 | .8407 | .8414 | .8420 | .8426 | .8432 | .8439 | .8445 |
| 70 | .8451 | .8457 | .8463 | .8470 | .8476 | .8482 | .8488 | .8494 | .8500 | .8506 |
| 71 | .8513 | .8519 | .8525 | .8531 | .8537 | .8543 | .8549 | .8555 | .8561 | .8567 |
| 72 | .8573 | .8579 | .8585 | .8591 | .8597 | .8603 | .8609 | .8615 | .8621 | .8627 |
| 73 | .8633 | .8639 | .8645 | .8651 | .8657 | .8663 | .8669 | .8675 | .8681 | .8686 |
| 74 | .8692 | .8698 | .8704 | .8710 | .8716 | .8722 | .8727 | .8733 | .8739 | .8745 |
| 75 | .8751 | .8756 | .8762 | .8768 | .8774 | .8779 | .8785 | .8791 | .8797 | .8802 |
| 76 | .8808 | .8814 | .8820 | .8825 | .8831 | .8837 | .8842 | .8848 | .8854 | .8859 |
| 77 | .8865 | .8871 | .8876 | .8882 | .8887 | .8893 | .8899 | .8904 | .8910 | .8915 |
| 78 | .8921 | .8927 | .8932 | .8938 | .8943 | .8949 | .8954 | .8960 | .8965 | .8971 |
| 79 | .8976 | .8982 | .8987 | .8993 | .8998 | .9004 | .9009 | .9015 | .9020 | .9025 |
| 80 | .9031 | .9036 | .9042 | .9047 | .9053 | .9058 | .9063 | .9069 | .9074 | .9079 |
| 81 | .9085 | .9090 | .9096 | .9101 | .9106 | .9112 | .9117 | .9122 | .9128 | .9133 |
| 82 | .9138 | .9143 | .9149 | .9154 | .9159 | .9165 | .9170 | .9175 | .9180 | .9186 |
| 83 | .9191 | .9196 | .9201 | .9206 | .9212 | .9217 | .9222 | .9227 | .9232 | .9238 |
| 84 | .9243 | .9248 | .9253 | .9258 | .9263 | .9269 | .9274 | .9279 | .9284 | .9289 |
| 85 | .9294 | .9299 | .9304 | .9309 | .9315 | .9320 | .9325 | .9330 | .9335 | .9340 |
| 86 | .9345 | .9350 | .9355 | .9360 | .9365 | .9370 | .9375 | .9380 | .9385 | .9390 |
| 87 | .9395 | .9400 | .9405 | .9410 | .9415 | .9420 | .9425 | .9430 | .9435 | .9440 |
| 88 | .9445 | .9450 | .9455 | .9460 | .9465 | .9469 | .9474 | .9479 | .9484 | .9489 |
| 89 | .9494 | .9499 | .9504 | .9509 | .9513 | .9518 | .9523 | .9528 | .9533 | .9538 |
| 90 | .9542 | .9547 | .9552 | .9557 | .9562 | .9566 | .9571 | .9576 | .9581 | .9586 |
| 91 | .9590 | .9595 | .9600 | .9605 | .9609 | .9614 | .9619 | .9624 | .9628 | .9633 |
| 92 | .9638 | .9643 | .9647 | .9652 | .9657 | .9661 | .9666 | .9671 | .9675 | .9680 |
| 93 | .9685 | .9689 | .9694 | .9699 | .9703 | .9708 | .9713 | .9717 | .9722 | .9727 |
| 94 | .9731 | .9736 | .9741 | .9745 | .9750 | .9754 | .9759 | .9763 | .9768 | .9773 |
| 95 | .9777 | .9782 | .9786 | .9791 | .9795 | .9800 | .9805 | .9809 | .9814 | .9818 |
| 96 | .9823 | .9827 | .9832 | .9836 | .9841 | .9845 | .9850 | .9854 | .9859 | .9863 |
| 97 | .9868 | .9872 | .9877 | .9881 | .9886 | .9890 | .9894 | .9899 | .9903 | .9908 |
| 98 | .9912 | .9917 | .9921 | .9926 | .9930 | .9934 | .9939 | .9943 | .9948 | .9952 |
| 99 | .9956 | .9961 | .9965 | .9969 | .9974 | .9978 | .9983 | .9987 | .9991 | .9996 |

# *APPENDIX* 5 NATURAL LOGARITHMS

**Table 3** Natural Logarithms of Numbers

| $n$ | $\log_e n$ | $n$ | $\log_e n$ | $n$ | $\log_e n$ |
|---|---|---|---|---|---|
| 0.0 | * | 4.5 | 1.5041 | 9.0 | 2.1972 |
| 0.1 | 7.6974 | 4.6 | 1.5261 | 9.1 | 2.2083 |
| 0.2 | 8.3906 | 4.7 | 1.5476 | 9.2 | 2.2192 |
| 0.3 | 8.7960 | 4.8 | 1.5686 | 9.3 | 2.2300 |
| 0.4 | 9.0837 | 4.9 | 1.5892 | 9.4 | 2.2407 |
| 0.5 | 9.3069 | 5.0 | 1.6094 | 9.5 | 2.2513 |
| 0.6 | 9.4892 | 5.1 | 1.6292 | 9.6 | 2.2618 |
| 0.7 | 9.6433 | 5.2 | 1.6487 | 9.7 | 2.2721 |
| 0.8 | 9.7769 | 5.3 | 1.6677 | 9.8 | 2.2824 |
| 0.9 | 9.8946 | 5.4 | 1.6864 | 9.9 | 2.2925 |
| 1.0 | 0.0000 | 5.5 | 1.7047 | 10 | 2.3026 |
| 1.1 | 0.0953 | 5.6 | 1.7228 | 11 | 2.3979 |
| 1.2 | 0.1823 | 5.7 | 1.7405 | 12 | 2.4849 |
| 1.3 | 0.2624 | 5.8 | 1.7579 | 13 | 2.5649 |
| 1.4 | 0.3365 | 5.9 | 1.7750 | 14 | 2.6391 |
| 1.5 | 0.4055 | 6.0 | 1.7918 | 15 | 2.7081 |
| 1.6 | 0.4700 | 6.1 | 1.8083 | 16 | 2.7726 |
| 1.7 | 0.5306 | 6.2 | 1.8245 | 17 | 2.8332 |
| 1.8 | 0.5878 | 6.3 | 1.8405 | 18 | 2.8904 |
| 1.9 | 0.6419 | 6.4 | 1.8563 | 19 | 2.9444 |
| 2.0 | 0.6931 | 6.5 | 1.8718 | 20 | 2.9957 |
| 2.1 | 0.7419 | 6.6 | 1.8871 | 25 | 3.2189 |
| 2.2 | 0.7885 | 6.7 | 1.9021 | 30 | 3.4012 |
| 2.3 | 0.8329 | 6.8 | 1.9169 | 35 | 3.5553 |
| 2.4 | 0.8755 | 6.9 | 1.9315 | 40 | 3.6889 |
| 2.5 | 0.9163 | 7.0 | 1.9459 | 45 | 3.8067 |
| 2.6 | 0.9555 | 7.1 | 1.9601 | 50 | 3.9120 |
| 2.7 | 0.9933 | 7.2 | 1.9741 | 55 | 4.0073 |
| 2.8 | 1.0296 | 7.3 | 1.9879 | 60 | 4.0943 |
| 2.9 | 1.0647 | 7.4 | 2.0015 | 65 | 4.1744 |
| 3.0 | 1.0986 | 7.5 | 2.0149 | 70 | 4.2485 |
| 3.1 | 1.1314 | 7.6 | 2.0281 | 75 | 4.3175 |
| 3.2 | 1.1632 | 7.7 | 2.0412 | 80 | 4.3820 |
| 3.3 | 1.1939 | 7.8 | 2.0541 | 85 | 4.4427 |
| 3.4 | 1.2238 | 7.9 | 2.0669 | 90 | 4.4998 |
| 3.5 | 1.2528 | 8.0 | 2.0794 | 95 | 4.5539 |
| 3.6 | 1.2809 | 8.1 | 2.0919 | 100 | 4.6052 |
| 3.7 | 1.3083 | 8.2 | 2.1041 | | |
| 3.8 | 1.3350 | 8.3 | 2.1163 | | |
| 3.9 | 1.3610 | 8.4 | 2.1282 | | |
| 4.0 | 1.3863 | 8.5 | 2.1401 | | |
| 4.1 | 1.4110 | 8.6 | 2.1518 | | |
| 4.2 | 1.4351 | 8.7 | 2.1633 | | |
| 4.3 | 1.4586 | 8.8 | 2.1748 | | |
| 4.4 | 1.4816 | 8.9 | 2.1861 | | |

# Answers to Odd-Numbered Problems

## CHAPTER 1

### *Section 1.1*

**1.** $-1$ **3.** $-7$ **5.** $-21$ **7.** $-10$ **9.** $-4$ **11.** $5/8$ **13.** $6/5$ **15.** $4/15$ **17.** $ac/b^2$ **19.** $(x/a)^3$
**21.** $1$ **23.** $\sqrt[3]{a}$ **25.** $7 + 2i$ **27.** $-6 + 17i$ **29.** $-7 + 3i$ **31.** $-1 - 2i$ **33.** $\frac{1}{2} - \frac{3}{2}i$ **35.** $\frac{2}{5} + \frac{11}{5}i$
**37.**  0 1 2 3 **39.** 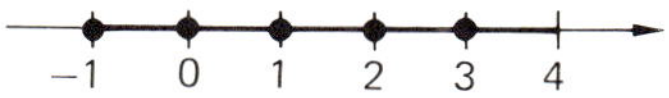 −1 0 1 2 3 4 **41.**  −1 0 1 2 3

**45.** $\dfrac{x + y}{2}$ is rational if $x$ and $y$ are.

### *Section 1.2*

**1.** $(9x - 2)(9x + 2)$ **3.** $(4x - 7y)(4x + 7y)$ **5.** $(4x + 7y)^2$ **7.** $(3x - 4y)^2$ **9.** $(2x - y)(4x^2 + 2xy + y^2)$
**11.** $(3x - 2y)(9x^2 + 6xy + 4y^2)$ **13.** $-\left(x + \dfrac{3 + \sqrt{17}}{2}\right)\left(x + \dfrac{3 - \sqrt{17}}{2}\right)$ **15.** $\left(x - \dfrac{3 + \sqrt{5}}{2}\right)\left(x - \dfrac{3 - \sqrt{5}}{2}\right)$

**17.** $\left(x - \dfrac{3 + \sqrt{29}}{2}\right)\left(x - \dfrac{3 - \sqrt{29}}{2}\right)$ **19.** $4\left(x - \dfrac{5 + \sqrt{119}i}{8}\right)\left(x - \dfrac{5 - \sqrt{119}i}{8}\right)$ **21.** $(3x + 1)(x + 2)$

### *Section 1.3*

**1.** $a^3 - 3a^2b + 3ab^2 - b^3$ **3.** $16a^4 + 32a^3b + 24a^2b^2 + 8ab^3 + b^4$ **5.** $729a^6 + 2916a^5b + 4860a^4b^2 + 4320a^3b^3 + 2160a^2b^4 + 576ab^5 + 64b^6$ **7.** $32x^5 - 320x^4y + 1280x^3y^2 - 2560x^2y^3 + 2560xy^4 - 1024y^5$
**9.** $(1/32)x^5 + (5/48)x^4y + (5/36)x^3y^2 + (5/54)x^2y^3 + (5/162)xy^4 + (1/243)y^5$ **11.** $x^5 - 5x^3 + 10x - 10x^{-1} + 5x^{-3} - x^{-5}$ **13.** $(x + 1)^5 - 5(x + 1)^4y + 10(x + 1)^3y^2 - 10(x + 1)^2y^3 + 5(x + 1)y^4 - y^5$
**15.** $32x^5 - 240x^4 + 720x^3 - 1080x^2 + 810x - 243$ **17.** $729 + 170.1 + 13.23 + 0.343 = 912.673$
**19.** $1 - 1 + 0.4 - 0.08 + 0.008 - 0.00032 = 0.32768$ **21.** $1 + 0.08 + 0.0024 + 0.000032 + 0.00000016 = 1.08243216$

## CHAPTER 2

### *Section 2.1*

1.

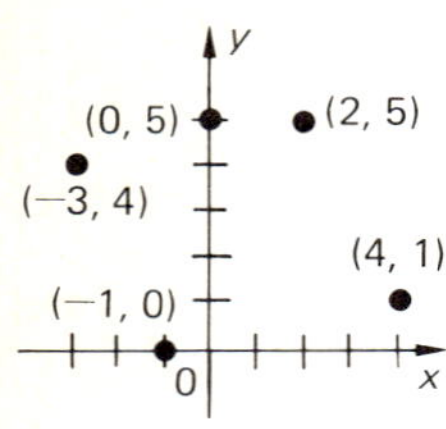


3. $y = 2, y = -2$

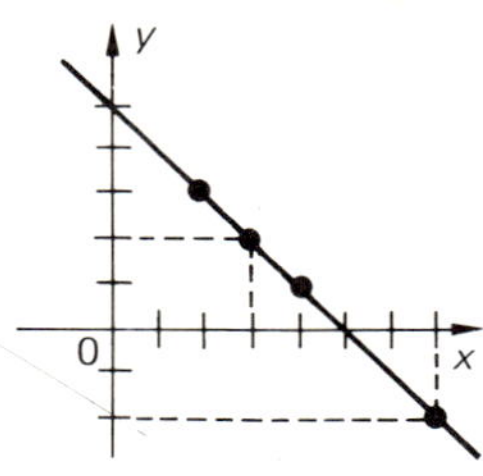


5. Function with reals as domain

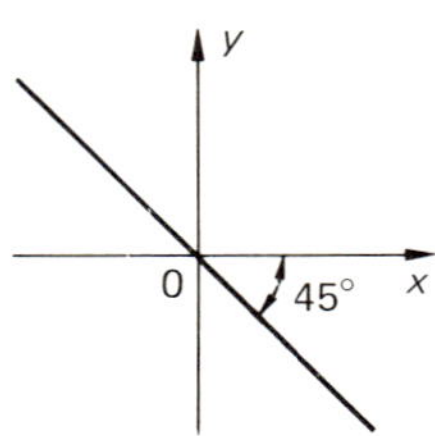


7. Function defined everywhere except at $x = 0$

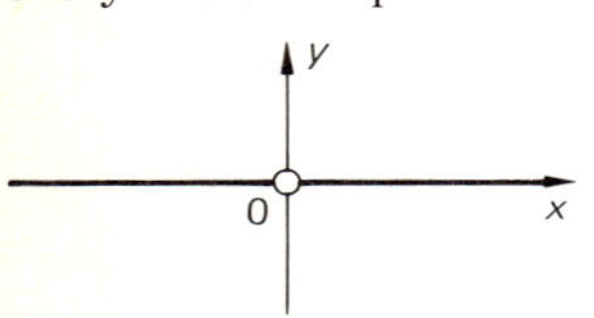


9. Relation

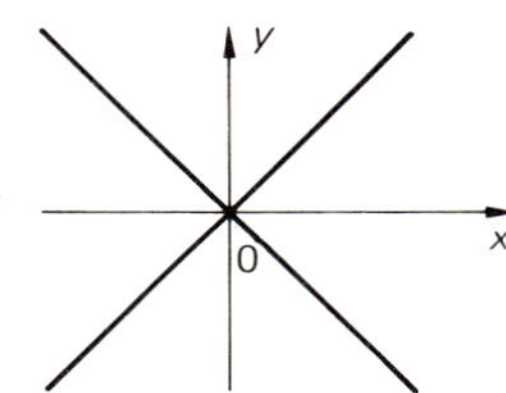


11. Relation

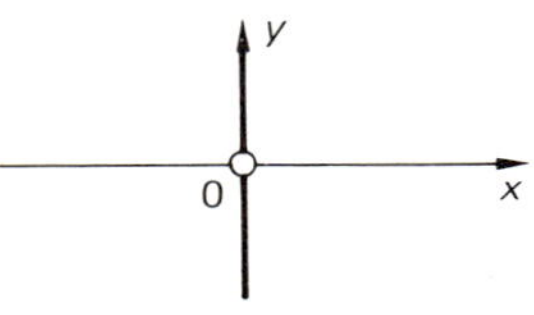


13. $f(0) = 2, f(2) = 4, f(5) = 7, f(-1) = 1, f(-3) = -1, f(a) = a + 2, f(a + b) = a + b + 2$ 15. $f(0) = 0, f(2) = 4, f(5) = 25, f(-1) = -1, f(-3) = 9, f(a) = a^2, f(a + b) = (a + b)^2$ 17. $f(0) = 0, f(2) = 2, f(5) = 20, f(-1) = 2, f(-3) = 12, f(a) = a^2 - a, f(a + b) = (a + b)^2 - (a + b)$ 19. $f(0) = 10, f(2) = \sqrt{96}, f(5) = \sqrt{75}, f(-1) = \sqrt{99}, f(-3) = \sqrt{91}, f(a) = \sqrt{100 - a^2}, f(a + b) = \sqrt{100 - (a + b)^2}$

21. Yes. Let the fingerprints be the independent variable.

23.

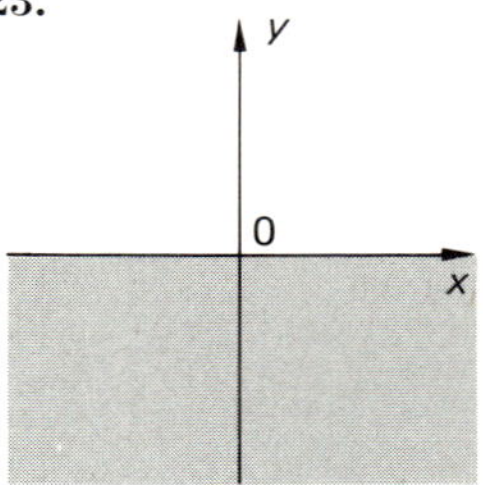


25.

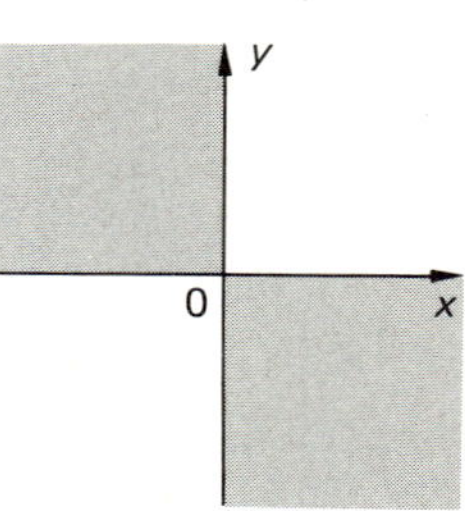


27. Year

### *Section 2.2*

1.

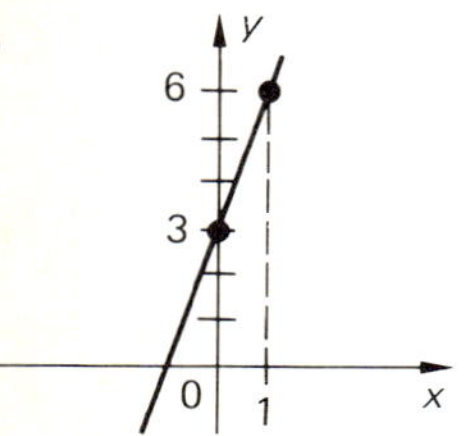


3.

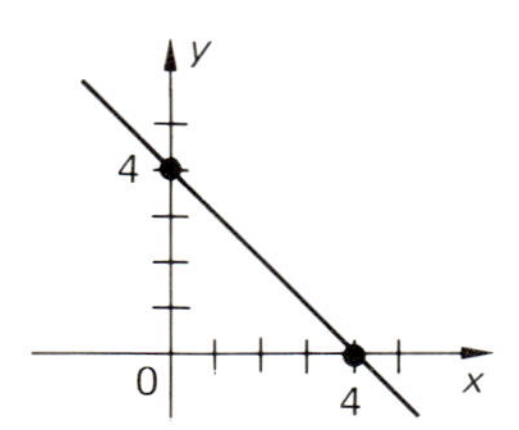


5.

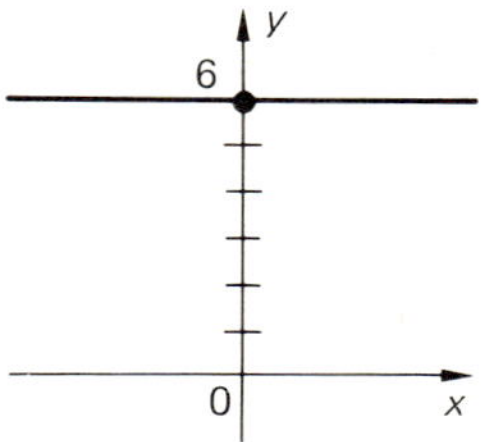


7. First line is five units above second. Lines are parallel.

**9.** $y = 2x + 3$

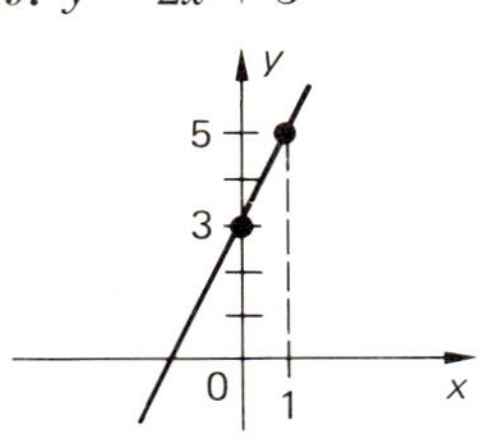


**11.** $y = 4x - 2$

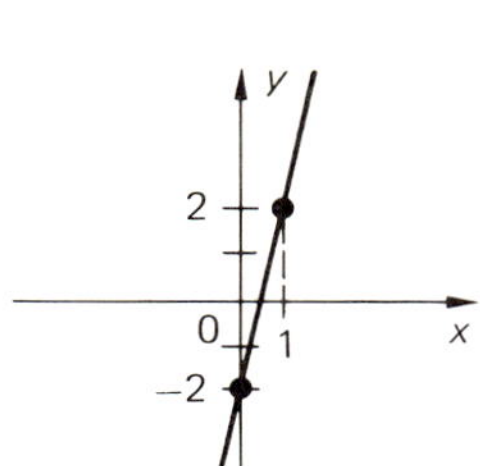


**13.** $y = x + 1$ **15.** $y = 3x - 14$
**17.** $y = 3x + 2$ **19.** $y = -x + 4$
**21.** $y = 2.40x + 1.60(72) = 2.40x + 115.20$ dollars/week
**23.** $y = 0.15x + 1.00$ dollars

### *Section 2.3*

**1.** (1, 2) **3.** (4, 0) **5.** $(\frac{3}{2}, \frac{1}{2})$ **7.** None **9.** None **11.** Line $\left(x, \dfrac{9 - 3x}{2}, \dfrac{x + 3}{2}\right)$ **13.** $(\frac{1}{2}, 2, \frac{5}{3})$
**15.** None

### *Section 2.4*

**1.** $L = \frac{1}{12}(T - 68) + 100$ **3.** $F = \frac{9}{5}C + 32$; 59°F **5.** $C = 13.95x + 0.11y$, $C = 13.95 + 0.11y$, $C = \$49.15$. **7.** No
**9.** $y = 0.14x$,
$y = 0.15(x - 1000) + 140$,
$y = 0.16(x - 2000) + 290$,
$y = 0.17(x - 3000) + 450$,
$y = 0.19(x - 4000) + 620$,
$y = 0.22(x - 8000) + 1380$.
**11.** $W = 3.75(h - 60) + 121$; 162.25 lb; 211 lb. **13.** $S = -7.7(w - 2) + 96$ **15.** \$3294.12
**17.** 34.75 years. The average lifespan of a grizzly is 31 years. **19.** 7 **21.** 62.9¢
**23.** 5 oz peanuts, 11 oz cashews **25.** $\frac{1}{14}$ yr or after approximately 26 days

### *Section 2.5*

**1.** (0, 0) **3.** (0, 7) **5.** (1, −8) **7.** $(\frac{1}{7}, \frac{111}{28})$ **9.** 2, −1 **11.** $\frac{1}{2}$, −2 **13.** $\frac{1}{2}$, $\frac{1}{2}$
**15.** $\dfrac{-3 \pm \sqrt{41}}{4}$ **17.** $\dfrac{1 \pm i\sqrt{47}}{6}$ **19.** $\dfrac{-1 \pm i\sqrt{15}}{2\sqrt{8}}$ **23.** (a) 55¢; (b) 65¢

### *Section 2.6*

**1.** $\sqrt{8}$ **3.** $\sqrt{20}$ **5.** $\sqrt{8}$ **7.** $\sqrt{109}$ **9.** $\sqrt{314}$ **11.** 2 **13.** $\sqrt{29}$ **15.** 10 **17.** $x^2 + y^2 = 4$
**19.** $(x - 2)^2 + (y + 1)^2 = 4$ **21.** $(x + 1)^2 + y^2 = 16$ **23.** $(x - 9)^2 + (y - 5)^2 = 36$ **25.** $x^2 + y^2 = 2$
**27.** $(x - 1)^2 + y^2 = 3$ **29.** $x^2 + (y + 1)^2 = 3 + 2\sqrt{2}$ **31.** Center = (1, −1), radius = 2 **33.** Center = (3, 4), radius = 1 **35.** Center = (−1, 0), radius = 1

### *Section 2.7*

**1.**

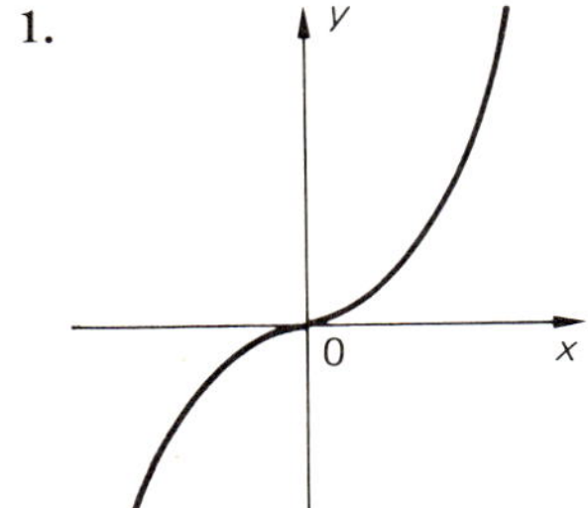

**3.**

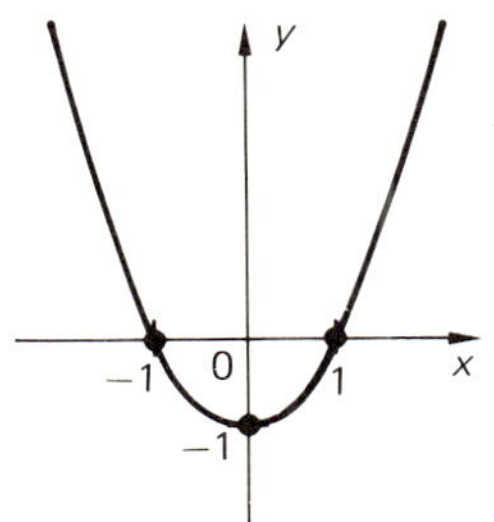


**5.**

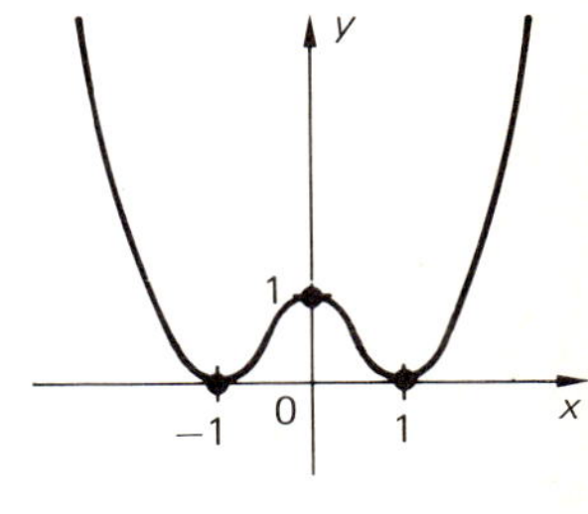

**7.**

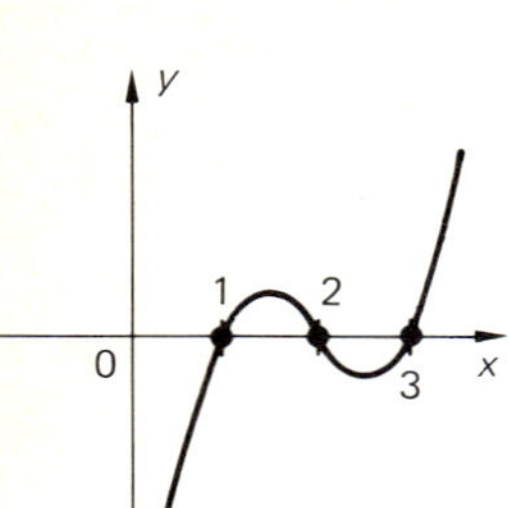


**9.**

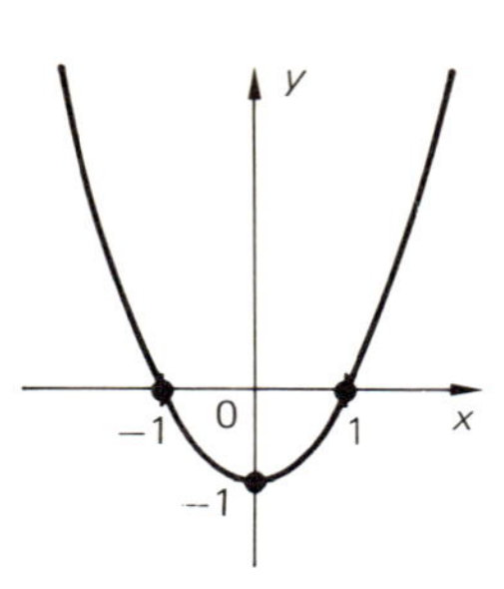


**11.** $y = 0$, $x = -1$

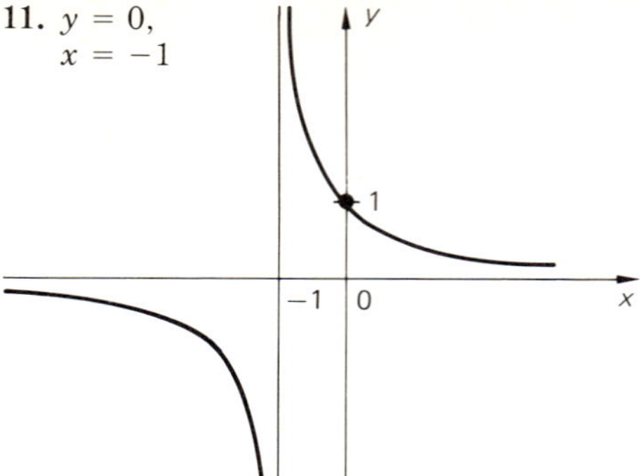


**13.** $y = 0$, $x = 2$

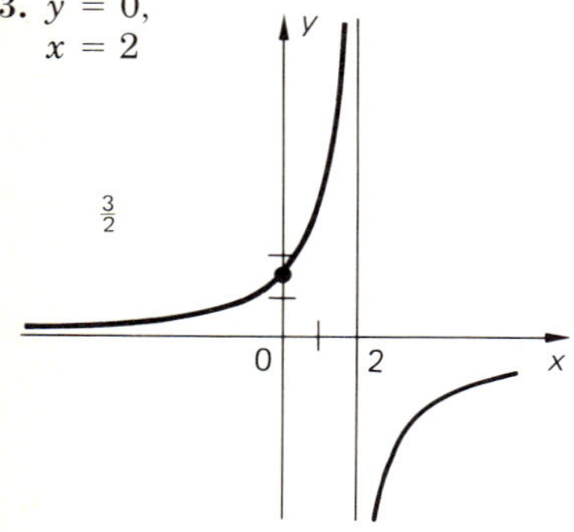


**15.** $y = 0$, $x = \pm 1$

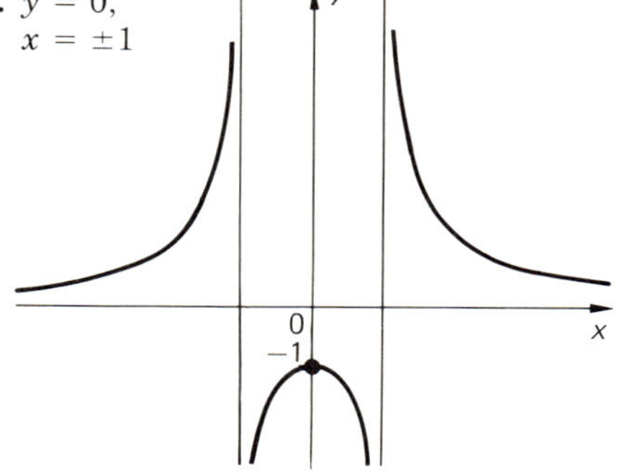


**17.** $x = 1$, $y = 1$

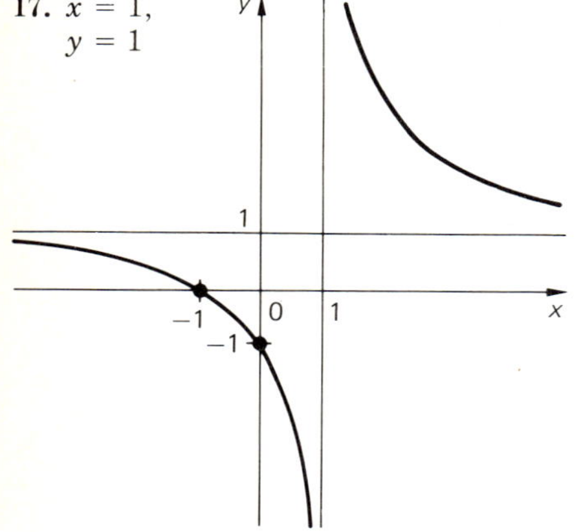


**19.** $x = 3$, $y = 2$

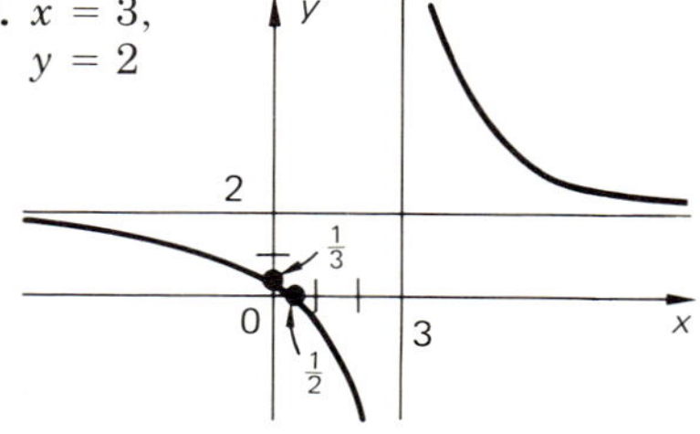


**21.** $x = 1$, $y = 0$

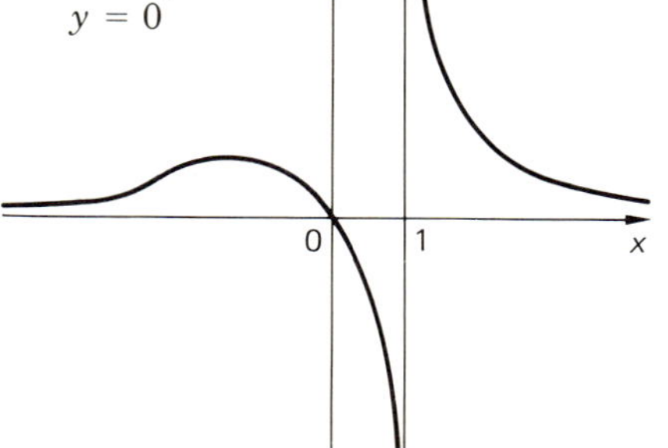

**23.**

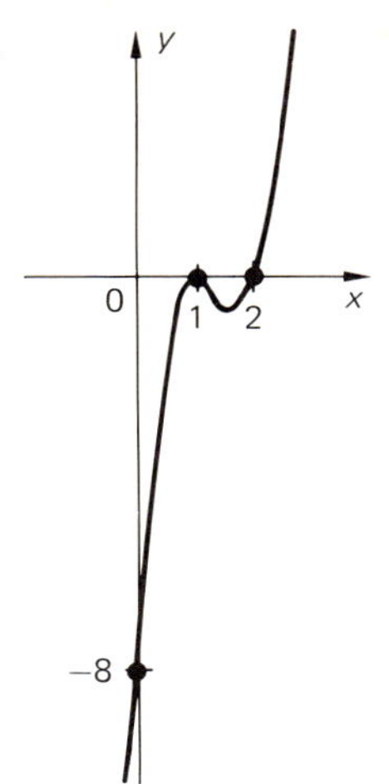


**25.** $y = (x-1)^2(x-2)^2(x-3)^2$ **27.** (24, 6) **29.** $(20, 6\frac{1}{4})$

*Section 2.8*

**1.** $y = 0.6x + 4.4$ **3.** $y = x + 5$ **5.** $y = 0.5x + 3.167$ **7.** $y = -0.25x + 4.25$ **9.** $y = -0.95x + 3.3$ **11.** $y = 0.723x + 1.314$ **13.** $y = 3.74x - 103.88$; 161.78; 210.42 lb **15.** $y = -7.76x + 111.35$; 7.91 weeks **17.** $y = 0.8376x + 0.6559$; 5.6815 kcal/min

## CHAPTER 3

*Section 3.1*

**1.** 9 **3.** 1 **5.** 1/5 **7.** 1 **9.** 1/9 **11.** 5 **13.** 1/27 **15.** 1 **17.** 9 **19.** 81

**21.**

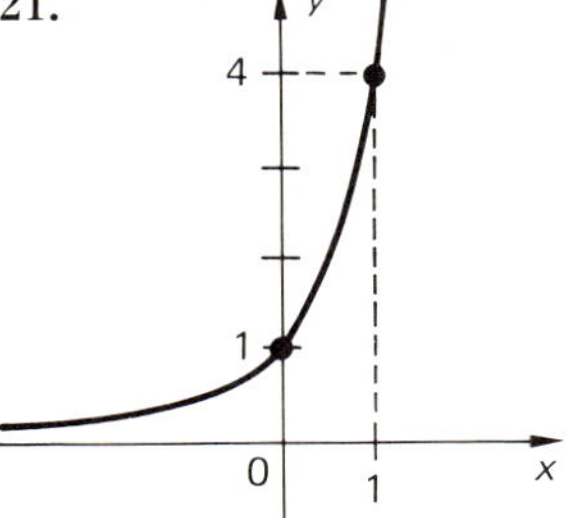


**23.**

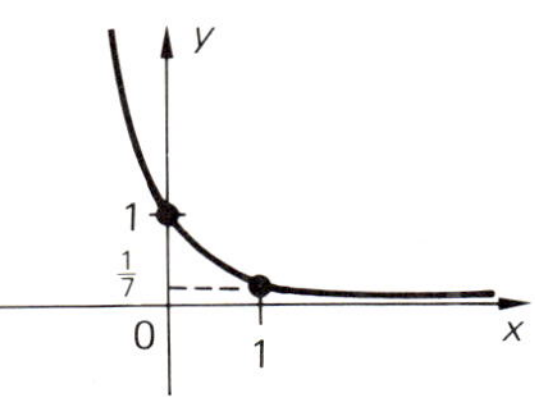


**25.**

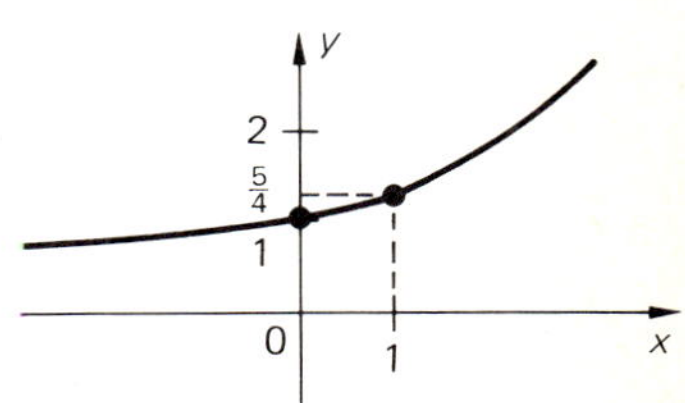


**27.**

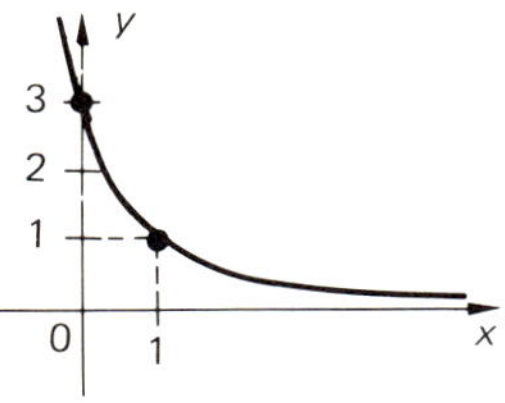


**29.**

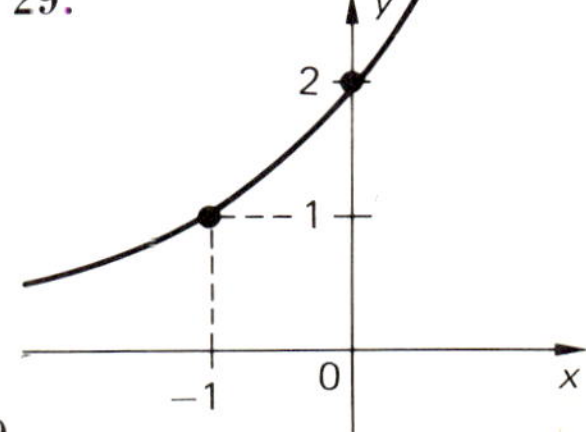


**33.** $(1.05)^{200} = 17292.58$g; $(1.05)^{500} = 3.9323262 \times 10^{10}$g **35.** \$5743.49

### Section 3.2

**1.** \$115.76 **3.** \$1331.00 **5.** \$1360.49 **7.** \$1552.92 **9.** \$8590.93 **11.** \$14,704.80 **13.** \$4099.55 **15.** \$9644.40 **17.** 10.904 dpm/gc **19.** 12.306 dpm/gc **21.** 2.209 dpm/gc **23.** 5360.2 years **25.** 16,820.2 years **27.** 734.0 years **29.** 18.14% **31.** 24,782,622 **33.** 61,391,861 **35.** 55,579,842 **37.** 83,919,963 **39.** (a) 99.9948%; (b) 94.6313%; (c) 82.5661% **41.** 103,432,062 **43.** 8,402,300 **45.** 14,972,339

### Section 3.3

**1.** 0.5922 **3.** 1.9877 **5.** 2.8156 **7.** 4.9542 **9.** 9.9400−10 **11.** 7.0170−10 **13.** 8.8463−10 **15.** 2.9996 **17.** 2.93 **19.** 33,700 **21.** 79.0 **23.** 4000 **25.** 0.0614 **27.** 475.33 **29.** 0.04426 **31.** 265.9 **33.** 0.0889 **35.** $6.11 \times 10^{-5}$ **37.** 298.9 **39.** 6289 years **41.** 13,140 years **43.** 46,090 years

### Section 3.4

**1.** $7(2)^x$ **3.** $7x^2$ **5.** $0.2(3)^x$ **7.** $0.2x^4$ **9.** $\pi x^3$ **11.** $3\pi^x$ **13.** $e^x$ **15.** $5.1(x)^{2.4}$

**17.** $y = 52\left(\frac{46}{52}\right)^{x/50}$; 40.7 sec **19.** $y = 1.33\left(\frac{1.08}{1.33}\right)^{x/50}$; 0.88 min = 52 sec

**21.** $y = 77{,}000{,}000\left(\frac{200}{77}\right)^{x/70}$; 229,220,000; 262,710,000; 301,090,000

**23.** $y = 1{,}030{,}000\left(\frac{220}{103}\right)^{x/20}$; 3,215,255 barrels **25.** $y = 60(x)^{0.78}$ **27.** $W = 0.14H^{1.68}$; 239 lb

### Section 3.5

**1.** $y = 3.03(1.99)^x$ **3.** $y = 1.95(3.19)^x$ **5.** $y = 1.01(2.71)^x$ **7.** $y = 3.14(2.72)^x$ **9.** $y = 3.08x^{1.98}$ **11.** $y = 3.19x^{1.98}$ **13.** $y = 50.9889(0.9979)^x$; 41.28 sec **15.** $y = 1.367(0.9952)^x$; 50 sec **17.** $y = 156{,}974(1.0804)^x$; 5,103,436 kWh **19.** $W = 0.1219H^{1.7023}$

## CHAPTER 4

### Section 4.1

**1.** $A + B = \begin{bmatrix} 9 & 8 \\ -1 & 10 \end{bmatrix}, A + 3B = \begin{bmatrix} 17 & 20 \\ -1 & 24 \end{bmatrix}, A - B = \begin{bmatrix} 1 & -4 \\ -1 & -4 \end{bmatrix}, AB = \begin{bmatrix} 20 & 44 \\ -4 & 15 \end{bmatrix}, BA = \begin{bmatrix} 14 & 26 \\ -7 & 21 \end{bmatrix}$; no

**3.** $A + B = \begin{bmatrix} 13 & 0 \\ 0 & 13 \end{bmatrix}, A + 3B = \begin{bmatrix} 25 & -4 \\ -2 & 27 \end{bmatrix}, A - B = \begin{bmatrix} 1 & 4 \\ 2 & -1 \end{bmatrix}, AB = \begin{bmatrix} 40 & 0 \\ 0 & 40 \end{bmatrix}, BA = \begin{bmatrix} 40 & 0 \\ 0 & 40 \end{bmatrix}$; yes

**5.** No $A + B$, $A + 3B$, or $A - B$; $AB = \begin{bmatrix} 30 & 6 & 6 \\ 62 & 18 & 22 \end{bmatrix}$; no $BA$; no

**7.** $A + B = (-1, -1, -1)$, $A + 3B = (-3, -7, -11)$, $A - B = (1, 5, 9)$; no $AB$ or $BA$; no **9.** No $A + B$, $A + 3B$, $A - B$, or $AB$; $BA = (5, 42, 63)$; no

**11.** No $A + B$, $A + 3B$, or $A - B$; $AB = (14)$, $BA = \begin{bmatrix} 0 & 0 & 0 \\ -2 & -4 & -6 \\ -6 & -12 & -18 \end{bmatrix}$; no

**13.** No $A + B$, $A + 3B$, or $A - B$; $AB = \begin{bmatrix} 97 \\ 38 \\ 99 \end{bmatrix}$; no $BA$; no

**15.** No $A + B$, $A + 3B$, or $A - B$; $AB = \begin{bmatrix} -12 & 40 \\ -25 & -45 \end{bmatrix}$, $BA = \begin{bmatrix} -17 & 58 & -38 \\ 16 & -34 & 49 \\ 3 & -30 & -6 \end{bmatrix}$; no

**17.** $A + B = \begin{bmatrix} 2 & 2 & 4 \\ 4 & 6 & 6 \\ 8 & 8 & 10 \end{bmatrix}$, $A + 3B = \begin{bmatrix} 4 & 2 & 6 \\ 4 & 8 & 6 \\ 10 & 8 & 12 \end{bmatrix}$, $A - B = \begin{bmatrix} 0 & 2 & 2 \\ 4 & 4 & 6 \\ 6 & 8 & 8 \end{bmatrix}$, $AB = \begin{bmatrix} 4 & 2 & 4 \\ 10 & 5 & 10 \\ 16 & 8 & 16 \end{bmatrix}$,

$BA = \begin{bmatrix} 8 & 10 & 12 \\ 4 & 5 & 6 \\ 8 & 10 & 12 \end{bmatrix}$; no

**19.** $A + B = \begin{bmatrix} 2 & 9 & 3 \\ 0 & 9 & 0 \\ 4 & 14 & 7 \end{bmatrix}$, $A + 3B = \begin{bmatrix} 4 & 23 & 3 \\ 0 & 25 & 0 \\ 4 & 32 & 9 \end{bmatrix}$, $A - B = \begin{bmatrix} 0 & -5 & 3 \\ 0 & -7 & 0 \\ 4 & -4 & 5 \end{bmatrix}$, $AB = \begin{bmatrix} 1 & 50 & 3 \\ 0 & 8 & 0 \\ 4 & 122 & 6 \end{bmatrix}$,

$BA = \begin{bmatrix} 1 & 9 & 3 \\ 0 & 8 & 0 \\ 4 & 14 & 6 \end{bmatrix}$; no

### *Section 4.2*

**1.** $\begin{bmatrix} 3 & -1 \\ 2 & 3 \end{bmatrix}\begin{bmatrix} x \\ y \end{bmatrix} = \begin{bmatrix} 1 \\ 8 \end{bmatrix}$ **3.** $\begin{bmatrix} 3 & -5 \\ 1 & 4 \end{bmatrix}\begin{bmatrix} x \\ y \end{bmatrix} = \begin{bmatrix} 12 \\ 4 \end{bmatrix}$ **5.** $\begin{bmatrix} 1 & 1 \\ 1 & -1 \end{bmatrix}\begin{bmatrix} x \\ y \end{bmatrix} = \begin{bmatrix} 2 \\ 1 \end{bmatrix}$

**7.** $\begin{bmatrix} 1 & 1 & 1 \\ 2 & 1 & -1 \\ 3 & -1 & 2 \end{bmatrix}\begin{bmatrix} x \\ y \\ z \end{bmatrix} = \begin{bmatrix} 6 \\ 3 \\ 4 \end{bmatrix}$ **9.** $\begin{bmatrix} 1 & 1 & 1 \\ 2 & 0 & -1 \\ 3 & -1 & 0 \end{bmatrix}\begin{bmatrix} x \\ y \\ z \end{bmatrix} = \begin{bmatrix} 0 \\ 3 \\ 3 \end{bmatrix}$

**11.** $\begin{aligned} 4x - 2y &= 8 \\ -2x + y &= 4 \end{aligned}$ **13.** $\begin{aligned} 6x - 15y &= 1 \\ -2x + 5y &= 2 \end{aligned}$ **15.** $\begin{aligned} 2x + 7y &= 0 \\ -3x &= 0 \end{aligned}$

**17.** $\begin{aligned} x + y + z &= 6 \\ 2x + y - z &= 3 \\ -4x - 2y + 2z &= 4 \end{aligned}$ **19.** $\begin{aligned} 2x + y + 3z &= 8 \\ -8y + 3z &= -9 \\ -6x + 7y &= 1 \end{aligned}$

**21.** $\begin{bmatrix} x \\ y \end{bmatrix} = \begin{bmatrix} 4 & -5 \\ -3 & 4 \end{bmatrix}\begin{bmatrix} 1 \\ 2 \end{bmatrix} = \begin{bmatrix} -6 \\ 5 \end{bmatrix}$, so $x = -6$ and $y = 5$.

### *Section 4.3*

**1.** $\begin{bmatrix} x \\ y \end{bmatrix} = \begin{bmatrix} 1 \\ 2 \end{bmatrix}$ **3.** $\begin{bmatrix} x \\ y \end{bmatrix} = \begin{bmatrix} 5 \\ -1 \end{bmatrix}$ **5.** $\begin{bmatrix} x \\ y \end{bmatrix} = \begin{bmatrix} 3 \\ -3 \end{bmatrix}$ **7.** $\begin{bmatrix} x \\ y \end{bmatrix} = \begin{bmatrix} 0 \\ 3 \end{bmatrix}$ **9.** $\begin{bmatrix} x \\ y \\ z \end{bmatrix} = \begin{bmatrix} -1 \\ 2 \\ -3 \end{bmatrix}$ **11.** $\begin{bmatrix} x \\ y \\ z \end{bmatrix} = \begin{bmatrix} 11 \\ 7 \\ 13 \end{bmatrix}$

**13.** $\begin{bmatrix} x \\ y \\ z \end{bmatrix} = \begin{bmatrix} 1/3 \\ 1/5 \\ 7/4 \end{bmatrix}$ **15.** $\begin{bmatrix} x \\ y \\ z \end{bmatrix} = \begin{bmatrix} 987 \\ 654 \\ 321 \end{bmatrix}$ **17.** $B = \begin{bmatrix} 0 & 1 \\ 1 & 0 \end{bmatrix}$, $C = \begin{bmatrix} 2 & 0 \\ 0 & 1/3 \end{bmatrix}$, $D = \begin{bmatrix} 1 & 0 \\ 2 & 1 \end{bmatrix}$

### *Section 4.4*

**1.** $\begin{bmatrix} -3 & 4 \\ 4 & -5 \end{bmatrix}$ **3.** $\dfrac{1}{54}\begin{bmatrix} 24 & -12 & 6 \\ -20 & 10 & 4 \\ 15 & 6 & -3 \end{bmatrix}$ **5.** $\dfrac{1}{72}\begin{bmatrix} -41 & 12 & 19 \\ 40 & -24 & -8 \\ 7 & 12 & -5 \end{bmatrix}$ **7.** $\dfrac{1}{129}\begin{bmatrix} 21 & 27 & -18 \\ -6 & 23 & -1 \\ -6 & -63 & 42 \end{bmatrix}$

**9.** $\dfrac{1}{84}\begin{bmatrix} 20 & 14 & -18 & 8 \\ -8 & 7 & 3 & 22 \\ 10 & -14 & 12 & 4 \\ -14 & 7 & 21 & -14 \end{bmatrix}$ **11.** $\dfrac{1}{294}\begin{bmatrix} -102 & 9 & 27 & 60 \\ 12 & -76 & 66 & 16 \\ 12 & 71 & -81 & 16 \\ 69 & 4 & 12 & -55 \end{bmatrix}$

**13.** $A^{-1} = \begin{bmatrix} -1 & 1 & -1 \\ 1 & -1/3 & 2/3 \\ 1 & -2/3 & 1/3 \end{bmatrix}, \begin{bmatrix} x \\ y \\ z \end{bmatrix} = \begin{bmatrix} -3 \\ 7/3 \\ 5/3 \end{bmatrix}$

**15.** No inverse. Solutions set consists of all vectors.

$$\begin{bmatrix} x \\ y \\ z \\ w \end{bmatrix} = \begin{bmatrix} \frac{1+t}{2} \\ \frac{3-t}{2} \\ t \\ 3 \end{bmatrix}$$

**17.** $A^{-1} = 1/3 \begin{bmatrix} -4 & -1 & 3 \\ -1 & -4 & 3 \\ 3 & 3 & -3 \end{bmatrix}, A^{-1}B_1 = \begin{bmatrix} 1 \\ 0 \\ 0 \end{bmatrix}, A^{-1}B_2 = \begin{bmatrix} -1 \\ -2 \\ 3 \end{bmatrix}, A^{-1}B_3 = \begin{bmatrix} -3 \\ -4 \\ 6 \end{bmatrix}, A^{-1}B_4 = \begin{bmatrix} -2/3 \\ -5/3 \\ 1 \end{bmatrix}$

**21.** $\begin{bmatrix} 2 & -3 \\ -3 & 5 \end{bmatrix}$ **23.** No inverse **25.** $\frac{1}{17}\begin{bmatrix} 4 & 3 \\ -3 & 2 \end{bmatrix}$ **27.** No inverse

**29.** $\frac{1}{82}\begin{bmatrix} 6 & -5 \\ -8 & -7 \end{bmatrix}$ **31.** $\frac{1}{2}\begin{bmatrix} 3 & -1 \\ 1 & 0 \end{bmatrix}$ **33.** No inverse **35.** $\frac{-1}{162}\begin{bmatrix} 0.4 & -1 \\ -2.1 & 1.2 \end{bmatrix}$

***Section 4.5***

**1.**

$$\begin{array}{c} \\ c \\ a \\ t \\ f \\ m \end{array} \begin{array}{c} \begin{array}{ccccc} c & a & t & f & m \end{array} \\ \begin{bmatrix} 0 & 1 & 1 & 1 & 1 \\ 1 & 0 & 1 & 0 & 1 \\ 1 & 1 & 0 & 0 & 1 \\ 1 & 0 & 0 & 0 & 1 \\ 1 & 1 & 1 & 1 & 0 \end{bmatrix} \end{array} = A,\ A^2 = \begin{bmatrix} 4 & 2 & 2 & 1 & 3 \\ 2 & 3 & 2 & 2 & 2 \\ 2 & 2 & 3 & 2 & 2 \\ 1 & 2 & 2 & 2 & 1 \\ 3 & 2 & 2 & 1 & 4 \end{bmatrix},\ A^3 = \begin{bmatrix} 8 & 9 & 9 & 7 & 9 \\ 9 & 6 & 7 & 4 & 9 \\ 9 & 7 & 6 & 4 & 9 \\ 7 & 4 & 4 & 2 & 7 \\ 9 & 9 & 9 & 7 & 8 \end{bmatrix}$$

**3.**

$$\begin{array}{c} \\ O \\ T \\ F \\ D \\ L \end{array} \begin{array}{c} \begin{array}{ccccc} O & T & F & D & L \end{array} \\ \begin{bmatrix} 0 & 1 & 0 & 0 & 1 \\ 1 & 0 & 1 & 0 & 0 \\ 0 & 1 & 0 & 1 & 1 \\ 0 & 0 & 1 & 0 & 1 \\ 1 & 0 & 1 & 1 & 0 \end{bmatrix} \end{array} = A,\ A^2 = \begin{bmatrix} 2 & 0 & 2 & 1 & 0 \\ 0 & 2 & 0 & 1 & 2 \\ 2 & 0 & 3 & 1 & 1 \\ 1 & 1 & 1 & 2 & 1 \\ 0 & 2 & 1 & 1 & 3 \end{bmatrix},\ A^3 = \begin{bmatrix} 0 & 4 & 1 & 2 & 5 \\ 4 & 0 & 5 & 2 & 1 \\ 1 & 5 & 2 & 4 & 6 \\ 2 & 2 & 4 & 2 & 4 \\ 5 & 1 & 6 & 4 & 2 \end{bmatrix}$$

**5.**

$$A^2 = \begin{bmatrix} 0 & 0 & 1 & 1 & 1 & 0 \\ 0 & 0 & 0 & 0 & 0 & 0 \\ 0 & 0 & 0 & 0 & 0 & 0 \\ 0 & 0 & 0 & 0 & 0 & 0 \\ 0 & 0 & 0 & 0 & 0 & 0 \\ 0 & 0 & 0 & 0 & 0 & 0 \end{bmatrix}$$

**7.**

$$A = \begin{bmatrix} 0&1&0&0&0&0&0&0&0&0&0&0&0&0&0&0&0&0 \\ 0&0&1&0&0&0&1&0&0&0&1&0&0&0&0&0&0&0 \\ 0&0&0&1&0&0&0&0&0&0&0&0&0&0&0&0&0&0 \\ 0&0&0&0&1&0&0&0&0&0&0&0&0&0&0&0&0&0 \\ 0&0&0&0&0&1&0&0&0&0&0&0&0&0&0&0&0&0 \\ 0&0&0&0&0&0&0&0&0&0&0&0&0&0&1&0&0&0 \\ 0&0&0&0&0&0&0&1&0&0&0&0&0&0&0&0&0&0 \\ 0&0&0&0&0&0&0&0&1&0&0&0&0&0&0&0&0&0 \\ 0&0&0&0&0&0&0&0&0&1&0&0&0&0&0&0&0&0 \\ 0&0&0&0&0&0&0&0&0&0&0&0&0&0&0&1&0&0 \\ 0&0&0&0&0&0&0&0&0&0&0&1&0&1&0&0&0&0 \\ 0&0&0&0&0&0&0&0&0&0&0&0&1&0&0&0&0&0 \\ 0&0&0&0&0&0&0&0&0&0&0&0&0&0&0&0&1&0 \\ 0&0&0&0&0&0&0&0&0&0&0&0&0&0&0&0&0&1 \\ 0&0&0&0&0&0&0&0&0&0&0&0&0&0&0&0&0&0 \\ 0&0&0&0&0&0&0&0&0&0&0&0&0&0&0&0&0&0 \\ 0&0&0&0&0&0&0&0&0&0&0&0&0&0&0&0&0&0 \\ 0&0&0&0&0&0&0&0&0&0&0&0&0&0&0&0&0&0 \end{bmatrix}$$

$A^3$ has a 1 in the top row, column 18, indicating that Robert Bruce had the strongest claim.

**9.** Louis IV d'Outremer has the stronger claim, since Gerberge also has claim to the empire.

**11.** $A = \begin{bmatrix} 0&1&1&0&0&0\\ 0&0&0&1&0&0\\ 0&0&0&1&1&0\\ 0&0&0&1&0&1\\ 0&0&0&0&0&1\\ 0&0&0&0&0&0 \end{bmatrix}$, $A^2 = \begin{bmatrix} 0&0&0&2&1&0\\ 0&0&0&1&0&1\\ 0&0&0&1&0&2\\ 0&0&0&1&0&1\\ 0&0&0&0&0&0\\ 0&0&0&0&0&0 \end{bmatrix}$, $A^3 = \begin{bmatrix} 0&0&0&2&0&3\\ 0&0&0&1&0&1\\ 0&0&0&1&0&1\\ 0&0&0&1&0&1\\ 0&0&0&0&0&0\\ 0&0&0&0&0&0 \end{bmatrix}$,

$A^4 = \begin{bmatrix} 0&0&0&2&0&2\\ 0&0&0&1&0&1\\ 0&0&0&1&0&1\\ 0&0&0&1&0&1\\ 0&0&0&0&0&0\\ 0&0&0&0&0&0 \end{bmatrix}$.

$A^2$ consists of food sources two links removed from the feeder, and so forth.

**13.** $A = \begin{bmatrix} 0&0&0&0&0&0&1\\ 1&0&0&0&0&0&0\\ 0&0&0&1&0&0&0\\ 0&1&0&0&0&1&0\\ 0&0&1&1&0&0&0\\ 0&1&0&0&1&0&0\\ 0&0&0&0&0&1&0 \end{bmatrix}$, $A^2 = \begin{bmatrix} 0&0&0&0&0&1&0\\ 0&0&0&0&0&0&1\\ 0&1&0&0&0&1&0\\ 1&1&0&0&1&0&0\\ 0&1&0&1&0&1&0\\ 1&0&1&1&0&0&0\\ 0&1&0&0&1&0&0 \end{bmatrix}$,

$A^3 = \begin{bmatrix} 0&1&0&0&1&0&0\\ 0&0&0&0&0&1&0\\ 1&1&0&0&1&0&0\\ 1&0&1&1&0&0&1\\ 1&2&0&0&1&1&0\\ 0&1&0&1&0&1&1\\ 1&0&1&1&0&0&0 \end{bmatrix}$,

There is one length-three chain shown on the main diagonal of $A^3$:

Nitrates → green plants → putrifying → nitrates.

$A^4 = \begin{bmatrix} 1&0&1&1&0&0&0\\ 0&1&0&0&1&0&0\\ 1&0&1&1&0&0&1\\ 0&1&0&1&0&2&0\\ 2&1&1&1&1&0&1\\ 1&2&0&0&1&2&0\\ 0&1&0&1&0&1&1 \end{bmatrix}$.

There are two length-four chains: $N_2$ → nitrogen → nitrates → denitrifying → $N_2$,

nitrates → green plants → animals → putrifying → nitrates.

The longest chain of distinct objects is
$N_2$ → nitrogen → nitrates → green plants → animals → putrifying → denitrifying → $N_2$

**15.** $\begin{bmatrix} 2100\\ 2700\\ 1200 \end{bmatrix}$ oz $= \begin{bmatrix} 131.25\\ 168.75\\ 75 \end{bmatrix}$ lb

**17.** $\begin{bmatrix} 320\\ 120\\ 360 \end{bmatrix}$

**19.** $\begin{bmatrix} 128\\ 320\\ 256 \end{bmatrix}$

**21.** Impossible to totally use ingredients since inverse matrix yields $\begin{bmatrix} 960\\ -840\\ 680 \end{bmatrix}$.

**23.** Lake 5 cannot be stocked optimally, since we get $\begin{bmatrix} 224.80\\ -145.32\\ 10182.84 \end{bmatrix}$ g.

**25.** $\begin{bmatrix} 791.65\\ 1111.00\\ 26.25 \end{bmatrix}$

**27.** Temporarily reverse traffic on $x_7$ for 300 VPH and set $x_1 = 600$, $x_2 = 200$, $x_3 = 600$, $x_4 = 400$, $x_5 = 600$, and $x_6 = 0$. **29.** $x_1 = 500$, $x_2 = 400$, $x_3 = x_4 = 700$, $x_5 = 500$, $x_6 = x_7 = 200$ **31.** $x_1 = 800$, $x_2 = x_4 = x_5 = x_9 = 900$, $x_3 = 500$, $x_6 = x_7 = 700$, $x_8 = 600$, $x_{10} = 1000$, $x_{11} = 400$, $x_{12} = 0$

## CHAPTER 5

### *Section 5.1*

**1.**

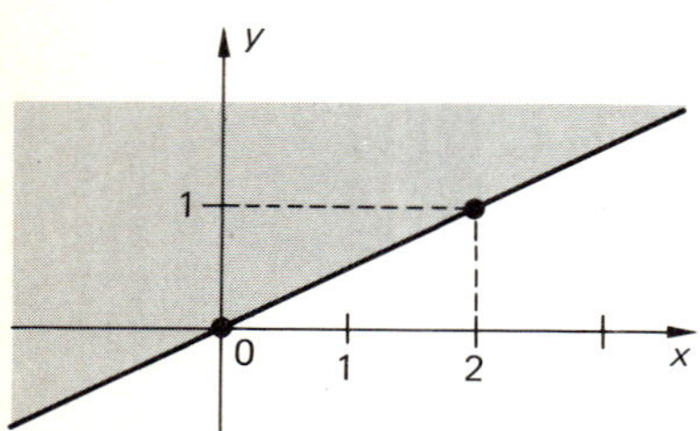


**3.**

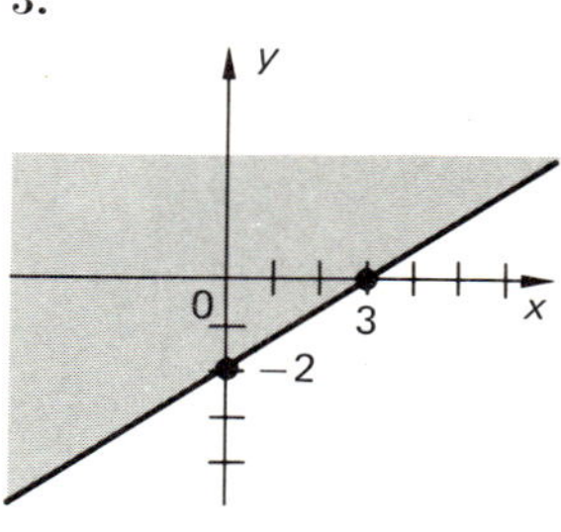


**5.**

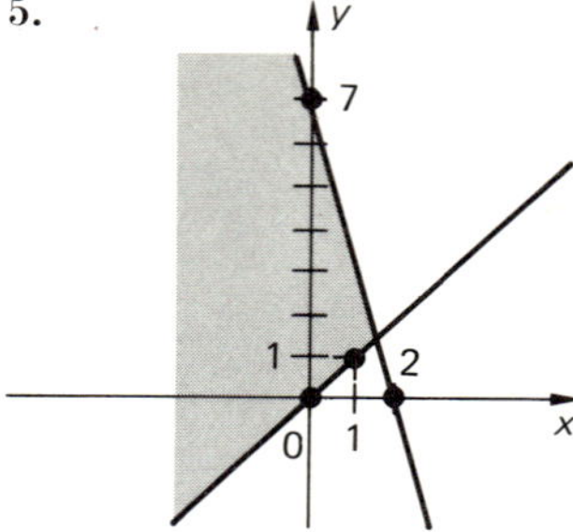


**7.**

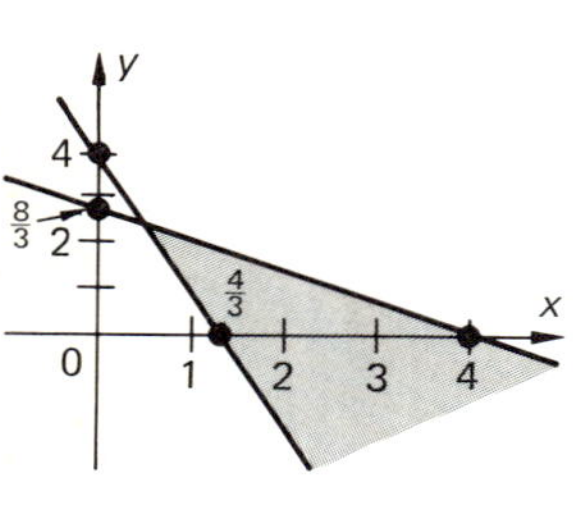


**9.**

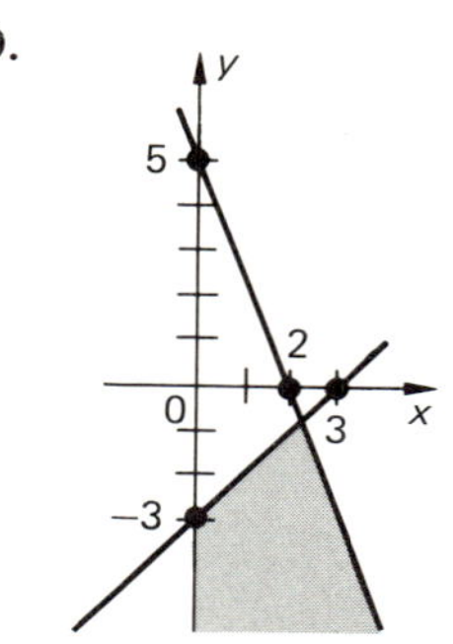


**11.**

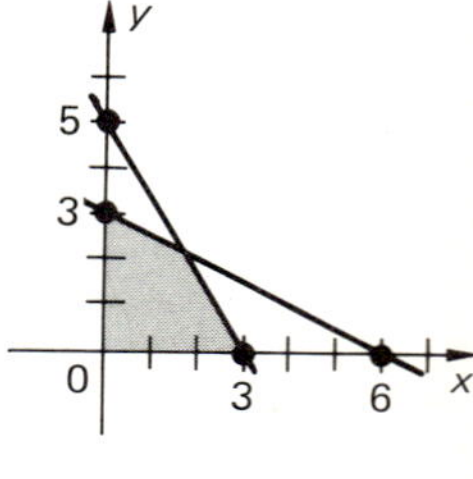


**13.** Empty set (no feasible set) **15.** Empty set (no feasible set)

### *Section 5.2*

**1–11.** Solutions of Exercises 1–11 are identical to those of Exercises 1–11 in Section 5.1. **13.** Maximum profit $P = \$360.00$ **15.** $x = 400$, $y = 1200$, $P = \$336.00$ **17.** $x = 100$, $y = 400$, $C = \$5800.00$ **19.** 94 kg bass and 456 kg bullheads **21.** 10 min on-line, 30 min batch, $C = \$3600$ **23.** 9 audits, no tax returns, $R = \$4500$ **25.** 9 min TV time, 4 one-page ads, exposure = 98,000,000 viewings

### *Section 5.3*

**1.** $x = 5/3, y = 0, P = 25/3$ **3.** $x = 3/4, y = 7/4, P = 5/2$ **5.** $x = 7/3, y = 0, P = 28/3$ **7.** $x = 0, y = 5/2, P = 15/2$; $(t_2 = -3/2)$ **9.** $x = 13/8 + t_2, y = 7/8 - t_2, -13/8 \le t_2 \le 0, P = 5/2$ **11.** $x = 0, y = 25/8, z = 5/4, P = 15/2$; $(t_2 = 11/8)$ **13.** $x = 1, y = 8/3, z = 5/3, P = 16/3$ **15.** $y = w = 0$, $2x + 4z = 7$, $x, z \ge 0, P = 7/2$ **17.** $x = 600 - 3t_2, y = 900 + \frac{9}{2}t_2, 0 \le t_2 \le 200, P = \$360.00$ **19.** $x = 400, y = 1200$, $P = \$336.00$ **21.** 94 kg bass and 456 kg bullheads **23.** 10 min on-line, 30 min batch, $C = \$3600$ **25.** 9 audits, no tax returns, $R = \$4500$ **27.** $x = 35.3, y = 0, z = 10183.1, P = 10218.4$ **29.** 0.1105 tons alfalfa meal, 0.5347 tons fish meal, 0.1057 tons soybean meal, 0.2492 tons filler cost the manufacturer \$77.94 per ton **31.** 255 prime rib, 170 steak and lobster, 85 chicken Kiev, \$1615

**33.** 258 prime rib, 166 steak and lobster, 86 chicken Kiev, \$1610 **35.** Let $x$, $y$, and $z$ be the number of pounds of sauce with meat, mushrooms, and meat and mushrooms combined, respectively. The profit equation is

$$\begin{aligned} P &= 0.7x + 0.7y + 0.82z - 0.4(0.8x + 0.85y + 0.7z) - 0.9(0.2x + 0.15z) - 1.2(0.15y + 0.15z) \\ &= 0.2x + 0.18y + 0.225z, \end{aligned}$$

so in order to optimize profits Mama Rosa should produce $2083\frac{1}{3}$ lb of sauce with meat, no sauce with mushrooms, and $3333\frac{1}{3}$ lb of combined, requiring 4000 lb of tomatoes, $916\frac{2}{3}$ lb of meat, and 500 lb of mushrooms.

### *Section 5.4*

**1.** $x = 5/3, y = 0, P = 25/3$ **3.** $x = 0, y = 2, P = 2$ **5.** No solution (feasible set is empty) **7.** $x = 0$, $y = 5/2, P = 15/2$ **9.** $x = 233\frac{1}{3}, y = 266\frac{2}{3}, C = \$5066.67$ **11.** Maximum problem is sensitive **13.** (a) 1500 lb mixed nuts, 500 lb cashew mix, $R = \$4000$; (b) 1500 lb mixed nuts, 500 lb cashew mix, $R = \$4375$; (c) 600 lb mixed nuts, 1400 lb cashew mix, $R = \$4350$; with mixed nuts: 50% cashews, 50% walnuts; cashew mix: 50% cashews, 50% peanuts; (d) 1500 lb mixed nuts, 500 lb cashew mix, $P = \$1905$; (e) 600 lb mixed nuts, 1400 lb cashew mix, $P = \$1880$

## CHAPTER 6

### *Section 6.2*

**1.** {1, 2, 3, 4, 5, 6, 8, 10} **3.** {7, 8, 9, 10} **5.** {1, 2, 3, 4, 5, 6, 7, 8, 9} **7.** {2, 4, 6, 7, 8, 9, 10} **9.** {7, 9} **11.** {1, 2, 3, 4, 5, 6, 8, 10} **13.** {all math majors taking an accounting course} **15.** {all students taking both an accounting and botany course} **17.** {all students not taking an accounting course} **19.** {all nonmath majors taking a botany course}

**29.**

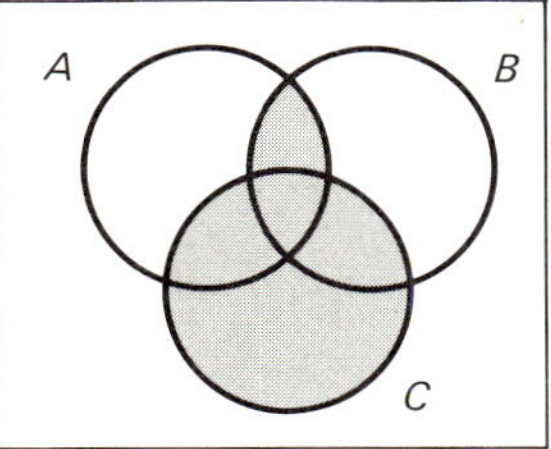


**31.**

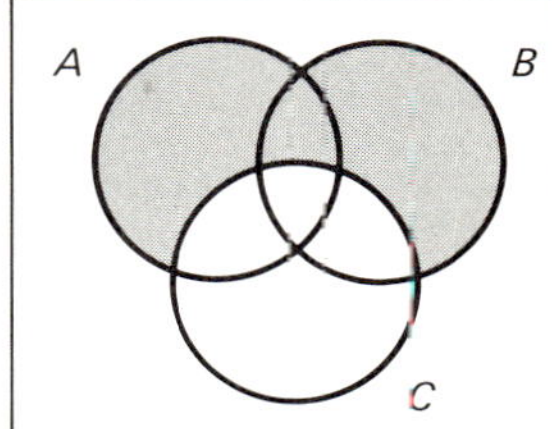


**33.**

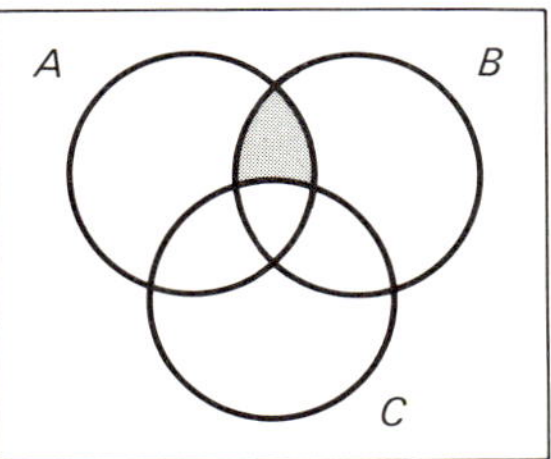


**35.**

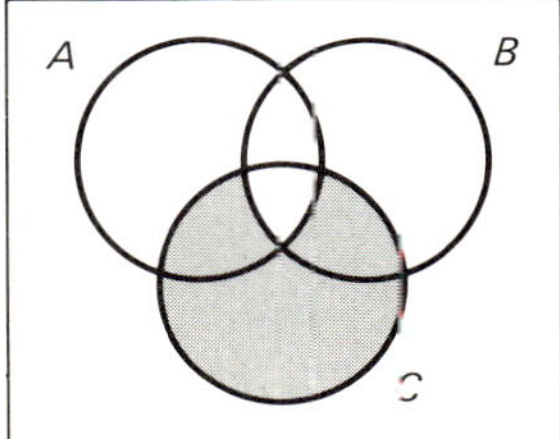


**37.** (a) 20; (b) 40 **39.** (a) 1; (b) 6; (c) 10

### *Section 6.3*

**1.** $S$ = {3H, 2H1T, 1H2T, 3T}; no **3.** No **5.** Yes **7.** $E'$ = {AABBA, ABABA, ABBAA, BAABA, BABAA, BBAAA, BBAAB, BABAB, BAABB, ABBAB, ABABB, AABBB} **9.** $p(\{\text{AAA}\}) = p(\{\text{BBB}\}) =$

$(0.5)^3 = 0.125$; $p(\{AABA\}) = p(\{ABAA\}) = p(\{BAAA\}) = p(\{BBAB\}) = p(\{BABB\}) = p(\{ABBB\}) = (0.5)^4 = 0.0625$; $p(\{AABBA\}) = p(\{ABABA\}) = p(\{BAABA\}) = p(\{ABBAA\}) = p(\{BABAA\}) = p(\{BBAAA\}) = p(\{BBAAB\}) = p(\{BABAB\}) = p(\{BAABB\}) = p(\{ABABB\}) = p(\{ABBAB\}) = p(\{AABBB\}) = (0.5)^5 = 0.03125$ **11.** $p(\{AAA\}) = 0.16666$; $p(\{BBB\}) = 0.08333$; $p(\{AABA\}) = p(\{ABAA\}) = p(\{BAAA\}) = 0.08333$; $p(\{BBAB\}) = p(\{BABB\}) = p(\{ABBB\}) = 0.04166$; $p(\{AABBA\}) = p(\{ABABA\}) = p(\{BAABA\}) = p(\{ABBAA\}) = p(\{BABAA\}) = p(\{BBAAA\}) = 0.04166$; $p(\{BBAAB) = p(\{BABAB\}) = p(\{BAABB\}) = p(\{ABABB\}) = p(\{ABBAB\}) = p(\{AABBB\}) = 0.02083$ **13.** Patient $P_1$ **15.** (a) $p(\mathrm{H} \cap \mathrm{S}) = 0.5$; (b) $p(\mathrm{H} \cap \mathrm{S}) = 0.2$ **17.** 8 **19.** Assuming a guess is made for each question, the probability is $p = 5/16$.

### Section 6.4

**1.** Yes **3.** $S = \{$AAA, BBB, AABA, ABAA, BAAA, BBAB, BABB, ABBB, AABBA, ABABA, ABBAA, BAABA, BABAA, BBAAA, BBAAB, BABAB, BAABB, ABBAB, ABABB, AABBB$\}$; no **5.** 5/18 **7.** 2/9 **9.** 9/19 **11.** 1/38 **13.** 1/26 **15.** 3/13 **17.** 4/815

### Section 6.5

**1.** 1/13 **3.** (a) 0.05882; (b) 0.24510; (c) 0.06373; (d) 0.12745 **5.** 3/16; 7/16 **7.** (a) $p = 2/9$; (b) $p = 28/729$; (c) $p = 1/4$ **9.** 0.4331 **11.** (a) $p = 248/465 = 0.5333$; (b) $p = 217/400 = 0.5425$; (c) $p = 136/256 = 0.5313$ **13.** $p = 0.03/0.0494 \approx 0.6073$ **15.** $p = 2/9 + 8/15 = 34/45$

### Section 6.6

**1.** 0.079, 0.921 **3.** 5/7 **5.** (a) 0.958; (b) 0.049 **7.** $p = 0.03/0.0494 \approx 0.6073$ **9.** (a) $p = 622/1057 \approx 0.5885$; (b) $p = 226/329 \approx 0.6869$ **11.** $p = 1/5$ **13.** (a) $p \approx 0.4211$; (b) $p \approx 0.1364$

### Section 6.7

**1.** (a) 60; (b) 840; (c) 72; (d) 181,440; (e) 6720; (f) 151,200; (g) 362,880; (h) 40,320; (i) 120; (j) 3,628,800 **3.** $P_{8,5} = 6720$ **5.** $C_{50,3} = 19{,}600$ **7.** $P_{40,3} = 59{,}280$ **9.** $P_{5,5} = 120$ **11.** $C_{25,10} = 3{,}268{,}760$ **13.** (a) $C_{13,6} \cdot C_{13,7} = 2{,}944{,}656$; (b) $C_{13,4} \cdot C_{13,4} \cdot C_{13,5} = 657{,}946{,}575$; (c) $C_{13,2} \cdot C_{13,3} \cdot C_{13,4} \cdot C_{13,4} = 11{,}404{,}407{,}300$ **15.** 1,000,000,000 **17.** 4; 0.75 **19.** 6 **21.** 6 **23.** 0.814 **25.** 0.941 **27.** 0.986 **29.** 0.777 **31.** 0.618

### Section 6.8

**1.** $C_{8,1} + C_{8,2} = 36$ **3.** $5 + 5^2 + 5^3 + 5^4 = 780$ **5.** $C_{3,1} \cdot C_{5,2} \cdot C_{6,2} = 450$ **13.** 0.497 **15.** $p(AA) = 169/200$, $p(Aa) = 22/200$, $p(aa) = 9/200$ for first generation; $p(AA) = 3479/3820$, $p(Aa) = 242/3820$, $p(aa) = 99/3820$ for second generation. The relative frequency of normal individuals increases.

### Section 6.9

**1.** $\dfrac{P_{9,9}}{P_{4,4} \cdot P_{5,5}} = \dfrac{9!}{4!\,5!} = 126$

**3.** $\dfrac{P_{9,9}}{P_{3,3} \cdot P_{3,3} \cdot P_{3,3}} = \dfrac{9!}{3!\,3!\,3!} = 1680$

**5.** $\dfrac{P_{6,6}}{P_{2,2} \cdot P_{2,2}} = 180$; $\dfrac{P_{10,10}}{P_{4,4} \cdot P_{3,3} \cdot P_{3,3}} = 4200$

**7.** $3\left(\dfrac{6!}{3!}\right) = 360$

# CHAPTER 7

## *Section 7.1*

1.

| Interval | Frequency | Cumulative frequency |
|---|---|---|
| 56.5-59.5 | 2 | 2 |
| 59.5-62.5 | 0 | 2 |
| 62.5-65.5 | 1 | 3 |
| 65.5-68.5 | 2 | 5 |
| 68.5-71.5 | 2 | 7 |
| 71.5-74.5 | 6 | 13 |
| 74.5-77.5 | 5 | 18 |
| 77.5-80.5 | 9 | 27 |
| 80.5-83.5 | 1 | 28 |
| 83.5-86.5 | 3 | 31 |

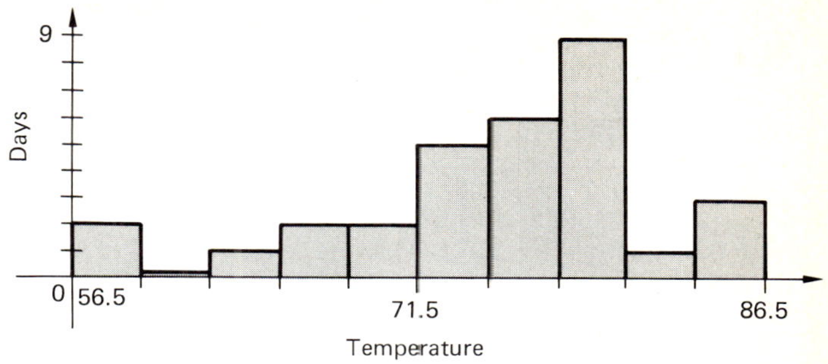

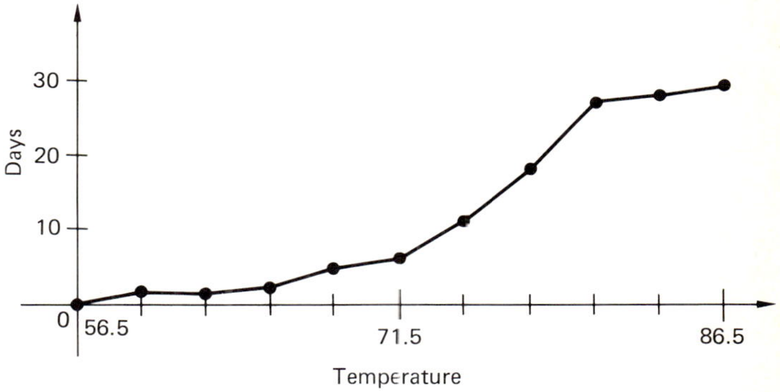

Days; 9; 0; 56.5; 71.5; 86.5; Temperature

Days; 30; 20; 10; 0; 56.5; 71.5; 86.5; Temperature

3.

| Interval | Frequency |
|---|---|
| 0-2,600 | 14 |
| 2,601-5,200 | 2 |
| 5,201-7,800 | 0 |
| 7,801-10,400 | 3 |
| 10,401-13,000 | 0 |
| 13,001-15,600 | 0 |
| 15,601-18,200 | 2 |
| 18,201-20,800 | 2 |
| 20,801-23,400 | 6 |
| 23,401-26,000 | 1 |

5.

| Digits | Frequency |
|---|---|
| 0 | 26 |
| 1 | 16 |
| 2 | 20 |
| 3 | 17 |
| 4 | 14 |
| 5 | 22 |
| 6 | 22 |
| 7 | 18 |
| 8 | 23 |
| 9 | 22 |

7. Since (T) would always be answered "no," a "yes" answer would mean that the respondent had engaged in premarital sex. Many respondents would answer evasively.

9. $p(\text{yes}) = \dfrac{p_0}{3} + \dfrac{1}{6}$; 40.5%

## *Section 7.2*

**1.** $\bar{x} = 57$; median $= 57$; range $= 11$; $s^2 = 12$ **3.** $\bar{x} = 44.875$; median $= 48.5$; range $= 24$; $s^2 = 61.1094$ **5.** $\bar{x} = 64.375$; median $= 64$; range $= 14$; $s^2 = 19.9844$ **7.** $\bar{x} = 56.55$; median $= 65.5$; range $= 92$; $s^2 = 735.4475$ **9.** $\bar{x} = 74.9355°$; median $= 76°$; range $= 28°$; $s^2 = 42.2539$ **11.** $\bar{x} = \$9498\frac{1}{3}$; median $= \$3775$; range $= \$25{,}100$; $s^2 = 84{,}806{,}080.57$ **13.** $E[x] = -1/37 \approx -\$0.027$, or an approximate cost of 2.7¢ per game **15.** $E[x] = \$2/3$ **17.** $E[x] = -\$15$ **19.** Pick all tomatoes on September 15 for $E[x] = \$1041.60$. **21.** Option (b) has an expected value of \$46,500.

### *Section 7.3*

**1.** $p(x_0) = 0.29630; p(x_1) = 0.44444; p(x_2) = 0.22222; p(x_3) = 0.03704; \mu = 1, \sigma^2 = 2/3$
**3.** $p(x_0) = 0.19753; p(x_1) = 0.39506; p(x_2) = 0.29630; p(x_3) = 0.09877; p(x_4) = 0.01235; \mu = 4/3; \sigma^2 = 8/9$
**5.** $p(x_0) = 0.07776; p(x_1) = 0.25920; p(x_2) = 0.34560; p(x_3) = 0.23040; p(x_4) = 0.07680; p(x_5) = 0.01024;$ $\mu = 2; \sigma^2 = 6/5$ **7.** $p(x_0) = 0.02799; p(x_1) = 0.13064; p(x_2) = 0.26127; p(x_3) = 0.29030; p(x_4) = 0.19354;$ $p(x_5) = 0.07741; p(x_6) = 0.01720; p(x_7) = 0.00164; \mu = 2.8; \sigma^2 = 1.68$ **9.** $p(x_0) = 0.38742;$ $p(x_1) = 0.38742; p(x_2) = 0.17219; p(x_3) = 0.04464; p(x_4) = 0.00744; p(x_5) = 0.00083; p(x_6) = 0.00006;$ $p(x_7) = 2.916 \times 10^{-6}; p(x_8) = 8.1 \times 10^{-8}; p(x_9) = 10^{-9}; \mu = 0.9; \sigma^2 = 0.81$ **11.** $p$(at least 90 show) $= 0.09945$; $p$(at least 85 show) $= 0.56832$ **13.** $p$(at least 21 show) $= 0.01487$
**15.** $p(x_0) = 1.024 \times 10^{-17}; p(x_1) = 5.0176 \times 10^{-15}; p(x_2) = 1.10638 \times 10^{-12}; p(x_3) = 1.44567 \times 10^{-10};$ $p(x_4) = 1.23966 \times 10^{-8}; p(x_5) = 7.28922 \times 10^{-7}; p(x_6) = 0.00003; p(x_7) = 0.00083; p(x_8) = 0.01531;$ $p(x_9) = 0.16675; p(x_{10}) = 0.81707$. Most likely is no defectives (10 successes).
**17.** $p(\text{successes} < 13) = 1 - [p(x_{15}) + p(x_{14}) + p(x_{13})] = 0.00304$ **19.** $p(x_0)' = 1 - (0.95)^{100} = 0.99408$, or 99.408% certain **21.** $p(x_0)' = 0.86602$

### *Section 7.4*

**1.** $p(|z| \le 1) = 0.6826$ **3.** $p(|z| \le 0.5) = 0.3830$ **5.** $p(|z| \le 1) = 0.6826$ **7.** $p(z \le -2) = 0.0228$
**9.** $p(3/4 \le z \le 9/4) = 0.2144$ **11.** $p(-4/3 \le z \le 0) = 0.4087$ **13.** $\bar{x} = 86.27; s^2 = 29.73;$ $p(sz + \bar{x} \le 90°) = p(z \le 0.6847) = 0.7532$; Temperatures not exceeding 90° occurred on 22 of the 30 days, with $22/30 = 0.7333$. Agreement is very close. **15.** $\bar{x} = \$4837.62; s^2 = 428186;$
(i) $p(sz + \bar{x} \ge \$5000) = p(z \ge 0.2482) = 0.4020$, or approximately 40.2% of the population;
(ii) $p(sz + \bar{x} \ge \$4000) = p(z \ge -1.2801) = 0.8997$, or almost 90%
**17.** $p(0.01z + 12.2 < 12) = p(z < -2) = 0.0228$ **19.** $p(6z + 75 \ge 90) = p(z \ge 2.5) = 0.0062$

### *Section 7.5*

**1.** Not a transition matrix **3.** Not regular **5.** Not a transition matrix **7.** Regular **9.** Not regular **11.** $p = 0.57004$ **13.** $p = 1/3$ **15.** $p = 0.4725$, two six-packs ($p = 88/225$)
**17.** $p(\text{RCA}) = 21/41; p(\text{Zenith}) = 17/41; p(\text{other}) = 3/41$

**19.** $A = A^3 = A^5 = \dots$, and $A^2 = A^4 = \dots = \begin{bmatrix} 1/3 & 0 & 2/3 \\ 0 & 1 & 0 \\ 1/3 & 0 & 2/3 \end{bmatrix}$

so matrix is not regular. However, the steady-state vector = (1/6, 1/2, 1/3).

### *Section 7.6*

Denote the entries of each payoff matrix by $(a_{mn})$.
**1.** $v = 4 = a_{12}$ **3.** $v = 2 = a_{13} = a_{23}$
**5.** Reduce to

$$\begin{pmatrix} 1 & -2 \\ -3 & 4 \end{pmatrix}$$

and play first row with probability $p = 0.7$, and first column with probability $q = 0.6$.
**7.** $v = 2 = a_{12}$ **9.** $v = 1 = a_{12} = a_{14} = a_{42} = a_{44}$
**11.** Reduce to

$$\begin{pmatrix} -2 & 4 \\ 3 & -5 \end{pmatrix}$$

playing first row with probability $p = 4/7$, and first column with probability $q = 9/14$.
**13.** Rutgers should end run 70% of the time and pass the other 30%; Cornell should blitz 10% of the time and play prevent defense the other 90%. Expected value is 4.3 yd per play.

**15.** Rutgers should end run 17/19 and short pass 2/19 of the time; Cornell should blitz 9/19 and play normal defense 10/19 of the time. Expected value is 3 18/19 yd per play. **17.** 5¢ per game
**19.** Richard selects a 1 and Charles a 6. The game is fair.
**21.** If your entries are

$$\begin{pmatrix} -a & b \\ c & -d \end{pmatrix}$$

with $a$, $b$, $c$, and $d > 0$, you should pick the first row with probability

$$p = \frac{c + d}{a + b + c + d}.$$

Then, regardless of the circumstances, your expected value is at least

$$\frac{bc - ad}{a + b + c + d}.$$

## CHAPTER 8

### *Section 8.1*

**1.** 109 **3.** 203 **5.** $-515\frac{1}{3}$ **7.** $30\frac{1}{21}$ **9.** $S_{185} = 17{,}575$ **11.** $S_{1000} = 500{,}500$ **13.** $S_{189} = 89{,}397$ **15.** $x_{15} = 16{,}384$ **17.** $x_{12} = 0.390625$ **19.** $x_{20} = 4.8324$ **21.** 65,534 **23.** 6648.513 **25.** (a) $A = \$1720$; (b) $A = \$2012.20$ **27.** (a) $A = \$1760$; (b) $A = \$2484.68$ **29.** (a) $A = \$8688$; (b) $A = \$10{,}425.09$ **31.** 1,073,741,823 gold coins **33.** $A = \$1448.66$ **35.** $r = 0.4$; rate of occurrence $= 1.66649 \times 10^{-5}$

### *Section 8.2*

**1.** $A = \$62.41$ **3.** $A = \$131.67$ **5.** $A = \$241.39$ **7.** $A = \$236.01$ **9.** \$209.76, \$220.13, \$230.67, \$241.39, \$252.26, \$10.63 **11.** \$273.29 **13.** \$4228.29 **15.** \$14,593.33 **17.** 212,196,200; 229,957,200; 211,993,560 **19.** $P_{10} = 0.7415$

**23.** $P_n = \dfrac{1 - (q/p)^n}{1 - (q/p)^N}, q = 1 - p$

### *Section 8.3*

**1.** 1 **3.** 1 **5.** 3/5 **7.** 1/2 **9.** Doesn't exist **11.** 1 **13.** 3/4 **15.** 5/6 **17.** 0 **19.** Doesn't exist **21.** 3 **23.** (20/3) **25.** (49/5) **27.** \$3,000,000; multiplier = 2.5

### *Section 8.4*

**1.** 2 **3.** 0 **5.** $-1/4$ **7.** 0 **9.** 2 **11.** 1/4 **13.** 0 **15.** 1 **17.** 1/2 **19.** $4x^3$

**21.** $\dfrac{1}{3x^{2/3}}$ **23.** $\dfrac{3\sqrt{a}}{2}$

### *Section 8.5*

**1.** 0 **3.** 0 **5.** 1/3, 1 **7.** None **9.** None **11.** 1 **13.** 0 **15.** 0, 2 **17.** No

**19.** $f(x) = \begin{cases} -0.02x + 0.7, & x \le 10 \\ -0.000364x + 0.50364, & x \ge 10 \end{cases}$

## CHAPTER 9

### *Section 9.1*

**1.** $y = 4x - 2$ **3.** $y = -5x - 3$ **5.** $y = -3x + 1$ **7.** $y = 24x - 32$ **9.** $y = 1$ **11.** $y = x$ **13.** $y = \frac{1}{4}x + \frac{1}{4}$ **15.** $y = -x + 2$ **17.** $y = 0$ **19.** $y = -x + 1$ **21.** \$300 **23.** (a) $-32$ ft/sec; (b) $16h - 64$ ft/sec; (c) $-64$ ft/sec

### *Section 9.2*

**1.** $f'(1) = 2$ **3.** $f'(2) = 5$ **5.** $f'(-1) = 12$ **7.** $f'(-2) = -32$ **9.** $f'(1) = 9$ **11.** $f'(0) = 0; f'(1) = 6$ **13.** $f'(x) = \dfrac{1 - x^2}{(x^2 + 1)^2}$ **15.** $f'(2) = 2; f'(3) = \frac{1}{2}$ **17.** $\sqrt{9.2} \approx \sqrt{9} + 0.2/2\sqrt{9} = 3.0333$; $\sqrt{9.2} = 3.0332$; $\sqrt{15.93} \approx \sqrt{16} - 0.07/2\sqrt{16} = 3.9913$; $\sqrt{15.93} = 3.9912$ **19.** $f(x) = (6 - 2x)(9 - 2x) = 4x^2 - 30x + 54$, and $f'(x) = 8x - 30$, so $f(0.88) - f(1) \approx (-0.12)f'(1) = 0.12(22) = 2.64$ in$^2$. (Actual saving is 2.6976 in$^2$.) The 500-page book with 1″ margins will occupy approximately 457 pages with 0.88″ margins.
**21.** (a) $\sqrt{5}$ sec; (b) $-32t$; (c) $-16, -32, -48, -64$ ft/sec, and $-32\sqrt{5}$ ft/sec at impact.
**23.** (b) The tangent line to the curve $y = f(x)$ at $x = x_n$ is $y = f'(x_n)(x - x_n) + f(x_n)$. Setting $y = 0$, we obtain $x = x_n - [f(x_n)/f'(x_n)]$, that is, $x_{n+1}$ is the $x$-intercept of the tangent line. The graph below indicates that the $x$-values $x_0, x_1, x_2, \ldots,$ approach the root $x_0 + h$. (d) $f(\sqrt{2}) = 0$

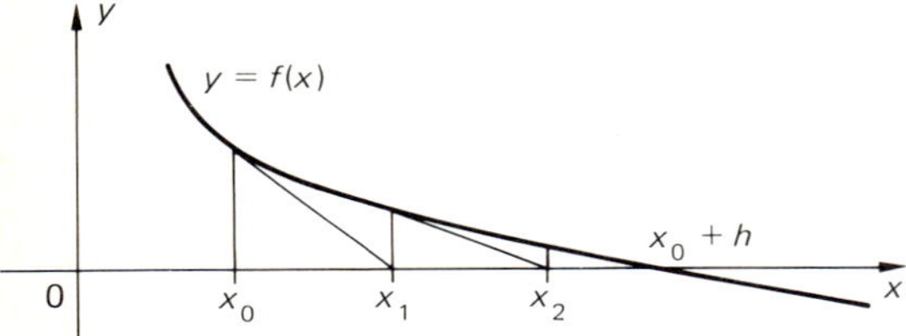


### *Section 9.3*

**1.** $f'(x) = 0$ **3.** $f'(x) = 0, y = 2/3$ **5.** $f'(x) = 171x^{170}, y = 171x - 170$ **7.** $f'(x) = 1/2\sqrt{x}$ **9.** $f'(x) = 0, y = 5$ **11.** $f'(x) = 6x^2 + 3$ **13.** $f'(x) = 85x^4 - 4x, y = -3$ **15.** $f'(x) = \frac{4}{3}x^{-2/3} + \frac{7}{2}x^{-3/2}$, $y = \frac{29}{6}x - \frac{47}{6}$ **17.** $2x - 1$ **19.** $\frac{5}{2}x^{3/2} - \frac{3}{2}x^{1/2} + 2x^{-1/2} + 2x^{-3/2}$ **21.** $4x^3 + 9x^2 + 6x + 3, y = 22x - 10$ **23.** $-2/(x - 1)^2, y = -2x - 1$
**25.** $\dfrac{8 + 42x - 6x^2}{(3x^2 + 4)^2}, y = \dfrac{44x - 79}{49}$ **27.** $\dfrac{-6x(x + 1)}{(1 + 2x + 3x^2)^2}$ **29.** $\dfrac{3x^4 + 4x^3 + 2x^2 - 4x - 1}{(3x^2 + 2x + 1)^2}$ **33.** $\dfrac{-f'}{f^2}$
**35.** (a) 6 sec; (b) $h'(t) = 80 - 32t, t = 2.5$ sec; (c) The rock is at the highest point of its trajectory.

### *Section 9.4*

**1.** $\dfrac{2x}{3(x^2 + 1)^{2/3}}$ **3.** $\dfrac{2x + 1}{2\sqrt{x^2 + x + 1}}$ **5.** $\dfrac{5x - 2}{\sqrt{5x^2 - 4x}}$ **7.** $\dfrac{2x - (1/2\sqrt{x})}{4(x^2 - \sqrt{x})^{3/4}}$ **9.** $\frac{5}{2}(x^2 + x + 2)^{3/2}(2x + 1)$

**11.** $\dfrac{1 + (1/2\sqrt{x + 1})}{2\sqrt{x + \sqrt{x + 1}}}$ **13.** $\dfrac{1 + (1/2\sqrt{x + 1})}{3(x + \sqrt{x + 1})^{3/2}}$ **15.** $\dfrac{2x + [(3x^2 - 1)/3(x^3 - x)^{2/3}]}{2\sqrt{x^2 + \sqrt[3]{x^3 - x}}}$

**17.** $\dfrac{\sqrt{x - 1} + (x + 1)/2\sqrt{x - 1}}{2\sqrt{(x + 1)\sqrt{x - 1}}}$

**19.** $\frac{1}{2}(x + \sqrt{x - \sqrt{x + 1}})^{-1/2}\{1 + \frac{1}{2}[x - \sqrt{x + 1}]^{-1/2}[1 - (1/2\sqrt{x + 1})]\}$
**21.** $15(x + \sqrt{x})^{14}[1 + (1/2\sqrt{x})]$ **23.** $14x(x^2 - 14)^6$ **25.** $3[x + (\sqrt{x} + 1)^2]^2[2 + (1/\sqrt{x})]$
**27.** $4\{x + [x^2 + (x^3 + 1)^2]^3\}^3\{1 + 6[x^2 + (x^3 + 1)^2]^2(3x^5 + 3x^2 + x)\}$ **29.** $\dfrac{9}{x^2}\left[\dfrac{x + 1}{x}\right]^8$

## Section 9.5

**1.** **(a)** $t = 5.7016$ sec; **(b)** $t = 2.5$ sec; **(c)** $h = 164$ ft; **(d)** $t = 0.4384, 4.5616$ sec **3.** **(a)** $t = 5.9628$ sec; **(b)** $t = 2\frac{1}{7}$ sec; **(c)** $h = 71.5$ m; **(d)** $t = 1.5896, 2.6961$ sec **5.** **(a)** $t = 8.1641$ sec; **(b)** $t = 2\frac{6}{7}$ sec; **(c)** $h = 138$ m; **(d)** $t = 6.5824$ sec **7.** **(a)** $t = 0.7417$ sec; **(b)** $t = 0$ sec; **(c)** $h = 80$ ft; **(d)** $t = 0.6572$ sec **9.** **(a)** $t = 6.5755$ sec; **(b)** $t = 2\frac{13}{16}$ sec; **(c)** $h = 226\frac{9}{16}$ ft; **(d)** $t = 1.524, 4.101$ sec **11.** **(a)** $t = 11.5314$ sec; **(b)** $t = 5.102$ sec; **(c)** $h = 202.551$ m; **(d)** $t = 0.5272, 9.6768$ sec **13.** 2 ft/sec **15.** **(a)** After 1 hr, 9 min, 11 sec = 1.1531 hr; **(b)** 60.1351 miles; **(c)** $-697.7146$ mph **17.** $dA/dr = 2\pi r$ **19.** $V' = 2\pi r^2$; $V' = \pi h^2$ **21.** Here marginal cost $C' = \$120.0004$/unit, while $R' = \$80$/unit, implying that $P' = -\$40.0004$/unit. She should not increase production, but instead should consider decreasing production and raising her prices.

## Section 9.6

**1.** $-5e^{-5x}$ **3.** $2(\ln 7)7^{2x}$ **5.** $\dfrac{8^{\ln x}\ln 8}{x}$ **7.** $\dfrac{-(2x)3^{\sqrt{4-2x^2}}\ln 3}{\sqrt{4-2x^2}}$ **9.** $2x^{2x}(\ln x + 1)$

**11.** $2^x x^{2^x}[(\ln x)\ln 2 + x^{-1}]$ **13.** $\dfrac{x}{(4+x^2)\ln 10}$ **15.** $\dfrac{2^{\ln x}\ln 2}{x}$ **17.** $e^{2^{3^{4^x}}}2^{3^{4^x}}3^{4^x}4^x\ln 2\ln 3\ln 4$

**19.** $\dfrac{3^{4^{\ln x}}4^{\ln x}\ln 3\cdot\ln 4}{x}$ **21.** $f'(20) = 2{,}546{,}389.1e^{0.284} \approx 3{,}382{,}707$ persons/year **23.** \$134.99

**25.** \$830.03 **27.** \$1575.61

## Section 9.7

**1.** $3\cos(3x+7)$ **3.** $\dfrac{-x\cdot\sin\sqrt{1+x^2}}{\sqrt{1+x}}$ **5.** $\sin x + x\cos x$ **7.** $3x^2\sec^2 x^3$ **9.** $2x\cot x - x^2\csc^2 x$

**11.** $-2e^x\csc^2(e^x)\cot(e^x)$

**13.** $\dfrac{-\sec^2(\csc\sqrt{1+x})\csc\sqrt{1+x}\cot\sqrt{1+x}}{2\sqrt{1+x}}$

**15.** $\sec(x\cot x)\tan(x\cot x)[\cot x - x\csc^2 x]$ **19.** $\frac{2}{125}$ rad/sec

**21.** If $x$ is the number of hours after 2 P.M., the equation determining the distance from the dock to the water level is

$$y = 3 + 2\cos\frac{2\pi x}{12.416667}.$$

Hence

$$y' = -\frac{4\pi}{12.416667}\sin\frac{2\pi x}{12.416667}\bigg|_{x=2} = -0.8581 \text{ m/hr}$$

so the water is rising at 0.8581 m/hr at 4 P.M.

## Section 9.8

**1.** $y(-3/10) = 6.55$ is the minimum; $y(1) = 15$ is the maximum **3.** $y(-1/2) = 5/4$ is the maximum; $y(1) = -1$ is the minimum **5.** $y(-1/\sqrt{2}) = -0.42888$ is the minimum; $y(1/\sqrt{2}) = 0.42888$ is the maximum **7.** $y(3) = 50$ is the maximum; $y(1) = -2$ is the minimum. There is a local maximum at $x = -2$. **9.** $y(-1) = -1/2$ is the minimum; $y(1) = 1/2$ is the maximum **11.** $y(-5/8) = -9/16$ is the minimum; $y(1) = 10$ is the maximum **13.** $y(1/14) = 12\frac{1}{28}$ is the maximum; $y(-3) = -54$ is the minimum

**15.** $y\left(\dfrac{-5-\sqrt{21}}{2}\right) = \dfrac{113+21\sqrt{21}}{2}$ is the maximum; $y\left(\dfrac{-5+\sqrt{21}}{2}\right) = \dfrac{113-21\sqrt{21}}{2}$ is the minimum

**17.** $y(e) = e^{-1}$ is the maximum; $y(1) = 0$ is the minimum **19.** $y(2.028757839) = 1.819705742$ is the maximum; $y(4.91318044) = -4.814469889$ is the maximum. The two values are the roots of $x + \tan x = 0$. **21.** **(a)** 55¢; **(b)** 65¢

**23.** Build the sides parallel to the stream 500 ft long, the other three sides (perpendicular to the

stream) $166\frac{2}{3}$ ft long. **25.** $10\sqrt{6/11} \approx 7.3855$ ft vertically and $\sqrt{66} \approx 8.124$ ft horizontally
**27.** $r = \dfrac{\sqrt[3]{147.87}}{2} \approx 2.64$; $h = \dfrac{8}{\pi}\sqrt[3]{147.87} \approx 13.47$ cm
**29.** 4 ft

***Section 9.9***

**1.** $y_x = t, y_t = x, y_{xx} = y_{tt} = 0, y_{xt} = y_{tx} = 1$ **3.** $y_x = te^{xt}, y_t = xe^{xt}, y_{xx} = t^2e^{xt}, y_{tt} = x^2e^{xt}$
$y_{tx} = y_{xt} = e^{xt}(1 + tx)$ **5.** $y_x = y_{xx} = e^x \ln t, y_t = y_{xt} = y_{tx} = \dfrac{e^x}{t}, y_{tt} = \dfrac{-e^x}{t^2}$

**7.** $y_x = \dfrac{e^{x/t}}{t}, y_t = \dfrac{-xe^{x/t}}{t^2}, y_{xx} = \dfrac{e^{x/t}}{t^2}, y_{tt} = e^{xt}\left(\dfrac{x^2 + 2xt}{t^4}\right), y_{xt} = y_{tx} = -e^{xt}\left(\dfrac{x + t}{t^3}\right)$
**9.** $y_x = y_{xx} = te^x, y_t = y_{xt} = y_{tx} = e^x, y_{tt} = 0$ **11.** $y_x = y_{xx} = e^x \sin t, y_t = y_{tx} = y_{xt} = e^x \cos t, y_{tt} = -e^x \sin t$ **13.** Test fails at $(t, 0)$ because $\Delta = 0$. A graph shows the line segment $\{(t, 0)|t > 0\}$ consists of minimums, while the line segment $\{(t, 0)|t < 0\}$ consists of maximums. The point $(0, 0)$ is a saddle point. **15.** Critical points at $(0, 0)$ and $(3, 3)$. A local minimum at $(3, 3)$ and a saddle point at $(0, 0)$. **17.** No critical points **19.** $y = 3.0854x + 1.5732$ **21.** $y = 0.5x - 3.48$ **23.** 50/9 lb of ground beef and 50/3 lb of potatoes **25.** $7\frac{1}{2}$ lb of ground beef and 75 lb of potatoes **27.** $15\frac{5}{8}$ lb of ground beef and $62\frac{1}{2}$ lb of potatoes **29.** A local minimum with $x = 4000/\sqrt{3}$ and $y = 10{,}000/\sqrt{3}$ units

## CHAPTER 10

***Section 10.1***

**1.** $\dfrac{x^3}{3} + x^2 - x + c$ **3.** $x^3 - 2x^2 + 7x + c$ **5.** $\dfrac{3x^8}{8} - \dfrac{4x^3}{3} + x + c$ **7.** $x^4 - x^2 + c$ **9.** $\dfrac{2x^{3/2}}{3} - \dfrac{3x^{4/3}}{4} + c$
**11.** $7e^x - 4\ln x + c$ **13.** $\dfrac{15^x}{\ln 15} - \dfrac{14^x}{\ln 14} + c$ **15.** $x - \dfrac{2x^{3/2}}{3} + c$ **17.** $e^t + c$ **19.** $\dfrac{6^z}{\ln 6} + c$
**21.** $4e^m - \dfrac{3m^{4/3}}{4} + c$ **23.** $2\sqrt{t} - 4\ln t + c$ **25.** $4\sin x + c$ **27.** $14\sin z + c$

***Section 10.2***

**1.** $\dfrac{e^{7x}}{7} + c$ **3.** $-\ln|4 - x| + c$ **5.** $\frac{1}{3}(\ln x)^3 + c$ **7.** $\sqrt{9 + x^2} + c$ **9.** $-\sqrt{9 - x^2} + c$
**11.** $\dfrac{1}{4}\ln\left|\dfrac{x - 2}{x + 2}\right| + c$ **13.** $\dfrac{1}{2\sqrt{3}}\ln\left|\dfrac{x - \sqrt{3}}{x + \sqrt{3}}\right| + c$ **15.** $\ln\left|\dfrac{x}{1 + x}\right| + c$ **17.** $\dfrac{\sin^2 x}{2} + c$ **19.** $-\dfrac{\cos^3 x}{3} + c$
**21.** $\ln\sqrt{\sec 2x} + c$ **23.** $\ln|\csc x - \cot x| + c$ **25.** $\sin^{-1}\left(\dfrac{x}{3}\right) + c$ **27.** $\dfrac{1}{3}\sin^{-1}\left(\dfrac{3x}{2}\right) + c$

***Section 10.3***

**1.** $e^{3x}\left[\dfrac{x^2}{3} - \dfrac{2x}{9} + \dfrac{2}{27}\right] + c$ **3.** $e^{2x^2}\left[\dfrac{x^2}{4} - \dfrac{1}{8}\right] + c$ **5.** $e^x[(x + 4)^2 - 2(x + 4) + 2] + c$
**7.** $2x\sqrt{4 + x} - \frac{4}{3}\sqrt{(4 + x)^3} + c$ **9.** $2x\sqrt{1 + x} - \frac{4}{3}\sqrt{(1 + x)^3} + c$
**11.** $\dfrac{x^2}{2}(\ln x - \frac{1}{2}) + c$ **13.** $\dfrac{x^3}{3}\left[(\ln x)^2 - \dfrac{2\ln x}{3} + \dfrac{2}{9}\right] + c$ **15.** $e^x[x\ln x - 1] + c$
**17.** $-\dfrac{x^3\cos 2x}{2} + \dfrac{3x^2\sin 2x}{4} + \dfrac{3x\cos 2x}{4} - \dfrac{3}{8}\sin 2x + c$ **19.** $x^2\sin x + 2x\cos x - 2\sin x + c$
**21.** $\dfrac{e^x}{2}(\cos x + \sin x) + c$ **23.** $\frac{1}{6}(3x + \sin 3x\cos 3x) + c$

### *Section 10.4*

**1.** 3 **3.** 13/3 **5.** $\frac{1}{3}(27 - e^3)$ **7.** $\frac{1}{4}(e^2 + 1)$ **9.** 0 **11.** 1 **13.** 1/3 **15.** $2 \ln 2 - 1$ **17.** ln 2
**19.** $2\sqrt{2}$ **21.** $1 + 1/\sqrt{2}$ **23.** $1/e$ **25.** 1 **27.** Diverges **29.** $2\int_0^{\infty} xe^{-x^2}dx = 1$ **31.** 2 **33.** $\pi$

### *Section 10.5*

**1.** 424 ft **3.** 344 ft **5.** $306\frac{2}{3}$ ft **7.** 420 miles **9.** 0.3787776 g **11.** 3.456 mg **13.** \$2000
**15.** \$27,500 **17.** $8\pi$ **19.** $\frac{\pi}{30}$ **21.** $\frac{1}{3}\pi r^2 h$ **23.** Let $\sin^2 x = (1-\cos 2x)/2$. Then volume $= \frac{\pi}{8}(e^{2\pi} - 1)$.
**25.** $\frac{2}{27}(37^{3/2} - 1)$ **27.** $\frac{1}{2}(e - 1/e)$ **29.** $\ln(\sqrt{2} + 1)/\sqrt{3}$ **31.** **(i)** $\frac{\pi h^3}{3m^2} = \frac{\pi r^2 h}{3}$; **(ii)** $\frac{2}{3}\pi r^3$;
**(iii)** $\frac{2\pi}{3}(8 \ln 2 - \frac{7}{3})$; **(iv)** $2\pi$; **(v)** $\pi\left(\frac{\pi}{2} - 1\right)e^{\pi/2}$

### *Section 10.6*

**1.** TR = 21.03125; exact = 21 **3.** TR = 0.7761296; exact = $\pi/4$ **5.** TR = 3.0606550 **7.** TR = 0.4593145; exact = $1 - \cos(1)$ **9.** TR = 0.1358096 **11.** TR = 382.9433123

### *Section 10.7*

**1.** $y = ce^{2x}$ **3.** $y = ce^{-3x}$ **5.** $y = ce^{2x} - 5/2$ **7.** $y = ce^{x^2/2}$ **9.** $y = x^x(1 + ce^{-x})$ **11.** $y = ce^{-x} + \frac{x^2}{2}e^{-x} + 1$
**13.** $y = ce^{\tan^{-1} x}$ **15.** $y = \frac{\sin x}{x^2}$ **17.** 21,831,743 **19.** 54,371,290 **21.** 942,304,355
**23.** 55 min, 46 sec **25.** 22 min, 22 sec **27.** 45 min, 48 sec **29.** $y^2 = e^x + c$ **31.** $y^{-1} = c - x^2/2$
**33.** $\ln y = e^x(x - 1) + c$ **35.** $y^2 = 2 \sin x + c$ **37.** Solve $s(t) = \frac{k}{c}\ln\left(\frac{h_0 + ct}{h_0}\right)$ at noon and 2 P.M., obtaining $(1 + 5c/h_0)^2 = (1 + 3c/h_0)^3$, implying that the snowfall began at 2:42:31 A.M.
**39.** $y(x) - y(x + 1) \approx -y'(x) = 1.5cx^{-2.5} = N(x)$; average income $\approx \int_{200,000}^{\infty} xN(x)\,dx = \$6,000,000,000$

### *Section 10.8*

**1.** $y = (ce^{-x} - 1)^{-1}$ **3.** $y = 3(4 + ce^{-3x})^{-1}$ **5.** $P = (1 - ce^{4x})^{-1}$ **7.** $y = (1 + ce^{-x^2/2})^{-1}$
**9.** $x = c_1e^{-3t}, y = c_2 - c_1e^{-3t} - 4t$ **11.** $x = c_1e^{-2t} + 2, y = c_2e^t - \frac{c_1}{3}e^{-2t} + 3$ **13.** $x = (c_1 + c_2t)e^{-3t}$, $y = -x' - 4x = -(c_1 + c_2 + c_2t)e^{-3t}$ **15.** $x = \frac{1}{3}(10e^{4t} - 7e^t), y = \frac{1}{3}(7e^t + 5e^{4t})$ **17.** **(a)** 50 min; **(b)** $x(50) = y(50) = 100/e$ lb; **(c)** $y(20) = 40/e^{0.4}$ lb; **(d)** less **19.** $x = 50(e^{-t/50} + e^{-3t/50}), y = 100(e^{-t/50} - e^{-3t/50})$

## APPENDIX 1

### *Section A1.1*

**1.** \$1220.19 **3.** \$1921.24 **5.** \$2159.19 **7.** \$3989.05 **9.** \$23,344.19 **11.** \$270,544.07
**13.** \$3558.90 **15.** 7.82%

### *Section A1.2*

**1.** \$628.89 **3.** \$39,863.31 **5.** \$104,044.32 **7.** \$118,587.28

9. $A = \frac{12P}{i}\left[\left(1 + \frac{i}{12}\right)^{12n} - 1\right]$, where $P$ is the monthly payment, $i$ is the annual interest, $n$ is the number of years.
11. Set the principal $= P/m$, interest $= i/m$, and number $= nm$ in Eq. (A1.12).

*Section A1.3*

1. \$449.86 3. \$411.07 5. \$402.61 7. \$615.13 9. \$1190.79

*Section A1.4*

1. \$2591.08 3. \$545.57 5. \$38,909.90 7. \$38,454.59 9. \$39.56 11. \$40.32 13. 55.54 payments less, saving \$15,082.44

# Index


Abscissa, 23
Absolute value, 10
Acceleration, 451
Additive identity, 150
Amortization, 380, 583
Amount, 370, 373, 578
Annual interest rate, 380, 578
Annuity, 377, 581
  future value, 377, 580
  present value, 580
Antiderivative, 504
Arbitrary constant, 504
Arc length, 543
Area function, 528
Arithmetic progression, 368
Asymptote, 75
Augmented matrix, 165
  procedure, 165
Axis, 23

Base
  of an exponent, 111
  of a logarithm, 109
Bayes's formula, 281, 284
Bell-shaped curve, 97
Belonging to, 247
Bernoulli experiment, 331
Bernoulli trial, 331
Binomial, 14
Binomial coefficient, 17
Binomial distribution, 332
Birthday problem, 240, 244
Branch of a tree, 242

Carrying capacity, 106
Cartesian coordinates, 22
Chain rule, 445–450
Change of variables, 509
Characteristic, 114
Circle, 69
Column dominance, 354
Column vector, 147
Combination, 289
Common logarithm, 111
Commutative multiplication, 156
Complement of a set, 249
Completing the square, 64
Complex arithmetic, 6
Complex number, 4
Compound interest, 99, 578
Composite function, 447
Conditional probability, 272
Confidence interval, 337
Consumer's surplus, 539
Contained in, 248
Continuity
  at a point, 404
  on a set, 404
Continuous compounding, 465
Continuous distribution, 340
Continuous function, 404
Converges, 389
Corner (feasible set), 207
Cosecant, 590
Cosine, 590
Cost function, 458
Cotangent, 590
Critical point, 476, 490
Cumulative frequency, 315
  grouped, 315
  polygon, 315

Definite integral, 525
Degrees, 591
Demand equation, 538
Demand function, 457
Dependent events, 276
Dependent variables, 30
Derivative, 423
  partial, 486
Determinant, 176
Difference quotient, 415
Differential equation, 552
Differentiation, 432
Direct influence, 180
Discount, 580
Discrete distribution, 339
Discrete model, 378
Discriminant, 66

Discontinuity at a point, 405
Disjoint set, 250
Disk formula for volume, 542
Distance formula, 69
Distribution, 332
  binomial, 332
  continuous, 340
  discrete, 339
  normal, 340
Diverges, 390, 534
DNA, 304
Domain, 27
Dominance
  column, 354
  row, 354
Dominant gene, 296
Dosage, 537
Double angle formulas, 592
Double logarithmic plotting, 128

Economic multiplier, 395
Element, 247
End-point of an interval, 477
Empty set, 248
Equally probable, 261
Equation, 37
  demand, 538
  differential, 552
  linear, 37
  simultaneous, 38
Equity, 382, 585
Event, 257
  dependent, 276
  independent, 275
  mutually exclusive, 258
Expected value, 324, 357
Experimental relative frequency, 260
Exponent, 7, 92
Exponential decay, 91
Exponential function, 94
Exponential growth, 91
Extrapolation, 51

Factoring, 12
Fair game, 360
Feasible set, 206
Final form of tableau, 217
First derivative test, 479
Fractions, 2
Free fall, 450
Frequency, 314
Frequency polygon, 315

Function, 22, 27
  area, 528
  composite, 447
  continuous, 404
  cost, 458
  demand, 457
  exponential, 94
  inverse, 110
  linear, 37
  polynomial, 72
  power, 128
  profit, 216
  quadratic, 59
  rational, 74
  trigonometric, 468, 590
Functional notation, 29
Fundamental theorem of calculus, 531
Future value, 377, 580

Game theory, 353
Gaussian elimination, 170, 171
Gene, 296
  dominant, 296
  recessive, 296
Genealogical table, 190
Genetic code, 304–309
Genotype, 296
Geometric progression, 372
Graph, 22, 24
Grouped cumulative frequency, 315
Grouped data, 315
Grouped frequency, 315
Growth rate, 103
  per individual, 103, 555, 564

Half-angle formulas, 592
Half-life, 101
Half-plane, 199
Heterozygous, 296
Histogram, 315
Homozygous, 296
Hybrid, 296
Hypotenuse, 69

Improper integral, 534
Indefinite integral, 505
Independent events, 275
Independent variable, 30
Indifference curve, 77
Indifference map, 78
Indirect influence, 181
Inequality, 9, 198

Infinite sum, 392
Inherited rate, 375
Initial condition, 554
Initial value problem, 554
Initial velocity, 451
Inspiration rate, 378
Instantaneous decay, 103
Instantaneous growth, 103
Integer programming, 213
Integers, 2
Integral, 505
  definite, 525
  diverges, 534
  improper, 534
  indefinite, 505
Integration, 505
  limits of, 525
  by parts, 518–528
Interest
  annual rate, 380, 578
  compound, 99, 373
  simple, 370
Interpolation, 51
Intersection of sets, 250
Inverse cosine, 533
Inverse of a function, 110
Inverse of a matrix, 161, 163
Inverse sine, 517
Invertible matrix, 163

Lagrange multipliers, 498
Lattice, 212
Law of cosines, 592
Law of sines, 592
Learning curve, 104
Least squares approximation, 80, 138
Least squares line, 80, 138, 495
Least squares method, 80, 138
Least squares theorem, 81, 86, 492
Limit
  of a function, 397
  of a sequence, 389
Line, 31, 37
  secant, 417
  tangent, 418
Line chart, 315
Linear equation, 37
Linear inequality, 198
Linear programming, 204, 215
Linear regression, 84
Local maximum, 476, 490
Local minimum, 476, 490

Logarithm, 109
  common, 111
  natural, 111
Logarithmic scale, 120
Logistic curve, 106, 563
Logistic growth, 106

Mantissa, 114
Margin, 416
Marginal cost, 416, 458
Marginal profit, 458
Marginal revenue, 458
Markov chain, 346
Markov process, 346
Matrix, 146
  addition, 149
  augmented, 165
  column, 147
  equation, 158
  inverse, 163
  invertible, 163
  multiplication, 150, 151
  payoff, 353
  row, 147
  square, 147
  transition, 344
Maximum problem, 228
Mean, 319
Median, 319
Mendel's laws, 296
Minimum problem, 229
Mixed partials, 490
Mixed strategy, 356
Mixed strategy rule, 358
Model, 246, 377
Multiplication principle of probability, 274, 280
Multiplication property of trees, 262
Multiplicative identity, 156
Mutation rate, 375
Mutually exclusive events, 258

Natural logarithm, 112
Newton's law of cooling, 556
Newton's method, 431
Normal distribution, 340

Odds, 240
Optimal value, 209
Ordinate, 23

Parabola, 59
Pareto's law, 559
Partial derivative, 486
Partial fractions, 514
Pascal's triangle, 15
Payoff matrix, 353
Perfect discrimination, 538
Permutation, 287
Plotting
  double-logarithmic, 128
  semilogarithmic, 119
Point-slope formula, 34, 37
Polynomial, 72
Population mean, 319
Population variance, 321
Power function, 128
Present value, 580
Principal, 370, 373, 578
Probability, 240
  conditional, 272
  multiplication principle of, 274, 280
Product rule, 439
Profit function, 216
Progression
  arithmetic, 368
  geometric, 372
Pure strategy, 353
Pythagorean theorem, 69, 590

Quadratic formula, 13, 65
Quadratic function, 59
Quadrature formula, 547
Quotient rule, 439

Radians, 591
Radicals, 8
Radioactive decay, 100
Radius, 69
Random sample, 312
Range, 27, 321
Ratio, 372
Rational functions, 74
Rational numbers, 2
Rational powers, 93
Real numbers, 3
Recessive gene, 296
Recursive equation, 385
Reduction formula, 521, 591
Regular Markov chain, 347
Relation, 27
Repeating decimals, 12, 394
Residence time, 567
RNA, 304
Roots
  of an equation, 13
  of a number, 7, 92
Row
  dominance, 354
  of a matrix, 146
  operation, 166
  vector, 147

Saddle point, 354, 491
Saddle point rule, 354
Sales curve, 104
Sample, 256
  mean, 319
  median, 319
  point, 322
  random, 312
  range, 321
  space, 256
  variance, 321
Scientific notation, 113
Secant, 590
Secant line, 417
Second derivative test, 482
Sequence, 368
Semilogarithmic plotting, 119
Sensitivity analysis, 233
Separation of variables, 552
Sequential counting principle, 286
Set, 2, 247
  complement, 249
  disjoint, 250
  empty, 248
  equality, 248
  feasible, 206
  intersection, 250
  union, 251
  Venn diagram, 249
Shell formula for volume, 547
Sigma notation, 540
Sigmoidal curve, 106
Signed numbers, 5
Simple interest, 369
Simplex algorithm, 217
Simplex method, 215
Simplex tableau, 215
Simultaneous linear equations, 38
Sine, 590
Slack variable, 217
Slope, 33
Slope-intercept formula, 33, 37
Specific infection rate, 566
Square matrix, 147
Standard deviation, 322

Standard normal distribution, 340
State
  of a process, 344
  vector, 346
Strategy
  mixed, 356
  pure, 353
Steady-state vector, 346
Subset, 248
Substitution methods, 509–518
Symmetry, 59
System of differential equations, 568
System of linear equations, 38

Table, 22
  genealogical, 190
Tableau
  final form, 217
  simplex, 215
Tangent, 590
Tangent line, 418
Tidal volume, 378
Total profit, 458
Total revenue, 458
Traffic problem, 185
Transition matrix, 344
Trapezoidal rule, 548
Tree, 241
Tree multiplication property, 262
Triangle inequality, 12

Uniform sample space, 261, 266
Union, 251
Unit circle, 69
Universal donor, 264
Universal recipient, 264
Utility, 78
Variable
  dependent, 30
  independent, 30
  slack, 217
Variance, 321
Vector
  column, 147
  row, 147
  state, 346
  steady-state, 346
Velocity, 414, 415, 451
Venn diagram, 249
Vertex
  of a feasible set, 207
  of a parabola, 60

Watson–Crick model, 304–309

$x$-axis, 23

$y$-axis, 23
$y$-intercept, 33